Biosemiotics

Volume 30

Combining research approaches from biology, semiotics, philosophy and linguistics, the field of biosemiotics studies semiotic processes as they occur in and among living systems. This has important implications and applications for issues ranging from natural selection to animal behaviour and human psychology, leaving biosemiotics at the cutting edge of the research on the fundamentals of life.

The Springer book series *Biosemiotics* draws together contributions from leading scholars in international biosemiotics, producing an unparalleled series that will appeal to all those interested in the origins and evolution of life, including molecular and evolutionary biologists, ecologists, anthropologists, psychologists, philosophers and historians of science, linguists, semioticians and researchers in artificial life, information theory and communication technology.

Peter N. Jackson

Aristotle on Meaning in the Living World

A Biosemiotic Perspective

Springer

Peter N. Jackson
Manchester, UK

ISSN 1875-4651 ISSN 1875-466X (electronic)
Biosemiotics
ISBN 978-3-032-00601-1 ISBN 978-3-032-00602-8 (eBook)
https://doi.org/10.1007/978-3-032-00602-8

This Springer imprint is published by the registered company Springer Nature Switzerland AG
The registered company address is: Gewerbestrasse 11, 6330 Cham, Switzerland

Preface

This work seeks to demonstrate the relevance of the philosophy of Aristotle to the modern discipline of biosemiotics. It seeks thereby to show biologists that there is meaning beyond science and to impress upon classicists the responsibility involved in safeguarding a treasure essential to the modern world. Both corrections are achieved by reference to the cutting-edge discipline of biosemiotics which finds and studies meaning as expressed by beings in the biological world.[1]

The problems we face today are primarily philosophical—perhaps even spiritual—and not merely practical or scientific. The value of biosemiotics is, in this context, not its formulation of a new, more efficient terminology for scientists to use but its wider insistence upon a philosophical re-framing for science. It addresses the meaning of meaning not by manipulating mental abstractions but by representing the real biological creations of nature that our minds reflect upon. This recognition of the derivativeness of the human mind and of its products, such as science, from the actual world we find ourselves in, goes hand in hand with a richer and deeper appreciation of such mind as a "*mens*" (from the Latin "*mēnsūra*") or "measure" in and of the world. As regards the application of this mind, we see that when Aristotle *created what we now know as the sciences* he quite deliberately and carefully drew out the limits and overlaps of their domains. It was precisely by recognising the "gaps" in our knowledge of the world that Aristotle could make room for a fuller "whole" as the base element in the richer conflicted reality in which we exist. It was by exploring the value of these gaps, means, wholes, and limits that Aristotle's philosophy exposed the great risk of Idealism, of reducing the world to abstract ideas and their scientific reductions, and opposed it with his own Realism, which logically explains and aesthetically interprets the real world through analysis and synthesis.[2]

[1] For an introduction to the world of biosemiotics, see https://www.biosemiotics.org/

[2] We will explore below this distinction between realism and idealism which has, as with that between subject and object, become confused and contradictory in the tradition. For an overview of Aristotle's own realism, see Paul Feyerabend "Realism" in *Conquest of Abundance: A Tale of Abstraction versus the Richness of Being* (Chicago, 1999).

As regards the nature and value of biosemiotics, a biosemiotician describes it as "the epitome of an interdisciplinary effort to bridge the sciences and the humanities in a way that is fully respectful of both domains—while simultaneously attempting to illuminate some of the mirror image 'blind spots' currently impoverishing them each—[and which] seeks to explicate a biologically based understanding of information-bearing processes that can account for how such processes emerge as 'sign relations' in the lived experience of animals and human beings".[3] Donald Favareau explains further on the problem biosemiotics is called to address that: "What changed in modernity…was the notion that 'signification' per se is ultimately a human construct only—an axiom whose corollary in science is the assumption that what appear to us to be 'meaning'-bearing relationships are, in fact, fully mechanistic and determinable (if only stochastically) events and processes, the precise event structures of which have simply not yet been scientifically ascertained".[4] The fruit of scientific reductionism is perhaps the guiding ideology of what we call "modernity". This reductionism reduces the world to the abstract products of the human mind (names and numbers) and then reduces the human mind to the manipulation schemes of science (logic and mathematics) as a previous age reduced both to 'God'. Biosemiotics is an attempt to restore the centrality of the world and to correct science itself on the basis that: "the contemporary teaching and conception of science is likewise 'severely, perhaps cripplingly, impoverished by the utter, frightening innocence of most of its practitioners about the [semiotic] order in which they and it are embedded'."[5]

Accordingly, both Aristotle and biosemiotics defend the reality of "meaning". This is not as trivial a distinction as might at first appear. For who would deny the existence of meaning in the world when it is so evident and omnipresent? The answer is that our "modernist" ideology has sought to deny this very thing. "Meaning" is from this perspective merely a mirage behind which stands the "higher-order" interactions of chance and/or necessity that it supposes the world to reduce to. By contrast, both Aristotle and biosemiotics press us to explain the biological world as a world full of meanings acted upon and expressed by engaged beings, rather than one captured and determined by abstract laws. Both perspectives are self-consciously *in the world* and do not, falsely, claim to be above it.

It is reasonable to hope to find beauty or wisdom or to gain access to our own civilisational origins from a classical source, from a thinker from the fourth century B.C. It is also clearly useful to encounter an insight different from our own, however primitive. This is all reasonable but erroneous if applied to Aristotle. Aristotle is the originator or creator of both science *and* philosophy and thoroughly epitomises their virtues. We should take Aristotle seriously and not as a relic. His thought is not only not exhausted; it has barely been touched. We see this particularly in respect to

[3] Favareau, Donald "Why this now? The conceptual and historical rationale behind the development of biosemiotics" *Green Letters: Studies in Ecocriticism* (2015) p 228.

[4] Favareau, Donald *Ibid.* p 235.

[5] Favareau, Donald *Ibid.* p 236.

his biology which comprises half his extant work. This biological corpus, which should be understood to sit at the very centre of his philosophy of "being" or "*ousia*", has only been rediscovered in recent years, and since given only a fraction of the attention it deserves. It was ignored or suppressed in the Christian era in some part on theological grounds. It can now be represented in less theological times, even if we have fresh dogmas to combat, if we are to clearly manifest and utilise its power and value. I have upon this basis sought to rethink Aristotle afresh as a philosopher of "biosemiotics" who bases his thought about the world on the existential fact of the organism within it, and the real-world relations that organise and inform it. This philosophical realism is the critical emphasis of Aristotle and it is not well represented by the later scholastic tradition. This is why I have sought to address Aristotle directly without confusing matters by introducing the later, derivative but divergent, neo-Aristotelian tradition.

This project is split into two books. This is the first, which reviews the science and philosophy of Aristotle and establishes its vitality and meaningfulness today in the face of a technological approach which prizes utility over value. The second shows that Aristotle's scientific and philosophical approach is supported by modern science itself and that it opens up the philosophical possibility for an approach to the world from the perspective of biosemiotics. I will not write in depth about biosemiotics in this first book, which will rather set the groundwork needed for my doing so in the volume that succeeds it, but will explain the philosophical case in principle as regards why biosemiotics should follow Aristotle's philosophical path in order to satisfactorily represent biological reality.

I would like to thank here the people who have expressed interest in my work and offered support, in the present and in the past. Kalevi Kull and Donald Favareau from the world of biosemiotics have shown great kindness towards me and given my research both an end and a means. Without them, I would have only one more wind egg in my life. I also extend my thanks to Rosalind Thomas and Bob Sharples from the world of classics at the University of London who have helped me in the past. For this work, Abraham Bos kindly used his expertise to review Greek transliterations from the Netherlands. Any opinions and errors in the text are, however, entirely my own. Thanks to Nikos Soueltzis for his excellent thoughts and for pointing me towards Jan Patočka. My old friend Simon Goff has generally provided me with much food for thought over the years. I also extend my thanks to Georgia Sermamoglou-Soulmaidi, John Casti, Frank McGillion, Stephanie Winder, and Claudia Zatta for their support and company. I thank an anonymous review for advice as regards the book's title. Christopher Coughlin at Springer Nature has been wonderfully supportive in guiding me towards presenting my wide, interdisciplinary project in as clear and manageable a format as I have been capable of providing. Thanks also to Thiru and Tessy in the publisher's team for their hard work in producing this book.

Manchester, UK Peter N. Jackson

Competing Interests The author has no competing interests to declare that are relevant to the content of this manuscript.

Table of Abbreviations Used

I have generally preferred to use English names over Latin so that the source is clearer to the reader. I have taken translations mainly from the Loeb and Oxford editions or, on rare occasions, from other editions and individual translations by scholars. Remarkably, the extant work listed below is only a small fraction of Aristotle's total work, an ancient catalogue of which is given by Diogenes Laërtius in Book V Chapter 1 of his *Lives of the Eminent Philosophers* (see webpage bit.ly/DiogLaertAristotleE). Other catalogues are provided by Hesychius and Ptolemy and both are recorded in V. Rose's *Fragments*.

Names of Groupings	**Names of Aristotelian Texts**	***Abbr. Used***
	Nicomachean Ethics / Ethica Nicomachea	***N.E.***
	Eudemian Ethics / Ethica Eudemia	***E.E.***
	Politics / Politica	***Pol.***
	Rhetoric / Rhetorica	***Rhet.***
	Poetics / Poetica	***Poet.***
	Magna Moralia	***MM***
	Physics / Physica	***Phys.***
	Metaphysics / Metaphysica	***Met.***
	On Coming-to-Be and Passing-Away / On Generation and Corruption / De Generatione et Corruptione	***GC***
	On the Soul / De Anima	***De An.***
	On Sense and Sensibles / On Sense and Sensibilia / De Sensu et Sensibilibus	***Sens.***
	On Memory and Recollection / De Memoria et Reminiscentia	***Mem.***
	On Sleep and Waking / De Somno et Vigilia	***Somn.***
	On Dreams / De Insomniis	***Insomn.***
Parva Naturalia	**On Divination by Dreams** / De Divinatione per Somnum	***Div.***
	On Longness and Shortness of Life / De Longitudine et Brevitate Vitae	***Long.***
	On Youth and Old Age [On Life and Death] / De Juventute et Senectute	***Juv.***
	On Respiration / De Respiratione	***Resp.***
	On Breath / **De Spiritu**	***Spir.***
	History of Animals / Historia Animalium	***HA***
	On the Parts of Animals / De Partibus Animalium	***PA***
	On the Movement of Animals / De Motu Animalium	***MA***
	On the Generation of Animals / De Generatione Animalium	***GA***
	On the Progression of Animals / De Incessu Animalium	***IA***
	Categories / Categoriae	***Cat.***
	On Interpretation / De Interpretatione	***De Int.***
	Prior Analytics	***Pr. An.***
Organon	**Posterior Analytics**	***Post. An.***
	Topics	***Top.***
	On Sophistical Refutations	***Soph. El.***
	Meteorology / Meteorologica	***Mete.***
	On the Heavens / De Caelo	***De Cael.***
	Fragments (Rose 3rd edition = R^3)	***Frag.***

"I have hitherto sometimes spoken as if the variations—so common and multiform in organic beings under domestication, and in a lesser degree in those in a state of nature—had been due to chance. This, of course, is a wholly incorrect expression, but it serves to acknowledge plainly our ignorance of the cause of each particular variation".

Charles Darwin *The Origin of Species* (1859) p 60

Introduction

Let us begin our introduction with Socrates and end it with Aristotle. What did the "Socratic" philosophical movement collectively achieve? It moved the focus of attention from the cosmos, the "outer" world, to the human being, the "inner" world. The initial impulse of this movement was provided by the humanism of Socrates—the "midwife" and "gadfly" of humanity—but it ended with the realism of Aristotle and did so through the idealism of Plato. To come to a consideration of Aristotle and his place in philosophy, we need to understand the relationship between these three movements of humanism, idealism, and realism and contemplate how such a remarkable spectrum of perspectives on the human condition was diffracted from a single "Socratic" light.

Plato's "Socrates" famously explained his humanism on his deathbed as that "my talking to you [cannot be] attribute[d] to sound, and air, and hearing [but only to] the true cause, which is, that the Athenians have thought fit to condemn me, and accordingly I have thought it better and more right to remain here and undergo my sentence (*Phaedo* 98c)". Plato, following Socrates, argues that we must not be content with finding meaning in the mechanisms of the world but must seek its "true cause" of which these mechanisms are only the worldly residue. We should not be distracted by the means of our science from concentrating upon the end-meanings of the world. Aristotle would undoubtedly not disagree with his teachers as regards the importance of looking through the world to the "true [four] cause[s]" which drive it but he does, we find, disagree with them as regards the role of what we call "science". He was himself as interested in the "becoming" understood through the investigations of science as in the "being" understood through the ideas of philosophy. He hence broke the world down into the various special sciences we know today—physics, biology, ethics, etc.—and separately created metaphysics, ontology, and logic as wider philosophical frameworks. A further important point to note here is that Aristotle defines and develops science at the same time as he develops a counterbalancing critique of science. This aporetic development of both problem and solution is typically Aristotelian.

Aristotle argues that we must understand the nature of the world by becoming intimate with its various beings. Aristotle's great philosophical entry point is to interrogate "being" (*ousia*) as Socrates' more foundational role is to establish within philosophy the basic principle of "know thyself" (*gnōthi seauton*).[1] For Aristotle, we ultimately "know ourselves" by understanding ourselves as beings within an environment of other beings and of being itself. Socrates' more directly humanistic insight is well explained by Michel Foucault as that: "in a sense, we may well say that the meaning and value of Socrates' death is at the heart of Western rationality"[2] and that: "Socrates appears as the person who, in taking care of others, is capable of teaching them to take care of themselves".[3] Socrates exemplifies the ethical stance of a teacher—and became an immortal paradigm of such through Plato—whereas Aristotle would later transform these images of good character into a science of human habit and behaviour in his *Ethics* which is itself more a continuation of his biological works than a pedagogical work on morality. The difference between them is radical in appearance and yet Aristotle's goal is to ground Socrates' insight and also, as we will see below, to provide it with further subtlety.

Aristotle critiques, transforms, and complements Socrates. He explains his own philosophy of realism by contrast with Socrates' as that: "Socrates…thought the virtues were rules or rational principles (*logoi*) (for he thought they were, all of them, scientific knowledge), while we think they *involve* a rational principle (*meta logou*) (*N.E.* VI 1144b 28-30)" (*cf. E.E.* I 1216b 11-15). It is through this concept of the "involvement" of rational principles in being that Aristotle transforms Plato's concept of the "participation" of beings in the Forms. The critical emphasis of Aristotle's revision here is that reality is not reducible to mind or mental abstractions but can only be understood as the expression of beings through such values and structures as "the good and beautiful". The world we encounter is not directly "*logos*" but emerges "*meta logou*" from the being and activity of beings. Values, meanings, goods, and even utility emerge from the world rather than being automatic, formal expressions of it. Aristotle adds the further important point that we must contextualise our humanity on this basis. We cannot reduce the world to our ideas but must regard our existence in the world as an expression of such ideas, emergent from our human situation. While he places stress in his logic upon the significance of definition (*horismos*) for all forms of thought, he does not *reduce the world* to definitions or ideas. This is one example of Aristotle's anti-reductionism. He also argues that the world is made up of form, mind, science, causation, mathematics, language, etc. but is not reducible to any one of them.

---

[1] Søren Kierkegaard explained: "He [i.e., Socrates] entered into the role of midwife and sustained it throughout; not because his thought 'had no positive content', but because he perceived that this relation is the highest that one human being can sustain to another. And in this surely Socrates was everlastingly right (*Philosophical Fragments* (Princeton, [*orig.* 1843] 1967) p 8)". On "the care of the soul" see also Jan Patočka *Plato and Europe* (Stanford, [*orig.* 1973] 2002) and the later work of Michel Foucault.

[2] Foucault, Michel *The Courage of Truth: The Government of Self and Others II* (Basingstoke, [*orig.* 1983–4] 2008) p 120.

[3] Foucault, Michel *Ibid.* p 159. He explains further that: "the issue is not the life and death of Socrates, it is love (*The Government of Self and Others* (Basingstoke, [*orig.* 1982–3] 2010) p 328)".

We will see below that Aristotle takes the critical ideas of all such thinkers as Socrates and Plato and refines and transforms their thought and guiding ideas (or ideologies) by contextualising them all within the world. This ability to absorb all positions and to transform them into a philosophical topology of being defines Aristotle's approach; our worldview must account for the world in all its aspects and representations. As regards how this general approach applies to his most immediate predecessors, he argues that: "Socrates…was busying himself about ethical matters and neglecting the world of nature as a whole but seeking the universal in these ethical matters, and fixed thought for the first time on definitions; Plato accepted his teaching, but held that the problem applied not the sensible things but to entities of another kind—for this reason, that the common definition could not be a definition of any sensible thing, as they were always changing. Things of this other sort, then, he called Ideas (*Met.* A 987b 1-8)". The critical emphasis here is that we cannot rely exclusively on universals or ideas. We see here, as throughout, that Aristotle is consciously involved in surveying *all* philosophical approaches in the world, as his *Politics* surveys all the political approaches of the societies around him. Aristotle's philosophy itself embodies his own synthetic and realistic approach to the world.

Aristotle himself ultimately completes Socrates' humanism in the sense that his *Ethics* concludes with the insight that the highest dimensions of the ethical life are the "contemplative insight" (*sophia*) exemplified by "Anaxagoras" and the "practical insight" (*phronēsis*) exemplified by "Pericles". Even Plato's paradigmatic "Socrates", his life and death, sits behind Aristotle's scheme as representing the possibility for a synthesis of the two sides of "mind" through the philosopher who is able and willing to apply his philosophical reflection and reason to the affairs of the human world. The important general principle of Aristotle's thought we find here is that an entity such as the human being can be looked at peripatetically, in many different ways and from many different perspectives. In respect to science itself, any given "being", in this case the "human being", cannot be subsumed in the science but it can be partially explained or represented by it. As regards human beings themselves, we find that they are exceptional in the way embodied by Socrates himself—as an exemplar of a human person—as exhibiting the possibility of individual greatness through a quotidian humanity. The genius of Socrates was that he could himself by being a spiritual "midwife" and a political "gadfly" become a catalyst for and exemplification of the peculiar qualities of our common humanity.[4]

"Socrates" (469–399 B.C.) claimed in respect to his relationship with his own philosophical heritage that he was spurred into philosophical reflection on the nature of "being" (*ousia*) and "mind" (*nous*) through "someone reading, as he said, from a book of Anaxagoras, that mind was the disposer and cause of all, and I was delighted at this notion". Plato depicts Socrates' philosophy as being inspired by and as a reaction to Anaxagoras' representation of "mind" (*nous*) as cosmic cause. He expands on this situation further as that: "What expectations I had formed, and how grievously was I disappointed! As I proceeded, I found my philosopher altogether

[4] See especially Pierre Aubenque *La prudence chez Aristote* (Paris, 1963).

forsaking mind or any other principle of order, but having recourse to air, and ether, and water, and other eccentricities. I might compare him to a person who began by maintaining generally that mind is the cause of the actions of Socrates, but who, when he endeavoured to explain the causes of my several actions in detail, went on to show that I sit here because my body is made up of bones and muscles [and] I am able to bend my limbs, and this is why I am sitting here in a curved posture (*Phaedo* 98c)". Socrates observes here that mechanicism is not naïve realism but is on the contrary a rejection of reality in favour of a system or technique. It is a top-down understanding of the world through the ideology of a detached "mind". We need to contend with the fact that the human mind itself imposes an ideological structure upon the world. Our language, mathematics, and logic limit our expression as well as enabling it, as Aristotle will explain.

Aristotle (384–322 B.C.) was also indebted to Anaxagoras (*c*. 500–428 B.C.) for both positive and negative inspiration. Aristotle's ultimate compliment was to take Anaxagoras' explanation of the eclipse as his own model for a good *scientific* explanation at *Posterior Analytics* 2.8 and to look to him for the basis of his *Meteorology*.[5] Aristotle commented on Anaxagoras' conceptualisation of "mind" (*nous*), as Plato's Socrates had previously, that: "When someone said that *Nous* is present—in nature just as it is in animals—as the cause of the cosmos and of all its order, he appeared a sober man among the random chatterers who preceded him. We know that Anaxagoras clearly held these views (*Met*. A. 984b 15-19)". He also, however, concurred with Socrates as regards the impossibility of building a philosophy from Anaxagoras' mechanistic worldview on the basis that: "The old *physiologoi*...did not see that the causes were numerous; they recognised only the material cause and the efficient cause (and even these they did not clearly distinguish), whereas they paid no attention to the formal cause and the final cause (*GA* V 778b 7-11)". Aristotle argues that the "sublunary" world of beings cannot be understood only as moving matter but must also be understood as a variegated world, as an ecosystem ("eco-" from "*oikos*" meaning "house"), of formal beings with their own place within their milieu. This is a clear and careful framing of reality which is both evidently true and philosophically profound and consequential.

As regards Aristotle own philosophy of cause (*aitia*), we first see that his "four causes" are an important statement of irreducible complexity which leads to an equally important description of the world as material potentiality (*dynamis*) realised in formal actuality (*energeia*/*entelecheia*). Through this elemental framing of complexity, Aristotle's wider philosophy and research can then unfold and reveal a more detailed and nuanced framing for causality and meaning. We do indeed find that there are four basic causes necessarily acting together in our world but also that there is *one* principle of causality operating on beings each with their *own* causality on the basis that: "The causes and the principles of different things are in a sense different, but in a sense, if one speaks universally and analogically, they are the same for all (*ta d' aitia kai hai archai alla allōn estin hōs, esti d' hōs, an katholou*

[5] See Patricia Curd *Anaxagoras of Clazomenae* (Toronto, 2007) p 234.

legē tis kai kat' analogian, tauta pantōn) (*Met.* Λ 1070a 31-32)". There is not only one reality but also many levels and dimensions of that reality and of our understanding and representation of that reality. In approaching the world through a careful but partial, i.e. scientific, understanding of it, we can achieve a solid sample of truth but will inevitably fail to capture this multi-faceted philosophical whole. The establishment of a rich layered and entangled world of meaning-making by and between beings (*ousiai*) occludes the alternative explanation of that world as an inert material world moved only by mechanical and/or stochastic mathematical laws.

When Anaxagoras invented the problem of "cause" he opened up the problem of the place and nature of "mind" and "meaning" in nature. Aristotle, following Plato (and his Socrates), critiqued Anaxagoras on the basis that he does not adequately explain "*nous*" (and hence human being and "Socrates") but that he rather "uses reason (*nous*) as a *deus ex machina* for the making of the world, and when he is at a loss to tell from what cause something necessarily is, then he drags reason (*nous*) in (*Met.* A. 985a 18-22)". Aristotle's student Eudemus explained this line of critique further as that: "despite having allowed for *Nous* [mind], Anaxagoras introduces the agency of chance [i.e., as a *deus ex machina*] for most things".[6] We see here the first example of the remarkable situation that if we reduce the world to causal necessity, we are led to simultaneously reduce it to chance. We find a proto-Cartesian dualism dividing the world into a free mind and a mechanical world. In his own day, Aristotle also argued that Plato's excessive formalism (necessity) was a premature reaction to Heraclitus' excessive flux (chance). His own alternative is represented by his logic which does not reduce the world to this dualism but rather unpacks and explains, as biosemiotics also seeks to do, our *bio*-logical world as the mental and worldly structures by means of which we engage with the world around us.[7] The tendency to break the world into "chance" and "necessity" does, however, still deeply inform philosophical thought in our day, and we will consider this situation in detail below.

Aristotle ultimately argues that "mind" is not (as with Anaxagoras) a mechanistic or practical affair and is not (as with Plato) an ideal or mathematical affair but is rather the matrix which transmits meaning through the world. Aristotle's "four causes" are hence not merely a scientific or logical method but are a philosophical framing for being(s). They provide us with a matrix of "realism" which expresses a real observed world in which beings have formal natures which move towards their ends according to their matters and movements. It is the "movement" (*kinēsis*) of a "being" (*ousia*) that brings "mind" (*nous*) into focus. Aristotle is similarly concerned with "soul" (*psychē*) in his *On the Soul* not as a detached spirit of personhood (as contrasted with Plato's *Phaedo*) but as the dynamic and embodied structuring of a living entity. Aristotle's study of "nature" (*physis*) moves our attention from the purely physical world and from the peculiarly human world to take in

---

[6] Simplicitus *Commentary on Aristotle's Physics* (London, [*orig. c.* 500 AD] 1992) 327.26

[7] See Walter Elsasser "A Form of Logic Suited for Biology" in Rosen, Robert (*ed.*) *Progress in Theoretical Biology*, Vol. 6 (Academic Press, 1981).

the living world of biological beings. He regards an entangled world in which beings exist and interact upon the basis of a natural platform of possibility. His "*nous*" is embodied in beings and also in being itself. His research projects all seek to draw out and explain the realised nature of these concepts. Aristotle ultimately retains all of Anaxagoras and Plato's insights but naturalises—and biologises—them in the real world. This philosophical picture is notably replicated in modern science by biosemiotics which similarly resists reductionism and seeks to understand "meaning making" through the "semiotic scaffolding" of a multiply determined and holistically constituted reality in which beings express themselves. I will indeed show in the subsequent volume to this world that this recognition of reality in modern science is by no means limited only to biosemiotics.

We ultimately find that Aristotle takes his efficient cause from Anaxagoras and his final cause from Socrates and he will also take his formal cause from Plato. What we see here is far more than a magpie approach to philosophy. We see not only the importance of synthesising ideas and systems but also the more profound emphasis that we should not understand the world through an ideology but through a nuanced philosophy of being(s). It is hence that Aristotle shows us through "being" that we cannot reduce "the well and beautifully disposed (*eu kai kalōs*) (*Met.* A 984b 11-12)" to "matter", "elements", or to "chance" or any other single "cause" such as "mind" (*nous*) or even "God",[8] He argues that "beauty" is not just a surface or chance attribute but is something meaningful in the sense that we must account for "a principle in things which is the cause of beauty, and the sort of cause by which motion is communicated to things (*tēn toiautēn hothen hē kinēsis hyparchei tois ousin*) (*Met.* A 984b 20-3)". The world as it appears, and as we find it before us, is a world of beauty, cause, and being which is generically of a nature to be expressed, channelled, and communicated. We need to create a rich world of ideas to express the rich world of beings before us.

Does Aristotle's belief in such concepts as "beauty", "good", "perfection", "virtue", etc. not represent a disqualifying anthropomorphism? Aristotle puts forward his position as that: "they are in error who assert that the mathematical sciences tell us nothing about beauty or goodness (*peri kalou ē agathou*) (*Met.* M 1078a 33-34)" and that: "The chief forms of beauty [are] order (*taxis*) and symmetry (*symmetria*) and definiteness (*to hōrismenon*) (*Met.* M 1078a36-1078b1)". Is it ridiculous to say that there is a beauty in the order of the world articulated by mathematics and of the symmetry and definiteness of beings as articulated most fully by biology? The concept of "symmetry" is deeply embedded in Aristotle's thought, most obviously in

[8] Aristotle's basic argument on "God" is that we can in some sense infer the "divine" but that: "each man wishes himself what is good, while no one chooses to possess the whole world if he has first to become someone else (for that matter, even now God possesses the good); he wishes for this only on condition of being whatever he is; and the element that thinks would seem to be the individual man, or to be more so than any other element in him (*N.E.* IX 1166a19-23)". Aristotle, in a typical argument, argues that we must live in accordance with and through our humanity and as individual human beings rather than seeking to become some sort of "divinity", a status that is by definition beyond us and different to us.

his biology.[9] The critical difference between the modern positivist position and the qualitative approach of Aristotle is that one finds in symmetry only a mathematical formula the other finds a principle through which the orderings of beings emerge through which they can articulate themselves so as to live and reach their ends. He explains this point in principle as that: "in the statement that 'coming-to-be is a channel [or leading] towards being' (*hoti hē genesis agōge eis ousian*) or that 'health is a balancing [or symmetry] of hot and cold' …[t]he words 'channel [or leading]' and 'balancing [or symmetry]' are equivocal (*homōnymos gar hē agōge kai hē symmetria*) (*Top.* VI 139b20-22)" and that: "'disposition', (*diathesis*) 'state' (*hexis*) and 'symmetry' (*symmetria*)… these terms cannot possibly exist anywhere else except in the things in relation to which they are employed (*Top.* IV 125a35-37)". This referencing of concepts to "beings" (*ousiai*) is essential to Aristotle's position generally and is equally true of all his other philosophical concepts such as of "proportion" and "middle" (or "mean") which denotes a "centre" or "channel" for "states" allowing us to produce a real picture of lived reality.

We find that biosemiotics naturally supports Aristotle in this approach. The biosemiotician Kalevi Kull was led in his mature, recent work to address "beauty" two and a quarter millennia later. We say better late than never! Kull introduces the concept of "semiotic fitting" as that: "Habit by itself is not good or bad, it is good or bad because of semiotic fitting…Semiosis is usually biased towards semiotic fitting, which stepwise leads towards perfection, and thus towards beauty".[10] He explains "perfection" by reference to Jakob von Uexküll, the great biological founder of biosemiotics, as that: "Perfection is not omnipotence, it only means the correct and complete use of all available means" (Uexküll, 1928: 137). For each species, perfection is their own. "Every organism can only be itself. But within itself it is perfect, […] unlike our human implements […] (Uexküll, 1928: 138)".[11] Aristotle himself prefigures this thought as that: "…'Each animal is thought to have a proper pleasure,

[9] Aristotle argues in principle regarding symmetry in biological beings that: "In fact, all of them [i.e. the organs] are double (*diphya*). And the reason for this is that the structure of the body is double (*diphyēs*), though its halves are combined under one source (*pros mian archēn*) (*PA* III 669b 18-20)". This duality is assumed to inform all aspects of organic being and hence: "In all animals the brain is double. Beyond this, at the far end, is the cerebellum as it is called; its form is different from that of the brain, as can be both felt and seen (*HA* I 494b31-4)". Aristotle adds that animals have a clear symmetry in respect to kidneys and lungs and he also tries, more disputably, to add the heart since its cavities make it bipartite (*PA* III 669b 22-26) and more obviously incorrectly the spleen as the "double" of the liver (*PA* III 669b 15-18 & 669b 36-670a 3). I note that Aristotle expresses doubts on this last matter just as he also doubts that the sense of touch can be easily explained in the terms of symmetry—though human beings do, of course, have two hands—as the other senses evidently can. In respect to physical symmetry Aristotle goes to extraordinary lengths to show the general "symmetricity" of the body by pointing out that even the tongue is divided (*PA* II 657a1-2) which Peck interprets (p 180 of his Loeb translation) as referring to forked tongues but which Lennox relates to the fact that the tongue in most animals has a clear "midline" (p 227 of his Oxford translation)).

[10] Kull, Kalevi "The Biosemiotic Fundamentals of Aesthetics: Beauty is the Perfect Semiotic Fitting" *Biosemiotics* (2022) p 1.

[11] Kull, Kalevi *Ibid.* p 2.

as it has a proper function'; *viz.* that which corresponds to its activity. If we survey them species by species, too, this will be evident; horse, dog, and man have different pleasures (*N.E.* X 1176a3-6)". This is an obvious principle but Aristotle builds upon it a profound philosophy in which such things as pleasure, mind, actuality, and beauty show us not just how "being" and "beings" transcend their material existence but exist for this transcendence. These "values" are not the add-ons of our human minds, as the Kantian position would suggest, but are embedded in a qualitative nature and reflected through humanity.

Kull approaches this subject by drawing out the biosemiotic concept of "*Umwelt*" (or speciate environment, milieu, or world) as an ideal fittedness on the basis that: "the correspondence between an organism and its umwelt is usually ideal"[12] and that: "By perfection I mean fitting multirelationally. Like how a perfect key is not only one that opens its lock smoothly; it should also be handy, suit the user's pocket, have no sharp edges that could damage the user, be durable, replaceable if destroyed, not too difficult to construct, recyclable, etc.".[13] A similar key analogy is also used by the philosopher of science Patrick Suppes to explain Aristotle's thought. Suppes holds up a key in order to "face up to the nature of the accuracy and speed of human perception, and also of many animals as well" through Aristotle's concept of "same form, different matter".[14] Suppes scientifically seeks to address here the famous philosophical "*Augenblick*". We see in all these cases that to engage with our lived reality is not anthropomorphism but a proper use of science to help fully and roundly explain the world in which we live. The reality considered here is not the flat mechanical world of the accountant or engineer but the rich, immediate, and ineffable reality that the philosopher, scientist, and theologian seek to appreciate.

Tim Ireland makes some interesting points as regards this "aesthetic" world. He goes back to philosophical and historical fundamentals and notes that: "The concept of aesthetics was originally coined by the philosopher Alexander von Baumgarten (1741–1762), who argued aesthetics is the study of the plenitude and complexity of sensations. When Kant took up the concept he drained it of its sensory plenitude, revising its significance to contemplation and judgement of beauty".[15] The principle of Baumgarten contra Kant is that: "art [is not] an intellectual pursuit [a]s opposed to an extension of nature" and this leads us back to Kull's point that: "understanding the nature of beauty requires biology".[16] We ultimately find that beauty is an absolute concern of biology as a peculiar potentiality of its forms and beings in their interrelations. It is not just an expression of "virtue" (*aretē*) or an "excellence"—these being the same in Greek so that a sharp knife manifests "*aretē*"—but is

---

[12] Kull, Kalevi *Ibid.* p 2.

[13] Kull, Kalevi *Ibid.* p 11.

[14] Suppes, Patrick "[Commentary on] Dagfinn Føllesdal" in Crangle, Colleen E. *et al* (*eds.*) *Foundations and Methods from Mathematics to Neuroscience* (Stanford, 2014) p 266.

[15] Ireland, Tim "Beauty: Synthesis of Intellect and Senses Commentary on the Biosemiotic Fundamentals of Aesthetics" *Biosemiotics* (2022) p 3.

[16] Kull, Kalevi "The Biosemiotic Fundamentals of Aesthetics: Beauty is the Perfect Semiotic Fitting" *Biosemiotics* (2022) p 3.

something even more fundamental. Kalevi Kull himself expands upon this point by quoting (among others) the cyberneticist and polymath Gregory Bateson that: "What is the pattern which connects all the living creatures? [...] By aesthetic, I mean responsive to the pattern which connects. [...] ultimate unity is aesthetic".[17] He also quotes the ethologist Irenäus Eibl-Eibesfeldt as that: "Our perception seeks and enjoys order, and this quest seems to be a general principle which derives, in part, from the limited capacity of our brain to process information. Our short-term memory seems to have the capacity to process 16 bits per second; less is perceived as boring, and more is stressful. In patterns, we attempt to discover regularities that allow us to form 'super-signs' to reduce the amount of information confronting us. [...] Thus, an aesthetic object must have a certain amount of order that should be neither too complex nor too simple".[18]

What do we find here? Is this philosophy or science? Kull's "semiotic fitting" is very different in meaning and tone that the reproductive "fitness" so beloved of neo-Darwinism. Indeed, Kull explicitly denies that "beauty" can be explained merely as a mechanism for sexual selection. The reality does not *reduce* to this law. There is no reason why we should seek to *reduce* our rich engagement with reality to such laws. It is perhaps this striving after meaning rather than solutions that characterises Aristotle's thought, biosemiotics, and the various modern scientific traditions that I will treat in the sister volume to this work.[19] Here, I note that Aristotle's fuller position as a scientist is that: "we should venture on the study of every kind of animal without distaste; for each and all will reveal to us something natural and something beautiful. Absence of haphazard (*tychontōs*) and conduciveness of everything to an end (*heneka*) are to be found in nature's works in the highest degree, and the end for which those works are put together and produced is a form of the beautiful. If any person thinks the examination of the rest of the animal kingdom an unworthy task, he must hold in like disesteem the study of man. For no one can look at the elements of the human frame—blood, flesh, bones, vessels, and the like—without much repugnance (*PA* II 645a 23-30)". Aristotle considers himself as a biological being from which emerges his own nature, emotions, drives, virtues, and motivations as a truly exceptional human being. It is on this basis that he considers reality through his "realism" rather than the disembodied abstractions that we have come to identify with the scientific project.

Aristotle concludes the above passage by noting that "the principal object of natural philosophy is not the material elements, but their composition (*peri tēs syntheseōs*), and the totality of the substance (*kai tēs holēs ousias*), independently of which they have no existence (*PA* II 645a 35-6)". The critical point made here is that "beauty" and "good" emerge from the complexity and synthesis of being(s). We can understand this position as a realistic reinterpretation of Plato's Theory of Forms

[17] Kull, Kalevi *Ibid.* p 5.

[18] Kull, Kalevi *Ibid.* p 13–14.

[19] For interesting discussions on aesthetics in this context, see Lancelot Law Whyte (*ed.*) *Aspects of Form* (London, 1951) and Philip C. Ritterbush *The Art of Organic Forms* (Smithsonian, 1968).

which understands a being as a *Gestalt* entity which is a distinct something (a "such" or "*toionde*") and a distinct individual (a "this" or "*tode ti*"). How does *this* being stand before us as one of many? I suggest that this concern with the beings of the real world is Aristotle's *true* (and original) form of empiricism. I add that whereas "theory" ("*theōria*") and "theoretical" means scientific "reduction" to the modern mind it originally meant philosophical "contemplation" about being(s) to the Greeks.[20] The approach(es) to the world considered above is(are) conceptual, philosophical, and traditional and not abstract, technical, and vanguardist and this is critical difference of viewpoint between realism and idealism.

How does Plato (*c.* 428–347 B.C.) further compound (according to Aristotle) Socrates' error of abstraction? Aristotle asserts that Plato's "third man" fallacy displaces the actual "being"—i.e., the "Socrates"—by transferring attention from "sensible things…to entities of another kind (*Met.* A 987b 5)".[21] He explains this problem with Plato further as that: "Only the name 'participation' was new; for the Pythagoreans say that things exist by 'imitation' of numbers, and Plato says they exist by participation, changing the name (*Met.* A. 987b 12-14)" and that: "The supporters of the ideal theory were led to it because on the question of the truth of things they accepted the Heraclitean sayings which describe all sensible things as ever passing away, so that if knowledge or thought is to have an object, there must be some other and permanent entities, apart from those which are sensible; for there could be no knowledge of things which were in a state of flux (*Met.* M 1078b 12-17)". Aristotle regarded Plato's "form" as a premature concession to Heraclitus' "flux", and believed that Plato's denatured notion of "participation" reduces the world to a reflection on mathematics.

Aristotle's own remarkable response to this conflict between the philosophies of form and flux was to accept and synthesise *both* positions, i.e. he argued that nature must be understood as containing "form" (*eidos*) which he redefined as "species" (*eidos*) and *also* as a material world which is constantly flowing or subject to constant movement and "change". We again see that Aristotle creatively builds upon his opponents, in this case Plato, even in his radical opposition. We can also usefully consider Aristotle's objection to Plato's idealism in the context of his objection to Anaxagoras. Aristotle argues that: "The statement that complete separation never

[20] See Nicholas Lobkowicz *Theory and Practice: History of a Concept from Aristotle to Marx* (Notre Dame, 1967) p 48–9.

[21] Aristotle puts forward the "third man" as an explanation of how abstractions and ideologies can obstruct and distract from our understanding of the world. He explains how both mind and language create alternative realities which confuse on the basis that: "a substance will be present in 'Socrates', who is a substance (*enyparxei ousia ousia*); so that it will be the substance of two things. And in general it follows that if "man" and all terms used in this way are substance, none of the elements in the formula is the substance of anything, nor can it exist apart from the species or in anything else; I mean, e.g. that neither "animal" nor any other element of the formula can exist apart from the particular species. If we look at the question from this standpoint, it is obvious that no universal attribute is substance; and it is also clear from the fact that none of the common predicates means "so-and-so" (*tode ti*), but "such-and-such" (*toionde*). Otherwise among many other awkward consequences, we have the "third man" (*Met.* Z 1038b 29–1039a 3)".

will take place is correct enough, though Anaxagoras is not fully aware of what it means…if separation took place, there would be a 'white' or a 'healthy' which was nothing *but* white or healthy…So his 'Mind' (*Nous*) is [metaphorically as] an absurd person aiming at the impossible (*Phys.* I 188a 6-10)". Aristotle goes to great lengths to rebut the idea that we can reduce the world to generic "atoms" or "elements" of being which are then mixed together somehow by "mind" or "chance". It also cannot be reduced to the "laws" of an abstract, disembodied "Mind". The world cannot be reduced to "simples"; it is always complex and "entangled" being and must be represented as such.

The core of Aristotle's philosophy is an emphasis upon the "*ousia*" (being). His emphasis is, as we will see below, similar with Heidegger's central concept of "*Dasein*". As regards "mind", Aristotle insists that the "mind" in this world is always the mind *of something*. As regards the cosmic "*Nous*" set out in his *Metaphysics* Book Λ, this is carefully left deliberately blank as representing the simple potentiality for meaning that we must employ and actualise in our sublunary existence. Aristotle makes it clear that we cannot know the "Mind" of the "Unmoved Mover" but can only recognise that there is a cosmic system, a closed circle of meaning that we can tap into. Aristotle hence famously argues as regards Plato's "Forms" that we cannot imagine that there is some person of the "Demiurge" in heaven churning out beings according to a divine archetype. He similarly argues that we cannot represent the world as a single "world-animal" since the beings of the world and their souls cannot be simply aggregated or unified on this basis.[22] He adds that we must recognise the existence of "mind" in nature, but should not reduce the world to mind and nor can we conflate "mind" with "soul" with Aristotle criticising Plato (as Leibniz will later criticise Descartes) for doing so at *De An.* I 407a. Aristotle ultimately argues that we cannot simply reduce the world to our mental representations or "ideas" of it and must rather approach it through its "beings". This is the basic principle of Aristotle's cosmogony or cosmology, and we need to expand upon it and justify it below.

As regards the fundamental point where Aristotle and Plato diverge in their ontologies, i.e. between realism and idealism, the nub is Plato's reduction of the world to geometrical atoms that are "invisible and characterless" which Aristotle critiques on the basis that: "If this body has weight or lightness, it will itself be one of the elements: if it has no impulse in any direction, it will be unmoved, i.e., a mathematical object (*De Cael.* III 305a 24-27)". His emphases upon the "movement" and "impulse" of something having matter are important and will be considered later. Here, though, let us note that Aristotle disagrees both with the framing of Plato's "Forms" and also with his "atoms" on the basis that neither explain the "growth" and "becoming" of being in the material world and hence: "Plato, it is true, investigated coming-to-be and passing-away [but he did not]…examine the conditions

[22] On the world-animal, see Dimitri El Murr "Platonic 'Desmology' and the Body of the World Animal (*Tim.*30c-34a)" in Salles, Ricardo (*ed.*) *Cosmology and Biology in Ancient Philosophy: From Thales to Avicenna* (Cambridge, 2021).

under which 'alteration' or 'growth' are attributable to things (*GC* I 315a 33-34)"; and see *Timaeus* 50e-55c. I will contend below that Aristotle's great innovation was to demand representation for the agency of beings (*ousiai*) and that this led him to transform Plato's ontology of mathematical structure into an ontology centred upon the example of biological beings (*ousiai*). Aristotle is concerned with the nature of "matter" (*hylē*) in creating a "being" (*ousia*) which provides an avenue to the problem of "the one and the many" alongside the problem(s) of the material nature of reality ("realism") and of the ideal and abstract quality of mind ("idealism").

I note that Andrea Falcon explains the above situation as a result of a remarkable sleight of hand by Plato around the term "body" (*sōma*). Falcon argues that: "Without a suitable definition of body that allows a pathway from mathematical entities to non-mathematical ones (and vice versa), the reconstruction proposed by Plato is impossible. It is the ambiguity of the Greek term *sōma* that renders the entire Platonic operation plausible. Moreover, this is an ambiguity that Plato exploits knowledgeably, using *sōma* to refer to the four bodies of the Empedoclean tradition as well as to the four regular polyhedra with which these simple bodies are associated".[23] Falcon identifies the problem that Aristotle identified and addressed as being the lack of an immediate pathway from mathematical constructions to material beings. Carl Friedrich von Weizsäcker, by contrast, presents us with the defence of Plato that: "The mathematical forms…are actually what Plato calls ideas. Now begins the descent, the theory of ideas in the strict sense. The ideas first make clear what mathematical forms are, these, what physical bodies are, and those, what sensory impressions of things are, thus always further into ideas released into multiplicity. The reason this philosophy appears as a conceptual notion lies in the difficulty of understanding the metaphor 'in the cave', the ontology of ideas, their way of being".[24] If we have a Pythagorean belief in the unreasonable efficacy of mathematics, the existence of bodies and beings becomes of relatively trivial importance.[25] Why should we waste our time messing around in the material world if that world is merely the end product of a formula or ideology?

Both perspectives are exclusively correct on their own terms. The patterns we observe in nature can be explained *either* mathematically or empirically. A.N. Whitehead explains as regards how we can Platonically and mathematically approach the "patterns" (*paradeigmata*) of nature (*physis*) that: "A truth-relation can be said to connect the objective contents of two prehensions when one and the same identical partial pattern can be abstracted from both of them. They each exhibit this same partial pattern, though their omitted elements involve the differences

[23] Falcon, Andrea *Aristotle & the Science of Nature: Unity without Uniformity* (Cambridge, 2005) p 54.

[24] Weizsäcker, Carl Friedrich von *The Ambivalence of Progress* (New York, [*orig*. 1977] 1988) p 132.

[25] See Eugene Wigner "The Unreasonable Effectiveness of Mathematics in the Natural Sciences" in *Symmetries and Reflections: Scientific Essays of Eugene P. Wigner* (Indiana, [orig. 1959] 1967) & Patrick Suppes "Why the Effectiveness of Mathematics in the Natural Sciences is not Surprising" *Interdisciplinary Science Reviews* (2011).

which belong to their diverse individualities. Plato uses the term 'participation' (*methexis*) to express the relation of a composite fact to some partial pattern which it illustrates. Only he limits the notion of the partial pattern to some purely abstract pattern of qualitative elements, to the exclusion of the notion of concrete particular realities as components in a composite reality. This limitation is misleading. Thus we will speak of a pattern as possibly including concrete particulars among its patterned elements".[26] Plato's (and Whitehead's) "patterns" are not trivial, poetical, or subsidiary. They are, rather, an attempt to represent the formal structures of the world.

Aristotle's original exercise in logic—making up his "Organon"—was his own further attempt to explain how we do make sense of the world through our thought upon it and its patterns. He also expands upon Plato's "great kinds" of "change, rest, being, sameness, and difference" at *Soph.* 257b, 258e-259a by arguing that an "element" is not limited to "any small unity which is useful for various purposes" but could also be applied to the genera themselves as elements of being (see *Met.* Δ 1014b3-13). We see here that good philosophy can reveal such obvious but easily neglected facts as that an "elemental" worldview does not necessarily mean that the world is built up as a magnification of little bits of matter. Elements can be used to explain a structured world at many levels: macro-, meso-, as well as micro-. Aristotle *agrees* that the world is "elemental" and that it is also structured and meaningful, but he also insists that this state of affairs is not adequately explained by the concept of the "participation" of beings in abstract and ideal "Forms" or their construction from supposed mathematical (or concrete) particles. He rather argues that we must consider the composition and integrity or "*synthesis*" of beings, of that which makes them the something they are. How are they whole and particular? To achieve this, we must lower our sight to the "beings" themselves, concentrate upon them, and not fall prey to unhelpful idealistic and ideological constructions that derive from our mind rather than from the world itself. The later positivist approach would show the error of this process of simplification by dismissing Aristotle's metaphysics and, essentially, philosophy itself as the result of "bad grammar".[27] The problem we face is simply how we interpret and represent the world around us through our concepts without those concepts themselves becoming reified. Aristotle's answer was to invent metaphysics and (classical) logic in order to achieve this end.

Plato and Aristotle provide us with alternative visions as regards how the world is constructed. In fact, Aristotle takes each of the elements of Plato's system and transforms them. Whitehead explains Plato's foundation for the whole of the

---

[26] Whitehead, A.N. *Adventures of Ideas* (Harmondsworth, 1933) p 278.

[27] Russell, Bertrand "Logical Atomism" in Ayer. A.J. (*ed.*) *Logical Positivism* (New York, [*orig.* 1924] 1959) p 36. Rudolf Carnap gives us a quintessential statement of the core "positivist" position that we need to substitute philosophy for science on the basis that: "the sentences of metaphysics are pseudo-sentences which on logical analysis are proved to be either empty phrases or phrases which violate the rules of syntax. Of the so-called philosophical problems, the only questions which have any meaning are those of the logic of science. To share this view is to substitute logical syntax for philosophy (*Logical Syntax of Language* (New York, 1938) p 9)".

Western philosophical tradition as "the interweaving of seven main notions, namely The Ideas, The Physical Elements, The Psyche, The Eros, The Harmony, The Mathematical Relations, The Receptacle".[28] Aristotle then takes each of these generic principles and transforms it from an "idealistic" (metaphorical) position to a "realistic" (literal) position. He converts, for instance, the "receptacle" from a formal, mathematical vehicle for matter into the idea that matter is "receptive" (*dektikon*) to form so that a being (*ousia*) is produced from and through its own matter and cycles through matter as a species is bound to its food source (see *De Cael.* III 305b 3-6). We will consider Aristotle's concept of "proper matter" (*oikeia hylē*) below. Regarding Plato's "Ideas", i.e. his Theory of Forms, Aristotle argues that it is an over-simplification (and in another sense an over-complication) of a world which is clearly made up from genera, species, and individuals. For Aristotle, the "species" (*eidos*) cannot meaningfully or satisfactorily be compared to a "day" or "sail" covering the particulars, a criticism of the "Theory of Forms" that Plato had himself put forward in his *Parmenides*. Aristotle expands upon and transforms Plato by picturing the world through a consideration of biological emergence rather than through mathematical formalism.

We find that Aristotle evidently builds upon Plato as he also builds upon Socrates and Anaxagoras. His own synthesis accepts but transforms Plato's philosophy. Most obviously, Aristotle produces "realism" from Plato's "idealism" and hence in a sense inverts his philosophical framework. Jan Patočka explains this movement beautifully as that Aristotle deliberately and explicitly "returns to the Cave"[29] in the sense that he argues systematically that we must find meaning in *this* world and not seek to locate meaning in *another* world. Returning to his critique of Socrates and his scientific ethics we can now understand why Aristotle insists that "the virtues [are not] rules or rational principles (*logoi*)...but *involve* a rational principle (*meta logou*) (*N.E.* VI 1144b 28-30)".[30] We do not merely find "rules" or "reasons" moving

---

[28] Whitehead, A.N. *Adventures of Ideas* (Harmondsworth, 1933) p 174. Whitehead represents this situation incisively and well as that: "Where Aristotle said 'observe' and 'classify', the moral of Plato's teaching is the importance of the study of mathematics. Of course, neither of them was so stupid as to dissuade from observation or, on the other hand, to deny the utility of mathematics. Probably Aristotle thought that the mathematical knowledge of his day was about as much as was wanted for the purposes of physical science. Any further progress could only minister to an unpractical curiosity about subtle abstractions (*Ibid.* p 179)".

[29] See Jan Patočka's chapter "The Transformation of the Ontological Project in Aristotle—The Return to the Cave" in *Plato and Europe* (Stanford, [*orig.* 1973] 2002).

[30] Aristotle explains "virtues" ("*aretai*" or "excellences") in the context of "ethics" (and "*hexeis*" or "havings") as that: "we feel anger and fear without choice, but the virtues are modes of choice or involve choice...For this reason also they are not faculties...all that remains is that they should be states of character (*hexeis*) (*N.E.* II 1106a 2-12)". Claudia Baracchi explains this situation well as that: "we may understand excellence [i.e., or '*arētē*' or 'virtue'] as a possession, property, or propriety of the soul, as a *hexis*—that which a soul 'has' and, when in action, *shows*...[and] the process [of learning] unfolds from a 'having' to a 'having', through the enactment of a 'having'...what is at stake is a certain transmission, or even translation, of a given 'having'. A certain activity is transferred from the outside to the inside, as it were. Someone learning takes something in and makes it one's own (*Aristotle's Ethics as First Philosophy* (Cambridge, 2008) p 117 & 109)".

through the world but "beings" (*ousiai*) which employ rules and reasons as they move through the world to their ends. This principle is set out in detail in Aristotle's *Ethics* which considers how individual human beings *develop* in the world in respect to their habits, virtues, ethics, and minds. The importance of this transformation can be understood by comparing Aristotle's *Politics* with Plato's *Republic*. We can then consider that Aristotle considers organic development in his *Generation of Animals* on a parallel basis in the sense that whereas one work considers behavioural change, the other considers biological change. We see the fundamental importance of biology for Aristotle's philosophical representation of the whole world in all its aspects.

Aristotle's ultimate philosophical burden is to explain why we cannot understand the *physical* world through material, mathematical, or any other "atomism" or "formalism" but only as a world which contains the living principle through which the "entelechy" of forms is embodied and expressed in matter as "beings" moving through the world. Against Plato, he insists that we must understand the world as the result of emergent "becoming" as well as of formal "being".

This work seeks to exhibit Aristotle within a "biosemiotic" context. It marries these two seemingly disparate approaches—i.e. that of Aristotle and of biosemiotics—precisely because they both attempt to understand how we find and represent "meaning" as expressed by biological beings and in the world generally. Both approaches seek to understand the world both philosophically and scientifically in order to flesh out what "meaning" itself actually signifies as regards our place in the world. We also find other traditions that seek to represent the world both philosophically and scientifically such as the *Lenbensphilosophie* and Gestalt Theory associated with phenomenology. Since these other traditions also recognise *bio*-logical "meaning", we will also consider and critique them below. We will also consider them in further detail in the subsequent sister volume to this work.

The objective of this work is not just to explain or even to defend Aristotle and/or Biosemiotics. It is, rather, to defend and define the role of philosophy, to demonstrate the limits of science, and to explore the error of conflating philosophy with or reducing it to science. The sister volume of this project will continue by showing that modern science itself refutes the reductionist philosophy that claims to speak for it. We will see throughout that whereas this reductionism may once have been an often noble and reasonable dream based upon realistic hopes in Enlightenment rationalism and scientific progress, today it is the nightmare of an ugly ideology that has captured the world and refuses to relinquish it despite its evident error. Reductionism is a form of destructive over-confidence and arrogance arising from a desire to reduce the world to *our* mind, *our* ideas. In this mode, we formulate a spiritual technology as a religion or a physical technology as a science and we then seek to reduce our lives to meet or fit that technology. Aldous Huxley rightly explains our situation as that: "We are living now, not in the delicious intoxication induced by the early successes of science, but in a rather grisly morning-after, when it has become apparent that what triumphant science has done hitherto is to improve the means for achieving unimproved or actually deteriorated ends".[31] The critique of technology

[31] Huxley, Aldous "Beliefs" in *Collected Essays* (New York, [*orig*. 1937] 1971) p 362.

in human life rightly occupies much philosophy from Aristotle all the way to Husserl, Heidegger, Ellul, Schumacher, Foucault, Camus, etc.

What does it mean to defend philosophy? Why would anyone want to conflate science with philosophy? The anti-philosophy of positivism famously reduces everything to science and dismisses the existence and meaning of metaphysis and ontology. Positivism may be palpable nonsense but, as we should hopefully have already learned in our life, useful nonsense has every chance of enjoying long life and success. The default philosophical position we serve today is that scientific specialists are the ultimate arbiters of meaning. We must challenge this "scientism" on behalf of both Aristotle and Biosemiotics. Either one or the other is correct, not both. The value of Aristotle as a thinker in this context is that this great genius concentrates upon the world itself and leaves "gaps" in his thinking when he cannot adequately represent it. That is ultimately why *his* "God" is a god of the gaps. His science and philosophy similarly address the world and our place within it but by doing so recognise their own limitations. It is by recognising our own limitations that we can recognise and appreciate the fuller world by which we are limited. I hope to demonstrate the value of this philosophical caution and subtlety in this work as a preliminary exercise to showing in its sister volume how modern science actually agrees with Aristotle rather than with the reductionist philosophers who idolise it and claim to speak for it.

The ultimate objective of this project is the reintroduction of "meaning" into our world in place of the cold and false scientific reduction of the world to the "anonymous observer" as set out by Kurt Riezler in his thought experiment presented as *Physics and Reality: Lectures of Aristotle on Modern Physics at an International Congress of Science* (1940). In respect to this recovery of the value of Aristotle's science, other excellent and highly recommended (and little recognised) works on Aristotle are Etienne Gilson *From Aristotle to Darwin and Back Again: A Journey in Final Causality, Species, and Evolution* ([orig. 1971] 1984), Richard Connell *Substance and Modern Science* (1988), René Thom *Semiophysics: A Sketch: Aristotelian Physics and Catastrophe Theory* (1990), and Paul Feyerabend *Conquest of Abundance: A Tale of Abstraction versus the Richness of Being* (1999). There is, of course, a vast literature on Aristotle but little on his biology and little contemplating how Aristotle can be relevant to us today rather than as a historical figure. The importance of biosemiotics is precisely that it provides us with a perspective that makes Aristotle's biology—long dismissed as the most antiquated area of his corpus—vital and relevant to us today.

This project has been broken down into two separate books. This first book considers Aristotle and his philosophy. It sets out Aristotle's philosophy as a vital and living philosophy which is likely as subtle, parsimonious but expansive, and correct a philosophical position as we can unearth. The second book will consider biosemiotics and modern science in the context of Aristotelian philosophy. I note that the critique I put forward as regards science is over its nature, place, and limits and is not, of course, an attack on science itself within its proper bounds. As regards the place of philosophy, I have already sought to show above through Plato or Aristotle that, however fundamental the differences between them, they together demonstrate the immense importance of philosophy as an examination of the structures of reality and a consideration of what they mean to us.

The first three chapters of this book set out the philosophical context for Aristotle in his day and for Aristotle in ours. We first consider how Aristotle's philosophy is a careful synthesis of the various schools of thought of the ancient world. We consider the significance of the fact that Aristotle's own synthesis does not merely represent a *quantitative* accumulation of perspectives but a *qualitatively* transformed philosophy. Philosophy as the quantitative reduction of the world to the force of a single principle—e.g. as form, matter, cause, etc.—is expanded in his hands into a rich qualitative topology of principles which animates and mirrors the real world. René Thom explains this situation, as we will see in more detail later, as that: "His [i.e., Aristotle's] revolt against Plato is that of the topologist against arithmetic imperialism, that of the apostle of the qualitative against the quantitative".[32] We then move on to consider the modern philosophical tradition in detail by comparison. We find that modern philosophy is broken down into a reductionist tradition and an anti-reductionist tradition. The anti-reductionist and synthetic tradition—as represented by Leibniz, Hegel, Whitehead, and Heidegger—is clearly in broad sympathy with Aristotle and his philosophy. The reductionist and analytic tradition—as represented by Descartes, Hume, Wittgenstein, and Rorty—is shown to be inadequate. It is shown that it is not possible to make the claim that philosophy has superseded Aristotle since reductionistic philosophy is clearly inadequate and anti-reductionist philosophy tends to be cluttered with excess baggage. Finally, we review the state of modern philosophy and modern philosophy of biology in outline, showing that there is ample room for improvement.

Having shown the nature of Aristotle's philosophy and its viability vis-à-vis alternatives, the next three chapters flesh out his scientific approach and philosophical vision. We find that Aristotle's logic and his philosophy is very much *bio*-logical in nature. His consideration of ethics and politics follows and extends his biology. His world covers both form and flux, being both a seemingly irregular and contingent world emerging through the possibilities open to beings and a structured and determined reality created through the actualisations of beings and the expression of their defining qualities and the virtues (*aretai*) which constitute their "fitness". Aristotle's concern with the *realisation* of beings as species (*eidē*) means that he develops his own theory of evolution as regards how beings come-to-be. Many people would, of course, say that it is Aristotle's error *not* to give an essential role to "chance" in the evolution of beings. We will see, however, when we come to consider the modern scientific literature in the sister volume of this book that this modern supposition about "chance" as a driver of form, function, and being is *not* generally accepted as valid. We should indeed rather suspect that the emphasis upon "chance" was always a philosophical premise brought to science rather than naturally emerging from its conclusions. Aristotle does in any case provide us with a

[32] Thom, René *Semiophysics: A Sketch: Aristotelian Physics and Catastrophe Theory* (Redwood City, 1990) p 166.

theory of evolution for filling a gap needing to be filled and I will make his case for doing so here.

The next three chapters show us the value of both Aristotle's science and philosophy for studying the meaning of the pathways taken by beings in an entangled or "sublunary" world. As individual human character is built up over life so that various dimensions and specialisms of character and self are built up upon the biological human being itself, so species of being—biological or otherwise—are built up through the peculiar channelling of natural cycles of growth and development. Aristotle's evolution is indeed contingent upon the place and history of a being in the world, but it is not in any meaningful sense reducible to the play of "chance". The final chapter rounds off philosophy in respect to how we approach meaning through it by the two methods, synthetic or analytic. We will see here the remarkable difference between philosophy which seeks to accept the meaning of being and of beings and to then interpret it and the philosophy which seeks to dismiss meaning and replace it with a sort of pragmatic technology. This is no small matter. The second tradition does nothing less than annihilate meaning, value, and humanity itself since it seeks to reduce the world to its abstractions which impose abstract structures upon the world. As regards the first tradition which does find meaning in the world, this has tended to fall into the stereotype of the "Gods" in Plato's Battle of Gods and has usually been willing to cede the physical world to its enemies, the "Giants". Aristotle exemplarily recognised the problem of ceding this territory, this reality.

The sister volume to this work takes its case forward by showing that Aristotle is not only valid philosophically but also scientifically. We will see that modern science, concluding in biosemiotics, thematically follows Aristotle and *not* the positivism that claims to speak for science. That modern science rejects scientism is critical for us to understand. Once this situation is cleared up and a new philosophical interpretation is needed, we find Aristotle helpfully waiting in the wings. He is not just a philosopher but also a scientist, he is indeed the founder of science, and he provides a model that can explain the whole of reality rather than just provide some valuable speculations about it.

Contents

Chapter 1
Aristotle in His Own Day

Abstract This chapter explores the value, complexity, and pitfalls of philosophy. The burden of this work is to show the value of Aristotle's own argument that science—and human life—needs to be framed by philosophy. On this basis, we first consider how philosophy creates worldviews. We then consider ancient philosophy in more detail. We see that ancient philosophy is marked by the reduction of the world to a series of ideas, e.g. number, flux, atoms etc., which then widens to reduce the world, with Plato, to ideas, and then, with Aristotle, to beings. I show that Aristotle's "realism" transcends the reductionism of his predecessors by grounding them and making them complex by showing how each ideology is *one* aspect or analogous dimension of a complex reality.

Keywords Aristotle's realism · Peripatetic synthesis · Syncretism · Anti-reductionism

A basic assumption of this work is that Aristotle's philosophy is powerfully salient in our own day. I will suggest later that we can learn from Aristotle's science and his biology but that is for later. Here, though, we are only considering Aristotle's philosophy. On this point we find that Aristotle has such followers within modern philosophy as Hegel, Heidegger, Patočka, and Feyerabend who all present his position beautifully as valid and meaningful, as we will see in more detail later. Strangely, though, we find that few Aristotle scholars really consider any such modern thought that takes Aristotle's thought seriously in the real world. Aristotle is trapped in the discipline of *ancient* philosophy. He is placed in a box, a coffin as it were. We see firsthand how the preconceptions of a discipline can negatively circumscribe the value of its subject matter.

Michel Foucault is a modern philosopher who lectured brilliantly and meaningfully on ancient philosophy in the later years of his life. He demonstrated by the very act of doing so how philosophy is distinct from scholasticism and is a

P. N. Jackson, *Aristotle on Meaning in the Living World*, Biosemiotics 30,
https://doi.org/10.1007/978-3-032-00602-8_1

meaningful aspect of human life and not the product of a talking shop. His exposition exemplified the movement of the mental life of philosophers through such examples as: "Whereas the Platonic movement consisted in turning away from this world in order to look towards another—even if souls, who, through recollection, have rediscovered and savoured the reality they have seen, are led more by force than by their own will back to this world in order to govern it—the Stoic movement defined by Seneca is completely different. It involves a sort of stepping back from the point we occupy. This liberation enables us to reach the highest regions of the world without, as it were, ever losing being out of sight. We reach the point from which God himself sees the world, we see the world to which we belong and consequently can see ourselves within this world."[1] He adds the further spiritual or meditative human possibility that: "the spiritual exercise of Marcus Aurelius tends towards a sort of dissolution of individuality, whereas the function of Seneca's spiritual exercise—with the subject's move to the world's summit from where he can grasp himself in his singularity—was, rather, to found and establish the subject's identity, its singularity and the stable being of the self it constitutes."[2]

I suggest that we can represent the three spiritual movements treated above in Fig. 1.1, as below:

Foucault argues that whereas Plato looks to a "higher" world for the informed meaning of all being and action, Seneca elevates the human being to a "higher" state from which point the self can understand and act in the world. Marcus Aurelius does

Fig. 1.1 Foucault's philosophical "movements"

[1] Foucault, Michel *The Hermeneutics of the Subject* (New York, [*orig.* 1981–2] 2005) p 276. See also his "Technologies of the Self" in *Technologies of the Self* (Amherst, [*orig.* 1982] 1988).

[2] Foucault, Michel *The Hermeneutics of the Subject* (New York, [*orig.* 1981–2] 2005) p 307. Marcus himself comments that: "Whether the world subsists by a fortuitous concourse of atoms, or an intelligent Nature presides over it, let this be laid down as a maxim, that I am part of a whole, governed by its own nature…I shall never be displeased with whatever is allotted me by that whole…Let us then employ properly this moment of time allotted to us by fate, and leave the world contentedly, like a ripe olive dropping from its stalk, speaking well of the soil that produced it, and the soil that bore it (*Meditations* X. 6 and IV. 39)."

not, according to Foucault, travel to a "higher" world but concentrates upon relieving us of the problems of our self in this world. What this shows us is that we do in a sense possess mental and spiritual bodies as well as physical bodies in the world. We see the world, and must see the world, from our own egoistic perspective, philosophical stance, and spiritual positionality. As individual human beings we cannot cover the whole world but must take a certain "spiritual" position and act within the world and understand it on this basis. In respect to this positioning, we find that we adopt a philosophy in the same way as we learn an occupation; we find and enter a world which empowers us at the same time as limiting us to its own confines. The philosophical "worldview"—a concept which originates with Kant and is explored by Martin Heidegger following Wilhelm Dilthey and Karl Jaspers—is a representational seeing of the whole "world" which, as we will see below, is treated by philosophers from Aristotle to Heidegger as a defining feature of humanity as opposed to animality.

Foucault also argues that the philosopher lives in the world at a certain "moment" in time which shapes his thought. He argues on this point that: "the Periclean moment…[is] situated…in the second half of the fifth century, [whereas] the…Platonic moment will be situated in the first half, even right at the start of the fourth century []. With the Platonic moment I have tried to locate we see what happens when the main part of parrhesiastic practice no longer takes place on the political stage, no longer principally on the political stage—understood in the strict, institutional sense of the term, with the Assembly, courts, and all those decision-making sites—but in philosophy."[3] I add that Aristotle, the tutor of Alexander *and* the defender of the *actual* Greek polis, heroically and tragically straddles these two "moments." Pierre Aubenque brilliantly explains this classical turning point as that: "The ideal of autarky, which had hitherto sought to satisfy itself by the mediation of the city, no longer recognizes any other way than the internal ascetic generating ataraxy or pure speculation as a refuge and as a substitute for an impeded action. This atmosphere of retreat or, as has been said, of "abstraction" is characteristic of Hellenistic philosophies."[4] Putting aside Foucault's own position on Aristotle and *his* "moment," to which we will return later, we see that the key emphasis for both Aristotle and Foucault (and Aubenque) is that philosophy should not be a spiritual consolation for a lost power to act in this world but an explanation of how power actually actualises, and actualises well or healthily, in the world.

Socrates is in a sense a touchstone for all philosophy. His refusal to reject the political authority of the demos of his polis and to allow his words to be written down and detached from living discourse makes him a paradigm for a philosophical refusal to detach oneself from reality. Foucault extends this principle and argues that human thought and action has been progressively "abstracted" and ossified

---

[3] Foucault, Michel *The Government of Self and Others* (Basingstoke, [*orig*. 1982–3] 2010) p 340. Pierre Manent adds in his important work to these spiritual and political "moments" the "Ciceronian moment" in his "Birth of the Nation" *City Journal* (2013) and the "Maastricht moment" in his *Democracy Without Nations? The Fate of Self-Government in Europe* (Wilmington, 2007) p 34.

[4] Aubenque, Pierre *La Prudence chez Aristote* (Paris, 1963) p 15–16.

throughout Western history through a movement from "memory" to "meditation" to "method," i.e., through the progressive transfer of meaning from inner to outer. He argues that the result of this abstraction, ossification, mechanisation, and detachment is the figure of "Descartes" who produced a system based upon "a flux of representations on which a work of analysis, definition, and description will be carried out."[5] "Method" is peculiarly identified with Descartes precisely because his philosophy reduces the world to an active ego applying method to a plastic and passive (or "extended") world. Francis Bacon sought to produce another method for putting nature "on the rack" from a completely different philosophical viewpoint.

Foucault emphasises the importance of "the philosopher." Plato is very different than Descartes; the two men both embody philosophy but very different philosophies from different worldviews. We cannot simply take the ideology and disconnect it from the thinker since it is the vision of a given human being. Foucault hence explains his own position as a philosopher, as a "Socrates," as that: "by placing myself at this Socratic moment of the end of the fifth century, I wanted to recover the point at which a certain relationship is established between this no doubt archaic, ancient, and traditional concern in Greek culture for a beautiful, striking, and memorable existence, and the concern with truth-telling"[6] although he sadly adds (with this lecture taking place in the last year of his life) that: "I am not able at present to lecture to you properly on this theme of the true life; maybe it will happen one day, maybe never."[7] Foucault clearly understands the significance of the person who has a world in his head as an exemplary fact of being, of nature, which needs to be explained rather than explained away. I note, however, that Foucault does not seem to adequately address Aristotle's explicit development of metaphysics and of the special sciences to push back against the disconnected ideological abstractions that he himself draws attention to and I will consider this situation later.

1.1 Ancient Philosophy

To appreciate the "Aristotelian moment" we need to look back at the tradition that produced him and understand how his "moment" represents a philosophical revelation that has spiritually transformed the world. It may seem tiresome to go into such detail as below on the intellectual context for Aristotle, but I believe that it is essential to understand what Aristotle achieved in his own time (and why he achieved what he did) as a platform for understanding the full significance of his thought in our own day. We will ultimately see that Aristotle formulated a systematic anti-reductionism in his own day which is eminently relevant in ours.

---

[5] Foucault, Michel *The Hermeneutics of the Subject* (New York, [*orig.* 1981–2] 2005) p 294.

[6] Foucault, Michel *The Courage of Truth: The Government of Self and Others II* (Basingstoke, [*orig.* 1983–4]) p 162–3.

[7] Foucault, Michel *Ibid.* p 163.

Whilst Aristotle is indeed truly a product of Socrates and Plato, I posit that he is also in a deeper but less obvious sense a culmination of and a syncretisation of *all* Greek thinking and culture. His thinking in all its manifestations represents a topology of being which is not merely a taxonomy of beings but a synthesis of thought which gives birth to a wholly new philosophy of "realism." The platform for his new philosophical insight is his critique of Plato and his "idealism." The nature of his opposition is, as Aristotle himself recognised, paradigmatically expressed by his decision to survey actual Greek political life in his *Politics* in express opposition to Plato's *Republic* which puts forward an *ideal* model of political existence; see *Pol.* II 1262b 8-25. The critical point for us to recognise here is that Aristotle's "common sense" empiricism or anti-idealism is not merely a collapse into mundanity or even, despite Aristotle's ground-breaking role in creating science and the sciences, a denial of philosophy in favour of science. The situation we encounter is that Aristotle *expands* philosophy to cover and include the whole of reality. He frames facts with ideas without reducing one to the other.

Let us therefore look at the cultural and intellectual ecosystem that produced Aristotle. I suggest that the figures set out in Fig. 1.2 below represent the dialectic between the "Gods" (numerologists) and "Giants" (elementalists) out of which Aristotle's "realism" emerged:

Pythagoras (570–495 B.C.) initiated the idea that we can and should look behind the material world to another mathematical world beyond. He posited that meaning emerged through a cosmic mathematical "music of the spheres," building a worldview from out of a supposed relationship between astronomical cycles and musical ratios, a theory that Aristotle discusses and dismisses, perhaps superficially, at *De Cael.* II 290b12ff.[8] Aristotle retains the principle that celestial objects regulate and control our sublunar world through their rhythms, a position supported by the evident importance of circadian and circalunar rhythms and by our concept "hours" being derived from the typically richer Greek ὥραι (*hōrai*) meaning periods in general, time of day, and seasons. He argues, however, that whereas our living world stands upon the principles of a mechanical universe which can and should be counted as a sort of "natural law" of the cycles of time, we cannot, as do Pythagoras and Newton, seek to *reduce* the living and organic to the mathematical and inorganic. Aristotle hence criticised the Pythagoreans on the basis that: "they omitted the more important principles (*kyriōteras archas*) …and supposed the principles that they did recognise to be inherent in all things alike (*De Cael.* II 285a 26-7)." He argued that the Pythagoreans only understood the world on their limited mathematical basis and not according to the variegated principles of its beings.

Plato, a follower of Pythagoras, famously placed the inscription above the door of his school "Let no one ignorant of geometry enter." Plato based his worldview upon mathematics but replaced Pythagoras' idea that the world is an "imitation" (*mimēsis*) of numbers (see *Met.* A 987b 12-13) with his own that it comes about through a "participation" (*methexis*) in Forms. Aristotle then performed the further

[8] On this subject see the useful article "Musica universalis" *Wikipedia* (online).

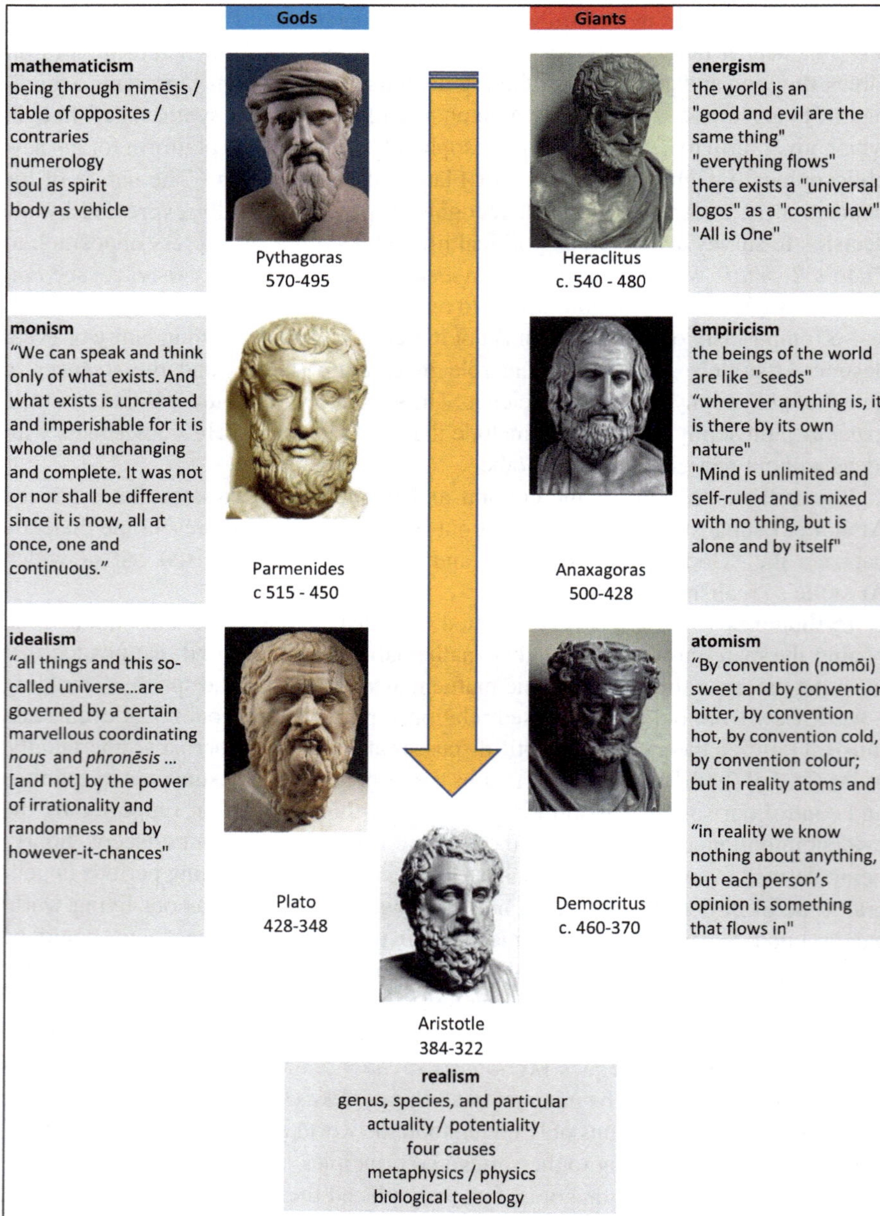

Fig. 1.2 Aristotle's philosophical influences

transformation of "*mimēsis*" in his *Poetics*, reflecting *his* worldview, through the observation that: "Imitation is natural to man from childhood, one of his advantages over the lower animals being this, that he is the most imitative creature in the world, and learns at first by imitation [and there is] the further fact: to be learning something is the greatest of pleasures not only to the philosopher but also to the rest of mankind, however small their capacity for it (*Poet.* 1448b6-15)." Aristotle grounded Pythagoras and Plato's idea that there are global structures that we somehow participate in and imitate through the observation that we form ourselves as (human) beings through our natural ability to "learn" through imitation and to then reflect upon the world.[9] We see from this the basic paradigm that Aristotle moderates his predecessors' ideas and transforms them from abstract hypotheses to metaphysical treatments of the observable, biological world.

Aristotle performs another meaningful transformation from mathematical idealism to biological realism as regards Pythagoras' Table of Opposites. The "elements" of this table that have come down to us (through Aristotle) are Limited/Unlimited; Odd/Even; Unity/Plurality; Right/Left; At Rest/In Motion; Straight/Curved; Light/Darkness; Good/Evil; Square/Oblong. Aristotle explains the table as follows: "the second column of contraries is privative, and everything is reducible to Being and Not being (*to on kai to mē on*), and Unity and Plurality (*hen kai plēthos*); e.g., Rest (*stasis*) falls under Unity and Motion (*kinēsis*) under Plurality. And nearly everyone agrees that substance and existing things (*ta onta kai tēn ousian*) are composed of contraries; at any rate all speak of the first principles as contraries (*tas archas enantias*)—some as Odd and Even [*viz.* Pythagoreans], some as Hot and Cold, some as Limit and Unlimited [*viz.* Platonists], some as Love and Strife [*viz.* Empedocles]; and the principles adduced by other thinkers fall entirely under these as genera (*hōs eis genē*) (*Met.* Γ 1004b 27–1005a 2)" (and see also *Met.* Λ 1075a 25 ff.). Aristotle's specific response to the Pythagoreans is that: "the Pythagoreans posited only two of these [characteristics of place], right and left, to the neglect of the other four, which are no less important; for above and below, front and back, are as distinctive from each other in all animals as right is from left. Sometimes the difference is in function only, sometimes in shape as well (*De Cael.* II 285a 12-16)." Aristotle argued that Pythagoras not only has a limited view of the world but also one which omits core principles for including living beings.

Aristotle puts forward a deeper critique of Pythagoras' Table of Opposites. He argues that Pythagoras stands at the root of the philosophy of his age and (1) observes that *all* subsequent thinkers take one "element" from his Table and follow it through as an ideology and (2) argues that he can himself survey *all* their principles from the vantage point of a "being" (*ousia*) within multiplicity.[10] How is Aristotle's own reduction to the principle of "being" (*ousia*) not also reductionist? Aristotle's "being" is itself necessarily complex and cannot be reduced to yes-no questions from a matrix of "positive" and "negative" attributes of a "being." This is the ultimate significance of the core Aristotelian principle that "the whole is more than the sum of the parts" with Aristotle's own operating dualism for this new worldview being the metaphysical distinction between "*energeia*" (actuality) and

[9] See Francis Wolff "The Three Pleasures of Mimēsis According to Aristotle's *Poetics*" in Bensaude-Vincent, Bernadette & Newman, William (*eds.*) *The Artificial and the Natural: An Evolving Polarity* (Cambridge Mass., 2007).

[10] See David Lefebvre "L'usage de la Série des Contraires en *Métaphysique* Γ 2, 1004B27–1005A6" in *Aristote Métaphysique Gamma* (Leuven, 2008).

"*dynamis*" (potentiality). Aristotle utterly transforms Plato's method of *diairesis* of branching yes-no logic into his own consideration of the bio-logic of a being and its expressions.[11] He similarly rejects Pythagoras on the basis that his approach represents a mathematisation of the world which falsely conflates a thing with its various mathematically measurable attributes.

We discern two essential points about Aristotle above. First, that he understands our intellectual world as a topology of "principles" (*archai*) and of "genera" (*genē*) which must be embodied in beings. Second, that he makes a real attempt to devise a syncretic solution which in this case (as we will see in detail below) includes his "categories" of being, his "four causes" and also, and radically, his transformation of Pythagoras' "contraries" into a framework of being which exists as "beings" existing within a "middle" (*to meson*) *between* "contraries." Aristotle's own solution is a triadic rather than dyadic picture as represented in *his* own table (of emotions/virtues) set out at *E.E.* II 1221a which begins with Irascibility, Spiritlessness, Gentleness (the mean); Rashness, Cowardice, Courage (the mean) etc. This utilisation of the "mean" or "middle" shows us how Aristotle sees mathematics as a platform and framework *for* a "being" and not as a simple and direct binary model for being. He does, however, explain that the forms of beings can be expressed by mathematics through his famous example of the "snub nose" as an embodiment of the "concave" and through his biological consideration of "symmetry."[12] Aristotle takes the position that the form and movements of a "being" (*ousia*) can usefully be represented mathematically but are not reducible to that mathematics.

Aristotle criticises how Pythagoras represents living beings in principle, as above. He also opposes the more specific Pythagorean hypothesis as regards the construction of living beings. Pythagoras argues that the soul is an external and material "spirit" which is "identical with the particles in the air (*De An.* I 404a 18)" and attempts to "try to explain what is the nature of the soul, without giving any details about the body which is to receive it; as though it were possible, as the Pythagorean stories suggest, for any soul to find its way into any body, <which is absurd,> for we can see that every body has its own particular shape or form (*hekaston idion echein eidos kai morphēn*). Such a theory is like suggesting that carpentry can find its way into flutes; [whereas in actuality] each craft must employ its own tools and each soul its own body (*De An.* I 407b 20-27)." Aristotle argues against Pythagoras that we cannot reduce the soul to "vitalism," i.e., to the idea that there is a bit of matter or spirit that adds the being to a being. He also insists upon the principle that the body or form of a being is not an add-on to the soul but is simply the physical dimension of a spiritual being. Aristotle therefore criticises Pythagoras both for excessive abstraction *and* excessive materialism. What matters is the whole being.

[11] On "*diairēsis*" see especially Martin Heidegger *Plato's* Sophist (Indiana, [*orig.* 1925] 1997) § 43-4.

[12] On "symmetry" see Bastit, Michel *Le principe du monde: Le Dieu du philosophe* (Paris, 2016) and on the "snub nose" see Suzanne Mansion "Τὸ σιμόν et la définition physique" *Naturphilosophie bei Aristoteles und Theophrast* (Heidelberg, 1969).

As regards Pythagoras' core theory—his numerology—Aristotle comments: "why are these numbers causes? There are seven vowels, the scale has seven strings, the Pleiades are seven, at seven animals lose their teeth (at least some, though some do not), and the champions who fought against Thebes were seven….These people are like the old Homeric scholars, who see small resemblances but neglect great ones (*Met.* N 1093a 13-27)." Even in this highly critical passage Aristotle discerns and draws out the limited value of Pythagoras' position. He observes the value of Pythagoras' philosophical innovation that there are indeed mathematical structures expressed in nature with his own thought depending upon such structures as "the one and the many," the four causes, the four elements etc. Books M & N of his *Metaphysics* explicitly consider the way in which "mathematical objects" (and "ideas") exist and are expressed in the world. He implicitly accepts the ontological significance of such principles of number as one (unity), two (duality), three (hierarchy, the mean), and four (proportionality) etc. C.S. Peirce grasps this vague but profound ontology as that: "One, Two, Three, are more than mere count-words like "eeny, meeny, mony, mi," but carry vast, though vague ideas."[13] Aristotle recognises this deep mathematical structuring but argues that we cannot reduce our world down to it.[14]

The philosopher Jan Patočka explains the problem of a mathematical envisioning of the world, and hence of Pythagoreanism, well as that: "Atomistics is a consistent arithmetism; it is close, in this respect, to Pythagoreanism, and its representatives arrive at significant geometric results on the basis of the atomic principle. The idea of the mechanical structure of bodies composed of elementary particles and that of the genesis of higher qualities on the basis of atomic structures (the great principle of atomistic explanations) are basically consequences of this arithmetism."[15] He explains the problem further as that: "Mathematisation presupposes the abstract view which fragments the unity of the movement into multiplicity. Mathematisation is therefore always something artificial, which does not grasp the internal character of the movement as such; something which, in the final analysis, compromises its originality on the level of ignoring the fact that movement does not primarily have a spatial meaning and orientation, but is rather an "upward" direction towards the fullness of being."[16] Aristotle's philosophy is, as we will see, formed on the negative basis that mathematics is an artificial quantitative analysis of a richer qualitative state of being which Aristotle will then positively seek to address.

---

[13] Peirce, C.S. "The Logic of Signs" in Favareau, Donald (*ed.*) *Essential Readings in Biosemiotics: Anthology and Commentary* (Dordrecht, [*orig.* 1910] 2010) p 142. For an interesting consideration of the Pythagorean significance of numbers see Harald Atmanspacher and Hans Primas "The Hidden Side of Wolfgang Pauli: An Eminent Physicist's Extraordinary Encounter with Depth Psychology" *Journal of Consciousness Studies* (1996).

[14] For a detailed consideration of this situation see Peter H. Nidditch "The First Stage of the Idea of Mathematics: Pythagoras, Plato, Aristotle" *Midwest Studies in Philosophy* (1983).

[15] Patočka, Jan *Aristote, Ses Devanciers, Ses Successeurs* (Paris, [*orig.* 1964] 2011) p 38.

[16] Patočka, Jan *Ibid.* p 210.

We find that Aristotle readily incorporates the principles of mathematics into his worldview but refuses to reduce it to them. In the sister volume to this work we will consider in some detail the alternative non-reductionist mathematics which Alain Boutot explains in a modern context as "a neo-Aristotelian mathematics."[17] Pythagoras, by contrast, paradigmatically represents a reductionist mathematics which breaks reality down into the "form" (structure) and "flux" (atoms) which, as Owen Goldin explains, is recycled in modern thinking by: "pancomputationalists such as Edward Fredkin [who] approach the Pythagorean account insofar as on their account all reality can in principle be expressed as one (very big) number, made up of discrete units…[and by the "structural realism" of] Floridi [who] share[s] with pancomputationalism the aspect of Pythagorean ontology that Aristotle finds so objectionable: positing structure or form with no substrate."[18] It should go without saying that the desire to reduce the real world to mathematical formulae, technologies, algorithms, and ideologies is as persistent today as ever.

Heraclitus (540–480 B.C.) famously perceived the world through the model of a flowing river ("*panta rhei*") whereas Aristotle would later see the world through the model of a foetus (and then as a "self-mover") which "flows" into and through a life according to its form, stages, and environmental circumstances. To explain this "flow" Heraclitus chose "fire" or, we could say, "energy," as his foundational material "element" whereas Thales had chosen "water" (and Anaximenes selected "air") as the "element" determining being. Theophrastus also observed that Anaximander invented the idea of the "infinite" (*apeiron*) as another single explanation of being.[19] Heraclitus' own emphasis upon "fire" or "energy" and its "flux" is an inspired form of materialism which still garners adherents today.[20] Aristotle

---

[17] Boutot, Alain *L'invention des formes: Chaos—Catastrophes—Fractales—Structues dissipatives—Attracteurs étranges* (Paris, 1993) p 185.

[18] Goldin, Owen "Aristotle, The Pythagoreans, and Structural Realism" *Philosophy Faculty Research and Publications* (2016) p 3–4. Goldin explains "structural realism" as a revival of positivism which is ""realism" because, on this account, science describes the world as it really is; "structural" because the realities identified are not kinds of stuff or objects, but structures (*Ibid.* p 3)."

[19] Theophrastus observed that: "Anaximander of Miletos, son of Praxiades, a fellow-citizen and associate of Thales, said that the material cause and first element of things was the Infinite (*apeiron*), he being the first to introduce this name of the material cause. He says it is neither water nor any other of the so-called elements, but a substance different from them which is infinite from which arise all the heavens and the worlds within them (*Phys*, Op. fr. 2; *Dox.* p. 476; R. P. 16)." I note that the philosophical usage of the term "*archē*" also comes from Anaximander.

[20] David Bohm reframes the Heraclitean "flux" as his "implicit order"; see his *Wholeness and the Implicate Order* (London, 1980). See also Werner Heisenberg *Physics and Philosophy: the revolution in modern science* (London, 1958) p 61. James Wayne Dye finds in Heraclitus a founder of "process philosophy" in his "Heraclitus and the Future of Process Philosophy" *Studies in Process Philosophy I* (1974) with this "Heraclitean" tradition of an informed flux clearly informing modern "information theory." Martin Heidegger from a very different philosophical tradition disagrees with Aristotle and defends Heraclitus' obscureness on the basis of its profundity: "He is the Obscure, because he thinks questioningly into the lighting (*Early Greek Thinking: The Dawn of Western Philosophy* (San Francisco, [*orig.* 1943–1954] 1975) p 123)."

argues, however, that we cannot refer the beings of the world to the matter of "fire" just as we cannot refer it to "water" even if human beings are indeed predominantly made up of water and are in a sense formed from energy. I note that Aristotle repurposes Anaximenes' emphasis upon "air" as "*pneuma*" but does not reduce being to this.

Aristotle rather contextualises matter by arguing that worldly being(s) are composed of all *four* "elements" or phase states of being, *viz.* earth (solid), water (liquid), air (gas), and fire (energy) and that they must positively extract and cycle their inputs (including glucose as energy[21]) according to their ousiological needs. He observes that the phase states of being move in a cycle of qualities based on energy levels—e.g., vapour—water—ice—water—vapour.[22] In respect to this situation Aristotle in a sense *begins* with Heraclitus by arguing on what we could call "energy" that: "fire *(to pyr)* is ever coming into being and rushing *(aei ginomenon kai rheon)* just like a river, but so speedily as to elude observation (*Juv.* 470a 3-5)." He continues, however, that we clearly cannot reduce "being" to "fire" on the basis that: "A flame *(hē phlox)* is a process of becoming, involving a constant interchange of moist and dry. It cannot be said to be nourished since it scarcely persists as one and the same for a moment (*Mete.* II 355a 9-11)" and hence that: "to say that the soul (*psychē*) is like fire (*pyr*) is like saying that the craftsman, or his craft, is the saw or the augur which he uses (*PA* II 652b 13-15)" (a critique he also makes of Pythagoras by reference to a carpenter and his tools). Aristotle here recognises that "life" *involves* "heat" (or "fire") whilst not being reducible to it, an emphasis he draws as that: "life (*to zēn*) and the presence of soul (*hē tēs psychēs hexis*) involve a certain heat (*meta thermotētos tinos estin*) (*Resp.* 474a25-6)"; and see also *Juv.* 469b16-20.

It could be said that Aristotle predicts the idea of "open systems" defined as any material system in which mass or energy can be lost to or gained from the environment. I rather suggest, however, that Aristotle goes way beyond a vague prediction of our "open systems" but rather breaks through in an original and incisive way to an understanding of how the living and the material relate to each other and how they differ. As regards their difference, Aristotle understands the essential "negative entropy" or "negentropy" of living beings and hence posits that a "fire" or "flame" (*phlox*) represents an "open system" whilst not being articulated as a being (*ousia*) since it does not itself have the capacity or need to be "banked up" over its lifespan; see *Juv.* 470a 13. To prove and support this finding he provides us with significant supporting research as regards how the principles of "heat," "movement," and

[21] Aristotle observes on the role of glucose in animal nutrition that: "all organisms are nourished by the sweet [i.e., glucose], either by itself or in combination with other savours (*Sens.* 442a 2)."

[22] Aristotle explains this principle of elemental change based on cycles of qualitative and material causality most fully in his *Meteorology*. The basic principle he builds upon here is that: "We have distinguished in the elements four causal factors whose combinations yield four elements: two of the factors are active, the hot and the cold, two are passive, the moist and the dry (*Mete.* IV 378b10-14)." He investigates the qualitative character of matter in his *Meteorology* and considers the full meaning of the concretion and boundedness of a being as its actualisation in his *Physics* and in his *On Generation and Corruption*.

"separation" affect being. He observes as regards ecosystems that: "the sea putrefies quickly when broken up into parts, but not as a whole (*Mete.* IV 379b 5-6)" with the operating principle here being that: "anything that is flowing or in motion is less apt to putrefy than a thing at rest (*Mete.* IV 379a 33-4)." Aristotle takes Heraclitus' vague and mystical idea that "everything flows," maintains its philosophical insight, and subjects the idea to rigorous scientific test. We will consider Aristotle and "open systems" in the sister volume to this work.

As regards Heraclitus, we find that Aristotle sought, in his *Meteorology*, not just to express an interesting maxim but to understand the reality of homeostasis in the context of an environment in which "everything, except fire, is liable to putrefy; for earth, water, and air putrefy, being all of them matter relatively to fire. The definition of putrefaction (*sēpsis*) is: the destruction of the peculiar and natural heat in any moist subject (*tēs en hekastō hygrō oikeias kai kata physin thermotētos*) by external heat (*hyp' allotrias thermotētos*), that is, by the heat of the environment (*tou periechontos*)…This explains why everything that putrefies grows drier and ends by becoming earth or dung. The subject's own heat departs and causes the natural moisture to evaporate with it, and then there is nothing left to draw in moisture, for it is a thing's peculiar heat (*oikeia thermotēs*) that attracts moisture and draws it in.[23] Again, putrefaction takes place less in cold than in hot seasons, for in winter the surrounding air and water contain but little heat and it has no power, but in summer there is more. Again, what is frozen does not putrefy, for its cold is greater than the heat of the air and so it is not mastered, whereas what affects a thing does master it (*to kinoun kratei*) (*Mete.* IV 379a 14-31)."[24] I represent Aristotle's position here as Fig. 1.3 below.

We see that Aristotle is concerned with how a "being" energetically engages with and masters the material elements of the world—air, water, earth, and fire—in order to maintain itself according to *its* lifecycle. The *Meteorology* is an interesting middle ground considering "open systems" and ecosystems which operate as entities but without being living beings.

Aristotle's basic principle is that there is an elemental cycling of being which is indeed like a river being filled and emptied in a flowing cycle which looks like a straight line on its surface. There is also another cycling of organic matter from

---

[23] Aristotle concludes his *On Youth and Old Age* with the thought that: "Everything that lives has a soul, and this, as we have said, cannot exist without natural heat. The assistance which plants get through food and the surrounding air is sufficient for the preservation of their natural heat. For the entry of food causes a cooling, just as it does in man when he first admits it; but fasting causes heat and thirst for the air when it is motionless always grows hot, but, moving as it does when food enters it, it becomes cool, until digestion is complete. If the surrounding air is excessively cold owing to the season of the year, when hard frosts occur, the plant withers (Juv. 470a 19-28)." Digestion represents not just a natural process but the core mechanism of an organic ecosystem.

[24] We see how Aristotle tests the same principle in different contexts by considering in his *On Sense and Sensibles* "why coldness and freezing render savours dull, and abolish odours altogether; for cooling and freezing tend to annul the kinetic heat (*to thermon to kinoun*) which helps to fabricate sapidity (*Sens.* 443b 16-17)." He is interested in why cold and freezing reduces the rate of the energic processes of nature.

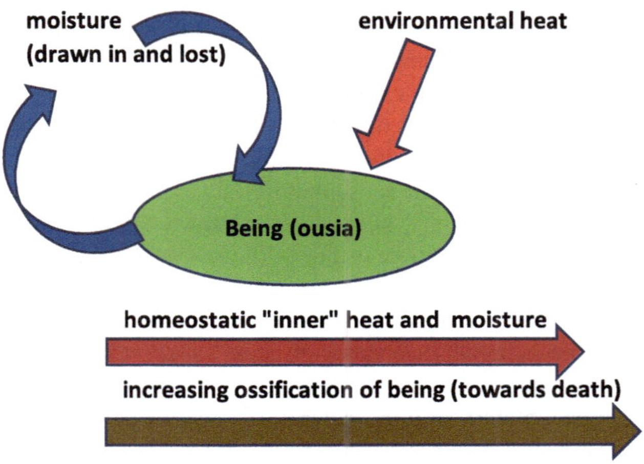

Fig. 1.3 Aristotle on the heat, moisture, and dryness of a being (ousia)

freshness to putrefaction. We can also add that there is a further organic cycling of being as a movement of biological reproduction through species on the basis that: "everything that is produced is something that is produced from something and by something, and that the same in species as it (*hapan to gignomenon gignetai ek tinos kai hypo tinos, kai touto tō eidei to auto*) (*Met.* Θ 1049b 28-9)." Aristotle observes further that this cycling operates through various other cycles of existence, e.g., the migration of birds, the seasons of our biosphere, etc.[25] His word for a meaningful cycle of movement, "*phora*," originally means the seasonal movement of a crop. His use of this term to refer to the cycles of the cosmos shows us how Aristotle understands the *metaphora* of nested cycles of meaning; see for e.g. *MA* 698b 8-18. Aristotle explains the evident difference between organic and inorganic cycles as the power for active selection by the organic being (*ousia*) which is required by the vital basic principle that: "food is acted upon by what is nourished by it, not the other way round (*De An.* II 416b 1-2)." The animal exists within natural and elemental cycles and channels these cycles from "outside" to its "inside" through a power of *action* in a world which *is acted upon*.[26] This is essentially Schrödinger's principle of "negentropy" that we will consider in our sister volume.

---

[25] Aristotle explains that: "Their [i.e., animals'] activities all have to do with mating or production of young, or with the supply of food, or are contrived against periods of cold and heat or the changes of the seasons. For all animals have an innate perception of change in respect of hot and cold, and just as among humans some move indoors during the winter while others who command extensive territory spend the summer in the cold parts and the winter in the warm sunny parts, so it is with those animals that are able to change their locations (*HA* VII 596b20-28)" and on seasonal bird migrations see *HA* VII xii–xvii passim and on the breeding seasons of animals see *HA* V viii–xii passim. The cyclic principle moves us from the heavenly cycles of astronomy through to the natural cycles of the biological world.

[26] Aristotle also argues that: "the hot substance prevails in the body [when] it induces growth, beginning from the centre (*poiei tēn auxēsin apo tou mesou*) along its own line of travel (*kata tēn autēs phoran*) (*PA* II 653a31-33)."

We find that, from Aristotle's perspective, Heraclitus' idea of a "universal *logos*" as a "cosmic law" is not explanatory.[27] Aristotle rather argues that the "*logos*" is always the *logos* of something—with the common *logos* being analogic (as *analogia*) with it being along these lines that Aristotle converts Plato's idea of "form" into his own idea of "species."[28] Aristotle's principle of "being" (*ousia*) is that nature is fundamentally ousiological in the sense that it is always "of something" and hence: "A habit is a habit of something, knowledge is knowledge of something, position position of something…Knowledge is thus of the knowable; the knowable is knowable by knowledge. Perception is of the perceptible, which is perceived by perception (*Cat.* 6b 4-38)." Aristotle's *bio*-logical conceptualisation of "*logos*" is as a circling, channelling, chreod, or conatus of being (*ousia*) through which an initial energy, or original piece of matter, first matures (and reproduces itself) and then breaks down in accordance with the natural lifecycle of the given being; see *Juv.* 469b–470a and his *On Longness and Shortness of Life*. It is in this sense of having a direction, structure, and distinct organic being that a "being" (*ousia*) has an *ergon* (function) and a *telos* (end). I note that there is nothing trivial or false about Aristotle's "circular" logic which gives us such thoughts as that: "mind thinks itself (*auton noei ho nous*) by sharing in the object of thought (*kata metalēpsin tou noētou*) (*Met.* Λ 1072b 20-1)"; the circularity here is the meaning circle of a being expressing and shaping the real world around it through its cycles and patterns of being and becoming.

As regards Heraclitus' famous maxim that "nature loves to hide itself" we find that Aristotle builds a working physics and logic which explicitly recognises that we must work from "what is more obscure by nature, but clearer to us" to "what is more clear and knowable by nature (*Phys.* I 184a 20-21)."[29] Aristotle accepts obscurity only in the sense that he builds the limits of our sense and knowledge into his worldview and he does, in fact, brilliantly and uniquely know when to leave "gaps" when we reach the unknowable, this being an intellectual modesty and talent that we should all reflect upon. Aristotle's response to Heraclitean obscurity is to produce an Aristotelian complexity theory which makes him object to all "one-sided theories (*ta monakhōs*) (*Met.* Γ 1012a 30)." Aristotle observes that reducing the world to matter, form, "God," freedom, evolution or any other "one-sided theory" is simply inadequate to explain our rich, entangled reality.[30] We rather need to understand our

[27] See A.-T. Tymieniecka "Inspirations of Heraclitus from Ephesus Fulfilled in Our New Enlightenment" *Analecta Husserliana* (2011) p 5–6.

[28] See Ernst Mayr *Towards a New Philosophy of Biology* (Cambridge Mass., 1988) p 57.

[29] On this important distinction see Suzanne Mansion ""Plus connu en soi", "plus connu pour nous." Une distinction épistémologique importante chez Aristote" *Pensamiento* (1979).

[30] Aristotle comments by reference to the "musical man" who must have a relation to (and capacity for) music and to a musical instrument so as to produce music that: "when we say that not-musical man becomes a musical man, both what becomes and what it becomes are complex (*Phys.* I 190a3-4)." This complex adoption by human beings of a specialised function and this ready ability to assimilate oneself both mentally and physically with a foreign object makes the "human being" a useful paradigm for a natural being. What other organisms achieve on a speciate level, human

"sublunary" world as a coming together of powers—form and matter, potentiality and actuality, idea (noosphere) and being (biosphere), being and becoming etc.—into nexuses of being which are entangled and informed into a single cosmos. This is Aristotle's response to Heraclitus' "*panta rhei*."

We ultimately see the inadequacy of Heraclitus by comparing him with Aristotle. We can represent this as the chemical/environmental cycle and the biological cycle in Fig. 1.4 below.

We see that Aristotle is concerned with the cycling of water and being in a very different way than Heraclitus. Indeed, he argues that: "The cause of this continuous process, as has been frequently remarked, is cyclical motion (*hē kyklō phora*), the only motion which is continuous (*synechēs*). Hence also the other things which change into one another, for instance, the simple bodies, by being acted upon or having power to act, imitate cyclical movement (*kata ta pathē kai tas dynameis, oion ta hapla sōmata, mimeitai tēn kyklō phoran*). For when Air comes-to-be from Water, and Fire from Air, and Water again from Fire, we say that coming-to-be has completed the cycle, because it has come back to its starting-point. Hence motion in a straight line is also continuous because it imitates cyclical motion (*hōste kai hē eutheia phora mimoumenē tēn kyklō synechēs estin*) (*GC* II 337a 1-8)." His "river" is a complex cycling of the bringing together of chemical elements into a species of being. Aristotle's idea that the straight line is an "imitation" of a circle is fascinating and presages Leibniz's later idea of a "conatus" having a "living force of rotation,"[31] as we will consider later.

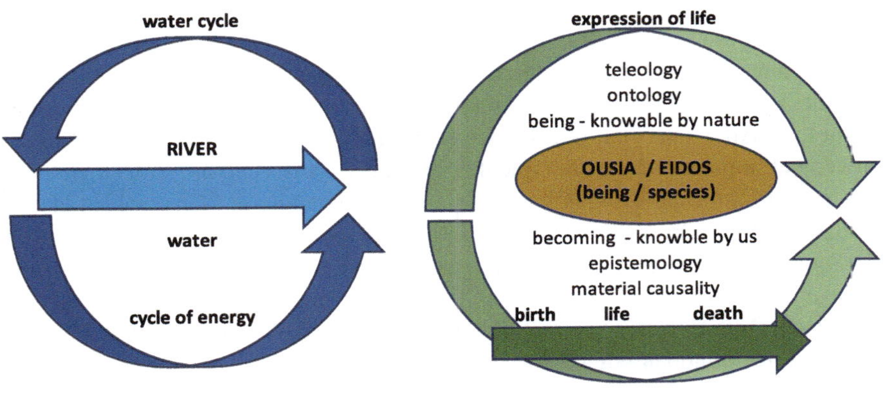

Fig. 1.4 Heraclitus and Aristotle's representative cycles of being

beings achieve on a cultural, professional, and individual level. The human being unpacks nature and creates novel forms and imitations through it. On music Mary B. Schoen-Nazzaro "Plato and Aristotle on the Ends of Music" *Laval théologique et philosophique* (1978).

[31] Leibniz, Gottfried "Letter to Johann Bernoulli, 18 November 1698" in *Philosophical Essays* (Indianapolis, 1989) p 169.

Aristotle ultimately takes a dim view of Heraclitus himself and of those who follow his philosophy. He comments that: "those…who introduce other's opinions, e.g. that "good and evil are the same thing," as Heraclitus says, refuse to admit the impossibility of contraries belonging at the same time to the same thing; not because they do not themselves believe this, but because on Heraclitus' principles one has to say so. The same thing is done also by those who take on the defence of one another's positions; their aim being to speak as would the man who stated the position (*Top*. VIII 159b 29-36)." Aristotle formulates his own science of logic in reply to this destruction of meaning which concretely explains how we do and can best formulate our thought regarding the objects of the world including through his principle of non-contradiction. He also adds along similar lines that: "Heraclitus [says] that "it is what is opposed that helps" and "from different tones comes the fairest tune" and "all things are produced through strife"; while Empedocles, as well as others, expresses the opposite view that like aims at like (*N.E.* VIII 1155b 4-8)" and in an insistently un-obscure mode he himself posits that perhaps the situation is as it appears so that: "friendship is the cause of good things, and strife of bad (*Met*. A 985a 5-7)." Aristotle refuses paradoxes in favour of clear statements of complexity and provides us with a logical assessment of what we can (and can not) know about the world in place of mystical obscurity.[32]

I conclude that Aristotle significantly gives the same response to the two very distinct philosophical traditions of Pythagoras (mathematicism) and Heraclitus (materialism): *viz*. we must work from the *ousia* (being) itself and its immanent being and movement in the world if we are to properly represent the world in which we exist.

Parmenides (515–450 B.C.) is a "God" (this distinction between philosophical "Gods" and "Giants" being a reference to Plato's Gigantomachia; see *Soph*. 246A-B) who is a recognised influence upon Plato. Parmenides supported Heraclitus' claim that "All is One" but rejected his claim that this "One" is subject to *constant* movement in favour of the principle that, in reality, there is *no* movement. He presents us with a formalism, a holism, and an absolute being or "One" which clearly predicts Plato and his "world-animal."[33] Plato did, however, successfully develop in his *Sophist* the principle of "not-being" which Aristotle then expanded into the principle of "*sterēsis*" as the absence of form which then comes-to-be through the movement from *dynamis* to *energeia* (and *entelecheia*). As with the Megarian idea that there is no hidden potentiality, i.e., so that a man *is* only a builder *when he is building*, and Zeno's paradoxes of movement, Parmenides is trapped into a thoroughly academic position by the paradox which supposedly bedevils all physics

[32] Aristotle mocks Heraclitus and his school as follows: "Cratylus, who finally did not think it right to say anything but only moved his finger, and criticised Heraclitus for saying that it was impossible to step twice in the same river; for *he* thought one could not do it even once (*Met*. Γ. 1010a 12-14)."

[33] See Dimitri El Murr "Platonic 'Desmology' and the Body of the World Animal (*Tim*.30c-34a)" in Salles, Ricardo (*ed.*) *Cosmology and Biology in Ancient Philosophy: From Thales to Avicenna* (Cambridge, 2021).

that "nothing comes from nothing." Aristotle avoids the problem by looking to the real world and studying the dynamics of movement as a becoming of being.

Aristotle replies to Parmenides (as also to Heraclitus) by reference to "being" and "beings" that: "a man investigating *principles* cannot argue with one who denies their existence. For if Being is just one, and one in the way mentioned [by Parmenides], there is a principle no longer, since a principle must be the principle of some thing or things (*Phys.* I 185a 2-5)." As regards "the principle of some thing or things" he argues further that: "that in which the motion takes place (the species) must be one and unintermittent, and that which is in motion must be one—not in an accidental sense (i.e., it must be one as…Coriscus who walks is one…) (*Phys.* V 227b32-33)." Aristotle's "one" is a certain continuous "being" (*ousia*) and not, as with Parmenides and Pythagoras, a mathematical principle. Aristotle transforms the Pythagorean contraries of Right/Left, At Rest/In Motion, and Straight/Curved by adding his own Actuality/Potentiality to produce the picture of a "being" (*ousia*) that: "if any one of their parts moves, another part must necessarily be at rest; it is on this account that animals have joints. For they use their joints as a centre, and the whole part in which the joints is situated is both one and two, both straight and bent, changing actually and potentially because of the joint (*MA* 698a17-21)."[34] Aristotle makes the important parallel in respect to the "movement of animals" between the continuity of the species and an individual walking; he demands in all cases that there be sufficient reason for sustained, stable, and *continuous* activity. His explanation of this situation will ultimately lead us to the stability theory necessary for "self-movers" and the "Unmoved Mover" of his *Metaphysics*.

As regards the meaning that Aristotle himself invests in the "one" we find that: "the actualisation must be one (*Phys.* III 202b 37)" and regarding the "two" that: "what is capable of desiring and what is capable of fleeing are not different, either from one another or from what is capable of perceiving, but their being is different (*alla to einai allo*) (*De An.* III 431a 12-14)." We see here the true meaning of Aristotle's principle of non-contradiction which is not just that something cannot be two different things at the same time but that it can *potentially* be many things but will *actually* be only one. This actuality is a nexus of being which is that of the being and its middle. Aristotle explains this situation in his *On the Soul* by considering that a being—such as Coriscus—has a central control over the "left and right" and over the "one and two" (the duality) of its moving being. I add that all of Aristotle's work is part of the same ontological exercise to study of the "middle" of the "being." Against Parmenides, he argues that the "one" does exist in nature but only as the principle of the identity of a "being" (*ousia*); i.e., the principle that "a principle must be the principle of some thing or things" and that: "everything that

[34] Aristotle explains further that: "every animal…must have in itself something that is at rest, in order to provide that which is moved with the origin of its movement, supported upon which it will move both as an integral whole and in its several parts (*MA* 698b 5-7)." Aristotle observes that a "being" must have a body and a manner or way of movement; animals are also "integral" and move literally as a whole through parts and a centre.

changes is something and is changed by something and into something (*Met.* Λ 1069b36–1070a1)."

Aristotle does indeed recognise that we exist in one cosmos but argues that this oneness of being is not sufficient to explain the plurality of beings that we encounter. His physics are on this basis not primarily concerned with how beings collide but with how they come-to-be and pass-away through unifying and concretising forces such that: "when bodies come into contact or are separated, their boundaries instantaneously become one at one time—when they touch, and two at another time—when they are separated (*Met.* B 1102b 1-3)." Aristotle's physics is a *qualitative* physics of the world which focuses upon the worldly nature of biological beings and capture the significance of their chemical constitutions as a "continuity" or "separation" of physical being, as we will consider in more detail later. Aristotle seeks to demonstrate a pregnancy in nature and an actuality concretised through *forces of continuity along a principle.* He finds a pathway in nature through "beginning" (*archē*), "middle" (*to meson, logos*), and "end" (*telos*). Aristotle hence rejects the monism and dualism of Parmenides who divides the world between divine meaning and its worldly shadow. He places a greater independent value in the entangled biological "sublunary" world in which we live whilst also recognising the potential divinity of the mechanical universe above and below us.

It is on account of his implicit dualism (along with his explicit monism) that Parmenides has been described, by Samuel Scolnicov, as the first Cartesian philosopher and hence: "Parmenides may be said to be the first Cartesian philosopher. He is the first to tackle the problem of method and to make truth dependent on it… And he is Cartesian also in the method he favours: an absolutely certain, undeniable, primordial intuition is attained, and consequences are deduced from it."[35]

Anaxagoras (500–428 B.C.) develops Parmenides' monism into an opposite empiricism. He asserts not that the cosmos is "One" but that each individual is one on the basis that "wherever anything is, it is there by its own nature (*Phys.* III 205b 4-5)." By reducing the world to what he calls "seeds" (*spermata*), he makes the "one" into an atom or monad.

Aristotle opposes Anaxagoras' form of empiricism with his own which contrarily argues that this particularity of nature *exists within a milieu* so that a seed is a principle that must occupy the right soil with this principle applying to all principles and all beings. He explains his opposition to Anaxagoras' naïve empiricism as that: "The theory of Anaxagoras that the principles are infinite in multitude was probably due to his acceptance of the common opinion of the physicists that nothing comes into being from not-being. For this is the reason why they use the phrase "all things were together" and the coming into being of such and such a kind of thing is reduced to change of quality…things come into being out of existent things, i.e., out of things already present, but imperceptible to our senses because of the imperceptibility of their bulk (*Phys.* I 187a27 187b 1)." Aristotle observes that Anaxagoras maintains Parmenides' monism on the basis that he simply asserts that everything

[35] Scolnicov, Samuel *Plato's Parmenides* (California, 2003) p 4.

simply emerges from the same primordial cosmic soup, through preformism and panspermia. In both cases, we find a simple brute actuality; the real is simply what appears before us.

The implication of this situation is explained by Patricia Curd as that Anaxagoras rejects "coming-to-be" and hence redraws Parmenides' position on the basis that: "The rejection of coming-to-be requires that any organism that appears in the world at any time—Kate, Max, and I, for instance—must have been already there from the start; the claim that seeds are entirely unlike requires that each must have been always present as the fully-fledged individual that it is during its period of active life."[36] Aristotle's critique is that dependence upon individual and infinite "seeds" only superficially resolves the issue of "being(s)." The idea of "everything in everything" does not provide any explanatory power as regards how or why individual and speciate things become in the world since organic beings observably do not exist on the basis of their predominating chemical elements but as concrete structures which feed off elements but are distinct structures from them. We again see the problem with conflating chemical and biological matter. The problem of "mind" also reappears. Aristotle argues that: "he [i.e., Anaxagoras] says that *nous* belongs to all animals, both the great and the small and the worthy and the less worthy. But mind does not appear to belong equally to all animals, nor even to all human beings (*De An.* I 404b 4-7)." As with matter, we cannot simply treat mind as a "cosmic law," as some sort of generic force of nature, but only as the mental expression *of something*.

Aristotle's ultimate position is that Anaxagoras' theory of infinite multiplicity is quite as vacuous as Parmenides' theory of the infinite "One." Aristotle's ousiology rejects *tout court* all such ideas as that form or mind happen to emerge from the predominating matter or the supposition that the macroscopic (or, rather, mesoscopic) world would make sense if only we could see its "seeds." Aristotle is himself not an empiricist (from "*empeiria*" (experience)) because he is only concerned with particulars, but he concerns himself with particulars because he believes that this is how meaning is ultimately encountered and accessed by human beings in our world.

I make note here of **Empedocles (494–434 B.C.)** whose philosophy builds upon Anaxagoras' and who was the first philosopher to explain nature as the product of four "elements" or what he calls "roots." Empedocles explains "change" and "movement" in the world as the result of the interplay of "Love" and "Strife" as opposed to Anaxagoras' "*Nous*."

Aristotle criticises this adapted position as follows: "Are we to say "Bone comes-to-be if the "elements" be put together in such-and-such a manner"? For, according to his own statements, nothing comes-to-be from their "fortuitous consilience," but only from their "consilience" in a certain proportion (*logōi tini*). What, then, is the cause of this proportional consilience? Presumably not Fire or Earth [i.e., simple elements]. But neither is it Love or Strife: for the former is a cause of "association"

36 Curd, Patricia *Anaxagoras of Clazomenae* (Toronto, 2007) p 173.

only, and the latter only of "dissociation." No: the cause in question is the essential nature of each thing (*estin hē ousia hē hekastou*)—not merely (to quote his words) "a mingling and a divorce of what has been mingled." And *chance* (*tychē*), not *proportion* (*logos*), "is the name given to these occurrences": for things can be "mingled" fortuitously. The cause, therefore, of the coming-to-be of the things which owe their existence to nature is that they are in such-and-such a determinate condition: and it is *this* which constitutes the "nature" of each thing—a "nature" about which he says nothing. What he [i.e., Empedocles] says, therefore, is no explanation of "nature" (*GC* II 333b 9-18)." Aristotle argues that he cannot find a sufficient explanation in Empedocles' theory of "Love" and "Strife" to account for the "*logos*" which expresses a materialised and definite "being" (*ousia*).

Aristotle ultimately critiques Empedocles' own theory of mechanicism as the association or mingling of "elements" on the basis that it is: "*Because* man is such and such, *therefore* the process of his formation must of necessity be such and such and take place in such a manner; which is why first this part is formed, then that. And thus similarly with all the things that are constructed by Nature (*PA* I 640b 1-4)." Aristotle argues that we clearly observe a "*logos*" (rationally representable pathways) operative in nature rather than "chance" (*tychē*). The paradigm for this principle is the human foetus (as per the *Generation of Animals*) and its controlled and continuous movement towards fulfilment as a competent functioning human adult (as per the *Ethics*). This paradigm is exemplified by the foetus (as opposed to the teratoma) and its possession of a whole organic developmental structure.[37] The schema revealed here is that organic beings possess a beginning (*archē*), an end (*telos*), and a middle (*to meson*) which is the expression of its becoming and being and is realised physically over its period of existence. It is hence that Aristotle places the problem or aporia of the nature of the foetus and its viable organic being and becoming, or development, at the centre of his philosophy.

Plato (428–348 B.C.) is, of course, the great idealist and the first idealist. Whilst the above thinkers had found some form of generic cosmic "*nous*" in the world, Plato also dug down to reveal a further "ideal" (and mythological) patterning. Plato introduced concrete patterns into our worldview but Aristotle rejected both of these framing patterns on the basis that they superimpose a system upon the world rather than explaining the world on its own terms. Aristotle's basic issue with Plato's Theory of Forms is well-known, i.e., that: "they [i.e., Plato's Forms] cause neither movement nor any change in them [i.e., sensible things]. But again they help in no way toward the *knowledge* of other things (for they are not even the substance (*ousia*) of these, else they would have been *in* them)…all other things cannot come from the Forms in any of the ways that are usually suggested. And to say that they are patterns (*paradeigmata*) and the other things share in (*metechein*) them is to use empty words and poetical metaphors. For what is it that works, looking to the Ideas? And any thing can both be and come into being without being copied from

[37] See William B. Hurlblut "Framing the Future: Embryonic Stem Cells, Ethics and the Emerging Era of Developmental Biology" *Pediatric Research* (2006).

something else, so that, whether Socrates exists or not, a man like Socrates might come to be. And evidently this might even be so even if Socrates were eternal (*Met.* M 1079b 15-31)." Aristotle is only content with a philosophical account that relates the working drivers of the specific being (*ousia*) being studied.

Aristotle argues that the Theory of Forms is neither necessary nor explanatory. It is even for him a misleading interjection which precludes the finding of an accurate solution. For Aristotle, the mundane world cannot satisfactorily be understood as a copy of the structures of a higher, ideal world. Plato's position and Aristotle's critique of it are both, however, far deeper and more interesting than this essential conflict. Plato's reduction of the world to Ideas goes hand in hand with his reduction of the world to mathematics, which is, in modern eyes, a more fruitful reductionism. Plato argued long before Galileo that the world is "written in mathematical language and its characters are triangles, circles and other geometrical figures."[38] Proclus explains Plato's conception of the soul on this basis as that: "Plato is right when he constructs the soul of all mathematical branches [kinds] and divides it numerically and binds it by proportions and harmonical ratios and places the erstwhile principles of figures in it…and makes the circles in it move in an intellectual motion. Thus, all mathematicals exist primarily in the soul…and the soul is the fullness of all mathematicals…The soul has its essence in these branches of mathematicals."[39] The soul is on this basis a combination of the geometric, the arithmetic, and the harmonic. Plato depicts the soul as a circle and his famous "divided line" breaks the world down into intelligibles, mathematicals, sensibles, and images. The abstract intelligibles feed through mathematics to inform the wholly derivative sensibles and images of the material world.

The soul is, however, more than a mathematical machine for Plato. He also maintained as regards the nature of "mind" that we should accept the view of "those who were before us" that "all things and this so-called universe…are governed by a certain marvellous coordinating *nous* and *phronēsis*," and we should reject the contrary view that they are directed "by the power of irrationality and randomness and by however-it-chances (*Philebus* 28d5-9)." Plato regarded the universe as a machine but one in service to what he would call "the Good." It is also notable that Plato's sophisticated "desmology" is a deep philosophical study into the "bonds" or connecting forces within and between beings, into how we can understand meaningful being or organism to exist.[40] Aristotle extends and expands Plato's conception of "the Good" at the beginning of his *Ethics* and argues that: "It is hard…to see how a weaver or a carpenter will be benefited in regard to his own craft by knowing this "good itself," or how the man who has viewed the Idea itself will be a better doctor or general thereby. For a doctor seems not even to study health in this way, but the health of man, or perhaps rather the health of a particular man; it is individuals

[38] The famous quote is from Galileo but this is the essential project of Plato's *Timaeus*.

[39] Quoted in Philip Merlan *From Platonism to Neoplatonism* 2nd ed. (The Hague, 1960) p 21.

[40] See Dimitri El Murr "Platonic 'Desmology' and the Body of the World Animal (*Tim.*30c-34a)" in Salles, Ricardo (*ed.*) *Cosmology and Biology in Ancient Philosophy: From Thales to Avicenna* (Cambridge, 2021).

(*kath' hekaston*) that he is healing (*N.E.* I 1097a 8-14)." For Aristotle, there is no "Good," only *a* good x or y. His *Ethics* ultimately considers human beings and how "*phronēsis*" and "*sophia*" manifest in this embedded context. It therefore takes Plato's generic principles and applies them specifically to "beings" (*ousiai*).

Aristotle makes clear his opposition to Plato's ideal "Forms" or "Ideas" and his idea of "the Good" but let us note that he himself clearly treads a fine line in doing so since he himself begins his own train of thought by observing that: "Every art and every investigation, and likewise every practical pursuit or undertaking, seems to aim at some good: hence it has been well said that the Good is That at which all things aim (*N.E.* I 1094a 1-2)." The seeming contradiction is mitigated, however, when we consider that Aristotle's ultimate point is that each being has its own good just as it has its own end (*telos*), form (*eidos*), matter (*hylē*) etc.; see *N.E.* X 1176a 3-6 and *De Cael.* II 292b 13-26. Aristotle similarly relates to the being (*ousia*) such concepts as work (*ergon*) or function which he will expand upon in other places as signifying the essence (*to ti ēn einai*) or soul (*psychē*) of a being. Aristotle is no stranger to abstractions or to values but they are for him emergent from beings and their milieux. It is hence that Aristotle's other categories, i.e., quality, whereness, whenness, relation, quantity, action, and being-acted-on, depend upon the essential principle of "*ousia*" rather than being formal and mathematical.

Aristotle's general point vis-à-vis Plato is that it is through the detailed "scientific" consideration of the objects of nature that we understand the being of the various objective beings of nature. For example, "mind" (*nous* and *dianoia*) is best studied by reference to individual human beings and species of animals who express it in a particular way even if it is also a potentially cosmic phenomenon.

Let us return briefly to reconsider A.N. Whitehead's statement that Western philosophy is, as footnotes to Plato, based upon "the interweaving of seven main notions, namely The Ideas, The Physical Elements, The Psyche, The Eros, The Harmony, The Mathematical Relations, The Receptacle."[41] We find in all cases that Aristotle takes on Plato's concept but then transforms it from idealism to realism. As regards Plato's concept of "Form" or "Idea" we find that Aristotle takes his own "formal cause" from Plato as he had taken the "material cause" from Heraclitus and the "efficient cause" from Anaxagoras. In this way, he transforms philosophy from an ideology considering some single cause of being, what he himself calls "one-sided ideologies," to what we will call a philosophical topology based upon our consideration of the various objects of being, with this being what we will call philosophical "realism." In more detail, we see that Aristotle takes Plato's basic question as regards the means for the transmission of form and idea in the world and reframes it as his own logic as expressed in the various books of his Organon. Exactly how a specific Idea manifests in a domain of being in the world is explained specifically and in detail in his various scientific studies.

Plato explains "Eros" through his famous concept of Platonic love through which love is purified from a carnal to an ideal form of love. The eros that is initially felt

[41] Whitehead, A.N. *Adventures of Ideas* (Harmondsworth, 1933) p 174.

for a person can with contemplation become an appreciation for the beauty within that person, or even an appreciation for beauty itself in an ideal sense. Eros can help the soul to "remember" beauty in its pure form. By contrast, Aristotle explores how the concept of "love" is first manifested biologically in animals as the evolution of care we observe in parents for their offspring in the animal kingdom; see *GA* III 753a8-16. He then puts great effort into the study of this attraction and care as it is manifested as "friendship" in human relationships in the final books of his *Ethics*. He argues, with remarkable perspicacity, against Aristophanes and Plato's conception of love as that the lover becomes as one with the object of their love; see *Pol.* II 1262b 8-25. He rather argues that "love" is not a simple assimilation to some other object but that we can and should define ourselves through our relationships with the various different objects of the world, e.g. as a musician becomes one through a love and aptitude for engaging with music; see *N.E.* IX 1168a 5-9. We do not look *through* people and objects to some sort of abstract love but find it expressed in our relationship with and appreciation for other people and objects.

By considering Plato's "Eros," we can appreciate the issue that arises for Aristotle in respect to his "Harmony." Aristotle does not seek to reduce the world to the principle of an ideal but argues not only that different beings have different principles but that they are themselves made up of a plurality of principles (*archai*). Most famously, a society should not be subordinated to an ideology as was the case with Sparta, but should rather be a healthy tapestry of different types of people, virtues, skills, and points of view.[42] As regards the "Psychē" we see that Aristotle transforms Plato on this point most directly in his *On the Soul*. In this work he argues that the "soul" is simply the dynamic living form of an organic being with a body.[43] Through this emphasis he disputes Plato's idea that the world is itself an "animal" with a soul and argues instead that the cosmos has its own order that we could describe as "natural law" but that we should pause before mythologising or anthropomorphising this cosmic system into a living entity like us. The soul is for Aristotle a specific term related to a living embodied entity. He does, however, apply "*nous*" as meaning "intuitive reason" (but not *phronēsis*) to the implicit meaning-making power of the universe as well as to the explicit meaning-making mind of human beings. As regards the "Physical Elements" we find the interesting transformation explained by Jan Patočka from Plato's *stoicheia* or "elements" to Aristotle's *archai* or "principles," one being static and simple and the other being dynamic and complex, and enabling the *entelecheia*. Patočka speculates that Plato was inspired to represent the world as mathematical elements by reference to the analogous building up of language out of letters and syllables.[44]

[42] See Jeremy Waldron "The Wisdom of the Multitude: Some Reflections on Book 3, Chapter 11 of Aristotle's *Politics*" *Political Theory* (1995) & Thomas Lindsay "Aristotle's Qualified Defense of Democracy through "Political Mixing"" *The Journal of Politics* (1992).

[43] On this matter see Abraham Bos "Why the Soul Needs an Instrumental Body According to Aristotle (*Anim.* I, 3, 407B13-26)" *Hermes* (2000).

[44] See Jan Patočka "Les Grammata et les Stoicheia dans la Philosophie Grecque Jusqu'à Aristote" in *Aristote, Ses Devanciers, Ses Successeurs* (Paris, [*orig.* 1964] 2011).

Plato's theory of "Mathematical Relations" is that there are abstract mathematical objects that exist independently of us on a higher plane. Aristotle contrarily argues that mathematics is derivative from being on the basis that: "(a) the objects of mathematics are not more substantial than corporeal objects; (b) that they are not prior in point of existence to sensible things, but only in formula; and (c) that they cannot in any way exist in separation (*Met.* M 1077b 11-14)." He argues further that mathematics exists as an attribute of being as one of the "common sensibles" of our "common sense" regarding which: "By these I mean shape (*schēma*), magnitude (*megethos*), movement (*kinēsin*) and number (*arithmon*) (*Sens.* 437a 9)" (and see also *De An.* II 418a 17-20). Mathematics are one dimension of how we represent our reality. I note that the brilliant mathematician René Thom defends Aristotle's philosophy as a qualitative topology of being based upon the mathematical principles of continuity (*synechēs*), succession (*ephexēs*), and separation (*chōristos*) that he calls "semiophysics" and that D.C. Schindler draws an interesting, related comparison between Plato and Aristotle as regards the "metaphysics of money."[45]

Plato's concept of the "Receptacle" is complex and fascinating. He explains his "*chōra*" as a space, a formless interval "which is to receive perpetually and through its whole extent the likenesses of all eternal beings (*Timaeus* 51a)." Aristotle, by contrast, argues that a "being" (*ousia*) is not the mere formal filling of a space but is itself a real something which is both formal and material or, in technical terms, it is a material being having a limit that is its shape or form with "place" on this basis being "the innermost motionless boundary of what contains it (*Phys.* IV 212a 21-2)." He speaks of the *periechon* or environment which surrounds the *periechomenon* or surrounded being on the basis that: "that which surrounds is on the side of the form, that which is surrounded is on the side of the matter (*to men periechon tou eidous einai, to de periechomenon tēs hylēs*) (*De Cael.* IV 312a 12-13)." Aristotle begins with Plato's observation about space, place, and being but then naturalises it as what it means to be a being-in-the-world. As regards the surface itself, the interface, he comments that: "for the limiting surfaces of the embracing and the embraced coincide (*en tautō gar ta eschata tou periechontos kai tou periechomenou*) (*Phys.* IV 211b 12)." We see here a philosophical account of the boundary conditions of a being that are so fundamental in modern physics.

Aristotle also reconfigures Plato's "Receptacle" as the "*to dektikon*" ("what is receptive") standing for the principle that the matter must be receptive of the form. He also carefully shows the lineage of the error of Plato's "receptacle" as that: "they [i.e., Empedocles and Democritus] speak of each element "inhering" and "being separated out," as if generation were emergence from a receptacle instead of from a material (*hōsper ex angeiou tēs geneseōs ousēs all' ouk ek tinos hylēs*), and did not involve change in anything (*De Cael.* III 305b 3-6)." Aristotle argues that there is no reality to these representations of being which are merely generic explanations of what might cause "being" without being a consideration of the real existence of

[45] See D.C. Schindler "Why Socrates Didn't Charge: Plato and the Metaphysics of Money" *Communio* (Fall 2009).

"beings." He interestingly uses "*angeion*" (container, vase, vessel, [vein, musical instruments]) to identify a "receptacle" rather than Plato who uses "*chōra*" (space, place), a difference which itself seems to characterise the dispute between a realistic and an idealistic perspective. Aristotle criticises even this concept of the receptacle as a container on the basis that it is simplistic and does not recognise the fact that beings are a *synthesis* of form and matter, surface and substance.

We clearly find from the above account a systematic distinction being made between Plato's "top down" and idealistic position and Aristotle's "bottom up" and realistic position. They differ on the basis that Plato reduces the world to "Forms" whereas Aristotle naturalises forms by adding the significance of matter and movement, i.e., the moving and material causes. In this way, Aristotle also naturalises the finality, as the final cause, which is a cosmic theme in Plato's scheme. Upon the supposition that we live in a naturalised world without a belief in "higher" powers, Aristotle's position seems to be stronger and an advance on Plato's. I comment, however, that both points of view are possibly correct on their own terms. If we see the world mathematically (or mythologically), we follow Plato and if we see the world empirically, we follow Aristotle. We could perhaps regard Plato and Aristotle as a spectrum of philosophical possibility which ultimately complement each other.

We find that although Aristotle is not himself a materialist, he defends matter against Plato's idealism as he also defends form against Heraclitus' materialism. His deceptively obvious conclusion is that a house is not an abstraction but is, like all beings, a combination of "form" and "matter"; see *Met.* H 1043a 15-21 and *Met.* H 1044a 28-30. He explains this hylomorphism in the following terms: ""Why is a saw like this?" "In order that it may have the essential character of a saw and serve for sawing." The purpose, however, could not be served if it were not made of iron (*Phys.* II 200a 11-13)." He adds regarding "matter" that it is more than the potentiality from which the actuality of "form" (*eidos*) and "being" (*ousia*) arises but that it plays the active role of the "mother" (*mētēr*) of nature.[46] Aristotle argues contra Plato that we cannot merely understand "matter" (and "place") as a mere filling for the "receptacle" of the "form"; see *De Cael.* III 305b 3-6. He rather insists that beings do rather have their own material place or milieu in the world and is willing to resort to the metaphor of a "mother" to explain this point, though this metaphor is also implicit in Plato's "*chōra*" on another basis.

Democritus (460–370 B.C.) is our last "Giant" and is the paradigmatic "atomist." Democritus' position on "mind" is that: "in reality we know nothing about anything, but each person's opinion is something that flows in (B9)" which Aristotle critiques on the basis that: "Democritus…says that the spherical atoms, as they move because it is their nature never to remain still, draw the whole body with them and so move it. But we shall ask whether these same atoms produce rest. How they

[46] Aristotle explains that: "the matter (*hylē*) is nearly, in a sense *is*, substance (*ousia*) …[and it is] a joint cause (*synaitia*), with the form, of what comes to be—a mother as it were…For what I mean by matter is precisely the ultimate underlying subject, common to all the things of nature, presupposed as their substantive, not incidental, constituent (*legō gar hylēn to prōton hypokeimenon hekastō, ex hou ginetai ti enyparchontos mē kata symbebēkos*) (*Phys.* I 192a 6-33)."

can do so, it is difficult, if not impossible, to say. In general the living creature does not appear to be moved by the soul in this way, but by some act of soul or will (*holōs d' ouch outō phainetai kinein hē psyche to zōon, alla dia prohaireseōs tinos kai noēseōs*) (*De An.* I 406b 20-26)" (see also *De An.* I 408b 14 *ff*).[47] Aristotle asks how atoms could have a complex potentiality which could account for the complex actuality of a body. His conclusion on Democritus' explanation is that "fire" (*pyr*) cannot explain being, nor can the movement or energy of "spherical atoms."

Democritus argued generally as regards "nature" that: "By convention (*nomō*) sweet and by convention bitter, by convention hot, by convention cold, by convention colour; but in reality atoms and void (*DK* 68B9)" and also that: "In reality we know nothing, for truth is in the depths (B117)." We are reminded here of Heraclitus' obscurity. Democritus explains regarding his unknowable depths that our understanding stops once "we can no longer see anything smaller (B11)" which Aristotle explains as being a conflation of "[i] what appears with [ii] what is true (*to alethes einai to phainomenon*) (*De An.* I 404a 29)." He explains the problem with this stress upon "appearance" (*to phainomenon*) through the example *contra* Democritus that the eye cannot work "as mere mirroring…that takes place in an eye due to the fact that the eye is smooth (*Sens.* 438a 7-8)" because this mirroring "exists not in the eye but in the observer; for the phenomenon is only reflection (*Sens.* 438a 8-9)."[48] We must understand that sight is not merely some mechanical process of the reflection of light but is a taking in of the world through the eye and its mechanism by a being, this being a psychic process very different from some superficial physical or material interaction.[49]

[47] Gregory Vlastos describes Democritus' basic position on the soul and body as being that there is: "a specific atomic cluster, dependent for its integrity upon another cluster (the body), and having the power to move the latter ("Ethics and Physics in Democritus" *Philosophical Review* (1945) p 92)."

[48] See Walter Burkert "Air-Imprints or Eidola: Democritus' Aetiology of Vision" *Illinois Classical Studies* (1977).

[49] Aristotle places great emphasis, both philosophical and scientific, on "sight" as a sense. Aristotle begins his *Metaphysics* with the statement that: "All men naturally desire knowledge. An indication of this is our esteem for the senses; for apart from their use we esteem them for their own sake, and most of all for the sense of sight. Not only with a view to action, but even when no action is contemplated, we prefer sight, generally speaking, to all the other senses. The reason for this is that of all the senses sight helps us to know things, and reveals many distinctions (*Met.* A 980a 22-28)." He develops the position later as that: "the object of a thing is its principle; and generation has as its object the end. And the actuality is the end, and it is for the sake of this that the potentiality is acquired; for animals do not see in order that they may have sight, but have sight in order that they may see. Similarly men possess the art of building in order that they may build, and the power of speculation that they may speculate; they do not speculate in order that they may have the power of speculation—except those who are learning by practice; and they do not really speculate, but only in a limited sense, or about a subject about which they have no desire to speculate (*Met.* Θ 1050a 8-15)." Aristotle, as always, stresses that movement arises from the being ("*ousia*") through its body and not merely *passively* by the action of environmental forces upon the being. He concludes regarding the critical feature of "sight" that it is "through the sensation connected with sight that we recognise the form which is in each thing (*Top.* II 113a 2-3)," hence showing the relationship between "sight" (and "soul") and "form."

The reduction of the world to appearance goes hand in hand with the reduction of the world to images and to atoms. It also suggests a reduction of the world to external representations, and this is what Aristotle charges Democritus with. Aristotle argues that: "If each animal and each of its parts is what it is in virtue of its shape and its colour, what Democritus says will be correct, since that was apparently his view, if one understands him right when he says that it is evident to everyone what "man" is like as touching his shape, for it is by his shape and his colour that a man may be told. Now a corpse has the same shape and fashion as a living body; and yet it is not a man. Again, a hand constituted in any and every manner, e.g., a bronze or wooden one, is not a hand except in name; and the same applies to a physician depicted on canvas, or a flute carved in stone. None of these can perform the functions appropriate to the things that bear those names. Likewise, the eye or the hand (or any other part) of a corpse is not really an eye or a hand (*PA* I 640b 30–641a 6)." Claudia Zatta generally explains Aristotle's critique of Democritus as "privileging animals' body structure and "surface" rather than "movement" (and life)."[50] We see here a fuller criticism of Democritus' conceptualisation of the "receptacle," as a shape filled in with matter as a vase contains various matters.

Democritus' reduction of the world to shapes and atoms resembles, on reflection, Plato's reduction of the world to forms and elements. Aristotle criticises Democritus on the same grounds as he criticises Plato, i.e., on the grounds of excessive literalness and not being able to imagine creative forces that we cannot simply picture.[51] Aristotle's ultimate critique of atomism in general is ultimately that it is "unsatisfactory" because it does not recognise the holistic principle that: "We have to state how the animal is characterised, i.e., what is the essence and character (*ti kai poion ti*) of the animal itself, as well as describing each of its parts; just as with the bed we have to state its form (*PA* I 641a 15-17)." He explains this holistic principle of living being as that its existence is "not without soul (*mē aneu psychēs*) (*PA* 641a 19)." He adds in his dedicated consideration of the foetus that: "of necessity the first principle must be present at the outset…since the growth and movement of the other parts are derived from it. That is why those people are wrong who, like Democritus, hold that the external parts of animals become distinct first, and then the internal ones. They might be speaking of animals carved out of wood or stone, the sort of things which have no first principle at all, whereas living animals all have such a principle, and it is inside them (*GA* II 740a 11-18)." As Aristotle used the necessary complexity of craft as an analogy for nature which refutes Heraclitus, he here argues that Democritus applies the analogy too literally as a sort of "demiurge" making things.

---

[50] Zatta, Claudia *Aristotle and the Animals: The Logos of Life Itself* (Abingdon, 2022) p 15.

[51] Aristotle repeatedly accuses Plato of over-literalness, most obviously by representing the gods as "eternal men" (*Met.* B 997b 11). He assumes that the "divine" cannot meaningfully be represented simply as a big man and that "forms" cannot be understood as heavenly paradigms which descend to reality. G.W.F. Hegel expresses this shift in application well when he explains that: "Final Cause is true and concrete, as opposed to the abstract Platonic Idea ("The Philosophy of Aristotle" *The Journal of Speculative Philosophy* ([*orig*. 1825-6] 1871) p 76)."

Aristotle criticises Democritus and *all* his peers for not recognising the central importance of the being (*ousia*) and of the becoming of beings (*ousiai*) in our living world. He observes that there is a constant temptation—and this pertains to our own day as to his—to reduce the world of living beings to a mechanically manipulable mass. Aristotle directs this criticism at atomism in his *Metaphysics*. He argues here that: "Leucippus…and his disciple Democritus hold that the elements are the Full and the Void (*to plēres kai to kenon*)—calling the one "what is" and the other "what is not." Of these they identify the full or solid with "what is," and the void or rare with "what is not" (hence they hold that what is not is no less real than what is because Void is as real as Body); and they say that these are the material causes of things (*Met.* A 985b 5-10)." He explains that Democritus reduces the existence of being(s) to such mechanical and material factors as "shape, arrangement, and position (*schēma te kai taxin kai thesin*); because they hold that what is differs only in contour, intercontact, and inclination (*rhysmō kai diathigē kai tropē*)," on which he comments that: "for motion (*kinēsis*), whence and how it arises in things, they casually ignored this point, very much as the other thinkers did (*Met.* A 985b 15-21)." The "motion" that all these thinkers casually ignore is the path or "meaning" engendered by our lives on this mortal coil.

Aristotle is perceptive and consistent in his criticism of Democritus' mechanicism and its mirror image as Plato's formalism. The "Gods" and the "Giants" are more similar than they might appear. They both seek universal, cosmic solutions for a biological reality that cannot be understood in such cosmic terms. It is ironic that none of these philosophers even address or approach "meaning" itself with it being hence that Aristotle concludes that: "Democritus…omitted to mention the final cause, and so all the things which Nature employs he refers to necessity (*eis anankēn*). It is of course true that they are determined by necessity, but at the same time they are for the sake of some purpose, some final cause (*hou heneka*), and for the sake of that which is *better* in each case (*GA* V 789b 3-6)." Aristotle argues that worldly being comes about through material necessity, i.e., in accordance with the permutations of how matter works in the world, but *also* through a functionality and optimality which creates the emergent meaning of the world that we experience around us. I suggest that this is a true representation of an "emergent" and "evolutionary" world as opposed to a mechanistic, mathematical, and materialistic world of statistical laws acting upon material "atoms."

I suggest further that Aristotle identifies through his critique of Democritus the problematic shallowness inherent in all our modern "*-isms*," most obviously "logical atomism," "empiricism," "idealism," "formalism," "operationalism" etc. They are *all*, as Aristotle well observes, part of the same fundamental error which combines "atoms" (materialism) with "mathematics" (idealism).[52] Aristotle properly identifies Democritus' (as also Plato's) failure to make contact with actual reality

[52] Robert Pasnau comments on the commonality between Democritus and Plato and their "atomisms" that: "even Democritus' arch-rival Plato accepted the need to eliminate all secondary qualities at the foundational level, remarking in the *Timaeus* (50e-51b) that the basic receptacle must be "invisible and characterless", and then going on to account for bodies in geometrical terms

not as a principled detachment from the world but as a shallow literalness masquerading as profound mysticism. The ultimate irony we face is that it is Aristotle, the sober and precise realist par excellence, who properly gives expression to the mysterious ineffable profundity of the spiritual depths and meaning of the world by insisting that their place cannot be filled by concrete representations.

Aristotle (384–322 B.C.) is, on the above account, the synthesiser of Greek thought who through his synthesis (meaning "seeing together") takes each of the elements to another level, this being a basic principle of holism. By accepting this complexity, Aristotle becomes the first "realist" who engendered, as Paul Feyerabend suggests, "the only reasonable empiricism that ever existed [whereas] Bacon...was an anti-empiricist plus an anti-rationalist ("rationalism" in the sense of Descartes) and in this respect the father of science."[53] This idea that Bacon gives us a false empiricism and, we can add, that Descartes gives us a misguided rationalism seems paradoxical since these figures are taken as the basis for our modern worldview which prides itself upon its (supposed) rationalism and evidence-based empiricism. We could, however, contend that it is Aristotle who is the *true* empiricist and the *true* rationalist and is so precisely because he is the *true* realist. It is hence that he is himself the "father of science," but one signifying an understanding of nature rather than technology.

Aristotle asserts in principle that we understand the metaphysical structures of reality through analogy (*ana*-logia; "a moving along the *logos*") and metaphor (*meta-phora*; a "carrying across" of meaning; see *Poet.* 1457b 6) and, in Hegelian terms, "speculatively." I will show below that Aristotle's analogical approach to the world as metaphysics is a wholly scientific approach to reality. His logic and his philosophy reasonably assume that our mind relates to and reflects the world, acts as its mirror, and in some way operates on common principles as it. He hence argues in his logic that we can approach some "*topoi*" (in varying degrees) from an *a priori* perspective and some from an *a posteriori* perspective. He explains in his *Rhetoric*—significantly, closely related to the *Topics*—that we must not only examine our subject-matter but must also interpret and internalise it so as to "bring [it] before the eyes."[54] That we must deliver this full expression and completion poetically, logically, rhetorically, mathematically etc. shows us that *nothing* is simple and that *everything* is complex and multi-dimensional in nature. Aristotle's scientific approach is not one of reduction to simples but is, rather, a complexity theory.

(53c-55c) ("Democritus and Secondary Qualities" *Archiv für Geschichte der Philosophie* (2007) p 103)."

[53] Feyerabend, Paul *For and Against Method* (Chicago, 1999) p 197.

[54] It is a vastly important point that we must re-create and re-present meaning—through our re-presentation in logic, mathematics, speech, sense—as well as receiving it from its origin. On this re-presentation see Sara Newman "Aristotle's Notion of "Bringing-Before-The-Eyes": Its Contributions to Aristotelian and Contemporary Conceptualizations of Metaphor, Style, and Audience" *Rhetorica: A Journal of the History of Rhetoric* (2002). On metaphor see also Fran O'Rourke "Aristotle and the Metaphysics of Metaphor" *The Boston Area Colloquium in Ancient Philosophy* (2005).

Aristotle posits that human "mind" must in some way operate on parallel or complementary principles with the world that it "sees" and "feels." His "reason" is not merely something that we apply to the world but is a symbolic (*sym-bolē*; "a coming together, a contract") meaning which combines inner and outer or, in biosemiotic terms, the "*Innenwelt*" with the "*Umwelt*." We will consider this biosemiotic "meaning" in more substantial detail in the sister volume to this work. Here, though, I emphasise that this "symbolic" meaning is not subjective but is based upon an objective relationship between the human being and the world. Aristotle argues regarding this relationship that: "We are…much better provided in relation to knowledge about the perishable plants and animals, because we live among them (*dia to syntrophon*)… Perishable things…take the prize in respect of understanding because we know more of them and we know them more fully…because they are nearer to us and more akin to our nature (*oikeiotera physis*) (*PA* I 644b 23–645a 25)." We share in the material world by being material beings and in the living world by circulating its matter within us, by being syn-trophic.[55] The living world is the milieu in which we live and we understand it best because we are naturally a part of the ecosystem of this biosphere.

Aristotle explains as regards the particular value of the study of human beings that they, and hence *we*, are notable not only on their own merits, as a separate genus of being defined by a whole new dimension of "mind." He adds the astute point that we can observe the nature of animal "life" and "mind" better from the perspective of our human scale in small animals such as birds, bees, ants etc. than in larger animals; see *HA* VIII 612b 20-32 and *Post. An.* II 99b 35–100a 4. We can literally stand over and study how they operate. The planets, by contrast, are the furthest removed from us so that all we can take from them is to discern the mechanisms of their movement. We can, however, take from this shape the principle that: "the sphere [of the planet] is at once the most useful shape for motion in the same place [whilst being] as different as possible from creatures which have organs of motion (*De Cael.* II 290b 2-6)."[56] Elemental matter is then explicable on the basis that it "imitates" cosmic matter by being linear and circular in motion—i.e., subject to fixed, mechanical transformations; see *Met.* Θ 1050b 28-30. Aristotle moves through the world peripatetically and considers what we can take from each of its dimensions.

For Aristotle, we need to "know ourselves" not merely for our own subjective benefit but to also to test and make our science and philosophy valid. Human beings are beings like any other "*ousia*" but are important as a working example because "man is, of necessity, the animal most familiar to us (*HA* I 491a 22-23)." Human

[55] On "*syntrophon*" see Claudia Zatta *Aristotle and the Animals: The Logos of Life Itself* (Abingdon, 2022) p 11-12.

[56] Regarding the "place" of the Earth Aristotle's position is that although he thinks it is observably the centre of the cosmos (see *De Cael.* II 296b 7-11) it is also the case that "our observations of the stars make it evident, not only that the earth is circular, but also that it is a circle of no great size (*De Cael.* II 297b 32-3)" and also that it may indeed be "perishable like any other formation of nature (*De Cael.* I 279b 13-14)." I suggest that the centrality of the Earth has little cosmological significance in Aristotle's thought.

beings are exceptional as exemplifying the channelling of "meaning" through the world as both "life" and "mind." The existence of meaning *in this case* is, or should be, indisputably evident. It is on this basis that Aristotle is an empiricist. Aristotle's empiricism is not anti-metaphysical as per our modern tradition but rather goes through and beyond the "*physis*" of the physical world to consider in what sense these principles manifest in the biological and human worlds. His empiricism as demonstrated by his *Ethics* and his *Politics* (and, indeed, by all his work) hence seeks to understand humanity as natural objects or beings and as an evolution from out of the physical world on parallel philosophical terms understandable by analogy (*kat' analogian*; along or through *logos*; see, for e.g., *PA* I 645b 1-11 and *Phys.* I 191a 8-12).[57] Aristotle's approach is to build up meaning by studying the world and *not* to reduce the world to mathematical laws.

1.2 Conclusions

It is perhaps worthy of consideration that Michel Foucault's bold and nuanced study of Greek thought, already touched upon, dismissed "the one who is called "the" philosopher, because he was no doubt the only philosopher in Antiquity for whom the question of spirituality was least important; the philosopher whom we have recognised as the founder of philosophy in the modern sense of the term: Aristotle. But as everyone knows, Aristotle was not the pinnacle of Antiquity but its exception."[58] With his "everyone knows" Foucault doth protest too much, methinks! Aristotle is, as I have shown above and as most people understand, quite reasonably understood as "the pinnacle of Antiquity." We can only claim a certain seemingly "modern" sophistication for Aristotle in the sense that his investigative method does not meditate upon or intuit the world but intellectually reconstructs it through a method which reviews and synthesises both thought about the world and the evidence of the world itself. Aristotle does not just tell us a tale of his own thought, his own idea or ideology, and experience as did his peers, but he rather surveys all the evidence before him and sets out a concrete explanation for it. As regards his ancient *bona fides*, however, we find that it is Plato's question of the *Meno* as regards how

[57] Aristotle argues regarding "analogy" that: "The causes and the principles of different things are in a sense different, but in a sense, if one speaks universally and analogically, they are the same for all (*Met.* Λ 1070a 31-32)" which Deborah Achtenberg explains as that: ""Analogy", or *analogia* in Greek, means "same *logos*" or "same relationship." To say that good is an analogical equivocal is to say that different things, persons, state of affairs, and so forth, are approximately called "good" because they are relata of a specified relationship *Cognition of Value in Aristotle's* Ethics (New York, 2002) p 46." See also Pierre Aubenque "Les Origines de la Doctrine de l'analogie de l'être: Sur l'histoire d'un contresens" *Les Études philosophiques* (1978).

[58] Foucault, Michel *The Hermeneutics of the Subject* (New York, [*orig.* 1981–2] 2005) p 17.

we "tie down" knowledge and truth that Aristotle resolves through his logic, metaphysics, and science.[59]

Foucault's problem with Aristotle may perhaps arise from Aristotle spoiling Foucault's narrative plan and his archaeological account of European thought as the passing through memory to meditation and then method. All three dimensions of human thought are explored and manifested by Aristotle. In his hands, all three retain a richness and purpose. His method is not a deflationary process of reductionism but an expansionary way to systematically approach the world. Aristotle's standard method is to look at the "common sense" understanding of any subject area ("*topos*" or place), to review "expert" opinion on the subject area, and to then go beyond these opinions, first dialectically and then philosophically. This passing through opinion (memory) and dialectic (meditation) to philosophy (method) is, of course, thoroughly Socratic and Platonic in nature. Aristotle gives us a reflective, philosophical, and modest method which constantly bears in mind that a human being is engaged with the world and seeking to pull out its meaning and value. According to this method we work *from* the "*topoi*" of being and must not on this basis expect the same proofs from a mathematician as from a rhetorician.[60]

Foucault claims that Aristotle is modern because he does not possess what he calls the "*de retour*" of spirituality. Foucault restricts the scope of spirituality to the traditions of inner meditations we find in Christianity itself and in its Stoic roots.[61] Aristotelian philosophical reflection is not included even though, as we have seen above, it is most definitely a "*de retour*" of human being upon himself as a natural being, implicitly sharing in the divine. Aristotle's spirituality is a deep reflection upon human mind and being which reflects upon its "good" as an expression of truth and excellence. This manly reflection upon spirituality expects striving for excellence with little consolation for suffering. This is the value that Aristotle brought into the world, and it surely encapsulates something of the glory of Greece. As the example of Ancient Greece led Marx to wonder how his "productive forces" could produce such a result, Foucault perhaps should have taken Aristotle as a counterexample for his theory of the "episteme" since Aristotle's thought clearly

[59] Michael Ferejohn explains that: "the central insight that drives the entire project of Aristotle's *Analytics* is the Platonic idea, briefly floated in the *Meno* 98 A, that what distinguishes knowledge from other types of true belief is that it is somehow "tied down" by the possession of an explanatory account, which I interpreted earlier to mean an account that explains the *truth* of what is known (*Formal Causes: Definition, Explanation, and Primacy in Socratic and Aristotelian Thought* (Oxford, 2013) p 66)." On this "desmology" see Dimitri El Murr "Platonic 'Desmology' and the Body of the World Animal (*Tim.* 30c-34a)" in Salles, Ricardo (*ed.*) *Cosmology and Biology in Ancient Philosophy: From Thales to Avicenna* (Cambridge, 2021).

[60] Aristotle argues that: "it is the mark of an educated man to look for precision in each class of things just so far as the nature of the subject (*hylē*) admits; it is evidently equally foolish to accept probable reasoning from a mathematician and to demand from a rhetorician scientific proofs (*N.E.* I 1094b23-27)."

[61] See Frederick Clifton Grant "St. Paul and Stoicism" *The Biblical World* (1915).

transcends his age. We cannot, indeed, properly treat any great philosophy on an environmental basis. Our lesson is to see that Foucault was perhaps so trapped by the problem of power that he could not fully account for the humanity that he otherwise so clearly expressed in himself and his thought.

Chapter 2
Aristotle in Our Own Day

Abstract This chapter considers modern philosophy in the context of Aristotle's critique of ancient philosophy. Modern philosophy is divided into a metaphysical philosophical tradition and a pragmatic anti-philosophical tradition. The latter is the philosophy that gives rise to reductionism, the reduction of meaning to the conclusions of science. In all cases we find the same fracturing and partiality of thought as previously witnessed—one perhaps consequential to the thought process itself—that was peripatetically reconstituted into a complex whole by Aristotle through his logic and metaphysics. I ultimately show that the split within our modern philosophical landscape between idealists and empiricists replays Plato's myth of the Battle of the Gods and Giants, written at the dawn of our intellectual history. The more things change, the more they stay the same…

Keywords Philosophical dualism · Anti-philosophy · The need for metaphysics · The Aristotelian tradition in modern philosophy

Let us take as our basic hypothesis that modern European thought swings radically between the extremes of religious eschatology and anti-religious nihilism due to its history. To expand upon this hypothesis, we can consider how this swinging between a deflationary and an inflationary perspective affects modern thought and its ability to satisfactorily represent the world. We can reasonably use Aristotle as a benchmark against which to assess our own thought. Let us hence begin this exercise by considering Aristotle's defence of the "definition" and then consider the reasons behind the modern attack upon such definitions.

What is the value of "definition" for Aristotle? Definition is the key to scientific demonstration on the basis that: "in all demonstration a definition of the essence is required as a starting-point (*pasēs gar apodeixeōs archē to ti estin*), so that definitions (*horismoi*) which do not enable us to discover the derived properties (*symbebēkota*), or which fail to facilitate even a conjecture about them, must

P. N. Jackson, *Aristotle on Meaning in the Living World*, Biosemiotics 30,
https://doi.org/10.1007/978-3-032-00602-8_2

obviously, one and all, be dialectical and futile (*De An.* I 402b 25–403a 3)." He gives us the worked example that: "A definition (*horos*) is a phrase (*logos*) indicating the essence of something (*to ti ēn einai sēmainōn*) (*Top.* I 101b 39)" with "man" being defined as "a mortal living creature receptive of knowledge (*Top.* V 128b 35-6)" on the basis that: "it is a property (*idion*) of man to be capable of learning grammar; for if a certain being is a man, he is capable of learning grammar, and if he is capable of learning grammar, he is a man (*Top.* I 102a 20-22)." We see that Aristotle looks at "man" and seeks to reach a definition as regards what differentiates him as a species and, in this case, as a genus of being. A human being is not defined by having the thought or language that other animals also possess to varying degrees but by reason of having the exceptional propensity to be able to learn grammar and have *structured and expressible representational thought.*

Beings are "defined" according to their differences, their niches, on the basis that: "in formulae (*logoi*) the first component, which is stated as part of the essence, is the genus, and the qualities (*poiotētes*) are said to be its differentiae (*diaphorai*) (*Met.* Δ 1024b 4-7)." Aristotle adds as regards the nature of these differences that: "Essence…will belong to nothing which is not a species of a genus, but only to a species of a genus; for it is these that are thought to be stated neither in virtue of participation, nor with an attribute (*pathos*), nor with an accident (*symbebēkos*). As for the others, if there is a name for each of them (*ean ē onoma*), there will be a formula signifying that this belongs to that, or instead of a simplified formula one that is more accurate; but there will be no definition (*horismos*) and no essence (*to ti ēn einai*) (*Met.* Z 1030a 11-17)." Aristotle puts forward the basic principle that only the "form" (*eidos*) taken as "species" and not the individual or genus will have an "essence" (*to ti ēn einai*) and hence have a "definition."[1] He continues by arguing that "definition" has many senses (*pleonachōs*)—as does "is" (*to estin*)—but that: "this point is clear: that the primary and unqualified definition, and the essence, belong to substances (*ousiai*). It is true that they belong equally to other things too, but not *primarily* (*Met.* Z 1030b 5-7)."[2]

Aristotle does *not* argue that "species" are immutable; *that* is Plato's contention, and it is obviously problematic to lazily conflate the two thinkers as is often done on

[1] On "essence" Aristotle explains that: "By substance without matter I mean the essence (*legō de ousian aneu hylēs to ti ēn einai*) (*Met.* Z 1032b 15)" and on "form" he explains further that: "By form I mean the essence of each thing and its primary substance (*eidos de legō to ti ēn einai hekastou kai tēn prōtēn ousian*) (*Met.* Z 1032b1-2)."

[2] Thomas Nagel tries to disprove Aristotle's idea of function (*ergon*) and definition through the example of a "combination corkscrew" which has multiple functions. This critique fails since it is clear that Aristotle does not argue that beings only have a *single* function. He observes that many or most parts of the body have several different functions and potentialities, e.g. *PA* II 659b 31-37, and that some species do indeed "dualise" between different genera of animals as the ape "dualises" between man and quadraped, see *HA* II 502a and *PA* IV 689b33-36. See Thomas Nagel "Aristotle on Eudaimonia" in *Essays on Aristotle's* Ethics in *ed.* Amelie O. Rorty (Univ. Of California, 1980). On Aristotle's "ergon argument" see Gavin Lawrence "The Function of the Function Argument" *Ancient Philosophy* (2001) and Deborah Achtenberg "The Role of the Ergon Argument in Aristotle's *Nicomachean Ethics*" *Ancient Philosophy* (1989).

this point. Aristotle's position on the difficulty of achieving precise definition and species boundaries is nuanced and reasonable, as is his insistence that beings do break down into species and do so because they exhibit differences (*diaphorai*) which necessitate a different name (*onoma*) and their own definition (*horismos*). Within this scope Aristotle explains the relationship between the "species" (or the "universal") and its members as that: "[since] the species is synonymous with its members….[we can engage in] the consideration of individual cases and the examination of the species to see whether the description fits…Such a procedure is useful against those who assume the existence of "ideas" (*Top*. VII 154a 18-20)." Once we have ringfenced the species, we can then review its particulars. Aristotle also explicitly argues that the "definition" is important to counter sophism on the basis that: "the starting point (*archē*) in dealing with all such people [as sophists] is definition (*horismos*) (*Met*. Γ 1012a 22)." Aristotle insists that we should not reify the world into "ideas" but are still able to understand and differentiate the structures and distinctions of species of being as they manifest in our world.

How is Aristotle's definition of definition denied by our modern thought? Karl Popper, one of the foremost philosophers of science of the modern age, argues that: "one of the prejudices we owe to Aristotle, [is] of the prejudice that language can be made more precise by the use of definitions."[3] He himself insists that: "In science…we never try to derive any information from the definition, or to base any argument upon it…we reach precision…by carefully phrasing our sentences in such a way that the possible shades of meaning of our terms do not matter."[4] Philosophy therefore boils down to the use of grammar. Popper further contends that all we need is "names," i.e., nominalism, on the basis that: "science does not use definitions in order to determine the meaning of its terms, but only in order to introduce handy shorthand labels."[5] Who, then, is "right" on "science" and "philosophy"; the ancient Aristotle or the modern Popper? Is it really true that we do not philosophically (or even scientifically) need definitions or consideration of essences, forms, or "shades of meaning"? Can we really dispense with philosophy and the subtleties of "meaning" and rest content with the "handy shorthand labels" of science?

I suggest here that it is Aristotle who is quite obviously *philosophically* right (and probably also obviously *scientifically* right) although I clearly need to support and expand upon this position below. Aristotle explicitly argued against the nominalism that seemed to underlie the bad philosophy in his own day (and which Popper and his colleagues propound in ours) that beings as species are more than mere labels or names on the basis that: "since in defining one exhibits either what the object is or what its name means, if definition is in no sense of the essence, it must be an expression meaning the same as a name. But this is absurd…All expressions

[3] Popper, Karl R. *The Open Society and Its Enemies vol. 2* (London, 1945) p 17.

[4] Popper, Karl R. *Ibid.* p 19.

[5] Popper, Karl R. *Ibid.* p 18. Popper explicitly identifies himself as a nominalist when he argues that: "scientific or nominalist definitions do not contain any knowledge whatever, not even any "opinion"; they do nothing but introduce new arbitrary shorthand labels; they cut a long story short (*Ibid.* p 14)."

would be definitions, because a name could be attached to any of them; so we should all converse in formulae, and the *Iliad* would be a definition (*Post. An.* II 92b 26-33)." As regards his own position, he insists that: "It is always possible to ask why (*dia ti*) (*Post. An.* II 92b 25)" and he argues that: "definition (*horos*) is a form of words (*logos*) which explains *why* a thing exists (*dia ti estin*) (*Post. An.* II 93b 39)" with the examples he offers being the "why" (*dia ti*) of thunder and of the eclipse. It is Aristotle who presents to us the standard generally accepted understanding of what science sets out to achieve, i.e., the establishment of the "tied down" cause rather than correlation. We find as regards Popper's attempt to conflate philosophy with science that it only replaces the prior equally dubious attempt to replace philosophy with religion.

We determine that Popper acts as a spoiler. He simply seeks to conflate science and philosophy by removing all meaning from the world apart from the description of the expert or scientist. The political subtext of his position—and this is significant—is that: "A social technology is needed whose results can be tested by social engineering."[6] As regards Aristotle's contrary position, he argues not only that we should seek to explain the meaning of the world rather than to produce a technology to manipulate it, but also that restricting our sight to abstractions such as names, language, and number is itself destructive. He argues in his aptly named *Sophistical Refutations* that: "[when] we use names as symbols in the place of things (*alla tois onomasin anti tōn pragmatōn chrōmetha symbolois*), we think that what happens also in the case of the names happens also in the case of the things (*epi tōn pragmatōn*)…[but] those who are unacquainted with the power of names are the victims of false reasoning (*Soph. El.* 165a 7-17)." Aristotle also explicitly warns us in his *Categories* of the tendency to reduce meaning to that which is expressed through *numbers and words*; see *Cat.* 4b 28-39. We rather find in reality that mathematics, language, and opinion can represent the world but can also, if we are not careful, displace and distort the world.

We do find, however, a counter-tradition in modern philosophy. Thinkers such as Martin Heidegger shows us both the vacuity of modern thinking and the depth exhibited by the Socratic tradition. His brilliant lectures on Plato's *Sophist* explore the thinking of both Plato and Aristotle and the relationship between them. He follows Plato's concern with the use of language that: "once a thing is put in writing, the composition, whatever it may be, drifts all over the place, getting into the hands not only of those who understand it, but equally of those who have no business with it (*Phaedrus* 275e)." He also follows Plato's concern with securing a *philosophical articulation* of meaning by closely translating him as that: "every λογος must συνεσταναι, hold together in itself, ωσπερ ζωον, like a living thing, which σωμα τι εχοντα, has a body, αυτον αυτου, with its own coherence, so that this ζωον is neither ακεφαλον, without head, nor without feet, and also has a middle and ends, ακρα, and everything is πρεποντα αλληλοις: all the parts are articulated, γεγραμμενα, in a

[6] Popper, Karl R. *Ibid.* p 222.

suitable way, among themselves and also in the context of the whole (*Phaedrus* 264c2ff)."[7] Heidegger shows us that Plato's great case study for the securing of meaning through definition is itself intended *to define philosophy as opposed to sophistry*.

Heidegger explains on this last point that: "a philosophy faces the tasks, on the one hand, of breaking through for the first time to the matters themselves (the positive task) and, at the same time the battle against idle talk. Both of these intentions are the genuine impulses of the spiritual work of Socrates, Plato, and Aristotle. Their struggle against rhetoric and sophistry bears witness to it."[8] I note that David Ambuel traces the roots of nominalism back to Parmenides.[9]

2.1 Modern Philosophy

I assert that it is simply ludicrous to suppose that philosophy can be dispensed with through the better use of grammar, even if we agree that the use of grammar defines humanity. The very idea is one of the most ridiculous gambits in history. That *we* fell for it should place a question mark over our powers of judgement and discernment. Understanding our own thinking will enable us to understand and appreciate Aristotle better, and hopefully to formulate a forward path. Let us on this basis review the main figures of modern philosophy under the provocative headings of "philosophy" and "anti-philosophy" in Fig. 2.1 below.

So, let us consider how Aristotle compares with the best modern thinking. Can our thought bear comparison with or even refute Aristotle's alternative position? We need to perform this exercise to show that Aristotle is not a dead letter and to understand the mental contortions we have got ourselves into so that we can then see if Aristotle can help us disentangle ourselves.

René Descartes (1596–1650) is correctly regarded as the father of modern philosophy. As regards why this is so we can consider his argument that: "the way in which I now understand the mind to be coextensive with the body [is as] the whole mind in the whole body and the whole mind in any one of its parts."[10] His "I think,

[7] Heidegger, Martin *Plato's Sophist* (Indiana, [*orig*. 1925] 1997) p 228.

[8] Heidegger, Martin *Ibid*. p 11.

[9] David Ambuel argues that: "The Eleatic denial of the image is bound with its implicit nominalism. In Parmenides this nominalism is implicit, because no theory of universals or conception of essence is distinctly formulated. Yet by the Eleatic assertion "Being is," no conceptual space is possible for an essence that bears some relation to something else, whether it be Platonic form, an Aristotelian universal, or some other model. That which is and that which can be thought are identical, completely merged in a featureless ontological landscape. The nominalist will have only particulars, each of them individual, and Parmenides will countenance only one (*Image and Paradigm in Plato's* Sophist (Las Vegas, 2007) p 171)."

[10] Descartes, René *Sixth Replies* CSM II §442 p 298. On this subject of mental extension see Robert Pasnau "Mind and Extension (Descartes, Hobbes, More)" *Studies of the History of Philosophy of Mind* (2007) and his "On Existing All at Once" in C. Tapp (*ed.*) *God, Eternity, and Time* (Ashgate, 2011).

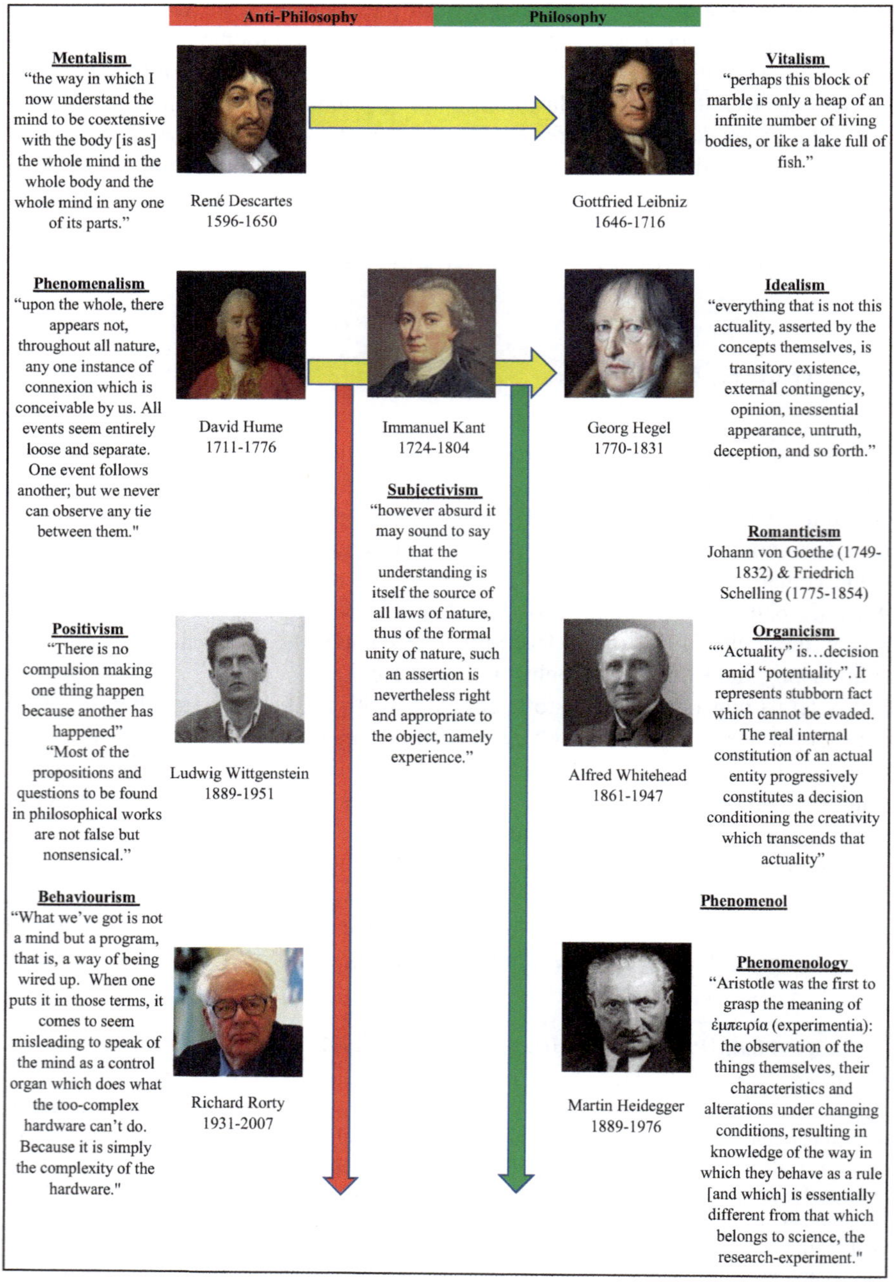

Fig. 2.1 Modern philosophical and anti-philosophical movements

therefore I am" can on this basis be interpreted as that "mind" occupies the world with matter being merely *an extension of mind*. Perhaps the most remarkable nuance here is Descartes' basing of his philosophy upon "mind" rather than "soul." Descartes' "ego" is not on this basis a true representation of the hard modern "self" and "individual" but has rather more in common with the belief (clearly derived from protestant Christianity) that the world is a reflection, product, and channelling of the "innate ideas" and/or "Will" of God. It is merely a step from seeing the world as the Will of God to regarding it as an interplay of abstract mathematizable forces. The "*res extensa*" of Descartes thereby gives way to the "billiard ball" physics of Isaac Newton who transformed Descartes' mechanistic *Principles of Philosophy* of 1644 into his own *Mathematical Principles of Natural Philosophy* of 1687 with the words "mathematical" and "natural" appearing in red on the cover of his work. I note that Paul Valéry rightly categorises Descartes as the first great "anti-philosopher."[11]

Gottfried Leibniz (1646–1716) is fittingly the opposite of a reductionist or "anti-philosopher" and is famously an opponent to both Descartes and Newton. He disputes Descartes' "I think" on the basis that: "Various things are thought by me. From this it follows not only that I am, but also that I am affected in various ways."[12] He explains further that: "It cannot be denied that Descartes has made some outstanding contributions and that it is his greatest merit to have revived the Platonic tendency to turn the mind away from the senses. Moreover, he has used advantageously the doubt of the Academic school. But, due to a certain inconsistency and to rash conclusions, he has all too soon gone beyond the goal, and has failed to determine what is certain from what is not. Thus he has advanced the absurd theory that extension is the essence of corporeal substance, and he has no sound understanding of the union of body and soul—all this because he has wrongly conceived of the nature of substance in general."[13] Descartes' error is, according to Leibniz, in respect to "substance" (*ousia*).

---

[11] See Catherine Wilson & Christiane Schildknecht "The Cogito Meant "No More Philosophy": Valéry's Descartes" *History of European Ideas* (1988).

[12] Leibniz, Gottfried "Critical Remarks Concerning the General Part of Descartes' Principles" in *Monodology and Other Philosophical Essays* (Indianapolis, [*orig*. 1692] 1965) p 26. Descartes himself counters Thomas Hobbes' objection that: "I am walking, therefore I am a walk" on the basis that: "I am not wholly certain of any of my actions, with the sole exception of thought (in using the word "certain" I am referring to metaphysical certainty, which is the sole issue at this point)…from the fact that I think I am walking I can very well infer the existence of a mind which has this thought, but not the existence of a body that walks (*Fifth Replies* §352, II p 244)." He explains his position (against Hobbes) as that: "I simply mean that all these modes of thinking inhere in me. I do not see how one can pretend that there is any doubt or obscurity here (*Third Replies* §177, II p 125)" (and see also "To Reneri for Pollot, April or May 1638" *CSM* III § 38 p 98). The problem arising from this attempt to hammer home a philosophical point regarding "mind" was the denial that animals have souls, and in a sense "life", on the basis that they do not have "minds."

[13] Leibniz, Gottfried "On the Improvement of Metaphysics" in *Monadology and Other Philosophical Essays* (Indianapolis, [*orig*. 1694] 1965) p 82.

What is the difference between Descartes and Leibniz on "*ousia*"? We can begin here with the observation that both Descartes and Leibniz reject the concept of material "atoms." Leibniz refused the idea of "atoms" on the basis that: "The atom which contains only a shaped mass of infinite hardness (which I hold not to be in conformity with divine wisdom, any more than the void is) cannot contain in itself all its past and future states, and even less all those of the entire universe."[14] René Descartes argued against "atoms" on the weaker grounds that: "We…know that it is impossible that there should exist atoms, that is, pieces of matter that are by their very nature indivisible…For anything we can divide in our thought must, for that very reason, be known to be divisible."[15] Leibniz argues that atoms are not explanatory because they cannot be the simple bottom-up basis of the complexity of the world and its beings whereas Descartes denies them for the strange reason that the world must conform to the assumptions of our thought. As explanation of this situation, we find that Leibniz accuses Descartes of thinking that "minds alone are monads"[16] whereas he himself argues that "monads" are "the true atoms of nature and, in brief, the elements of things."[17]

Leibniz himself describes "monads" as being like organic cells which gather power through their placement on the basis that: "there are also infinite degrees of life in the monads, some dominating more or less over others. But when a monad has organs that are adjusted in such a way that, through them, there is contrast and distinction among the impressions they receive, and consequently contrast and distinction in the perceptions that represent them [in the monad] (as, for example, when the rays of light are concentrated and act with greater force because of the shape of the eye's humours), then this may amount to sensation, that is, to a perception accompanied by memory—a perception of which there remains an echo long enough to make itself heard on occasion. Such a living thing is called an animal, as its monad is called a soul. And when this soul is raised to the level of reason, it is something more sublime, and it is counted among the minds."[18]

---

[14] Leibniz, Gottfried "Letter to Arnauld Nov/Dec 1686" *Philosophical Essays* (Indianapolis, 1989) p 81.

[15] Descartes, René *Principles of Philosophy* (1644) § 51, *CSM* I p 231. He disputed "atomism" with Pierre Gassendi on the grounds that: "The fact that they [i.e., innate ideas] may not accord with your suppositions about the nature of things, or with the atomic conception of reality invented by Democritus and Epicurus, is merely an extraneous feature which changes nothing; in spite of this they undoubtedly conform to the true nature of things established by God….I do not, incidentally, concede that the ideas of these [geometrical] figures ever came into our mind via the senses, as everyone commonly believes (*Fifth Replies CSM* III § 381 p 262)."

[16] Leibniz argues that: "the Cartesians have failed badly, since they took no account of the perceptions that we do not apperceive. This is also what made them believe that minds alone are monads and that there are no animal souls or other entelechies ("Monadology" (1714) § 14 in *Philosophical Essays* (Indianapolis, 1989) p 82)."

[17] Leibniz, Gottfried "Monadology" § 3 in *Philosophical Essays* (Indianapolis, [*orig*. 1714] 1989) p 213.

[18] Leibniz, Gottfried "Principles of Nature and Grace Based on Reason" in *Philosophical Essays* (Indianapolis, [*orig*. 1714] 1989) p 208.

Leibniz makes the "monad," which clearly bears some resemblance to Aristotle's "*ousia*," the basic element of being. His atoms of being are therefore qualitative rather than quantitative in the sense that: "It is only *atoms of substance*, that is to say unities which are real and absolutely without parts, which can be the sources of actions, and the absolute first principles of the composition of things, and as it were the ultimate elements into which substantial things can be analysed. They might be called *metaphysical points*; there is about them *something vital* and a kind of *perception*, and *mathematical points* are their *points of view* for expressing the universe."[19] Along these lines Leibniz also rehabilitates Aristotle's concept of "*entelecheia*" and explains regarding "first entelechy…[that t]his same substantial principle is what is called soul in living beings and substantial form in other corporeal objects. In so far as it constitutes, joined to matter, a truly unified substance, that is, a unit by itself (unum per se), it is what I call a Monad."[20] This emphasis upon cells as the basis for being and also upon hierarchical fields of being reminds us not only of Hans Driesch's famous sea urchin experiments but also foreshadows later "evo devo" biology. Driesch also revived the idea of "entelechy."

The problem with Descartes' intuitive position from Leibniz's metaphysical perspective is that: "the Cartesians, since they acknowledge no active principle substantial and capable of modification in the body, are compelled themselves to reject all action, and to transfer it to God alone, a far-fetched mechanical view, which is not philosophical."[21] Leibniz himself rather pictures the world as being *full* of creation in the sense that: "perhaps this block of marble is only a heap of an infinite number of living bodies, or like a lake full of fish."[22]

Could we say that Descartes is a revival of Plato and Leibniz of Aristotle? There are certainly grounds for this perspective. Descartes explicitly speaks about his philosophical debt to Plato and Leibniz explicitly speaks about his philosophical debt to Aristotle. We find, for instance, that Leibniz explicitly builds his concept of the "monad" upon Aristotle's thesis of the "alterations" (*alloiōsis*) between "*ousiai*"

---

[19] Leibniz, Gottfried "New System, and Explanation of the New System" *Philosophical Writings* (London, [*orig*. 1695–6] 1973) p 121.

[20] Leibniz, Gottfried "What is Nature?" in *Monadology and Other Philosophical Essays* (Indianapolis, [*orig*. 1698] 1965) p 106. Leibniz explains further that: "I distinguish: (1) the primitive entelechy or soul; (2) the matter, namely, the primary matter or primitive passive power; (3) the monad made up of these two things; (4) the mass [massa] or secondary matter, or the organic machine in which innumerable subordinate monads come together; and (5) the animal, that is, the corporeal substance, which the dominating monad in the machine makes one ("From the letters to de Volder" in *Philosophical Essays* (Indianapolis, [*orig*. 1703] 1989) p 177)."

[21] Leibniz, Gottfried "Appendix" in *New Essays Concerning Human Understanding* (London, [*orig*. 1702] 1896) p 703.

[22] Leibniz, Gottfried "Letter to Arnauld April 30, 1687" *Philosophical Essays* (Indianapolis, 1989) p 89. Leibniz explains further in respect to the role of the "divine" that: "Prime matter is merely passive, but not a complete substance: to it must be added a soul or a form analogous to a soul, a first entelechy that is, a certain tendency or primitive power of action, which is the law inherent in this substance, impressed on it by divine decree ("What is Nature?" in *Monadology and Other Philosophical Essays* (Indianapolis, [*orig*. 1698] 1965) p 107)."

(this being a key difference with Descartes, as we see above).[23] We also find, however, that Leibniz rejects Aristotle on the basis that: "No primitive entelechy whatsoever can ever arise or be destroyed naturally, and no entelechy ever lacks an organic body. As far as my consideration of these matters goes, these things could not be otherwise; they are not derived from our ignorance of the formation of foetuses, but from higher principles."[24] Leibniz rejects Aristotle precisely due to the latter thinker's emphasis upon "the formation of foetuses" rather than "higher principles" and Leibniz adds further on this basis that: "I believe with Plato and even with the Schoolmen, and with all of those who find meaning in the passage of St. Paul (*Romans* 2:15) when he states that the law of God is written in our hearts."[25] Leibniz insists upon a framing of Platonic mythology and Christian theology.

Leibniz's core disagreement is with Aristotle is therefore with Locke's idea of the "blank slate" which is often conflated with Aristotle's. He explicitly argues against Aristotle's own idea that the principles of beings are *in* them as "*archai*" which is a *bio-logical* principle he describes as "barbaric."[26] Both Leibniz and Descartes (as also Locke) argue that we must see the world from the perspective of "higher principles"—i.e., "God"—and Leibniz hence (also) asserts that: "Plato, in my opinion, surpasses all others, for he brings us to hope for a better life with good reason, and he is closest to Christianity."[27]

I have already noted above Nikolai Berdyaev's sharp observation that we moderns may be less Christian than the Angelic Doctor in our spiritual lives whilst at the same time being more "Christian" than him in our philosophy.[28] Nicholas Jolley

---

[23] Leibniz argues that: "To account for this heterogeneity, that is, diversity of qualities and hence ἀλλοίωσις or alteration, which Aristotle has not sufficiently explained, we can refer to the various degrees and directions of the tendencies which are modifications inherent in the monads. This also makes it clear, I think, that something else must be assumed in bodies than a uniform mass, the mere displacement of which does not explain any change ("What is Nature? Reflections on the Force Inherent in Created Things and on Their Actions" (1698) in *Monadology and Other Philosophical Essays* (New York, 1965) p 110)."

[24] Leibniz, Gottfried "Letter to de Volder 20 June 1703" *Philosophical Essays* (Indianapolis, 1989) p 176.

[25] Leibniz, Gottfried "Preface to the New Essays (1703–05)" in *Philosophical Essays* (Indianapolis, 1989) p 292.

[26] Leibniz dismisses as "barbaric" a "return to chimeras, to *archae*, to certain plastic intelligences that attend to the formation of the foetus ("Against Barbaric Physics" in *Philosophical Essays* (Indianapolis, [*orig*. 1686] 1989) p 31)." I note that Leibniz's opposition to the "blank slate" is directed against the philosophy of John Locke; see "Preface to the New Essays (1703–05)" in *Philosophical Essays* (Indianapolis, 1989) p 292. Aristotle does *not* argue that mind exists as an outgrowth of sense but, in a position with much similarity with Jean Piaget's, that it develops through a process of natural biological and experiential maturation.

[27] Leibniz, Gottfried "Letter to Molanus (?), On God and the Soul, c. 1679" in *Philosophical Essays* (Indianapolis, 1989) p 241.

[28] Pierre Manent explains that we move in a landscape of Christianity like "fish in water" and that: "The observation applies to everyone, or almost everyone, in the nineteenth century—to the liberals as well as the Saint-Simonians, to Chateaubriand as well as Quinet, to Tocqueville as well as Michelet…this metapolitics that determines politics, is actually Christianity "realising" itself (*An*

compounds our confusion by observing that: "Philosophers in the period [i.e., in the Early Modern period] were not really turning their back on the tradition as a whole; rather they were casting down Aristotle in order to raise up Plato."[29] Dennis des Chene explains the consequences of this situation as that: "What in the twentieth century appeared as the "problem of emergence" has its parallel in the Aristotelian tradition as the problem of the education of forms: if the forms of plants and animals do not exist potentially in matter, where do they come from? The standard answer [for later "Aristotelian" thinkers but not for Aristotle himself], for higher animals and humans at least, is that they come from the heavens or from God. Descartes did not solve the problem of education. He gets rid of it. There are no souls in animals or plants, and thus no education of forms. In the vocabulary of the present day: there are no emergent properties in the world of Descartes."[30] The same criticism applies to Leibniz, on explicitly theological grounds. Our modern thought actually moves *away* from reality, evolution, and nature (in a meaningful sense) rather than towards it.

Thomas Hobbes (1588–1679) and **Baruch Spinoza (1632–1677)**. Let us here combine the "God" Spinoza with the "Giant" Hobbes in order to consider the ridiculous modern collapse of thought into "chance" and "necessity." Thomas M. Lennon explains this situation as that: "Wittingly or not, philosophers have tended either to collapse the contingent into the necessary, or to give up the necessary altogether and to regard the real as a colossal accident. As instantiations of this tendency in the seventeenth century one thinks of the necessitarianism of Spinoza and the pantychism of Hobbes."[31] We have now formalised the problems of "chance" and "necessity" raised by the ancient philosophers considered previously. The explanation of world through chance and mechanism is foundational to modern thought, exemplified by neo-Darwinism.

What is the significance of this dichotomy between "chance" and "necessity"? We can point here to the thinking of the Nobel-laureate biologist Jacques Monod (who himself claims to be a Cartesian) and his seminal lectures on *Chance and Necessity*. Monod argues as a biologist that we are all as "beings" the product of an "enormous lottery presided over by natural selection, blindly picking the rare winners from among numbers drawn at random"[32] and as a philosopher that: "Modern societies, woven together by science, living from its products, have become as dependent upon it as an addict on his drug. They owe their material power to this

Intellectual History of Liberalism (Princeton, [*orig.* 1987] 1995) p 82–3)." Even Karl Marx regarded his political philosophy as a development through Christianity.

[29] Jolley, Nicholas *The Light of the Soul: Theories of Ideas in Leibniz, Malebranche, and Descartes* (Oxford, 1990) p 11.

[30] Des Chene, Dennis *Spirits and Clocks: Machines and Organism in Descartes* (Ithaca, 2001) p 154–5.

[31] Lennon, Thomas M. *The Battle of the Gods and Giants: The Legacies of Descartes and Gassendi, 1655–1715* (Princeton, 1993) p 51.

[32] Monod, Jacques *Chance and Necessity* (London, 1971) (subtitled as "a philosophy for a universe without causality") p 138.

fundamental ethic upon which knowledge is based, and their moral weakness to those value-systems devastated by knowledge itself, to which they still try to refer."[33] This is the significance of the duality of "chance" and "necessity" which inexorably leads to Bentham's "panopticon" and latterly Dawkins' "robot vehicles blindly programmed to preserve the selfish molecules known as genes."[34]

There is something strange and twisted about this mixture of desperate chance and hard machine. Even the routine lashing together of the rationalism (and associated "method") of Descartes and the empiricism (and associated "method") of Francis Bacon is observably incoherent at the same time as being a modern dogma.[35] All such rationalist ideas as that we just need to be led by "the facts" or by "clear and distinct ideas" are ultimately naïve and false. Aristotle's position is, by contrast, vastly more subtle and true. Aristotle ultimately realises that philosophy is not just a talking shop for misguided ideologies but a grasping of being measured by truth against the world as we experience it. "The Philosopher" ironically understands best the pratfalls and pitfalls of philosophy.

David Hume (1711–1776) perhaps represents the apogee (or nadir) of anti-philosophy. He seeks to invalidate the principle of power or "*dynamis*" (and of nature or "*physis*") on the basis that: "upon the whole, there appears not, throughout all nature, any one instance of connexion which is conceivable by us. All events seem entirely loose and separate. One event follows another; but we never can observe any tie between them. They seem *conjoined* but never *connected*…the necessary conclusion *seems* to be that we have no idea of connexion or power at all, and that these words are absolutely without any meaning, when employed either in philosophical reasonings or common life."[36] We have already seen this attempt to deny potentiality and to depend upon activity and appearance made by Democritus but it is also a feature of the Megarian philosophy which argued that a builder is only such when he is actually building.

Hume significantly takes Isaac Newton as his authority on *philosophical* matters. He argues that: "It is entirely agreeable to the rules of philosophy, and even of common reason; where any principle has been found to have a great force and energy in one instance, to ascribe to it a like energy in all similar instances. This indeed is Newton's chief rule of philosophising."[37] Hume's is not a move towards Aristotle's analogical method which extends related meaning to beings but towards a scientific reductionism which claims to consolidate any such meaning. The principles that we use for the purposes of physical mechanics—i.e., Newton's physics—can be applied to the *whole* world on the basis that: "the experimental reasoning itself, which we possess in common with the beasts, and on which the whole conduct of life depends,

[33] Monod, Jacques *Ibid.* p 164.

[34] Dawkins, Richard *The Selfish Gene* (Oxford, [*orig*. 1973], 1989) p xxi.

[35] Jorge Secada comments on this situation that: "I believe, in fact, that Descartes would have preferred to remain an Aristotelian rather than end up a "new" empiricist (*Cartesian Metaphysics: The Late Scholastic Origins of Modern Philosophy* (Cambridge, 2000) p 185)."

[36] Hume, David *An Enquiry Concerning Human Understanding* (1757) §58.

[37] Hume, David *Ibid.* §163.

is nothing but a species of instinct or mechanical power, that acts on us unknown to ourselves."[38] Hume's ultimate solution is to reduce the world to atoms (as well as forces) on the basis that: "time, as it exists, must be composed of indivisible moments…if each moment, as it succeeds another, were not perfectly single and indivisible, there would be an infinite number of coexistent moments, or parts of time; which I believe will be allow'd to be arrant contradiction."[39] Time becomes the atomic events or moments of circumstance rather than being the temporal pathway of the timescale or lifespan *of many somethings*. Hume's banal interpretation of time is simply that it is that which is measured by the hands of a clock.

Michael J. White describes this atomistic position well as an ""at-at" ontology." He explains Bertrand Russell's "logical atomism" in this context as that: "Russell gives a lucid (and tendentious) statement of the positional or "at-at" ontology of motion…[as that] "Motion consists merely in the fact that bodies are sometimes in one place and sometimes in another, and that they are at intermediate places at intermediate times"…[which White glosses as that] This positional ontology of local motion has been very widely disseminated. And, when it is presented to them, it seems to be rather easily internalised by most people in contemporary Western culture. Perhaps the facts of cinematography have helped make the idea that motion is ultimately composed of "freeze-frame slices" seem a very natural idea. The positional ontology of motion is not a *correct* view according to Aristotle…The positional or "at-at" conception of motion is certainly intimately related to the development of point-set topology. Aristotle possesses a *kinematical* conception of motion that appears, in many respects, to be very similar to a standard, contemporary conception: that is, he conceives of motion as occurring in smooth stretches rather than by jumps or instantaneous displacements. However, he obviously does not share the common, contemporary positional ontology."[40] Aristotle simply and completely rejects a "billiard ball" view of the world.

Georg Henrik von Wright explains the objective and subtext of this "at-at" ontology and Aristotle's rejection of it further by observing that: "The goal "pulling from the future" could be replaced by the will (to reach the goal) "pushing towards the future.""[41] Hume's innovation is to collapse the world not into the necessary power and concurrence of "God" as with Descartes' picture of a divine ego in a plastic world but into the supposedly contingent happenings of matter through which the being itself pushes forward through time and space. He rejects the idea of an "end" and hence of an "entelechy" (and "final causes") because it requires a living being (an *ousia*) rather than a merely "pushing" thing. A Newtonian world of forces and atoms rather than of beings becomes superficially viable on this basis. As regards why we need to take this reductionist position into account, A.N. Whitehead cor-

[38] Hume, David *Ibid.* §85.

[39] Hume, David *A Treatise of Human Nature* (1740) § II "On the infinite divisibility of space and time."

[40] White, Michel J. "Aristotle on the Non-Supervenience of Local Motion" *Philosophy and Phenomenological Research* (1993) p 150–4.

[41] Wright, Georg Henrik von *Explanation and Understanding* (Ithaca, 1971) p 89.

rectly, but sadly, comments that the "Hume-Newton situation…[is]…the primary presupposition for all modern philosophic thought. Any endeavour to go behind it is, in philosophic discussion, almost angrily rejected as unintelligible."[42] Most modern thinking simply leads us to the same insistent but dubious conclusion: *viz.* to a sort of "logical atomism" which explains the world as the "chance" of the being itself *and* the "necessity" of the world in which it exists.[43]

Immanuel Kant (1724–1804) was famously woken from his "dogmatic slumber" by David Hume. He surely counts as the representative philosopher of modernity since he is the masterful, audacious blender of Humean empiricism and Cartesian rationalism (and Newtonian science) into the strange but brilliant cocktail that passes for Enlightened modern thought. It is hence that Kant stands—along with Aristotle—at the centre of the table of philosophers put forward as Fig. 2.1.

Kant's great innovation was the subjectivism expressed in principle as that: "however absurd it may sound to say that the understanding is itself the source of all laws of nature, thus of the formal unity of nature, such an assertion is nevertheless right and appropriate to the object, namely experience."[44] Kant argued that *we create the laws of nature through our reflection upon the world.* He explained further that: "As a rational being, and thus as a being belonging to the intelligible world, the human being can never think of the causality of his own will otherwise than under the idea of freedom; for independence from the determining causes of the world of sense (which reason must always ascribe to itself) is freedom. With the idea of freedom the concept of autonomy is now inseparably combined, and with the concept of autonomy the universal principle of morality, which in idea is the ground of all actions of rational beings, just as the law of nature is the ground of all appearances."[45] Kant's emphasis upon freedom, autonomy, and duty clearly lead him to a mentalistic and subjective world which we need to consider below vis-à-vis Aristotle's objective perspective.

Many books could, and indeed have, been written on the contrast between the philosophical positions of Aristotle and Kant. Here, I will just draw some crude, overall distinctions between the two (which will be built upon later on). Aristotle sought to objectify the world, including humanity and its various "parts," in order to understand it. Humanity is taken as an exemplary case study for being ("*ousia*"), though Aristotle's biological programme demonstrates that it is only one "*eidos*" of

---

[42] Whitehead, A.N. "Nature Lifeless" in *Modes of Thought* (New York, 1938) p 135.

[43] See Stefanie Rocknak "Locke, Hume, and the Treacherous Logos of Atomism: The Eclipse of Democratic Values in the Early Modern Period" *Notre Dame Philosophical Reviews* (online, 2015). Michel Bastit speaks of the "astounding weakness of the Humean conception of causality" and argues rather that: "the dispositional powers come from the thing itself, *ex autou pragmatos*, as Aristotle would have said ("Causality: From Contemporary Scientific Conceptions of Causality to the Classical Four Causes" *Angelicum* (2014) p 58)." See also the wonderful work on Iris Murdoch as regards the errors of modern philosophy.

[44] Kant, Immanuel *Critique of Pure Reason* (Cambridge, [*orig*. 1781] 1998) A127.

[45] Kant, Immanuel *Groundwork of the Metaphysic of Morals* (Cambridge, [*orig*. 1785] 1998) 4.452-3.

many. Aristotle's *Ethics* seeks to survey the "habits" (*ethē*) of man in order to understand his spiritual and behavioural structuring in parallel with his biological structuring. He does not really consider morality as such but rather considers good action according to the nature and objectives of a given human being on a *situational* basis. Kant's position is, by contrast, subjective rather than objective, as we have already seen. His concentration is upon man and his moral world and is *universalistic* with his imperative being: "Act only according to that maxim whereby you can, at the same time, will that it should become a universal law.[46]" Kant is concerned with realising human freedom as a moral civilisation rather than with unpacking the human power of creation to reveal the springs of nature beneath. See Kurt von Fritz's "Aristotle's Anthropological Ethics and its Relevance to Modern Problems" (1981) as regards this contrast.

Kant only places human beings in the world by reducing the world to "mind." Josef Pieper explains this approach as that: "According to Kant man's knowledge is realised in the act of comparing, examining, relating, distinguishing, abstracting, deducing, demonstrating—all of which are forms of active intellectual effort. Knowledge, man's spiritual intellectual knowledge (such is Kant's thesis) is activity, exclusively activity."[47] Whereas Aristotle's metaphysics and his categories were centred upon how we understand and engage with "being" (*ousia*), the centre of Kant's argument is the liberatory idea that we create the world by interpreting it through its "appearances" to us. Kant therefore still frames his world through the Cartesian picture of mind and ego acting upon matter. For Kant, human beings are not parts of nature but spiritual creatures seeking to transcend nature. His "analytic" approach represents itself as our active enlightenment applied to a world defined by its "passivity" as a plastic and extended thing.[48] Aristotle, by contrast, considers the world not as an intuitive revelation but as a synthesis of thought based upon an analysis of the world. The need for a synthesis of reality leading to truth is the difference between Aristotle's realism and Kant's (and Plato's) idealism.

Kant does, however, himself provide us with a critique of the "Cartesian" (or "Platonic") position. He himself explains that: "Plato himself a master of this science [of geometry], was overcome by enthusiasm [when he saw] that the original character of things is such that it can be discovered without any experience whatever, and that the mind is able to derive the harmony of beings from their supersensible principle; (to [these beings] we must add the properties of numbers, with which the mind plays in music). It was this enthusiasm that lifted Plato above empirical concepts to ideas that he thought could be explained only by an

[46] Kant, Immanuel *Ibid.* 4:421.

[47] Pieper, Josef *Leisure the Basis of Culture* (London, [*orig*. 1947] 1965) p 26.

[48] Brian Ellis explains on this point that: "The view that things in nature are essentially passive, and obedient to nature's laws, was very widely shared by philosophers of all persuasions in the eighteenth century, as indeed it has been ever since. It was accepted not only by Descartes, Newton, and Hume, but also by Locke and Kant, and therefore by the founding fathers of all of the major philosophical traditions of western Europe. Let us call this still-dominant world-view "passivism." Passivism has both religious and secular versions (*The Philosophy of Nature: A Guide to the New Essentialism* (Chesham, 2002) p 2)."

intellectual community [between us and] the origin of all beings. No wonder he turned away from his school everyone who was ignorant of geometry; for what Anaxagoras inferred from objects of experience and their connection in terms of purposes, Plato meant to derive from the pure intuition residing within the human intellect [*Geist*]…Surely it is pardonable if, as the result of a misunderstanding, this admiration gradually increased to the point of fanaticism."[49] Kant, as also Plato, perceived the limitedness of idealism but was, perhaps, prevented from overcoming such inadequacies by the very parameters of his own (idealistic) philosophical approach.

What is Plato's misdirection according to Kant? Kant argues that mathematics has only a formal purposiveness as constructed by us for our subjective purposes. He argues that we must also recognise an external, natural or "real, [i.e., material]" purposiveness in nature which possesses its teleology in itself. Kant admits this "real" in his final critique, his *Critique of Judgment*, as that there are "existing things that must be given empirically if they are to be cognised and are not a mere presentation in myself determined according to an *a priori* principle. Hence the latter (empirical) purposiveness is [not formal but] *real*, and hence is dependent on the concept of a purpose."[50] This admission of the "real" leads Kant (belatedly) towards Aristotle, towards the innate purpose of beings, and towards the principle that we must "modestly restrict ourselves to the expression that says no more than we know—*viz.*, purpose of nature."[51] We also see, however, alongside this "Aristotelian" Kant, a purely "Kantian" Kant whose main thrust is towards an understanding of such "unconditioned objects" of thought as are *not* emergent from the world such as (according to Kant) "soul," "world," and "God." Kant's hypostatisation is to transcend the real world by pursuing these higher objects. Kant hence clearly worked from what Leibniz called "higher principles" rather than from "the formation of foetuses"; his innovation was to subjectivise these principles in a different way than Hume.

Hegel is to my mind correct in criticising Kant's "critical" philosophy as a process of whittling away objectivity and giving the burden to a subjectivity which relegates the world and effectively reduces itself to psychologism. Kant elevates his sources into a full and meaningful metaphysics but is reduced by them to a woolly theologism (spiritualism) alongside a woolly scientism (mechanicism). By comparison, we find the judiciousness of Aristotle who carefully constructs a physics (and metaphysics) of being which is rigorously designed to exhibit the

[49] Kant, Immanuel "Critique of Teleological Judgment" in *Critique of Judgment* (Indianapolis, [*orig*. 1790] 1987) § 363. He argues further that: "Plato…was…the father of all fanaticism [aroused] *by means of philosophy* ("On [the] Dignified Tone Recently Adopted in Philosophy", Ak. VIII, p 398)."

[50] Kant, Immanuel *Critique of Judgment* (Indianapolis, [*orig*. 1790] 1987) § 364.

[51] Kant, Immanuel *Ibid.* § 382. Kant interestingly also proceeds through the problems of monism and empiricism that we have treated above. For Parmenidean monism he argues against Spinoza and on mechanicism he refers to Democritus and Epicurus, a position he regards as "manifestly absurd"; see *Ibid.* § 72-3.

purposiveness which exists in the (animal) world and to explain it on its own terms explicitly rejecting any compromise which reduces it in some measure to either machine (science) or to "God" (mythology).[52]

Georg Hegel (1770–1831) is likely the greatest systematic philosopher and metaphysician in the Western philosophical canon, excepting Aristotle himself. Indeed, it is critical to observe how this philosopher manages to reintroduce Aristotle afresh into the tradition (as Leibniz did previously and Heidegger did subsequently) by repurposing, rethinking, and distorting him in accordance with his times.

Hegel profoundly understands Aristotle when he observes that: "He [i.e., Aristotle] gets the sensuous phenomenon before him in its entire completeness, and omits nothing, be it ever so common. All sides of knowing enter his mind, all interest him; all are handled by him with depth and exhaustiveness...[and] Aristotle...abandons a determination only when he has traced it to another sphere wherein it retains no longer its former shape...[and] sometimes Aristotle does not aim to reduce all to unity, or at least to a unity of antithetic elements; but, on the contrary, to hold fast each one in its determinateness, and thus to preserve it."[53] This is a beautiful statement of the value of Aristotle's topological and analogical approach to the world. As regards what it means to Aristotle to "save the phenomena," Hegel comments that: "Aristotle's conception of nature is...nobler than that of today, for with him the principal point is the determination of the end as the inward determination of natural things. Thus he comprehended nature as life."[54] Hegel very precisely identifies the nub of Aristotle's "physics" as being in respect to "the determination of the end as the inward determination of natural things," i.e., regarding "determination," "the end," "inward," and "natural things." He represents Aristotle as the epitome of the philosopher on the basis that: "The ultimate aim and business of philosophy is to reconcile thought with reality."[55]

Hegel builds his own philosophy upon Aristotle's brilliant (but little understood) assessment of "abstraction" (*abs-traho* = "I detach from") which regards how human beings extract "forms," "ideas" and "meaning" from the objects of the world and internalise them as the "form[s] in the soul (*Met.* Z 1032b 2)." He also picks up on Aristotle's little-appreciated concept of "*symphysis*" or "growing together" on which Ivan Il'in comments that: "In the very nature of this "growing-together" is hidden the ideational root of all of "speculative" philosophy."[56] Hegel himself

---

[52] For a profound critique of Kant see Hegel's *Logic* (Oxford, [*orig.* 1830] 1975).

[53] Hegel, G.W.F. "The Philosophy of Aristotle" *The Journal of Speculative Philosophy* ([*orig.* 1825–6] 1871) p 73–75. On the advantage of Aristotle over Plato Hegel argues that the "Final Cause is true and concrete, as opposed to the abstract Platonic Idea (*Ibid.* p 76)."

[54] Hegel, G.W.F. *Lectures on the History of Philosophy* (New York, [*orig.* 1840] 1974) II p 157.

[55] Hegel quoted in Nicholas Lobkowicz *Theory and Practice: History of a Concept from Aristotle to Marx* (Notre Dame, 1967) p 157.

[56] Il'in, Ivan *The Philosophy of Hegel as a Doctrine of the Concreteness of God and Humanity Vol. I* (Evanston, [*orig.* 1918] 2010) p 141. See also the translator's introduction, Ibid. p xlvi. Bodies are themselves a manifestation of this concretion of being and Il'in adds on this situation that: "the universal "penetrates," "saturates" (*durchdringt*) its specific parts. By no means should this term be

explains the "Spirit" (*Geist*) animating this approach as that: "Just as the totality of life is contained to an equal degree in the nature of the polyp and the nature of the nightingale and the lion, so too the world spirit revealed in each shape (*Gestalt*) its absolute self-feeling (*Selbstgefühl*)—be it more obscure or more developed—and enjoyed its own essence and its own self in each nation, in each ensemble of mores and laws."[57] Hegel's philosophy of "shapes" (*Gestalten*) is clearly of the same milieu as Goethe's biological "morphology" (from "*morphē*" (shape)) and Schelling's "speculative physics" of self-organisation.[58] He also adds, however, the "shapes" of human history which is an evolutionary approach that broadly goes beyond, wisely or not, Aristotle's very carefully limited ontology, metaphysics, politics, and science.

How does Hegel critique or distort Aristotle in accordance with the spirit of his age? Hegel argues that: "Aristotle always moves in the speculative…he seems always to be philosophising only on the individual, the special, and not to arrive at what is absolute, universal or God…the one Absolute, the Idea of God, appears in Aristotle's Philosophy, but as a particular somewhat, side by side with the others."[59] He moves beyond Aristotle because he himself believes that: "To think the empirical world is…to change its empirical form and to transform it into something universal"[60] and that: "everything that is not this actuality, asserted by the concepts themselves, is transitory existence, external contingency, opinion, inessential appearance, untruth, deception, and so forth."[61] Hegel's position here is indeed compatible with Aristotle's argument in respect to how we form ideas and universals but goes beyond it towards something more reificatory and more Platonic. As regards the ideological extension of Hegel's position, we find that Kant saw the acme of

taken in a metaphorical or figurative sense. A substance penetrates into its attributes, saturates them, and itself represents their totality (*Ibid.* p 101)." I suggest that this movement towards absoluteness is represented in Aristotle's concept of "complete difference (*teleia diaphora*)" through which a being or species becomes fully itself, fully realised, and autonomous; see *Met.* I 1055a13-17. On Aristotle's "*symphysis*" see *Met.* Δ 1014b 22-26 and *Met.* Z 1040b 13-16.

[57] Hegel quoted in Il'in, Ivan *Ibid.* p 207.

[58] On Schelling see Arran Gare "Overcoming the Newtonian paradigm: The unfinished project of theoretical biology from a Schellingian perspective" *Progress in Biophysics and Molecular Biology* (2013) and on Goethe see Dana Pauly "Goethean Science: A Phenomenological Study Of Plant Metamorphosis" *Biodynamics* (2000) and Werner Heisenberg "The Teachings of Goethe and Newton on Colour in the Light of Modern Physics" in *Philosophical Problems of Quantum Physics* (Woodbridge, [orig. 1941] 1979).

[59] Hegel, G.W.F. *Ibid.* p 77–8. On the nature of "speculative" philosophy see *Hegel's* Logic (Oxford, [*orig.* 1830] 1975) § 26-32.

[60] Hegel quoted in Ivan Il'in *The Philosophy of Hegel as a Doctrine of the Concreteness of God and Humanity Vol. I* (Evanston, [*orig.* 1918] 2010) p 35).

[61] Hegel quoted in Il'in, Ivan *Ibid.* p 190. Il'in comments further that "The absolute idea is itself a "method" (*Ibid.* p 190)."

human thought as being the "moral law within me" and that Hegel saw it as being the realisation at the end of history of the peculiar freedom of (German) Christianity.[62]

Interestingly, Hegel himself identifies the mantraps of idealism that are incumbent upon reducing the world to "the Absolute" by reference to the "absolute Substance" of Spinoza, who "defrauds the principle of difference or finitude," but not in respect to his own "absolute Spirit."[63] Ivan Il'in explains Hegel's position here as that: "The Concept "contains the value of the object," its "inner." The Concept constitutes the "object's own self. Therefore, one can say that "thought is the principle and essence of the world""[64] with Hegel's own core principle being that: "the logical becomes nature, and nature becomes spirit."[65] Hegel does not remain "speculative" along with Aristotle but seeks to progress to something else, something "absolute" (*absolvo* = "untied, detached"). Hegel on this basis takes detached and absolute positions on such things as Philosophy, Science, Logic, Concept, Dialectic, Spirit, Right, God etc. (all capitalised) which serve Hegel's putative directionality of a "Spirit" as "God" moving (through a "Phenomenology of Spirit") through "World History."

Johann von Goethe (1749–1832) and **Friedrich Schelling (1775–1854)** deserve special mention in our condensed survey of philosophy. These thinkers represent a "romantic" denial of mechanicism through their *Naturphilosophie* which presages the later *Lebensphilosophie* of the later phenomenologists. The recognition of "nature" and "life" by philosophy is so obviously necessary that it is shocking and troubling to find a trenchant refusal to recognise such facts, and this *denial of meaning* is important in itself. I make note of the *ideological* and *religious* basis for this refusal and comment that the idea that we can and should reduce the world to physical matter and to physics is a strange dream emerging from the success of Newtonian

---

[62] Hegel's whole philosophy ultimately drives towards the "conclusion" that: "It was first the Germanic peoples, through Christianity, who came to the awareness that every human is free by virtue of being human, and that the freedom of spirit comprises our most human nature. This awareness arose first in religion, in the innermost region of Spirit (*Introduction to The Philosophy of History* (Indianapolis, [*orig*. 1840] 1988) p 21)." Hegel typifies the hidden tendency of modern thought explained by Pierre Manent as "Christianity "realising" itself."

[63] Hegel argues that: "Spinoza puts substance at the head of his system [i.e., as God], and defines it to be the unity of thought and extension, without demonstrating how he gets to this distinction, or how he traces it back to the unity of substance. The further treatment of the subject proceeds in what is called the mathematical method… The so-called atheism of Spinoza is merely an exaggeration of the fact that he defrauds the principle of difference or finitude of its due. Hence his system, as it holds that there is properly speaking no world, at any rate that the world has no positive being, should rather be styled Acosmism (*Hegel's Logic* (Oxford, [*orig*. 1830] 1975) § 151, p 213–4)."

[64] Hegel quoted in Il'in, Ivan *The Philosophy of Hegel as a Doctrine of the Concreteness of God and Humanity Vol. I* (Evanston, [*orig*. 1918] 2010) p 165.

[65] Hegel quoted in Il'in, Ivan *Ibid.* p 200. Il'in explains that: "In this new process, a new subject is created, capable of receiving the Divine revelation. This subject exists in the element of bifurcation, or other-being, and possesses "con-sciousness." And it is to this subject, enriched and burdened with the form of other-being, that the essence of Divinity is revealed anew (*Ibid.* p 200)."

physics which remains a modern dogma.[66] Goethe critiques this ideology by reference to his aesthetic Theory of Colour which he put forward in express opposition to Newton's analytic treatment of colour. Werner Heisenberg in a brilliant lecture explains the philosophical significance of this conflict between Newton and Goethe as a choice between the "continuous movement of modern science towards an abstract control of nature, removed from an appreciation based on common experience"[67] and "that other reality, important and full of meaning for us. In that reality events are not counted but weighed, and past events not explained but interpreted."[68] We can hence choose between abstract and embodied interpretations of nature.

Arran Gare explains the parallel, equally remarkable legacy of Friedrich Schelling as that: "For Schelling, "Speculative Physics" is required to produce the concepts necessary to comprehend the self-constructive activity by which nature has evolved…He argued that we are part of nature, so our comprehension of nature is simultaneously nature comprehending itself through us and our comprehension of ourselves as products of nature…[and] such knowledge presupposes living nature and only emerges from it, and we can know nature through reflection because we are part of, have emerged from, and are practically engaged within it."[69] Gare shows upon this basis that the various positive trends in modern science such as field theory, morphology, and systems theory which I argue are clearly products of Aristotle's thought are simultaneously products of Schelling's thought. How can they be counted as both? The simple fact is that Aristotle and Goethe/Schelling both insist that we must approach the world from the perspective of "nature" and "life" rather than through a mathematical or physical theory. It is therefore natural for both schools of thought—"realism" and "romanticism"—to be parallel or alternate influences on later developments.

Do we need Aristotle if we have a more modern or immediate source for the philosophico-scientific developments which have themselves influenced modern biosemiotics? I suggest that Aristotle is important in himself as a direct influence upon this modern thought and that he brings his own distinct philosophical and scientific perspective to the table which is different from that of Goethe and Schelling and equally, or more, explanatory and penetrating. I will explain the value of Aristotle's science later on in this volume and its influence in more detail

---

[66] A. N. Whitehead correctly observes that we live in a: "Hume-Newton situation…[which is]…the primary presupposition for all modern philosophic thought. Any endeavour to go behind it is, in philosophic discussion, almost angrily rejected as unintelligible ("Nature Lifeless" in *Modes of Thought* (New York, 1938) p 135)." Most (or all) the thinkers considered later in this work recognise this "situation" and seek to break out of it. On the problem of "physicalism" see Tim Crane & D.H. Mellor "There is No Question of Physicalism" *Mind* (1990).

[67] Heisenberg, Werner "The Teachings of Goethe and Newton on Colour in the Light of Modern Physics" in *Philosophical Problems of Modern Physics* (Woodbridge, [*orig.* 1941] 1979) p 60.

[68] Heisenberg, Werner *Ibid.* p 68.

[69] Gare, Arran "Overcoming the Newtonian paradigm: The unfinished project of theoretical biology from a Schellingian perspective" *Progress in Biophysics and Molecular Biology* (2013).

in its sister volume. Here I note that Aristotle, broadly speaking, does not give us the same brilliant insights on the human experience as can be harvested from the "romantics" and the later "existentialists" but he also does not manifest the drawbacks and distractions of their emphases. "*Sturm und Drang*" is a valuable approach in some circumstances but not, perhaps, in the context of metaphysics or ontology. Heidegger's idea of "thrownness" in the world captures something of the mystery, genius, and pain of the human condition and Aristotle also maturely captures both this wonder and this suffering in his own work to some degree.[70] Aristotle restrains himself, however, from speculation since his aim is to explain how we fit in to reality so that he can explain reality—hence his "realism"—rather than to explore how we envisage ourselves as human beings in the world—hence "romanticism."

Ludwig Wittgenstein (1889–1951) represents the (much needed) charismatic side of the "linguistic turn" in philosophy. The reduction of the world to words, terms, names in a sense reminds us of Karl Popper's reduction of the world to "bold ideas" and "handy shorthand labels."

Setting off from Hume, Wittgenstein asserts that: "There is no compulsion making one thing happen because another has happened"[71] and he ultimately ends up with the positivist idea that: "Most of the propositions and questions to be found in philosophical works are not false but nonsensical."[72] We see on this basis how Wittgenstein embodies the *anti*-philosophy explained by Alain Badiou as that: "It is typical of antiphilosophy that its purpose is never to *discuss* any philosophical theses…since to do so it would have to share its norms (for instance, those of the true and false). What the antiphilosopher wants to do is to situate the philosophical desire in its entirety in the register of the erroneous and the harmful. The metaphor of sickness is never absent from this plan, and it certainly comes through when Wittgenstein speaks of the "nonsensical.""[73] The emphasis upon sickness and therapy is certainly a central theme of modern thought.

---

[70] Aristotle ultimately recognises that nothing is guaranteed in our (short) life and that anything actually attained is always only achieved in moments and following toil and strain; see *Politics* Book VIII. In his striving for happiness and wisdom a man seeks, and is genetically given to seek, to overcome and transcend his "composite nature (*sytheton*) (*N.E.* X 1177b 29)," i.e., to seek to transcend his body. Regarding this composite nature Aristotle in a fragment (*Protrepticus* 10b) compares the relationship of the soul with the body to a prisoner of war bound to a corpse and on this struggle for being see *Met.* Θ 1050b 26-28. He also adds in his *Ethics* that a great benefit of philosophy is in distracting the wealthy and politicians from engaging in destructive forms of war and politics; see *N.E.* X 1177b.

[71] Wittgenstein, Ludwig *Tractatus Logico-Philosophicus* (1922) 6.37.

[72] Wittgenstein, Ludwig *Ibid.* 4.003. Rudolf Carnap explains more prosaically that: "the sentences of metaphysics are pseudo-sentences which on logical analysis are proved to be either empty phrases or phrases which violate the rules of syntax. Of the so-called philosophical problems, the only questions which have any meaning are those of the logic of science. To share this view is to substitute logical syntax for philosophy (*Logical Syntax of Language* (New York, 1938) p 9)."

[73] Badiou, Alain *Wittgenstein's Antiphilosophy* (London, 2011) p 77.

Badiou also explains through Wittgenstein how *anti*-philosophy can result in a collapse of meaning into mysticism, as we have previously seen in respect to Heraclitus and Democritus. We can here consider Wittgenstein's statements that: "There are, indeed, things that cannot be put into words. They *make themselves manifest*. They are what is mystical"[74] and that: "The sense of the world must lie outside the world. In the world everything is as it is, and everything happens as it does happen: *in* it no value exists – and if it did exist, it would have no value. If there is any value that does have value, it must lie outside the whole sphere of what happens and is the case. For all that happens and is the case is accidental. What makes it non-accidental cannot lie *within* the world, since if it did it would itself be accidental. It must lie outside the world."[75] We find another combination of "chance" and "necessity" exhibited in Wittgenstein's own metaphysical (or theological) "nonsense." We see that when we move away from engaging with and interrogating the world, we end up in an existentialist quagmire in which "meaning" can exist only as some extra-worldly thing.

The "linguistic turn" was always, as Aristotle himself predicted, "absurd" since nominalism is itself absurd since we cannot have "an expression meaning the same as a name"; see *Post. An.* II 92b 26-33. Robert Pasnau describes this approach as a collapse into lexicography.[76] It is ironic, though fitting, that a literalist and formalist philosophy based only upon language cannot even rise to the level of a purely descriptive representation of the world. It produces a world with no "meaning" *in* it.[77]

Alfred North Whitehead (1861–1947) was a founder of computational logic who later in life created a "process philosophy" which sought to refute the monster he had himself in part raised up. He put forward, inadequately as we will see later, a philosophy of "process" as a "philosophy of organism." He argued regarding "process" or what Aristotle calls "becoming" that: "There is a rhythm of process whereby creation produces natural pulsation, each pulsation forming a natural unit of historic fact. In this way, amid the infinitude of the connected universe, we can discern vaguely finite units of fact. If process be fundamental to actuality, then each ultimate individual fact must be describable as process. The Newtonian description of matter

[74] Wittgenstein, Ludwig *Tractatus Logico-Philosophicus* (1922) 6.522.

[75] Wittgenstein, Ludwig *Ibid.* 6.41.

[76] Robert Pasnau explains the problem as that: "Epistemology is prone to lexicology because it lacks other sorts of stable grounds on which to build…Without an idealised epistemology…all we can expect to produce is a more-or-less detailed report on our actual linguistic practices—epistemology as lexicology [and regarding an "idealised epistemology" we find that] In the end, an idealised epistemology points towards a normative, prescriptive rather than descriptive enterprise ("Epistemology Idealised" *Mind* (2014) p 1–4)."

[77] Aryeh Kosman explains this situation as that: "By a formal (or trivial) answer… I mean the kind of answer (or nonanswer) that evaders and philosophers are fond of giving: *What's in your bag?—its contents. Where are you headed?—to my destination. What's a human being?—something that is being human.* Here an "answer" is in fact merely the formal description of whatever sort of thing might be mentioned in a proper and informative reply: *the contents of my bag* simply means *what's in my bag; my destination* means *where I'm headed* (*The Activity of Being: An Essay on Aristotle's Ontology* (Cambridge Mass., 2013) p 154)."

abstracts matter from time. It conceives matter "at an instant." So does Descartes' description. If process be fundamental such abstraction is erroneous."[78] He argued regarding "being" that: ""Actuality" is…decision amid "potentiality." It represents stubborn fact which cannot be evaded. The real internal constitution of an actual entity progressively constitutes a decision conditioning the creativity which transcends that actuality."[79]

Whitehead's philosophy is notably unapologetic and direct. It is historically informed by the tradition and is explicitly metaphysical. Whitehead explicitly refuses to regurgitate the apologetics needed to maintain a "Hume-Newton situation…[which is]…the primary presupposition for all modern philosophic thought. Any endeavour to go behind it is, in philosophic discussion, almost angrily rejected as unintelligible."[80] I take Whitehead to be a model for what modern philosophy could have been, his work being both irrefutably lucid and fundamentally solid. His concentration upon "process" or "becoming" is still, however, a form of reductionism. It is indeed salutary to concentrate on the efficient cause and to stress its necessity—this being also Aristotle's predominant emphasis—but this world and its beings can be understood only through a matrix of *four* causes, not through an emphasis upon one of these.

Richard Rorty (1931–2007) stands at the end of the road as the quintessential *anti*-philosopher who angrily rejects any alternative to the "Hume-Newton situation." Rorty describes his own position as "epistemological behaviourism."[81] The basis of his position is that: "we do not start with visual images. We do not "start" with *anything*. We are just trained to make reports—some perceptual, some introspective—as part of our general training in uttering true sentences, our learning of the language."[82]

Rorty explains his philosophy as being based upon "uttering true sentences" on the basis that: "The nominalist…construes "finding the nature of X" as just finding the most useful way to talk about the things which have traditionally been called "X"—a way which need not employ any term coreferential with "X.""[83] His argument that all we have to do to perform philosophy is utter "true sentences" disre-

[78] Whitehead, A.N. "Forms of Process" in *Modes of Thought* (New York, 1938) p 88–89.

[79] Whitehead, A.N. *Process and Reality* (New York, 1929) p 68–69. He adds that: "Various occasions are thus comparable in respect to their relative depths of actuality. Occasions differ in importance of actuality (*Religion in the Making* (Cambridge, 1926) p 108)."

[80] Whitehead, A.N. "Nature Lifeless" in *Modes of Thought* (New York, 1938) p 135.

[81] Rorty explains that: "Epistemological behaviourism…might simply be called "pragmatism"…it is the claim that philosophy will have no more to offer than common sense (supplemented by biology, history etc.) about knowledge and truth (*Philosophy and the Mirror of Nature* (Princeton, 1979) p 176)." See also Hilary Putnam's *Ethics without Ontology* (Cambridge Mass., 2004).

[82] Rorty, Richard "Mind as Ineffable" in Richard Q. Elvee (*ed.*) *Mind in Nature; Nobel Conference XVII* (San Francisco, 1982) p 71.

[83] Rorty, Richard *Ibid.* p 79. On Rorty see Roger Scruton's "Richard Rorty's Legacy" in *Open Democracy* (online, 12 June 2007) and Olavo de Carvalho's "Mr Rorty and his Fellow Animals" (online, 2011). For scepticism regarding Rorty's philosophical perspective see the other contributions to the 1982 Nobel conference "Mind in Nature" and for a functionalist defence of Aristotle

gards or shows profound ignorance of the whole history of philosophy in respect to the nature of truth, being, and representation. We rather find, according to Rorty, that since the name "X" need not be "coreferential," i.e., referring to a something or "*ousia*" which is *called* "X," any "truth" must only be present in the grammar itself. Rorty therefore reduces the world to a grammatical framework on the basis that: "What we've got is not a mind but a program, that is, a way of being wired up. When one puts it in those terms, it comes to seem misleading to speak of the mind as a control organ which does what the too-complex hardware can't do. Because it is simply the complexity of the hardware."[84] I note that it is important to recognise the cross-over here between a subjective/linguistic error and an objective/technological error.

What does Rorty mean by "the complexity of the hardware"? Rorty explains as regards the substance of his world that: "Functionalism comes down to saying that anything you want to say about persons will have an analogue in something you can say about computers, and that if you know as much about a person as a team consisting of the ideal design engineer and the ideal programmer know about a computer, then you know *all* there is to know about the person."[85] Rorty explains further as regards how we reduce the world to an "analytic" representation that: "If Kant had gone straight from the insight that "the singular proposition" is not to be identified with "the singularity of a presentation to sense" (nor, for that matter, to intellect) to a view of knowledge as a relation between persons and propositions, he would not have needed the notion of "synthesis." He might have viewed a person as a black box emitting sentences, the justification for these emissions being found in his relation to his environment (including the emission of his fellow black boxes). The question "How is knowledge possible?" would then have resembled the question "How are telephones possible?" meaning "How can one build something which does that?""[86] Kant would, of course, have done nothing of the sort. Indeed, no serious philosopher would.

We again see the attempt to dismiss the need for "synthesis." Rorty ultimately insists that we have no relation with the world through sense or intellect but only through analytic propositions. He concludes as regards how we approach the real world that: "It may seem weird to say that there might turn out to be no living bodies, or that there might turn out to be no minds. It was of course weird to say it turned out that the earth was not at rest. It seems to be that what we need to explain is not the truth of a proposition, but the inclination of human beings to assert the

---

see Martha C. Nussbaum & Hilary Putnam "Changing Aristotle's Mind" in *Essays on Aristotle's* De Anima (Oxford, 1992).

[84] Rorty, Richard "Mind as Ineffable" in Richard Q. Elvee (*ed.*) *Mind in Nature; Nobel Conference XVII* (San Francisco, 1982) p 114.

[85] Rorty, Richard *Ibid.* p 74.

[86] Rorty, Richard *Philosophy and the Mirror of Nature* (Princeton, 1979) p 152.

proposition."[87] None of this is philosophically explanatory or useful. It is, however, useful to investigate *why* Rorty himself claims to hold *this* proposition.

Rorty himself explains his motivation as being based upon: "the conviction that seventeenth-century science discovered that everything could be explained by atoms and the void, and that philosophy has a moral duty to preserve this insight…[and] if there was supposed to be answers to such general questions as "How is abstraction (recognition, constancy) possible?" [we] can say, there is *no* answer, except the pointless remark that nature has evolved suitable hardware to get the job done."[88] Rorty's reduction of the world to computers somehow claims to defend "the seventeenth century." How can this be? The "seventeenth century" did not generally support the philosophy of "atoms and the void" in the first place, e.g. Descartes and Leibniz both rejected both "atoms" and the "void." The philosophy of Gottfried Leibniz, the pre-eminent philosopher of the seventeenth century, is based upon the principle (with which Descartes himself probably concurred) that: "the mind is not only a mirror of the created universe, but is also an image of God. But this belongs to rational substances alone"[89] whereas the object of Rorty's attack is indeed precisely this seventeenth century idea that there is a "mirror of nature." Rorty's attack on the "mirror of nature" is, yet again, a rejection of a complex, two-way world and an assertion of a simplistic, reductionistic one.

Rorty gives us above a crude restatement of Isaac Newton's principle "*hypotheses non fingo*" alongside the use of Darwin's theory of evolution as a blanket alibi. He demonstrates how modern thought has collapsed from Descartes' Cogito and of Leibniz's monadology into an anti-philosophical pragmatic technologism. He denies the very existence of Descartes' "ego" on the basis that: "if assertions are justified by society rather than by the character of the inner representations they express, then there is no point in attempting to isolate *privileged* representations."[90] Rorty seeks to suppress the "*ousia*" and, ultimately, the person. His supposed defence of the seventeenth century is simply an assertion of scientific reductionism as a *mathesis universalis* outlined by Descartes as that: "there must be a general science which explains all the points that can be raised concerning order and mea-

---

[87] Rorty, Richard "Mind as Ineffable" in Richard Q. Elvee *(ed.)* *Mind in Nature; Nobel Conference XVII* (San Francisco, 1982) p 90.

[88] Rorty, Richard *Philosophy and the Mirror of Nature* (Princeton, 1979) p 217 and 235. Robert Rosen explains that this reference to "hardware"—to which we could add "software", "wiring", "hardcoding" etc.—is not a harmless metaphor but, rather, a deeply entrenched intellectual obfuscation according to which: "the organisation is "wired in" to the physics of the system from the outset…[which] in turn conveys a strong, but utterly mistaken, impression that the physics is entailing that wiring, rather than (in a precise sense) the reverse (*Life Itself: A Comprehensive Inquiry Into the Nature, Origin, and Fabrication of Life* (New York, 1991) p 140)."

[89] Leibniz, Gottfried "Metaphysical Consequences of the Principle of Reason" in *Philosophical Writings* (London, [*orig.* 1712] 1973) p 177.

[90] Rorty, Richard *Philosophy and the Mirror of Nature* (Princeton, 1979) p 174. "The crucial premise of this argument is that we understand knowledge when we understand the social justification of belief, and thus have no need to view it as accuracy of representation. Once conversation replaces confrontation, the notion of the mind as Mirror of Nature can be discarded (*Ibid.* p 170)."

sure irrespective of the subject-matter, and...this science should be termed *mathesis universalis*."[91] Aristotle had already explained philosophically why sciences cannot and should not reduce in this way, and our modern science has itself, as we will see in the sister volume to this work, shown us that this ideological dream is toxic and its scientific claim false. Whilst Descartes and Leibniz should be excused for their seventeenth century dream, Rorty has no excuse for his twentieth century error. He is willing to sacrifice the "inner representations" of humanity for the sake of a dubious ideology.

If we reflect on this situation we see, strangely enough, that Rorty's "defence of computers" is actually a strange and extreme politico-theological position. Rorty's denial of "nature" goes hand in hand with the denial of value (privilege) and of inner representations in the name of some sort of collective will. According to this world-view, the world is however we (whoever *we* is) say it is. It is really a radical form of Protestantism or a secularised version of such.

Martin Heidegger (1889–1976) represents the end position of an utterly opposing philosophical pathway. In marked contrast with Rorty, Heidegger's work is painfully built up from a series of reflections on the philosophical tradition from the pre-Socratics, Plato, and Aristotle to Leibniz, Kant, Hegel, and Nietzsche. The significance of this contrast is that one represents philosophy as insight built up from understanding (perhaps in the Aristotelian tradition) whereas the other represents philosophy as a light of intuition, as a Cogito (perhaps in the Platonic tradition).

Aristotle is a foundational influence upon Heidegger who argued regarding his own phenomenological philosophy that: "Phenomenology radicalised in its ownmost possibility is nothing but the questioning of Plato and Aristotle brought back to life: *the repetition, the retaking of the beginning of our scientific philosophy*."[92] For Heidegger, Aristotle is a phenomenologist and phenomenology is Aristotelian (and Platonic). Michael Bowler explains as regards the significance of this relationship that: "Heidegger rejects any philosophy's claim to be primordial science if it is not situated in life. The lack of primordiality in the philosophies of Rickert, Husserl, and Natorp is indicated by the fact that one finds at the foundations of their philosophies a living element that they simply do not account for. Thus, according to Heidegger, the renewal of philosophy as primordial science requires exhibiting how philosophy is located in life as well as how it can be productive of life."[93] Heidegger's is a philosophical consideration of life and of the primordial options open to it.

Bowler adds that: "Phenomenology, Heidegger demonstrates, is not just a hermeneutically naïve appeal to the things themselves, as if it were a matter of recapturing or approximating some lost original position. It is the self-address of factical life. Heidegger's pervasive claim in this essay is that philosophy *is* life, that is the self-articulation from out of itself of life."[94] The interesting question here is quite

[91] Descartes, René *Rules for the Direction of the Mind, CSM* I (*c*. 1628) § 378 p 19.

[92] Heidegger, Martin *History of the Concept of Time* (Indiana, [*orig*. 1925] 1985) p 136.

[93] Bowler Michael *Heidegger and Aristotle: Philosophy as Praxis* (London, 2008) p 92.

[94] Bowler, Michael *Ibid.* p 12.

how we understand the "consciousness" of life and on this point Bowler argues that Heidegger rejects Husserl's idea that we can slice away the world and leave a stream of consciousness that can ultimately form as a final expression of true philosophy a "world consciousness."[95] The fine point of this argument seems to be that Heidegger argues that the world can only be reduced to a content-*ed* "life" rather than to an abstract-*ed* "consciousness." The distinction between working from soul *or* mind seems to emerge back into view.

Heidegger himself traces back the problem of reductionism, of reducing the world to our representations of it, to Descartes on the basis that: "It is in the metaphysics of Descartes that, for the first time, the being is defined as the objectness of representation, and truth as the certainty of representation."[96] We thus objectify the world and logicise our mind. Heidegger adds that this reduction of the world to representation (*eidōlon*) rather than substance (*ousia*) and its manifestations is a form of sophism on the basis that: "sophistry is characterised by an unconcern with substantive content, an unconcern in a quite determinate sense, not one that is haphazard…We could determine their [i.e., the sophists'] unconcern in a better way by calling it emptiness, a *lack* of substantive content."[97] We find a contrast between two concentrations, substantive philosophy and abstracted technique.

Heidegger continues to argue that our modern thought is generically a form of sophism which needs to be corrected through a reestablishment of our philosophy on fundamentally "Greek" grounds through which we can undertake "the retaking of the beginning of our scientific philosophy." The question regarding the precise nature, problem, and value of sophism goes back to Socrates and is treated in the modern world by Heidegger among others. Nikolai Berdyaev explains sophism well as a necessary accompaniment of a sophisticated age.[98] Georges Sorel correctly categorises Descartes' philosophy as that of an age of the salons.[99] Edmund Husserl

[95] Husserl, Edmund *The Crisis of European Sciences and Transcendental Philosophy: An Introduction to Phenomenological Philosophy* (Evanston, [*orig.* 1936] 1970) p 264.

[96] Heidegger, Martin "The Age of the World Picture" in *Off The Beaten Track* (Cambridge, [*orig.* 1938] 2002) p 66.

[97] Heidegger, Martin *Plato's* Sophist (Indiana, [*orig.* 1925] 1997) p 159. Heidegger argues that: "ειδος [*eidos*] means the outward look of something, i.e. that ontological determination which presents something as what it is. Ειδωλον [*eidōlon*], on the other hand, refers to a mere outward look; it is not nothing, but it is such that it merely appears to be so and so (*Ibid.* p 239)."

[98] Nikolai Berdyaev explains that: "I believe that the cultures of all times and of all peoples have passed through a period of enlightenment. The cultural development of all people is subject to a certain cyclical movement. This affinity between cultural processes would seem to indicate the organic character of their development. Thus, Greek culture, one of the greatest known to mankind, had its age of enlightenment essentially analogous to that experienced by mankind in the eighteenth century. The age of the sophists was in its way the flowering of Greek culture; but in spite of its specifically Hellenic features it had many particularities in common with the age of the eighteenth century enlightenment. As the eighteenth-century enlightenment was to do, and as every enlightenment tends to do, the age of Greek enlightenment attacked and discredited what was sacred in the "historical" and all the organic and traditional elements of history (*The Meaning of History* (London, 1936) p 5)."

[99] See Georges Sorel *The Illusions of Progress* (Berkeley, [*orig.* 1910] 1969) p 87.

considered sophism and argued that its antidote was the philosophy of the Greeks who created a "science of the universe, of the all-encompassing unity of all that is…[which is] philosophy, the one science, [which] branches into many particular sciences. In the breakthrough of philosophy in this sense, in which all sciences are thus contained, I see, paradoxical as it may sound, the primal phenomenon of spiritual Europe."[100] We find here a grounded science facing off against an ungrounded science.

This birthing of science from philosophy is most particularly represented by the project of Aristotle within the Greek tradition even if Husserl himself relates more to the Platonic tradition. The break-out, or break-away, of science is also, however, expressed by the parallel ungrounded, technological tradition of the sophists which also emerged from this "Greece." Whether we approach the world from a Husserlian perspective as a radical and original interpretation of the Cartesian ego or from a Heideggerian perspective as an Aristotelian study of being and beings, the phenomenological approach does itself clearly represent a revolution against the enlightened sophism of the "Hume-Newton situation." It is an assertion of "being" and of "life" against the ascendancy of "technology."[101] This is the schism of Western thought into vitalism and mechanicism that we have observed above and will need to consider below.

2.2 Conclusions

How do we understand our philosophical tradition? We can, as always, return to Plato, to whom we all stand as footnotes, and here consider one of Plato's myths, his repurposing of the Gigantomachia, the battle between the "Gods" and "Giants," in

[100] Husserl argues in full that: "[in] the ancient Greek nation in the seventh and sixth centuries B.C… there arises a *new sort of attitude* of individuals towards their surrounding world. And its consequence is the breakthrough of a completely new sort of spiritual structure, rapidly growing into a systematically self-enclosed cultural form; the Greeks called it *philosophy*. Correctly translated, in the original sense, that means nothing other than universal science, science of the universe, of the all-encompassing unity of all that it is. Soon the interest in the All, and thus the question of the all-encompassing becoming and being in becoming, begins to particularise itself according to the general forms and regions of being, and thus philosophy, the one science, branches into many particular sciences. In the breakthrough of philosophy in this sense, in which all sciences are thus contained, I see, paradoxical as it may sound, the primal phenomenon of spiritual Europe ("Philosophy and the Crisis of European Humanity" in *The Crisis of European Sciences and Transcendental Phenomenology* (Evanston, [*orig*. 1936] 1970) p 276)." See also Paul Feyerabend's *Conquest of Abundance: A Tale of Abstraction versus the Richness of Being* (Chicago, 1999).

[101] See Martin Heidegger's *The Question Concerning Technology and Other Essays* (New York, [*orig*. 1952–62] 1977) and Edmund Husserl's "The Vienna Lecture" in *The Crisis of European Sciences and Transcendental Philosophy: An Introduction to Phenomenological Philosophy* (Evanston, [*orig*. 1936] 1970).

his *Sophist*. Let us ask: how is our modern thought and its exponents "footnotes" to Plato?

Plato depicts his mythological battle as an "interminable battle [which] is always going on between the two camps…on this issue…over being [substance] (*ousia*) (*Sophist* 246A)" regarding which "One party [i.e., the Giants] is trying to drag everything down to earth out of heaven and the unseen, literally grasping rocks and trees in their hands; for they lay hold upon every stock and stone and strenuously affirm that real existence belongs only to that which can be handled and offers resistance to the touch. They define reality as the same thing as body, and as soon as one of the opposite party asserts that anything without a body is real, they are utterly contemptuous and will not listen to another word …[whereas the Gods are] very wary in defending their position somewhere in the heights of the unseen, maintaining with all their force that true reality consists in certain intelligible and bodiless Forms (*Sophist* 246A-B)." This battle is represented in Fig. 2.2 below.

The power and accuracy of Plato's myth should be evident. We have indeed seen over the millennia a constant "battle" between physicalists and idealists. The use of the imagery of war to express this "interminable argument…over *ousia*" is itself inspired since it brings home the real violence of this difference in worldviews. We also actually find, as so often in real war, that in the cold light of day the two sides are, ridiculously, far less different than they might themselves imagine.[102] Both

Fig. 2.2 Extract from a sixteenth Century representation of the Gigantomachy

[102] Aristotle makes the important point of emphasis that both "sides" end up with much the same thing: "In a way these thinkers [physiologoi] too [i.e., along with the Pythagoreans/Platonists] are saying that everything that exists is numbers, or evolved from numbers; they may not show it clearly, but nevertheless that is what they mean (*De Cael*. III 303a8-11)."

"sides" in this case actually reduce the world to their fixed ideas about it without recognising complexity; they are *both* reductionistic. Regarding this reductionism, Plato himself concludes that the true philosopher should *not* follow "the champions either of the one or of the many forms of the doctrine that all reality is changeless [i.e., the "Gods"], and he must turn a deaf ear to the other party who represent reality as everywhere changing [i.e., the "Giants"] (*Soph.* 249C-D)." The obvious targets of this critique are Parmenides and Heraclitus and yet it is interesting that Plato himself criticises his own Theory of Forms, that "true reality consists in certain intelligible and bodiless Forms."

Does modern thought supersede and correct the Greeks? I have argued above that we have not advanced *from* Plato's Gigantomachia, created at the dawning of philosophy, but have proceeded *within* it. We could even say that the modern world has embodied Plato's battle through its central conflict between religion and science. Our scientific technology has today simply replaced the similarly problematic spiritual technology of dogmatic religion as the basis for our worldview. Reductionism is a paradigmatically modern condition. It is the dreadful, systematic reduction of philosophy and its spirit to the results of science and technology. Its opposite, the new "Gods," have produced a patchily profound but ultimately disengaged thinking which really cedes worldly meaning to the "Giants." Its lack of connection to the world makes this thought opaque. In the necessary exercise to meaningfully address our human condition, our best friend is Aristotle. He shows us through his philosophy and his ethics and politics that there is a reasonable alternative to the mythologies that we have been constrained to follow. He does so not by adding an alternative mythology but by carefully parsing the world before us—which is much the same then as now—and showing us how its goods manifest into reality.

Chapter 3
Our Philosophical Context

Abstract From the preceding survey of philosophy, we move to consider the philosophical situation we find ourselves in. We see in more detail how Cartesian dualism and his "*epoché*" valorised the simplification and abstraction of reductionism, for both better and worse. For a time this has allowed us to concentrate upon reaping the fruits of mathematical science at the expense of a full engagement with our rich, complex human situation. Despite the benefits of this reductionism it has left us with the problem today not of developing philosophy but of recovering it. To this end, the reality of structure needed to be reintroduced through the structuralism of Saussure, the phenomenon through the phenomenology of Husserl, and the sign through the semiotics of Peirce. We see in all cases excellent developments of thought towards complexity and meaning. This recovered thought is, however, still fractured and partial. We then consider how this partiality informs the philosophy that frames modern biology. The main problem we face is the conflation of the limited theoretical assumptions used for the practise of scientific research with full philosophical insight and truth.

Keywords The problem of modern philosophy · Recovering Greek philosophy · Peirce · Husserl · Saussure

"*Cogito ergo sum*." We could take this famous statement as regards the relationship between "thinking" and "being" as Descartes' answer to Socrates' complaint that: "my talking to you [cannot be] attribute[d] to sound, and air, and hearing [but only to] the true cause, which is, that the Athenians have thought fit to condemn me, and accordingly I have thought it better and more right to remain here and undergo my sentence (*Phaedo* 98c)." Descartes' point of departure is that the ego and mind is simply *the* central essential fact we must include when we address the world around us.

Descartes transforms how we engage with the world through his "I think" and asserts that: "a man who walks across a room shows much better what motion is than a man who says "It is the actuality of a potential being in so far as it is

P. N. Jackson, *Aristotle on Meaning in the Living World*, Biosemiotics 30,
https://doi.org/10.1007/978-3-032-00602-8_3

potential," and so on."[1] Descartes here explicitly moves to reject Aristotelian metaphysics and its core idea of the "actuality" and "potentiality" of beings through which meaning enters the world through a being acting in a world that mirrors and enables it through its own natural potentiality. He rather seeks to see the world intuitively and not through a metaphysical framework. The touchstone for his "new" approach is Plato's Theory of Ideas on the basis that: "according to Plato, Socrates asks the slave boy about the elements of his geometry and thereby makes the boy able to dig out certain truths from his own mind which he had not previously recognised were there, thus attempting to establish the doctrine of reminiscence. Our knowledge of God is of this sort."[2] This belief in an intuitive and divine mind (and soul) is what he sets out to demonstrate in his *Meditations*, the full title of which is *Meditations on First Philosophy, in which the existence of God and the immortality of the soul are demonstrated.*

The result is that Descartes asks us to: "Compare my assumptions with the assumptions of others. Compare all their *real qualities*, their *substantial forms*, their *elements* and countless other such things with my single assumption that all bodies are composed of parts [which are as infinitely divisible as "wax"]."[3] The perils of such simplification should hopefully be immediately evident. We will, however, see below that the Cartesian solution, which frees the human mind and ego and renders the world itself plastic and malleable to that ego, pervades modern thought and action for both better and worse. We should conclude from this that there is a radical issue with our modern thought.

Since we are here speaking of philosophy not as a talking-shop but as a palate of worldviews let us begin by contemplating quite how Descartes reconfigured the world around him. We find, as previously with Socrates, that Descartes brought philosophy back to the fundamental question regarding the nature of the human condition. What does it mean to be a thinking human being and how does our worldview transform how we see the world? Descartes insisted upon the human ego and created a method through which the mind could supposedly control the world mathematically, i.e., through the famous Cartesian coordinates through which modern man represents the world around him. He commented upon this "analytic" approach that: "Those long chains composed of very simple and easy reasonings, which geometers customarily use to arrive at their most difficult demonstrations, had given me occasion to suppose that all the things that can fall under human knowledge are interconnected in the same way."[4] The wider philosophical significance of this Cartesian approach is well explained by Emily Grosholz as that: "according to Descartes, everything important about a subject-matter can be captured in an abstract relational system, transparent to reason, whose terms are mere

[1] Descartes, René "Letter to Mersenne 16th October 1639" §597 *CSM* III p 139.

[2] Descartes, René "Letter to Voetius May 1643" §167 *CSM* III p 222–3.

[3] Descartes, René "Letter to Morin, 13 July 1638" § 200 *CSM* III p 107.

[4] Descartes, René *Discourse on the Method* (1637) §19 *CSM* I p 120.

place-holders without intrinsic interest."[5] It is this reduction of the world to mathematical abstraction that Aristotle challenged in his day and which makes him, as we will see, relevant in ours.

3.1 Consequences of the Cartesian Reduction

The critical result of the Cartesian philosophical transformation is to remove the "beings" from being to leave only "mind" and its abstractions, *res cogitans* and *res extensa*. Descartes' elevation of "mind" and "reason" does not merely raise up the "ego" of "man"—potentially to a Comtean "religion of man"—but also gives rise to the question of "idealism" as regards how "ideas," represented in Descartes' philosophy as the "innate ideas" of "God," shape or produce the world.

Ladislav Kvasz explains the slippery slope of this move as that: "Galileo studied ordinary objects of everyday experience, and represented only some of their aspects in a mathematical form. Descartes replaced all ordinary objects by his extended things, and identified them with space. Finally Newton showed that this Cartesian identification is a mistake, and he took as the characterisation of *the thingness of the things* not extension but hardness (i.e., forces)."[6] Along with the reductive development of science we see a corresponding reduction of philosophy—of our mental framing of reality—through which "beings" are finally reduced to "forces." This desubstantiation of being(s) is a subject we will return to repeatedly below. Kvasz explains the consequence of the ultimate Newtonian reduction as that: "Determinism and randomness are two aspects of the same reality. Determinism is the ontological side and probability the epistemological side of the same world. According to Laplace, the world is absolutely deterministic, but to the human mind, it is opened only in a probabilistic way…In his *Scholium generale*, Newton characterised the absolute space as *Sensorium Dei*. Therefore the possibility of its mathematicisation originates in God's perfection. To humans, only the relative, empirical space is accessible."[7] We see here not only Cartesian dualism but also the already observed splitting of our picture of the world into chance and necessity and into flux and form.

---

[5] Grosholz, Emily *Cartesian Method and the Problem of Reduction* (Oxford, 1991) p 17. She explains further that the Cartesian system "tends to conflate the already flattened subject-matter of geometry with that of number, in an identity which suppresses one or the other…[whereas in reality] Points, lines, and surfaces are emphatically different kinds of things. While points serve as boundaries for lines, and lines for areas, points do not compose lines, nor lines areas; higher dimensional objects cannot be reduced to the lower dimensional objects (*Ibid.* p 17–8)."

[6] Kvasz, Ladislav "Heidegger's Interpretation of Mathematical Science in the Light of Husserl's Concept of Mathematization in the *Krisis*" *Philosophia Naturalis* (2013) p 346.

[7] Kvasz, Ladislav "The Invisible Link Between Mathematics and Theology" *Perspectives on Science and Christian Faith* (2004) p 112 p 114–115. See also his "Galileo, Descartes, and Newton—Founders of the Language of Physics" *acta physica slovaca* (2012). For an overview of the problem inherent in the Cartesian-Newtonian worldview (which is critical to this work) see

Ernst Cassirer explained the significance of "mathematical reason" in this context as that: "Mathematical reason is the bond between man and the universe; it permits us to pass freely from the one to the other"[8] but with this power coming along with the "curse" that: "They [i.e., modern philosophical systems] strive, so to speak, to turn the apparent curse of the new cosmology into a blessing. Giordano Bruno was the first thinker to enter upon this path, which in a sense became the path of all modern metaphysics…In Bruno's doctrine infinity no longer means a mere negation or limitation. On the contrary, it means the immeasurable and inexhaustible abundance of reality and the unrestricted power of the human intellect."[9] What are the consequences of Cassirer's "curse"? The expansion of humanity and its mind to "infinity" involves the rejection of the bounded body or being (*ousia*) which is, as we will see later, an opposite position to Aristotle's emphasis upon the bounded being and his rejection of the unbounded or uninformed idea. I stress here that this dispute is not merely one of surface ideas but of basic approach to the world.

The consequence of philosophically replacing bounded nature with infinite abstraction is explained by Nicolas Berdyaev as that: "With Nicolas of Cusa, with Descartes and in part with Spinoza, with Leibniz, Locke, Berkeley and Hume, a new orientation of philosophical thought begins…Medieval scholastic philosophy was fundamentally Greek; it did not pass beyond the bounds of ancient thought; it was a philosophy of the object, that is to say it was cosmocentric. Modern philosophy, on the other hand, has become a philosophy of the subject; it is anthropocentric and its centre of gravity is transferred to man. But this means that Christian emancipation from the power of the objectified world over man had not yet made its way into thought in the Middle Ages, whereas in modern times Christianity does not enter into thought, it carries on its work unseen within it and leads to the autonomy of man and of his thought."[10] The subjective ego or will naturally goes hand in hand with an infinitely extended and plastic world. Berdyaev adds the useful emphasis that modern thought, including that of Descartes himself, is a "Christian emancipation from the power of the objectified world" and we will consider the consequences of this situation slightly later.

The transformation of philosophy into (hidden or assumed) ideology is perhaps a genuinely modern development. It is a gaming, an apologetics, which admittedly follows on from, but transforms, the eristics of the sophists. This modern gaming is observed in Descartes but with the irony of this situation (and the risks of this approach) being that Descartes' game was turned against him in order to make him seem to validate the philosophical positions that he expressly sought to refute. For

---

esp. the essays by David Bohm, Theodosius Dobzhansky, and John Lewis in John Lewis (*ed.*) *Beyond Chance and Necessity* (London, 1974).

[8] Cassirer, Ernst *An Essay on Man* (New York, [*orig*. 1944] 1970) p 18.

[9] Cassirer, Ernst *Ibid.* p 16–17.

[10] Berdyaev, Nicolas *The Beginning and The End* (New York, [*orig*. 1941] 1957) p 11. He explains further that: "St Thomas Aquinas was, of course, much more of a Christian than Hegel was, but in its theme and in its line of direction his philosophy was less Christian than Hegel's, not to speak of Kant's (*Ibid.* p 12)."

instance, Descartes presented his philosophy from a position of "hyperbolic doubt" not because he was a radical sceptic himself but because he wanted to refute the sceptics and show that even their strongest arguments could not disprove the existence of the "soul" or of "God." This is indeed the express point of his *Meditations*. Descartes himself categorised scepticism as "a disease" which he wants to "explain how it can be cured"[11] through exposure on the basis that: "I wanted to show the firmness of the truths which I propound later on, in the light of the fact that they cannot be shaken by these metaphysical doubts."[12] His monastic friend Fr. Marin Mersenne prudently objected to his philosophical attempt to *prove* God by gently asking him whether "you see how, in your desire to champion the truth, you may end up proving too much, and thus overturn the truth rather than build it up"[13] whereas the less sympathetic Jesuit Pierre Bourdin inquired rather more astringently as to the feasibility of a "new and unheard of hope of deriving certain results from what is uncertain."[14] Descartes' theology was, ironically, too radical for both these men of the cloth.

In the age that produced Pascal's wager, we find that Descartes ultimately lost his own. In seeking to refute the Sceptics he injudiciously employed a rhetorical device, i.e., doubt, which could easily be twisted to serve their agenda. His method let him down. By paring down the ontological structuring of the world to what Edmund Husserl called "a little tag-end of the world, as the sole unquestionable part of it for the philosophising Ego"[15] he achieved only the most Pyrrhic victory. The tragic end result of his proof of "God" through the "ego" was only that once the extended world had been ceded, the remaining "little tag-end" could then be assailed as a final "hard problem." Looking back ruefully, we perhaps see, as Bernard Bosanquet explained, the need to keep hold of the whole world come what may on the basis that: "Philosophical consideration explains, and on the whole justifies, the verdict of a really serious and careful empirical outlook and the usage of common sense. And that I am convinced is the typical position of traditional philosophy…Its demand for precision is so intense, and its valuations are drawn from so universal and crucial a survey, that its opponents can be shown, I think, in quite definite and assignable respects, not to see at all what sort of requirements they are up against…Failing this, the attempt to philosophise becomes a game of play."[16]

I ultimately posit that Descartes' "idealism" is an unwise acceptance and application of the attitude of religious apologetics whereas Aristotle's "realism" comes from an understanding that it is a false economy to defend the world by sacrificing

[11] Descartes, René *Third Replies* CSM II §172 p 121 (and see Antoine Arnauld's perceptive comments in his *Fourth Objections* CSM II §215 p 151).

[12] Descartes, René *Third Replies* CSM II §172 p 121.

[13] [Descartes, René] *Second Objections* CSM II §127 p 91.

[14] [Descartes, René] *Seventh Objections* CSM II §484 p 327.

[15] Husserl, Edmund *Cartesian Meditations: An Introduction to Phenomenology* (Dordrecht, [orig. 1929] 1999) p 24.

[16] Bosanquet, Bernard "Science and Philosophy" *Proceedings of the Aristotelian Society* (1914–15) p 21.

it. Aristotle genuinely used philosophy as metaphysics and logic as a framework and not as apologetics or as a game. This is a fundamental difference of philosophic approach.

3.2 The Problem of Philosophy

I will argue below that modern philosophy is a victim of its own history and, in a sense, of its own success in making itself disappear, the most obvious manifestations of this direction being nihilism and postmodernism. The radical scientific and technological progress of the nineteenth century was achieved by displacing not only religion but also philosophy. Philosophy was on this basis stripped down to a positivism that reduced the world to the laws of physics and mathematics, this being what is known as "reductionism." Matter of fact empirical research (preferably performed in a laboratory) was also sometimes allowed. The reduction of philosophy to science was justified on the basis that metaphysics is simply lack of clarity, bad use of grammar, sophism. Philosophy is reframed as an apparent glitch in a scientific model which can be excised by a technical explanation that does not allow vagueness to creep in. This position analyses the world into rational mathematical mind operating on plastic inert matter.

The Cartesian schema and schism persist through the various archetypal -isms of modern thought. On the one side, we find such objective -isms as materialism, positivism, operationalism, functionalism, and, most characteristically, behaviourism and on the other side we have such subjective -isms as subjectivism, relativism, psychologism, and existentialism. Taken as a whole, Jerry Fodor and Massimo Piattelli-Palmarini argue that we find -isms or ideologies on all sides which they call "one-factor theories." The living world is philosophically reduced through this framing on the basis that: "Skinner's account of learning and Darwin's account of evolution are identical in all but name"[17] and, in more detail, that: "there are "single level" theories—of which Newtonian mechanics is perhaps the extreme example…[and] ET [evolutionary theory] and OT [operant conditioning: behaviourism] are one-factor theories of their respective domains. According to the former, selection is overwhelmingly decisive in shaping the evolution of phenotypes; according to the latter, reward is overwhelmingly decisive in determining the conditioning of psychological profiles."[18] In all cases, the world is reduced to *one* certain given factor or law, normally mechanistic, materialistic, and environmental.

I posit here and will explain below, and in more detail in the sister volume to this work, that the solution to the swamp of modern philosophy is Aristotle allied with biosemiotics. I will show that Aristotle provides us with a synthetic, complex, and

---

[17] Fodor, Jerry & Piattelli-Palmarini, Massimo *What Darwin Got Wrong* (London, 2011) p xvii and p 109.

[18] Fodor, Jerry & Piattelli-Palmarini, Massimo *Ibid.* p xxi and 9.

thus humanistic philosophy which consciously counters any form of reductionism through its framing of being as a philosophy of "beings." As regards biosemiotics, it is critically important as an implied statement that meaning is not a mental or purely human dimension but is embedded in and emergent from the physical and biological world. The mind, sense, and creativity we find in human nature, if we look at it naturalistically, must in some sense be a natural phenomenon, necessarily derived from and emerging from the natural world. The biological basis of our "sub-lunary" world is essential both to Aristotle's philosophy and to biosemiotics as a science. Both Aristotle and biosemiotics assert in their respective areas that the world they address is living and real and not inert and abstract. This "realism" and anti-reductionism is an essential part of this side of the argument and counters the "idealistic" and reductionist side which argues that there is a science of being which can be produced a priori. We can perhaps legitimately follow either side of the argument—essentially Platonism or Aristotelianism—but it is important to understand the fundamental existence of these two very different worldviews with all their implications and consequences.

The great gambit of positivism was to argue that philosophy is itself simply redundant, rather like phlogiston theory or phrenology. Metaphysics is famously dismissed as a sophism resulting from the use of "bad grammar."[19] We can reduce the world to science—essentially to Descartes *mathesis universalis*—if we frame it scientifically without vagueness, through "clear and distinct ideas."[20] This project continues today and is, in fact, perhaps even more embedded in our lives and thought, though expressed more through a dead controlling hand of technocratic managerialism than through expectant trust in the white heat of technological progress.[21]

A counter-position to this progressive technologism does, however, emerge. The spiritual father of this reaction is Edmund Husserl and his phenomenology. The structuralism founded by Ferdinand de Saussure who originated what is known as "semiology" represents a further break. Phenomenology and structuralism are often

---

[19] Rudolf Carnap gives us a quintessential statement of the core "positivist" position that we need to substitute philosophy for science on the basis that: "the sentences of metaphysics are pseudo-sentences which on logical analysis are proved to be either empty phrases or phrases which violate the rules of syntax. Of the so-called philosophical problems, the only questions which have any meaning are those of the logic of science. To share this view is to substitute logical syntax for philosophy (*Logical Syntax of Language* (New York, 1938) p 9)."

[20] Karl Popper argues that: "In science…we never try to derive any information from the definition, or to base any argument upon it…we reach precision…by carefully phrasing our sentences in such a way that the possible shades of meaning of our terms do not matter (*The Open Society and Its Enemies vol. 2* (London, 1945) p 19)."

[21] Aldous Huxley explains that: "We are living now, not in the delicious intoxication induced by the early successes of science, but in a rather grisly morning-after, when it has become apparent that what triumphant science has done hitherto is to improve the means for achieving unimproved or actually deteriorated ends ("Beliefs" in *Collected Essays* (New York, [*orig*. 1937] 1971) p 362)." He precisely diagnoses our modern complaint as the result of "the philosophy of meaninglessness [as] an instrument of liberation."

paired together though they are in many ways opposites, as reinventions of Descartes' *res cogitans* and of his *res extensa* respectively. I will try to show below that both movements recognise the problems of inherited modernity whilst being unable to fully overcome them. Charles Sanders Peirce and his new philosophical science of "semiotics" perhaps goes further than they towards expanding out the meaning of the world. The problem we encounter here, however, is that his semiotics can itself become reified or ossified into a "code" of meaning explaining the world through "sign vehicles" rather than "beings." I note here that the ultimate trap open to biosemiotics is manifested by its collapse into "code biology" which represents the world as an accumulation of algorithms and hence loses itself in "information theory." The obvious utility of "code biology" and "information theory," and of "biosemiotics" if it wants, is to allow a desired reduction of reality to computers and computer code. Ultimate acceptance provides quite an allure for a critic. Recognising the problems of modernity and its reductions is possible but overcoming them remarkably difficult.

Let us begin this story with **Ferdinand de Saussure (1857–1913)** and his structuralism which explains much about our struggle to break out from the inherited constraints of our thought. Saussure's own great breakthrough was to point out that language should be approached primarily as a system (as "langue") and not merely as practical use (as "parole"). He recognised that language is meaningful as a structure and a culture itself and cannot be reduced to the philological study of the history of words and their usage. Innate structure is more determining than contingent history. This insight led Saussure to insist that the word or sign is itself arbitrary and is a token like money. John Sturrock expands upon the economic metaphor invoked here as that: "Economics…is the structural study par excellence, as Saussure himself recognized by citing it as a model for his own synchronic linguistics. An economy is the ideal example of a functioning whole all of whose parts interact and depend utterly on one another. It has a history and historical explanations may be offered for particular features of its current condition; but the proper economic explanation of that condition is synchronic, to be determined by observation of the various very complex relations which constitute the economy…the Saussurean concept of linguistic "value"…is basically an economic one."[22] Saussure also compares language to a chessboard in which the pieces move according to the rules of the game.[23]

Saussure's emphasis upon the "whole" and the "structure" was, of course, salutary and no less a figure than Jacques Derrida claims Aristotle, with only some justification, as the founder of this movement.[24] Perhaps the problem of this movement, its own destructive reduction, is *its ideological reduction of the world to the "language" of structure*. Admitting structure to our representation of the world is neces-

---

[22] Sturrock, John *Structuralism* 2nd ed. (Oxford, [*orig*. 1986] 2003) p 71.

[23] See Jean-Marie Benoist *The Structural Revolution* (London, [*orig*. 1975] 1978) p 3.

[24] See Mauro Senatore "Teleotheology: Derrida and the Aristotelian Foundations of Structuralism" *Philosophy Today* (2019).

sary and informative, reducing the world to structure is excessive and misleading. We see here the unhealthy emphasis upon language itself which—along with an equally unhealthy emphasis upon number—will bedevil modern philosophy. Aristotle argued by contrast that reducing the world to language, mathematics, money, or any other means of expression is itself a mistake and we will consider this critique later. Here I note that the reduction of the world to a market based upon economic relations and to a game based upon a prefigured structure of rules—such as chess—are both metaphors for reality that are repeatedly reified within modern philosophy.

As regards structuralism in its full development, we have a paradigm in the shape of the French anthropologist Claude Lévi-Strauss (1908–2009) who argued that the structures of kinship systems could be understood as a "periodic table" on the basis that: "By drawing up an inventory of all the customs that had been observed, all those imagined in myths, those also evoked in the games of children and adults, and the dreams of healthy and sick individuals and psychopathological behaviours, one would be able to draw up a periodic table like that of the chemical elements, in which all real or merely possible customs would appear grouped in families, and in which we would simply need to recognize those which societies have in fact adopted."[25] We find in Lévi-Strauss a recrudescence of the old ambitious Cartesian dream of a *mathesis universalis*. We can organise the whole world into the tables of a relational database. The subtext of Lévi-Strauss' perspective is also important. He argued that kinship relations could be understood as a "language" and as regards language he argued that: "[language is] a change from a stage where nothing had meaning to a stage where everything had."[26]

We find, as Sturrock observes, that: "Greatly influenced by Marx, Lévi-Strauss has that thinker's obsessive awareness that everything that is social or cultural is man-made and might therefore be remade. The truly natural is by definition unintelligible, but the cultural, which is a departure from the natural, *is* intelligible and thus capable of being integrated within some systematic explanation."[27] Identifying meaning only with sources you can access is misleading. The identification of meaning with the cultural is erroneously anthropocentric from an Aristotelian position. We also find, however, a paradoxical attempt to reject the culture of humanity altogether. Lévi-Strauss' ideology led him to stress that the structured lives of "primitive" man was of equal value and complexity to the culture of his "civilised" cousins. The associated Annales school of historiography also emphasised the mundane to the extent of "integrating historiography with geography, geology and economics."[28] And we can go further. Lévi-Strauss' structuralism explicitly sought to abolish the (human) "subject" and "ego" in favour of the natural and societal (and

[25] Claude Lévi-Strauss quoted in John Sturrock *Structuralism* 2nd ed. (Oxford, [*orig.* 1986] 2003) p 60.

[26] Claude Lévi-Strauss quoted in Jean-Marie Benoist *The Structural Revolution* (London, [*orig.* 1975] 1978) p 69.

[27] Sturrock, John *Structuralism* 2nd ed. (Oxford, [*orig.* 1986] 2003) p 56.

[28] Sturrock, John *Ibid.* p 69.

anthropological) structures around him. Michel Foucault's different ideological convictions—his evident distrust of all forms of corporate power—also led him to seek to abolish man on the basis that: "Man is an invention whose recent date, and whose approaching end, is clearly evident from the archaeology of our thought."[29] Good sense would go even further off the rails with the deconstructions performed by post-structuralism. C.S. Lewis explains the generic urge to "abolish" man in his jewel of a masterpiece *The Abolition of Man* (1943).

Structuralism is, then, a salutary restoration of the system, the structure, and the whole to our representation of the world. It is, however, itself strangely corrupted by ideological commitments to downplay content to such an extent as to seek to abolish man himself and his "ego." Its politics is to produce a humanism by reducing humanity itself to an object or mass. The emphasis upon structures necessarily de-emphasises the content of being and of individual beings. We hence return to Aristotle's fundamental argument against Plato and his formalism. Aristotle argues that if we reduce being in the world to "form," we must simultaneously reduce beings in the world to "flux." This problem is most obvious, to we humans in any case, in respect to how we represent human beings.

It is hence that the bedrock of Aristotle's complaint against Plato is found in his own *Politics* in direct response to Plato's *Republic*. Aristotle argues here that politics is a *bottom-up* phenomenon of human activity within a constitutional democratic framework within which "the many are better judges than a single man (*Pol.* III 1281b 9)," even than the few good men.[30] He adds that private property and individual families are necessary for allowing the fullest expression of meaning, love, and value in human life. A *top-down* one-size-fits-all ideology of human politics is, for Aristotle at least, destructive and dehumanising. *Contra* Plato, Aristotle critiques the paradigmatic model for proto-fascism and proto-communism provided by Sparta in which the state possesses one central ideological objective, in this case war, that the whole society is structured to serve; see especially Darrel Dobbs' article "Aristotle's Anticommunism" (1985) on this subject.[31] The problem *we* face is that our commitment to ideological and technological systems generically crushes our humanity, each in their own way.

[29] Michel Foucault quoted in John Sturrock *Structuralism* 2nd ed. (Oxford, [*orig.* 1986] 2003) p 145.

[30] See Jeremy Waldron "The Wisdom of the Multitude: Some Reflections on Book 3, Chapter 11 of Aristotle's Politics" *Political Theory* (1995) and Thomas Lindsay "Aristotle's Qualified Defense of Democracy through "Political Mixing"" *The Journal of Politics* (1992).

[31] Aristotle explicitly criticises Plato's use of Sparta as a model on the basis that it "began to decline when they had won an empire, because they did not know how to live in peace (*Pol.* II 1271b 4-5)." Aristotle argues that we cannot base a society upon a single "idea" such as military valour but that it must be a mix of the various human virtues and activities. Aristotle does, however, argue that a society is based upon its constitution and hence that "this community is the constitution (*koinōnia d'estin hē politeia*)"; see *Pol.* III 1276b 28-31. We are, in other words, bound by and shaped by our political system but not in an ideological Platonic or Hegelian sense.

Edmund Husserl (1859–1938) sought out, in marked contrast to the structuralists, an "egological"[32] understanding of the world with human intentionality being the problem that "encompasses the whole of phenomenology."[33] He revived the question as regards how objective knowledge of the "real" world is possible with his conclusion being that once we pass through the conscious act we come to a "residue"[34] that must be understood as a transcendentally pure consciousness. We must question both the idea that we pass through the physical world and that there is something like *pure* mind beyond.

Jan Patočka nicely illuminates the fundamental nature and aim of phenomenology as that: "the fundamental problematic of phenomenology" is "the question of the ultimate ground of the appearance of that which appears."[35] Husserl himself explains, in more detail, that: "Phenomenology…does not discuss states of animal organisms (not even as belonging to a possible nature as such), but perceptions, judgements, feelings *as such*, and what pertains to them *a priori* with unlimited generality, as *pure* instances of *pure* species, of what may be seen through a purely intuitive apprehension of essence, whether generic or specific. Pure arithmetic likewise speaks of numbers, and pure geometry of spatial shapes, employing pure intuitions in their ideational universality. Not psychology, therefore, but phenomenology, underlies all clarifications in pure logic (and in all forms of rational criticism). Phenomenology has, however, a very different function as the necessary basis for every psychology that could with justification and in strictness be called scientific, just as pure mathematics, e.g. pure geometry and dynamics, is the necessary foundation for all exact natural science (any theory of empirical things in nature with their empirical forms, movements etc.)."[36] The objective given here is that of looking through the images of the world to something *pure* beyond.

We see the Cartesian influence upon Husserl in his intent "to give firm clarity to notions and laws on which the objective meaning and theoretical unity of all knowledge is dependent"[37] with this influence being made explicit in his assertion that: "Phenomenology…lays bare the "sources" from which the basic concepts and ideal laws of *pure* logic "flow," and back to which they must once more be traced, so as to give them all the "clearness and distinctness" needed for an understanding, and for an epistemological critique, of pure logic."[38] Patočka explains this situation as that Husserl "renews Cartesian metaphysics with the Kantian concept of the

[32] Quoted by Shigeru Taguchi "Ur-Ich in Husserl" in N. de Warren, T. Toadvine (*eds.*) *Encyclopedia of Phenomenology* (Springer, 2023) p 2.

[33] Quoted by Karl Schuhmann "Intentionality and the Intentional Object in the Early Husserl" in John J. Drummond & Otfried Höffe (*eds.*) *Husserl: German Perspectives* (New York, 2019) p 141.

[34] Lohmar, Dieter "Ego and Arch-Ego in Husserlian Phenomenology" in R. Breeur and U. Melle (*eds.*) *Life Subjectivity & Art: Essays in Honor of Rudolf Bernet* (Springer, 2012) p 284.

[35] Patočka, Jan "What Is Phenomenology?" in John J. Drummond & Otfried Höffe (*eds.*) *Husserl: German Perspectives* (New York, 2019) p 85.

[36] Husserl, Edmund *Logical Investigations vol. 1* (Abingdon, [*orig.* 1900–1] 2001) p 176.

[37] Husserl, Edmund *Ibid.* p 166.

[38] Husserl, Edmund *Ibid.* p 166.

transcendental."[39] He also comments on Husserl's argument that we must proceed from: "being as consciousness and being as something which becomes "manifested" in consciousness, "transcendent" being"[40] that: "Only the immanence [of *res cogitans*] was emphasized; the sense of the *res extensa* faded into being an object, and the relation of the two is turned into one of intentionality rather than a causal-objective one."[41] He concludes that: "One can thus characterize the Husserlian standpoint in the following way: it is the continuation of the metaphysical pursuit of the concept of being without a clear, explicit differentiation of the ontological difference between being and beings. In this difference, however, lies the proper sense of the *epoché*, when conceived generally."[42]

Can we get away with eliding the difference between "being" and "beings"? Can we collapse the *res extensa* into the *res cogitans*? As we will see below, Aristotle would emphatically argue not. For him, we can neither reduce world to mind nor beings to being. We do not look through the real world of beings to some sort of ideal world of being or mind. This is precisely his critique of Plato and, even if Husserl himself rejected the Platonic epithet, we must still regard him as an idealist in the tradition of Plato, Descartes, and Kant. Husserl rejected the "mythical" or "metaphysical hypostatisation" of Plato, but it is moot whether his concept of "eidetic intuition"—or "ideating abstraction" or "intuition of the universal"—and of "eidetic variation" really solve the issues raised by Plato in his *Theaetetus*.[43] Dieter Lotmar explains Husserl's proposed solution to these problems of knowledge and representation as that: "If, in free transformations in perception and phantasy, we imaginatively transform [*umfingieren*] the object in question in indefinitely many possible variants and observe in all of them the same element, then we can be sure that this element must also be present in further cases. That is to say, this element is necessary, *a priori* in the sense characteristic of eidetic intuition…[and i]n eidetic variation…[w]hat guarantees the universality of the seen invariants is thus the consciousness of the free "and so on freely" [*und so weiter nach Belieben*] that is effected along with all the variations."[44] Husserl simply provides us with a philosophical expression of empiricism which suffers, unlike Aristotle's, from being ungrounded in the "beings" of the physical world.

It is not evident that Husserl improves upon or advances from Aristotle's metaphysics and his logic. His argument that: "Each species, if we wish to speak of it, presupposes a meaning, in which it is presented, and this meaning is itself a

[39] Patočka, Jan "What Is Phenomenology?" in John J. Drummond & Otfried Höffe (*eds.*) *Husserl: German Perspectives* (New York, 2019) p 99.

[40] Patočka, Jan *Ibid.* p 96.

[41] Patočka, Jan *Ibid.* p 96.

[42] Patočka, Jan *Ibid.* p 98.

[43] Quotes from Dieter Lohmar "The Phenomenological Method of Eidetic Intuition and Its Clarification as Eidetic Variation" in John J. Drummond & Otfried Höffe (*eds.*) *Husserl: German Perspectives* (New York, 2019) p 110.

[44] Lohmar, Dieter *Ibid.* p 115 and 120.

species"[45] is simply recycled Aristotle but with an excessive residual Platonism that supposes "a *class* of "universal objects" or species."[46] It is hence moot whether Jan Patočka is right to argue that: "Pure phenomenology, we see, takes the place of traditional metaphysics."[47] It is, however, for our purposes, more significant that Husserl revives the various questions of the nature of being and mind—and of quite how we "tie down" the world as objective knowledge—first posed by Plato and Aristotle. What I take to be the shortcomings of Husserl's "egological" approach are significant only for demonstrating the deconstructed and circumscribed nature and the recycled errors of even the very best modern thought.

Husserl's "*Krisis*" correctly identified the fact that modernity is a problem as well as a blessing. Patočka sets out our situation on this score as that: "this whole period [of modernity] down to the present day can be designated as a triumph of construction. In repeated advances, modernity has replicated the foundational process called subject-object division, the mathematical outline of being, the progressive specialization from the point of view of an "objective" structure, and the reducing [*Zurückführung*] of every as yet "nonobjective" remnant to an objective ground. Technical civilization is nothing other than the setting to work of this project, which gradually comes to appear self-evident because it has become important, even indispensable, for life."[48] He adds that: "Husserl's late work *Crisis* has shown that science and traditional metaphysical philosophy begin with constructed concepts and never push beyond to something like pure, unbroken experience."[49] The problem we face is how to expose and to counter the preconceptions of "technical civilisation" which has become "indispensable for life" on the basis that even a critic such as Husserl cannot avoid believing that he can and should establish a *pure* logic, mathematics, or science of the world and to "push beyond to something like pure, unbroken experience." Husserl proves his point by himself being trapped by modernity into an elevation of idealistic construction which is simultaneously rejected and deconstructed.

C.S. Peirce (1839–1914) and his science of "semiotics" provides us with a further development, obviously relevant to this work. Whereas Ferdinand de Saussure's "semiology" argued that signs are arbitrary tokens of language, Peirce maintained that signs are the very language of being in the sense that: "the entire universe is perfused with signs, if it is not composed exclusively of signs."[50] His mission to understand the world through a "science" of semiotics led him to identify 66 distinct kinds of sign which increase in their correspondence to reality over the three classes of "symbol," "index," and "icon." Peirce explained himself as that: "I define a Sign

[45] Husserl, Edmund *Logical Investigations vol. 1* (Abingdon, [*orig*. 1900–1] 2001) p 231.

[46] Husserl, Edmund *Ibid.* p 231.

[47] Patočka, Jan "What Is Phenomenology?" in John J. Drummond & Otfried Höffe (*eds.*) *Husserl: German Perspectives* (New York, 2019) p 95.

[48] Patočka, Jan *Ibid.* p 87.

[49] Patočka, Jan *Ibid.* p 86.

[50] Peirce, Charles *Collected Papers* (Cambridge Mass. 1931–1958) 5.448n.

as anything which is so determined by something else, called its Object, and so determines an effect upon a person, which effect I call its Interpretant, that the latter is thereby mediately determined by the former. My insertion of "upon a person" is a sop to Cerberus, because I despair of making my own broader conception understood."[51] John Michael Krois explains this "sop to Cerberus" as that: "What he meant was, Semiosis is not a matter of psychology, even though human thought is semiosis."[52] We do not through this framing reduce the world to structure or to a supposedly pure logic but the problem we now face is in defining the wider basis of "semiosis" and of meaning making in the world.

Peirce reinvented the world as one of "semiosis," of sign and signification. This new semiotic worldview full of meaning(s) was an obvious advance on Descartes' *res extensa* lacking all meaning, as were the structuralist and phenomenological worldviews. A new layer or dimension of meaning had been introduced. We could, however, contend that the risk incumbent in these revolutions is in respect to the revolutionary, iconoclastic, and radical aspects of their founders' thought and of the dominating "idea" or "mythos" they propound. One example of Peirce's radicality is that he posits an evolution of the cosmos as a generalised gradual emergence of what he calls "concrete reasonableness" from an initial condition of entropic disorder. He pictures this evolution as that: "In the beginning—infinitely remote—there was a chaos of unpersonalised feeling, which being without connection or regularity would properly be without existence. This feeling, sporting here and there in pure arbitrariness, would have started the germ of a generalising tendency. Its other sportings would be evanescent, but this would have a growing virtue. Thus, the tendency to habit would be started; and from this, with the other principles of evolution, all the regularities of the universe would be evolved."[53] "Existence" only becomes a phenomenon when we have "personalised feeling." Peirce on this basis envisions an evolutionary development through a cosmic movement from Firstness (pure feeling), which by way of Secondness (actualities tending to repeat themselves) enters over time into Thirdness (relational lawfulness, conceived of as relatively stable forms of habit).

Let us consider the problem inherent in the iconoclasm of philosophy by reviewing Peirce's relationship with Aristotle. We find that Peirce takes on Aristotle's concept of "entelechy" but then burdens it with the production of an "ideal sign which should be quite perfect" through a unification of "the matter denoted" and the "very form signified by it."[54] He then speculates further that: "[t]he entelechy of the Universe of being, then, the Universe *qua* fact, will be that Universe in its aspect as a sign, the "Truth" of being. The "Truth," the fact that is not abstracted but complete,

[51] C.S. Peirce "Letter to Lady Welby (1908)" quoted in John Michael Krois "More than a Linguistic Turn in Philosophy: the Semiotic Programs of Peirce and Cassirer" *Sats—Nordic Journal of Philosophy* (2004) p 32–3.

[52] Krois, John Michael "More than a Linguistic Turn in Philosophy: the Semiotic Programs of Peirce and Cassirer" *Sats—Nordic Journal of Philosophy* (2004) p 33.

[53] Peirce, Charles *Collected Papers* (Cambridge Mass. 1931–1958) 6:33.

[54] Coppock, Patrick "Pragmaticism and Semiotics" *The Whitehead Encyclopedia.*

is the ultimate interpretant of every sign."[55] Peirce's reified semiosis depends upon a reified universe and a reified concept of truth or, we should say, of "Truth." We can yet again potentially discern in Peirce the problem of idealism, of the over-articulation of "mind" moving hand in hand with an under-articulated world. In respect to Aristotle's idea, *his* priority, we find that Peirce to some degree minimises or dismisses it. His semiotic meaning valorises the category of "relation" over that of "substance" (*ousia*)—as did Saussure's semiology—on the basis that it is a "Nantucket of thought" inappropriate for modern science.[56]

Peirce provides us with a powerful image and idea of the "sign." He also, however, runs the risk of reducing beings to the meaning relations *between* those beings. The apparent problem here is how we can valorise a concept like "sign," "consciousness," "evolution," or, with Aristotle, the "being" (*ousia*). We should note that "sign" is a Greek concept of "*sēmeion*" subtly recognised and perhaps invented by Aristotle with the related "semen" appropriately being Latin for "seed." The fundamental question we must address is quite in what sense these seeds of meaning *are*. Can we understand beings through their relations rather than in themselves? Donald Favareau argues that Peirce does not fall into the obvious trap of reducing the world to signs and hence to the relations between beings on the basis that: "signs are not things for Peirce, nor are they ideas. They are, like "gifts", only the analytical terms we use to denote the enactively situated terminals in a presently-being-accomplished action between agents, objects, and meanings, (the properties of each of which contribute to the shaping of the resultant particular "sign relation", or physically realizable interpretant)."[57] These "signs" as "gifts" bear some similarity with Aristotle's concept of "actuality" (*energeia*) in which a being is qualitatively and substantively actualised through its worldly expression or manifestation. Although "actuality" seems a more inclusive and less loaded term for explaining manifested meaning than "signs," I note that we encounter here simply the different limitation(s) provided by the idea of "*actualitas*"/"*en-ergeia*" or of the "function inside" being acted out outside. The idea of "consciousness" or of shared knowing is another important concept pressing at the very limit of the generalising possibilities of language.

Do we need to waste our time with such philosophical subtleties as "actuality" (*energeia*)? Surely something that is just *is*. Is it not progress that the speculations of "ontology" have been replaced with the useful study of knowledge-forming that is "epistemology"? This is most certainly *our* mythology. René Descartes hence asserted that: "a man who walks across a room shows much better what motion is than a man who says "It is the actuality of a potential being in so far as it is potential", and so on."[58] Francis Bacon famously described Aristotle's claim that "the

[55] Peirce, Charles *Collected Papers* (Cambridge Mass. 1931–1958) 2.304.

[56] Krois, John Michael "More than a Linguistic Turn in Philosophy: the Semiotic Programs of Peirce and Cassirer" *Sats—Nordic Journal of Philosophy* (2004) p 19.

[57] Donald Favareau (personal correspondence).

[58] Descartes, René "Letter To Mersenne 16th October 1639", *CSM* III §597, p 139.

hairs of the eye-lids are for a quickset and fence about the sight" as "impertinent".[59] He rather argued that: "When man contemplates nature working freely, he meets with different species of things, of animals, of plants, of minerals; whence he readily passes into the opinions that there are in nature certain primary forms which nature endeavours to educe…[But in] nature nothing really exists besides individual bodies [true particles], performing pure individual acts according to law."[60] We similarly find that modern thinkers have sought to rationalise Peirce by quietly putting aside his idea that "the entire universe is perfused with signs, if it is not composed exclusively of signs"[61] and arguing that we can say nothing about the "entire universe" but that semiosis is, as they say, "coextensive with life."[62] This rationalisation or retrenching of (bio)semiotics to the scope of biology seems sensible, but is it?

Is this standardising and rationalising of philosophical thought useful? I say, anything but. The reduction of insight to what can be tested, standardised, and communicated reduces original philosophy to academic philosophy, creativity to the requirements of the pragmatic maxim, and individual insight to managed groupthink. We ultimately see *the ossification of philosophy into science* by means of which complex and nuanced insight is reduced to standardised reuseable formulae. We need not to find a way of reducing philosophy to science but of keeping both forms of insight in their own proper fields. It is hence that we should *follow* Peirce in such speculations as that: "The evolutionary process is…not a mere evolution of the existing universe, but rather a process by which the very Platonic forms themselves have become, or are becoming developed"[63] and that: "The one intelligible theory of the universe is that of objective idealism, that matter is effete mind, inveterate habits becoming physical laws."[64] Peirce is here considering the very real ideas of traditional philosophy as we have considered them above going all the way back to its origins which focussed upon determining the very nature of "mind" (*nous*), "habit" (*hexis*, *ethos*), "law" (*nomos*), "nature" (*physis*) etc. We can disagree with Peirce's interpretation or prefer another, but we should also recognise that the "truth" of "reality" will certainly be something like this, something approaching the ineffable, and *not* something like an accountant's ledger book or a controlled chemical experiment.

Even with Peirce we find the collapse of meaning into relations, processes, or appearances we have already encountered in ancient philosophy. In our own time this reductionism is, for example expressed by W.V. Quine's as that: "To be assumed

[59] Bacon, Francis "On the Advancement of Learning" in *Bacon's Essays and Advancement of Learning* (London, [*orig.* 1605] 1925) p 265.

[60] Bacon, Francis *Ibid.* I.66, II 2, II 8.

[61] Peirce, Charles *Collected Papers* (Cambridge Mass. 1931–1958) 5.448n.

[62] Sebeok, Thomas A. "*Biosemiotics: Its Roots, Proliferation, and Prospects.*" *Semiotica*, (2001).

[63] Peirce, Charles *Collected Papers* (Cambridge Mass. 1931–1958) 6.194. On Peirce's concept of evolution see Josiah Royce & Fergus Kernan "Charles Sanders Peirce" *The Journal of Philosophy, Psychology and Scientific Methods* (1916).

[64] Peirce, Charles *Collected Papers* (Cambridge Mass. 1931–1958) 6.25.

as an entity is, purely and simply, to be reckoned as the value of a variable"[65] and that: "Our acceptance of an ontology is, I think, similar in principle to our acceptance of a scientific theory, say, a system of physics: we adopt, at least insofar as we are reasonable, the simplest conceptual scheme into which the disorganised fragments of raw experience can be fitted and arranged. Our ontology is determined once we have fixed upon the over-all conceptual scheme which is to accommodate science in the broadest sense."[66] Karl Popper properly encapsulates Quine "in a nutshell" by quoting him as follows: "The bodily states exist anyway; why add the others?"[67] Richard Rorty, already, considered, refers this reductionist approach to "pragmatism" by describing his own position as: "Epistemological behaviourism…[which] might simply be called "pragmatism"…it is the claim that philosophy will have no more to offer than common sense (supplemented by biology, history etc.) about knowledge and truth."[68] The objective in all cases is to flatten philosophy and meaning and make it computable, something that can be scientifically or ideologically represented and manipulated. We cannot necessarily attribute this reductionism to Peirce who first invented "pragmatism" but called his own philosophy "pragmaticism" to dissociate it from the various errors of other variants of pragmatism. We cannot, however, fully dissociate him from the straight path on a very slippery slope from philosophy to scientism that we discern here.

I will take the opportunity here to add a few preliminary thoughts on another great philosopher, A.N. Whitehead. Whitehead, the father of "process philosophy," develops "process" from the Platonic principle of "participation" through which an "occasion" is produced as a "synthesis" of "actuality" produced by the relationship between "eternal objects." He argues that: "every occasion is a synthesis of all eternal objects under the limitation of gradations of actuality…as a process"[69] and that this renders "Plato's general point of view" into "a philosophy of organism."[70] He gives a brilliant account of the evolution of meaning through morphology in his chapter "Expression" in his *Modes of Thought* (1938) that we will consider in more detail later. We again see, however, that the word "process" taken *unphilosophically* allows Whitehead's followers to reduce his profound philosophical thoughts to a reduction of the world as scientific method and process. His talk about "eternal

---

[65] Quine, W.V. "On What There Is" in *From a Logical Point of View* (Cambridge, Mass., 1953) p 13.

[66] Quine, W.V. *Ibid.* p 18.

[67] Popper, Karl "Knowledge: Objective and Subjective" in *Knowledge and the Body-Mind Problem: In defence of interaction* (London, [*orig.* 1969] 1994) p 8. On the deficiencies of Quine's philosophy see George Engelbretson "Quine on Aristotle on Identity" *Crítica Revista Hispanoamericano de Filosofía* (1985), Jaakko Hintikka "Three Dogmas of Quine's Empiricism" *Revue Internationale de Philosophie* (1997), and Peter Hacker "Passing by the Naturalistic Turn: On Quine's Cul-de-Sac" *Philosophy* (2006).

[68] Rorty, Richard *Philosophy and the Mirror of Nature* (Princeton, 1979) p 176. See also Hilary Putnam's *Ethics without Ontology* (Cambridge Mass., 2004).

[69] Whitehead, A.N. *Science and the Modern World* (Cambridge, [*orig.* 1925] 1929) p 217–8.

[70] Whitehead A.N. *Process and Reality: An Essay in Cosmology* (New York, [*orig.* 1927–8] 1978) p 39.

objects" can on this basis be quietly shelved along with his argument that evolution is a power of creativity realised as deepened consciousness and complexity in place of an interpretation of this evolution as the emergence of self-organisation to maintain the species' survival. This reduced interpretation is not *philosophically* valuable. It is not sufficient to explain the meaning of the world but merely lays out the theoretical framework needed for a scientific research programme upon it. I note that Husserl also has a theory of signs from his own point of view which begins with the symbols and functions of mathematics and moves to a more nuanced but confusing and unclear emphasis upon the "act-character [which] is a descriptive trait in the sign-experience which, stripped of intuition, yet understands the sign."[71]

I add that we see a similar irreducibility in respect to *all* great philosophers. We find a personal dimension to their thought in a comparable way as we find one for the great composers in the sense that we cannot confuse or conflate Schubert with Bach or Shostakovich. We ultimately see here the meaning and nature of philosophy. It *cannot* be conflated with science. I wish to make several important summary comments here. The above highlighted philosophers are all philosophers *I like*. They have revitalised philosophy by restoring the *aporiae* first set out by Plato and Aristotle. The fruits of their labour are the emergence of such new thinking as that of biosemiotics, which *I also like*. What I have sought to show above is, however, how inadequate these thinkers and the schools based upon them are. This is the problem of philosophy. Even the best thought is only ever partial and it is only by understanding this partiality that we can appreciate the richness of the whole.

3.3 The Problem of Biology

I argue that biology and a consideration of the nature of humanity is, or should be, at the very centre of philosophy. It is only through a very strange thought process that the world can be reduced to mind or matter, derecognising beings and, more particularly, human beings. The idea that we can find a medium other than "beings" (*ousiai*), whether this be language, mathematics, science, logic, or sign theory, runs throughout modern philosophy. I will show in this work that Aristotle not only places biology at the heart of his peripatetic project but also insists that human "meaning" must emerge from biological function and structure and then somehow from the plastic potentiality of matter itself. We are, on this basis, meaningful and creative beings that somehow channel and express in some specific sense a meaning inherent in the world itself. We do not need to invent a just-so story or mythology to explain the human situation but only to recognise and represent the reality in which we somehow find ourselves.

This is the naturalistic Aristotelian position that I will set forth in this book and seek to promote as the most fruitful and reasonable philosophical approach to the

---

[71] Husserl, Edmund *Logical Investigations vol. 1* (Abingdon, [*orig*. 1900–1] 2001) p 210.

world in general and to the biological and human worlds specifically. The standard alternative is the neo-Darwinian position paradigmatically explained by the 1969 lectures of the Nobel-prize winning biologist Jacques Monod published in English as *Chance and Necessity*. This work is useful because it is often taken as being the landmark philosophical explanation of neo-Darwinism, along with Richard Dawkins' *The Selfish Gene*. It is important to stress here that we are not considering the biology of this work which is of some quality but since superseded. We are not even considering the fact that the biology—and science in general—has moved on in a direction that invalidates Monod's position. I will treat this story in the sister volume of this work on modern science. What we need to consider here is the *philosophical* significance of this work which retains its power in demanding that we reduce philosophy to science even whilst it puts forward a science that does not justify this philosophical reduction.

Jean-Marie Benoist explains this situation from an interesting perspective in his 1975 philosophical explanation of structuralism entitled *The Structural Revolution*. Through Benoist we can consider the nature both of Monod's original position and of Benoist's structuralist critique of that position. We can begin here with Benoist's own definition of structuralism as that: "An analysis is structural if, and only if, it displays the content as a model, i.e., if it can isolate a formal set of elements and relations in terms of which it is possible to argue without entering upon the significance of the given content."[72] Note the analysis without synthesis, the terms used such as "formal," "elements," and "relations," and the attempt to dismiss "the significance of the given content." As regards how Monod's biological representation of the world meets this structuralist definition, Benoist quotes Monod as that: "we shall arbitrarily choose to define the essential teleonomic project as consisting in the transmission, from one generation to another, of the content of the invariables characteristic of the species"[73] which he himself quixotically expands upon as that: "Once the lures of teleology and the demons of finalism have been exorcised, the cellular code can for Monod be expressed simply in terms of an autonomous morphology, of reproductive invariables and of teleonomy. The concept of teleonomy which Monod presents and develops in the course of his work is a key concept that, in spite of the finalist appearances of its equivocal prefix, is articulated around two fields of structural causality and a revisited Cartesian mechanism."[74] A bacteriologist reduces life to the replication of a "cellular code."

How convincing is Monod's philosophical position? How advanced is Benoist's cutting-edge philosophy? I suggest that both are disappointing. It is clear that Monod's philosophy is tacked on to a science that does not really support it. Indeed, Benoist himself makes this point. He complains that there are "two" Monods. The first Monod is "the invincible man of science, whose genius lies in having been able to treat the living world as a language through combining in the most fertile and

[72] Benoist, Jean-Marie *The Structural Revolution* (London, [*orig*. 1975] 1978) p 8.

[73] Jacques Monod quoted in *Ibid*. p 90–1.

[74] Benoist, Jean-Marie *Ibid*. p 90.

rigorous way the gains of information theory and the discoveries of molecular biology. It is this side of the man that exhales the presocratic freshness of a reinvented chance and necessity."[75] It is interesting that this Monod is described as "presocratic." The second Monod is the rather tiresome amateur existentialist who famously argues from a Cartesian perspective that: "Modern societies, woven together by science, living from its products, have become as dependent upon it as an addict on his drug. They owe their material power to this fundamental ethic upon which knowledge is based, and their moral weakness to those value-systems devastated by knowledge itself, to which they still try to refer."[76] Benoist's critique of this strange idea is that: "Instead of deriving these ethics from the knowledge of a transcendental critical epistemology without a subject/substance that would bestow meaning, Monod presents them as an ethical and ontological wager realised by a kind of phenomenological subject."[77] Benoist bemoans the fact that Monod does not go all the way to the total expurgation of the human subject, i.e., through a "critical epistemology without a subject/substance."

Benoist ironically critiques Monod's science-worship as a misguided existentialism by producing the same sort of misguided existentialism himself. He explains that: "Having skidded off the Kantian and Bachalardian road where the sagacity of his scientific analysis should have assured him a rightful place, the ethical philosophy of Jacques Monod swells the ranks of all those personalisms and phenomenologies that tirelessly continue to fuel the old anthroponarcissitic dream whose last somersault one had hoped momentarily was Sartrian existentialism."[78] It is clear from this statement simply that Benoist's position is as un-objective as Monod's whilst both make the same false claims to scientific objectivity. Ultimately, Benoist argues alongside his literal reduction of the world to "Bachalardian" mathematics that we must abolish humanity and being—what he describes as a "substantialist view of nature"[79] and as the "substantialist trap"[80]—and achieve "the decentring of man, who would no longer think himself as the hierarchised and teleological conclusion to the evolutionary process, as its *end*, but who would have to consider the emergence of that admirable machine, the brain, as the product of that game of roulette set in motion by chance alone."[81] We see how both Monod and Benoist insist that they are arguing from a grand objective position of "science" and yet produce a disconnected, subjective, radical philosophy which simply attacks "substance" and the "subject" in the name of "chance."

How do we explain this appeal to a Bachalardian philosophy of "non" and its rejection of "substance" and humanity? We must first recognise that this position is

[75] Benoist, Jean-Marie *Ibid.* p 98.

[76] Monod, Jacques *Chance and Necessity* (London, 1971) p 164.

[77] Benoist, Jean-Marie *The Structural Revolution* (London, [*orig.* 1975] 1978) p 97.

[78] Benoist, Jean-Marie *Ibid.* p 97.

[79] Benoist, Jean-Marie *Ibid.* p 87.

[80] Benoist, Jean-Marie *Ibid.* p 211.

[81] Benoist, Jean-Marie *Ibid.* p 93.

not merely an eccentric outlier which will conclude with the odd, barely intelligible extremes of post-structuralism. Rather, the attack on substance clearly goes back to Descartes, as we have already seen, and perhaps even to Plato. Even C.S. Peirce attacks "substance" as a "Nantucket of thought" inappropriate for a world centred upon modern science.[82] Indeed, it is the fixation upon reducing philosophy and all meaning to science that seems to unite even all the thinkers mentioned above. It is this shoehorning of reality into "science" that collapses complex reality into "form" (structure) and "flux" (information) on the assumption well-articulated by Monod as that: "From the reign of pure chance, it enters into that of necessity, of the most implacable of certitudes."[83] This supposed transition from chance to necessity makes no immediate sense, scientific or otherwise, but simply represents the latest variant of Cartesian dualism, itself derived from Plato's Theory of Forms. It is a confusion arising in part from the conflation of philosophical examination with scientific method.

Aristotle, by contrast, consciously refuses this breakdown into material mechanisms and a disembodied, existential freedom which can be understood as "chance." He rather surveys the world on a "holistic" basis as a qualitative topology of interacting and entangled beings. This scientifico-philosophical approach to the living world can also be expressed as biosemiotics.

3.4 Conclusions

A hard assessment of modern philosophy shows us not how far we have come but how problematic our thought actually is. The collapse of philosophy into the solipsistic dualism of an objective world and a subjective ego goes back to Descartes. This dualism has long been understood as unsatisfactory and yet its simplification has yielded the quite incredible fruits of modern science and technology. Its allure therefore remains perennial *despite its obvious philosophical error*. This erroneous power is at the heart of the phenomenon of reductionism. Reductionism is in this sense the expression of Cartesian physicalism and additional "ego" and it is in another sense an expression of Platonic idealism. It is the reduction of the world to a certain medium—normally mathematics and/or language—and it is notable that Plato worked from the example of language to reduce the world to a language of mathematics.[84] As regards the elements of this "language," Plato presents us with mathematical "atoms" alongside his "Forms." Democritus also famously presents us with "atoms and the void."

---

[82] Krois, John Michael "More than a Linguistic Turn in Philosophy: the Semiotic Programs of Peirce and Cassirer" *Sats—Nordic Journal of Philosophy* (2004) p 19.

[83] Jacques Monod quoted in Jean-Marie Benoist *The Structural Revolution* (London, [*orig*. 1975] 1978) p 92.

[84] See Jan Patočka's chapter "Les Grammata et les Stoixeia dans la Philosophie Grecque jusqu'a à Aristote" in *Aristote, Ses Devanciers, Ses Successeurs* (Paris, [*orig*. 1964] 2011).

We ultimately find that our modern philosophy is built upon the same conceptual building blocks as prevailed at the very origin of our thought. We can from this perspective widen out our problem as a problem of abstraction itself as the use of technology, and of sophism as the use of political power. I will claim below that Aristotle identifies this fundamental problem of philosophy itself, i.e., its degeneration into over-simplifying sciences and -isms, and counters it through his "realism" which surveys the whole of reality and the whole of our mental landscape and synthesises them. He does so, most notably in his *Politics*, from a bottom-up perspective of beings rather than, as in Plato's *Republic*, as a top-down ideological construction. The most obvious top-down plans for human beings and for being in general are religious and this influence, whether accepted or rejected, has indeed compounded our problems today.

The central claim made in this work is that Aristotle is notable for his determined emphasis upon the primacy of "beings" (*ousiai*). These complex units can be studied in their various dimensions through different sciences. The being is always, however, more than the sum of its parts and the addition is not trivial. Complexity points to the richness of reality and shows the error of reductionism.[85] Aristotle on this basis "saves the phenomenon" and elevates it in its being. The alternative is the reductionism which reduces being(s) to our explanation of them. The final conclusion of this reductionism and one of its basic assumptions is that humanity can in some way be explained away. This deconstruction of being is most obviously effected through post-structuralism. Jacques Derrida seeks to dispel what he calls a "metaphysics of presence" and reduces the world to language and its signs and dismisses not only the idea of an author but also of a book, which becomes merely a text.[86] This attack on "being," on "substance," is not, however, a mere perversity on the fringe of European thought. We constantly find, as Kvasz previously explained, a collapse of thought into the most reduced "laws" possible, stripping away the richness of reality. Even Occam's razor is, on reflection as dangerous as it is useful.

The recognition of the error of our modern philosophical position has led to the painful emergence of a "new" philosophy, a philosophy resurrected only due to the evident failure of religion and science to replace it. We have seen above that even though this "new" philosophy—whether phenomenology, structuralism, or semiotics—recognised the problems arising from a worldview recognising only science (or only theology) as a valid framing for knowledge, they were unable to break out of this worldview themselves. The true problem we face is to establish what philosophy itself *is* and that it is distinct from science. No one reaches more reasonable and considered conclusions upon this difference than Aristotle, as we will see below.

---

[85] See Paul Feyerabend's *Conquest of Abundance: A Tale of Abstraction versus the Richness of Being* (Chicago, 1999).

[86] See "Post Structuralism" in John Sturrock *Structuralism* 2nd ed. (Oxford, [*orig*. 1986] 2003).

Chapter 4
Aristotle and Meaning in Life

Abstract This chapter considers how Aristotle understands the biological fact of (human) life and death philosophically. Aristotle has a complex definition of life which is both philosophical and biological. He explains the "lifecycle" of an organism as a "period" of its being and a "cycling" of the species. Aristotle extrapolates from the physical and psychical pathways of individual lives to the physical and psychical "evolution" of species. Aristotle's theory of evolution is a sort of living optimisation and expression of biological being in its milieu. The contention pressed here is that Aristotle's is a more realistic and observable theory of evolution than the neo-Darwinian alternative. We find, furthermore, that this picturing of meaning and of its structuring and refinement is found anew in semiotics and in biosemiotics. Neither assume that general meaning or biological meaning can be attributed to "chance and necessity" rather than being approached on its own terms as the expression of life itself.

Keywords Aristotle's theory of evolution · Psychical and physical evolution · Aristotle and biosemiotics · Aristotle and biological cycles and periods

We have encountered above the Heideggerian claim that: "philosophy *is* life." Let us ask: how and in what sense can and should philosophy explain "meaning" and "life"? Further, how could any discipline tasked with making sense of the world itself fail (or refuse) to recognise such evidently essential aspects of our experienced reality as the existence of meaning(s) in life?

The problem as regards the meaning of life goes back to Plato's comment in his *Phaedo*, a dialogue we have already looked at, that: "if things would not reciprocate, one to the other, going around as it were in a circle (*kyklō*), but if coming-to-be were on a straight line (*eutheia*), going from one state to the opposite, and did not bend back to the other side, and did not make turns, then ultimately, you know, everything would end up in the same state, would suffer the same fate, and thus would stop coming-to-be…In the end, by necessity, everything would be dead, and nothing would be alive (*Phaedo* 72bc)." Plato rests his case regarding the meaning of life itself on the unexpected (but profound) idea that we must envisage the world

P. N. Jackson, *Aristotle on Meaning in the Living World*, Biosemiotics 30,
https://doi.org/10.1007/978-3-032-00602-8_4

along the paradigm or model of a circle rather than as a straight line. I have already noted on this point that Ernst Cassirer's "curse" of modernity is its emphasis upon infinity over boundedness, abstraction over being, line over circle, and I add Walter Burkert's further explanation that: "the final and inevitable result of linear progress [is] nihilism, the big black hole."[1] The abstracted thought of modernity gives us *both* progress *and* nihilism.

I suggest that the problem as regards how to recognise "being" (*ousia*) and "idea" (*idea*) within their domains is fundamental to Aristotle, following Plato. Aristotle follows Plato's cryptic but important above emphasis with his own argument that: "The cause of this continuous process, as has been frequently remarked, is cyclical motion (*hē kyklō phora*), the only motion which is continuous (*monē gar synechēs*). Hence also the other things which change into one another, for instance, the simple bodies, by being acted upon or having power to act (*kata ta pathē kai tas dynameis*), imitate cyclical movement (*mimeitai tēn kyklō phoran*). For when Air comes-to-be from Water, and Fire from Air, and Water again from Fire, we say that coming-to-be has completed the cycle, because it has come back to its starting-point. Hence motion in a straight line is also continuous because it imitates cyclical motion (*hōste kai hē eutheia phora mimoumenē tēn kyklō synechēs estin*) (*GC* II 337a 1-8)." Aristotle's remarkable point is that a straight line is a pathway because it "imitates" cyclical motion as a cyclical chain of cause-and-effect. For an example, we most obviously think about life both as a straight line from birth to death and also as a "cycle" of being which follows a cyclical pattern.

The path and cycle of life is an obvious example and a clear paradigm. Aristotle's thought and intention is, however, more expansive than this simple questioning of the significance of the foetus as a model of meaning, as we might expect. For the wider principle we can take any action which follows a process (*kinēsis*) to achieve an end (*telos*) since the end exists in the origin (*archē*) with the middle (*to meson*) being its creative expression.[2] The path of the process is the staging of a cycle; the straight line (*eutheia*) is also a circle (*kyklos*). Eric Voegelin explains the philosophical implications of this situation as that: "The eschatological interpretation of history results in a false picture of reality… Specifically, the Gnostic fallacy destroys the oldest wisdom of mankind concerning the rhythm of growth and decay which is the fate of all things under the sun."[3]

---

[1] Burkert, Walter "Impact and Limits of the Idea of Progress in Antiquity" in *The Idea of Progress ed.* Arnold Burgen et al. (Berlin, 1997) p 35–6. Albert Camus also argues that the critical point of difference between Greeks and moderns is that: "The difference between the two is the difference between a circle and a straight line (*The Rebel* (London, 1953) p 157)."

[2] Claudia Baracchi explains this point as that: "In its highest manifestation…the end is not an outcome separate from the activity leading to it (we should especially avoid a naïve temporal understanding of finality here), but, rather, the activity itself. The end is manifest in and as the action, from the start. It already informs the unfolding of the activity, of a certain way of living (*Aristotle's Ethics as First Philosophy* (Cambridge, 2008) p 97–8)."

[3] Voegelin, Eric *The New Science of Politics* (Chicago, 1952) p 166 (and see also H.B. Acton's "Tradition and Some Other Forms of Order" in Proceedings of the Aristotelian Society (1952–3)).

4.1 Aristotle and the "Period" of a Lifecycle

Aristotle provides us with food for thought with his concept of the period ("*peri-hodos*" "around the road");[4] on this subject see especially his *On the Longness and Shortness of Life*. He conceptualises time in general as a linear series of "nows" regarding which he argues that "that which is determined (*to horizomenon*) either way by a "now" seems to be what we mean by time (*Phys*. IV 219a 30)" though "The present "now" is not part of time at all (*Phys*. IV 218a 7)," i.e., the "now" is a creative point which actualises as reality. He explains the "*periodos*" itself as a potential framing for being on the basis that: "any finite portion of time is a portion of revolution (*periphora*), but is not a revolution (*Phys*. IV 218b 3)" and that: "if time, then motion, inasmuch as time is but an aspect or affection of motion (*ho chronos pathos ti kinēseōs*) (*Phys*. VIII 251b 27-28)." We see that both the point and the line are part of but distinct from the circle. Time fills a period and supports the movement or cycle of a being ("*ousia*") through multiple movements along a linear path (or "straight line"). These distinct paths and movements take place within the context of an "entanglement" or "environment" meaning "en-circling" (from *en + viron*) in the sense that the entangling of movements of discrete beings passing through proximate space is encompassed by our total "sublunary" world.[5]

We also find that: "time is regarded as the rotation of the sphere, inasmuch as all other orders of motion are measured by it, and time itself is standardised by reference to it (*tautē metrountai hai allai kinēseis kai ho chronos tautē tē kinēsei*). And this is the reason for our habitual way of speaking; for we say that human affairs and those of all other things that have natural movement and become and perish seem to be in a way circular, because all these things come to pass in time and have their beginning and end as it were "periodically" (*kata tina periodon*); for time itself is conceived as "coming round"; and this again because time and such a standard rotation mutually determine each other. Hence, to call the happenings of a thing a circle is saying that there is a sort of circle of time (*ta gigomena tōn pragmatōn kyklon to legein esti tou chronou einai tina kyklon*); and that is because it is measured by a complete revolution (*kyklophoria*), and the whole measurement of a thing is nought else but a defined number of the units of its measurements (*Phys*. IV 223b 22–224a 2)." In respect to the biological reproductive period Aristotle adds that: "In all cases,

[4] See also Aristotle's interesting treatment of "method" ("*meta-hodos*" "through the road") by reference to a race-course through which, according to Plato, we can reach the end of the course in two directions, from two perspectives; see *N.E.* I 1095a 31–1095b 4.

[5] Aristotle famously gives the example of a man who goes to the marketplace for the sake of buying some goods and happens to meet his debtor who pays him back; this is not, as Pascal Massie explains, a manifestation of chance but a contingent causality through a circulating movement of beings by means of which: "Luck is then a simulacrum of teleology; a random conjoining that nevertheless *mimics* causality ("The Irony of Chance: On Aristotle's *Physics* B, 4–6" *International Philosophy Quarterly* (2013) p 23)." For Aristotle's position that "no designed result is due to chance" see *Post. An.* II 94b 36–95a 9 and on his view that chance exists meaningfully only in the "sublunar" (organic) world see *Phys*. II 196b 5.

as we should expect, the times of gestation and formation and of lifespan aim, according to nature (*kata physin*), at being measured by "periods" (*hoi chronoi kai kuēseōn kai geneseōn kai tōn biōn metreisthai*). By a "period" I mean day and night and month and year and the times which are measured by these (*GA* IV 777b 18-20)."[6]

How does Aristotle regard the specific "*periodos*" of human life? The first thing to observe is that he deploys two words for "life": "*zōē*" meaning the fact and force of life, and "*bios*" meaning the structured or contented life or lifespan. "*Bios*" shows us that life is not just "*zōē*" since even for plants it represents an arc—birth, fruition, and death—which cannot be understood as a linear force. The critical distinction here is that the "*telos*" of biological life is not located as simple "cause and effect" in its physical end which is death but in its "middle," its maturity or "*akmē*." On this "middle" Aristotle dates (for men) the physical prime of life to 30–35 years and the mental prime to 49 years with natural death being 70 years; see *Rhet*. II 1390b 6-10. Human life itself breaks down (following Pythagoras) into 7-year phases or stages.[7] On this situation see Carl Jung's essay "The Stages of Life" (1933).

I represent how Aristotle frames human life in Fig. 4.1 below. Life is broken down into stages of 7 years and runs from (inception as) 0 to 70. Biological maturity occurs at 30–35 and mental maturity at 49–50.

For Aristotle, the peculiarity of the human situation is that man has two selves combined in one "*synholon*" having a physical self and a spiritual self, body and soul, different though parallel structures with different "primes."[8] Aristotle recognises our "spiritual" self as manifested in the "master craftsman" but never allows this dimension of being to be reduced to an abstraction (as disembodied mind) or an ideology (as disembodied spirit). Perhaps because we are by nature so "close" to

---

[6] Aristotle adds that "in the case of the animals and trees the seasons are the turning-points of their lives more <than in the case of man>, and so when there is a change of season, then they follow suit and grow or shed their feathers or hair or leaves. In man, however, the spring, summer, autumn, and winter of his life are not seasons according to the calendar but seasons of his own age; so that, as these do not go through the cycle of change, neither do the conditions which depend on them; although the cause which controls the change of conditions is a similar one in his case too (*GA* V 784a 14-22)."

[7] Michel Foucault comments that: "He [i.e., Pythagoras] divides man's life in this way: (quoting Diogenes Laërtius) "A child for twenty years, a youth for twenty years, a young man for twenty years, and an old man for twenty years" (*Lives of Eminent Philosophers* Book VIII 10)" and (quoting Seneca quoting Demetrius) that: "every seventh year imprints a new character on the life of a man" (*The Hermeneutics of the Subject* (New York, [*orig*. 1981–2] 2005) p 232). For Aristotle's own reference to these stages see *Pol*. VII 1335b 33-36.

[8] Aristotle explains in his *Politics* that the physical prime and the mental prime do not overlap on the basis that: "the period [of conception] must be limited to correspond with the mental prime; and this in the case of most men is the age stated by some of the poets who measure men's age by periods of 7 years,—it is about the age of fifty (*Pol*. VII 1335b 33-36)" and he adds further about the cycle of "generation" that: "the succession of the children to the estates, if their birth duly occurs soon after their parents marry, will take place when they are beginning their prime, and when the parents' period of vigour has now come to a close, towards the age of seventy (*Pol*. VII 1335a 33-36)."

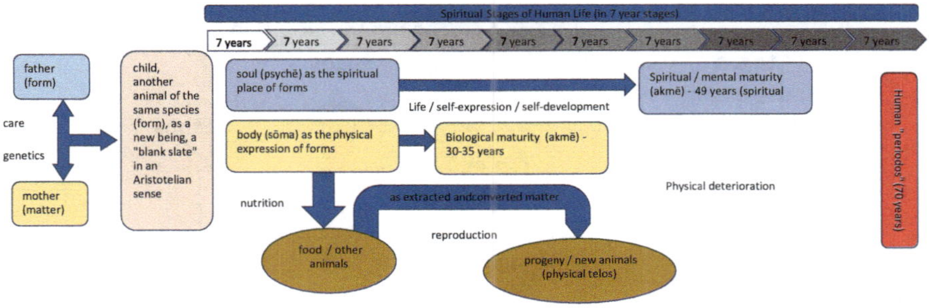

Fig. 4.1 Aristotle's stages and cycle of life

human being, we can clearly see (if we want to) that this being reveals a paradigmatically complex and qualitative structure that cannot be reduced to an idea. And the more we consider the world, the more we see that the complexity of our human nature is typical of the world and that neither can be reduced. This appreciation of complexity is one thing we can take from the human example.

Aristotle explains this spiritual structuring of human being further as that: "there are admittedly three things by which men are made good and virtuous, and these three things are nature (*physis*), habit (*ethos*) and reason (*logos*). For to start with, one must be born with the nature of a human being and not of some other animal; and secondly, one must be born of a certain quality of body and soul (*poion tina to sōma kai tēn psychēn*). But there are some qualities that it is of no use to be born with, for our habits make us alter them: some qualities in fact are made by nature liable to be modified by the habits in either direction, for the worse or for the better (*Pol.* VII 1332a40–1332b3)." Let us represent this position as below in Fig. 4.2 and then consider what it means:

We could say that Aristotle is stating the obvious here, i.e., that the human situation comprises of *three* main inputs; *physis*, *ethos,* and *logos*, and it is indeed positive if he is stating the obvious. Our own problem is the opposite, *viz.* how to come back to an ideological position which is not so obvious, which reduces the world to "nature" *or* to "nurture." We should recognise as regards this situation that Aristotle's position is not more simplistic than ours but rather insists upon an irreducible triadic complexity as opposed to our binary alternative. I note as regards the peculiar quality of "*logos*" as regards human beings that Aristotle explains in his *Politics* that language (*logos*) sets man apart from other animals and he expands upon this principle in his *Ethics* as that man alone lives in accordance with reflective, independent, and *right* (or wrong) reason (*orthos logos*); and see *Met.* A 980a 28–981a 6. I add as regards how Aristotle represents the divine or transcendent etc. that he presents us with the concept of "*periousia*" as a superfluity of energy and "*ousia*"; see *Top.* III 118a 8. Aristotle represents transcendence as a surpassing evolution of "movement" and the "supersensible" as being precisely that, a development of mind

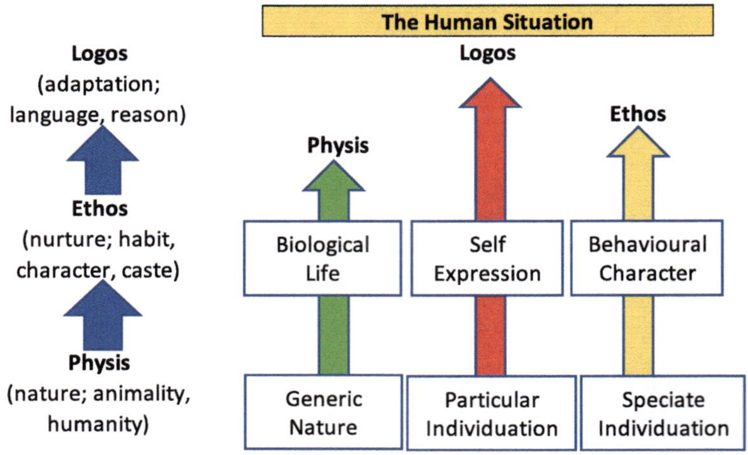

Fig. 4.2 Aristotle's tripartite representation of the human situation

through the refinement and transformation of sense, both on an individual and a speciate basis.[9]

We also find that Aristotle represents the movement or passage of a being (*ousia*) through the world through his famous four causes, which are not a basic principle or model but a key component of his philosophy, the nuances and logic of which he seeks to explain through all his works. The "being" necessarily possesses its own matter and form—these two go together as the being of a given "species"—and its own power of movement to *its* ends. I represent this ontological matrix, this situation in Fig. 4.3 below.

This four-cause causality is expressed (as three causes) by material interactions and (as four causes) by human life and action. Plant and animal behaviour is somewhere in between. We see that we must understand the world as moving,

[9] Aristotle's concept of "movement" is not that of dynamics or ballistics but of the natural expression of being(s). An outline of this argument is that: "the common feature that characterises them all [i.e., natural things] seems to be that they have within themselves a principle of movement (or change) and rest—in some cases [1] local (*kata topon*) only, in others [2] quantitative, as in growth and shrinkage (*kat' auxēsin kai phthisin*), and in others again [3] qualitative, in the way of modification (*kat' alloiōsin*) (*Phys*. II 192b13-16)" His position on physical movement in space is that it is an actualisation of other kinds of movement in the sense that: "local movement (*phora*) [i.e., movement in space as opposed to change of being] characterises things in proportion to the fullness of nature they have realised (*Phys*. VIII 261a18-19)." On the "supersensible" Aristotle argues that there is a *scala naturae* based upon the realisation and extension of "sense" to "mind" on the basis that a life expressed by mind and its associated bodily movement is the "aim of all things, or of all things that possess sensation or reason; or would be, if they could acquire the latter (*ei laboi noun*) (*Rhet*. I 1362a23-24)"; and see also *Post. An*. II 99b 35–100a 4 and *Top*. V 137b 24-26. Aristotle gives as examples of this higher expression of "mind" both the principle of "care" and such facts as that: "man alone, so to speak, among animals perceives and takes pleasure in the odours of flowers and such things (*Sens*. 444a 32-34)" (and Aristotle also notes the emergent principle in his biological works that animals are attracted to certain plants for medicinal reasons).

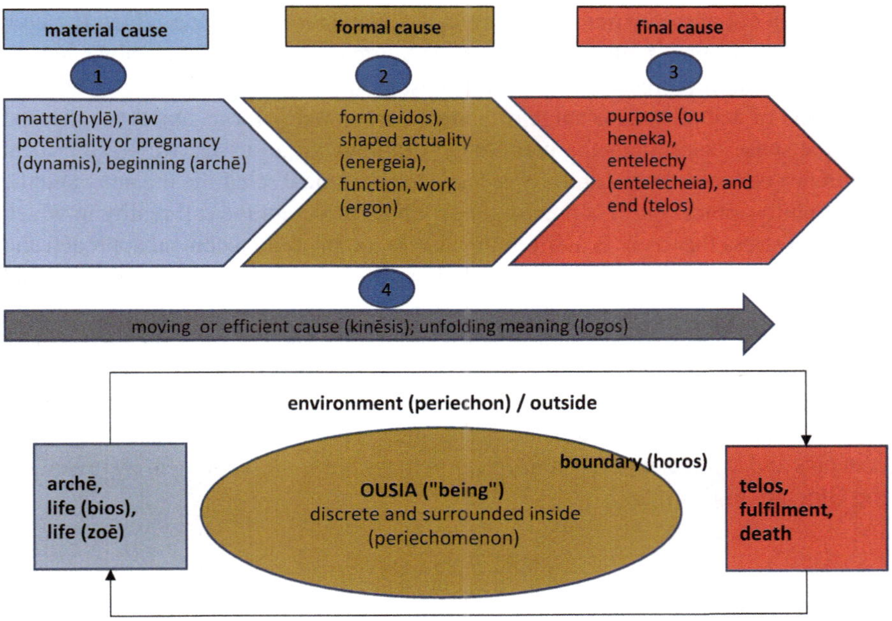

Fig. 4.3 The human being moving through the world through the "four causes"

physical beings with an inner, an outer, a boundary and boundary conditions, a space, place, environment, and a direction. The human example shows us that material or simple cause-and-effect is not sufficiently explanatory to explain nature tout court. Indeed, the human example shows us the evolution of causality itself in the world from what we could describe as a simple quantitative interaction to what is unavoidably a complex qualitative entanglement and expression. We can posit on this basis an evolution or development of causality itself which cannot realistically be understood as an accumulation of chance events in the sense that a foetus clearly differs from a teratoma and then, at a further remove, from the chemicals that comprise them.

Aristotle produces a philosophy of nature—his *Physics*—which is certainly highly abstract, but which is not mathematical and quantitative but analogical and qualitative. It seeks to represent the structures of nature in a way that explains the becoming of the being of its various beings. It *physically* represents "reality" in a realistic and biological way. This creates a whole different way of looking at the world technically that I would like to draw attention to here. We find a depth of meaning in the Greek terminology (as stressed by Heidegger) exemplified by "*hōra*" meaning season, period, or fitting time which possesses a scope not expressed by our mechanical expressions of time. A further example is that the original meaning of "*phora*" is "crop" and only later "movement" with the root of this term being explained in Aristotle's comment that: "in the generations of men there is a kind of crop (*phora*) as in the fruits of the field; sometimes, if the race is good, for a certain period men out of the common are born in it, and then it deteriorates (*Rhet.* II 1390b 24-27)." The critical emphasis of Aristotle's organic conception of movement as "*phora*" is that it represents not a contentless or mechanical motion but a surge of being and of actuality. This surge of being takes on many forms and strategies and in some sense determines the being which follows. In all cases, we see a critical difference between terms carefully reduced to a technical scientific meaning and the richer philosophical but less precise terms which represent the full reality in which we live. I stress that there is a different value to the modern, technical approach and the older, richer, common-sense tradition; they have different purposes and objectives.

Aristotle's philosophy explains nature and reality and does not seek to create a terminology and a technology for reducing that reality to itself. It is on this basis useful to compare Aristotle's expression of "*phora*" with Kant's reductionistic "phoronomy" as a desubstantiated kinematics. This contrast between full Greek concepts and abstract modern concepts is perhaps made most apparent by reviewing the Glossary to this work.

4.2 Aristotle's Theory of Physical Evolution

It is important to recognise that a central objective of Aristotle's entire philosophical project is to naturalise Plato's philosophy of "form" as a model for the representation of "being" into an explanation for the "becoming" of "beings." His magisterial *Generation of Animals* explains this process of becoming or reproduction in respect to animals and his *Ethics* extends it to explain development or "becoming" in a purely human behavioural context. His fascinating treatise *On Coming-to-be and Passing-away* considers material transformation on a generic material basis. The transformation of elements and the development of beings is thoroughly accounted for in Aristotle's extant works. We need, however, to reconstruct Aristotle's general position on the evolution or emergence of species from his work as a whole.

The most explicit statement of the transformation and "evolution" or "emergence" of species is provided by Aristotle's theory of the *scala naturae*. The *scala naturae* can be understood as Aristotle's theory of evolution and it has no relation to the idea that there is a divine, fixed great chain of being with his own account being that: "Nature proceeds little by little from things lifeless to animal life in such a way that it is impossible to determine the exact line of demarcation, nor on which side thereof an intermediate form should lie. Thus, next after lifeless things in the upward scale comes the plant, and of plants one will differ from another as to its amount of apparent vitality; and, in a word, the whole genus of plants, whilst it is devoid of life as compared with an animal, is endowed with life as compared with other corporeal entities. Indeed, as we just remarked there is observed in plants a continuous scale (*synechēs*) towards the animal (*HA* VII 588b 4-12)." It is a shame that Aristotle's *On Plants* is not extant since it would be interesting to know how he would have explained his hypothesis that plants evolve their vitality towards animality.

Aristotle notes as regards this "scale" that the sponge is an intermediate form between plant and animal by commenting that "it has sensation of a sort (*HA* I 487b 9)" even if "the whole thing is similar to a lung (*HA* V 549a 8)." The biosemiotician Jesper Hoffmeyer expands upon this observation that: "the earliest known animals, the sponges, may have been around for more than 600 million years, but the evidence for determining this decisively is scarce. They possessed no nerve cells at all, and no body symmetry either…In the sponges, we have only faintly developed exosemiotic activity—understood as semiotic interaction between different organisms—but endosemiosis, i.e. semiotic interactions between the myriad of cells and tissues inside of an organism, is already complex and sophisticated. The main activity of sponges consists in pumping up water from the bottom and expelling it from the top of the organism (much like a chimney), and in capturing bacteria and other small nutrient particles from the internal flow of water. Water flow is controlled by various combinations of wholly or partially closing the organism's intake pores, and by varying the beat of the small whips (flagella) that are located in the cellular membranes of interior cells. Doing so requires a stringent synchronization of flagellar movements up along the "chimney" (because the whips must operate in synchrony), and since the sponge has no nerve cells, this synchronization must be assured through chemical semiosis, i.e., by the diffusion of signal molecules (as sign vehicles, or carriers of signs) through the body fluids."[10] Exactly how an animal begins with "chemical semiosis" and then transcends this is at the heart of our, and in a sense Aristotle's, philosophical investigation or *zētēsis*.

In the context of the sponge, let us add that the amoeba is another brilliant example for demonstrating an elemental principle. Richard Connell explains the principle that: "only at rest or in death does it round up" that: "upon its death an "amoeba" obeys the laws of physics in the sense that its shape after death is determined solely by surface tension forces; the ordinary flattened shape of the living amoeba has disappeared. From this it is plain that cell shape is a property determined by the function or operation of the living whole itself; and once the whole loses its proper functional character, it loses its characteristic shape also."[11] We find an expansion of and vindication of several important Aristotelian concepts from the above two case studies of elemental life. We see from the sponge that organisms begin with the basic functions of nutrition and reproduction along (in animals) with self-movement.

[10] Hoffmeyer, Jesper "Semiotic individuation and Ernst Cassirer's challenge" *Progress in Biophysics & Molecular Biology* (2015) p 611.

[11] Connell, Richard J. *Substance and Modern Science* (Houston, 1988) p 153. Gregory Bateson adds regarding the amoeba that: "An amoeba will, for a considerable period of time, move more when it is hungry. Its energy expenditure is an inverse function of energy input ("Pathologies of Epistemology" in *Steps to an Ecology of Mind* (New York, 1972) p 482)." Karl Lorenz notes that: "the amoeba does not devour all tiny objects but, with rare exceptions, only those which serve as food (*Behind the Mirror: A Search for a Natural History of Human Knowledge* (Methuen, 1977) p 53)" and more generally that: "Many years ago Curt Richter discovered that when rats were given the various nutriments they required in a number of individual dishes, even to the extent of breaking proteins down into their amino acids, they took from each dish just what they needed for a balanced diet (*Ibid.* p 91)."

The development of the organism's relationships with the outside world and its more sophisticated external shape comes later. The amoeba shows us the basic principle that the "form" is imposed upon the living whole in opposition to mere physics. We find that the organism—sponge or amoeba—actively selects from its environment and is hence fundamentally an active actor rather than merely the passive subject of a material environment. Aristotle expresses this fundamental principle of activity (*energeia*) as that: "food is acted upon by what is nourished by it, not the other way round (*De An.* II 416b1-2)." We do not just absorb food by osmosis but select it according to our needs.

Moving from the sponge to the human being, Aristotle argues that it is pointless to consider the *origin* of human beings rather than the *nature* of human beings on the basis that: "it is probable that primitive mankind, whether sprung from the earth or the survivors of some destructive cataclysm, were just like ordinary foolish people…so that it would be absurd for us to abide by their notions (*Pol.* II 1269a 5-8)." He argues further for an emergence of humanity on the basis that: "The whole land of the Egyptians, whom we take to be the most ancient of men, has evidently gradually come into existence and been produced by the river (*Mete.* I 352b 20-23)." He speculates further that his own polis civilisation has developed from simpler communities such as family and village (*Pol.* I. 1252b 27-34) and that the Greek tribes could be descended from first begetters such as Hellen (Hellenes) and Ion (Ionians) (see *Met.* Δ 1024a 33-34). Finally, he speaks morphologically about the origin of man by considering that human beings "dualise" with apes on the basis that: "The ape is, in form (*morphē*), intermediate (*epamphoterizein*) between the two, man and quadruped, and belongs to neither (*mēdeterōn*), or to both (*amphoterōn*), and consequently he has no tail, qua biped, and no buttocks, qua quadruped (*PA* IV 689b33-36)." Aristotle recognises that man is related to the ape morphologically. He understands how the "form" of man (and of the ape) is fitted to their lives. An organism's soul is fitted to its body which is in turn fitted to its environment. It is not, however, reducible to either or both.

Aristotle importantly recognises the question of origins but places little firm emphasis on it. He argues that there is no real meaning in the question of origins, but that the species is evidently contingent upon the natural process of generation and materialisation in the world and can come-to-be and pass-away as all other natural objects aside from the cosmic bodies (and the chemical elements) which are "eternal." Aristotle's core paradigm as regards being and becoming is that the human race will only exist if it is reproducing on the basis that: ""while the race of man lasts" means "while the generation of them goes on continuously (*hē genesis synechēs autōn*)" (*Met.* Δ 1024a 29-31)." He hypothesises further that: "If all animals ceased to exist, there would be no knowledge at all, though there might in that case, notwithstanding, be still many objects of knowledge…If the animal itself is destroyed, then perception is also destroyed. But perceptibles yet will remain, such as body, heat, sweetness and bitterness and everything else that is sensible (*Cat.* 7b 33–8a 6)." For Aristotle, species clearly both emerge and go extinct. It is hence strange and shocking that Aristotle's emergent position on species and biological being has routinely been conflated with the essentialist approach of Plato that he expressly sought to counter. Even Linnaeus' taxonomy has, apparently, been similarly misrepresented.[12]

Aristotle's theory of speciation is explicitly *not* that of a fixed hierarchy of being(s). It is, rather, a simple observation as regards the increasing complexity of function (*ergon*) as is clearly found in nature. Aristotle argues that the only function (*ergon*) of a plant is to reproduce itself and to take nutrition in order to grow and reproduce. It is on this basis, for simplicity, that most plants are hermaphrodite.[13] Moving on to consider animals he comments that they are necessarily more complex than plants since they must *find* their food (and their mates) and hence have a stomach, a "bag" which acts like an "earth" that they carry around; see *PA* II 650a 21-27. He suggests further that this complexity in respect to nutrition is mirrored in a complexity as regards reproduction such that: "of those animals that generate, one will emit semen into itself, one into another individual. The former is known as "female," the latter as "male"—though in some animals male and female are not found (*HA* I 489a 9-13)." His assumption is that sexual reproduction is somehow better because it enables greater complexity and consciousness and we will consider this hypothesis in more detail below.

We find as regards Aristotle's definition of a separate species is that it cannot interbreed with other species since it has its own period of gestation. The technical term he uses for this is "complete difference" (*teleia diaphora*) which is the principle that: "everything which is different (*to diapheron*) [in species] have not common matter and cannot be generated into or out of each other (*Met.* I 1054b 28-30)." As regards the emergence of "new" species, he explains these on the basis that: "there is a proverb that Libya ever bears something new. For it is thought that owing to the lack of rain animals meet at the water-holes and mate with each other even if not the same breed, and bear to term if their periods of gestation are the same and if their sizes are not far off each other's; they become mild to each other because of their need for drink (*HA* 606b VII 20-24)." Aristotle considers both hybrids and mutations in his *Generation of Animals* and he assumes that "new" species come about as a sort of blending between species so that a mule is between a horse and a donkey.[14] He observes as regards the mule, and generally, that such cross species are

[12] See Mary Pickard Winsor "Linnaeus' biology was not essentialist" *Annals of the Missouri Botanical Garden* (2006).

[13] Aristotle argues that: "A plant, in its essence (*ousia*), has no function (*ergon*) or activity (*praxis*) to perform other than the production of its seed (*spermatos genesis*); and since this is produced as the result of the union of male and female, nature has mixed the two and placed them together, so that in plants male and female are not separate (*GA* I 731a25-27)." Aristotle observes that animals are "like divided plants (*GA* I 731a 22)." See Rosamond Kent Sprague "Plants as Aristotelian Substances" *Illinois Classical Studies* (1991).

[14] Aristotle explains the situation of hybrids on the basis that: "something happens contrary to nature (*para physin*), e.g. the production of a mule by a horse. (And even these cases are similar; for that which would be found to be common to horse and ass, the genus next above them, has not received a name, but it would doubtless be both, in fact something like a mule). Obviously, therefore, it is quite unnecessary to set up a Form (*eidos kataskeuazein*) as a pattern (*hōs paradeigma*)...the begetter is adequate to the making of the product and to the causing of the form in

sometimes infertile owing, in his view, to their "periods" of gestation being different. He argues that these "periods" are significant for procreation and also that there is a need for "common matter." Aristotle clearly understands that "matter" is "energy" with each type of being having its own energy, and this is why he places such emphasis upon "food" (*trophē*). We are again returned to consider the meaning of a psychic or energic representation of being.

The principle we see here is that, for Aristotle, the evolution from plant to man is as much a spiritual evolution of consciousness—or, in Aristote's term, of "soul"—as it is a physical evolution of bodily form.[15] His explanation is emergent and naturalistic and does not depend upon any idea of divine creation. It is descriptive rather than dogmatic so that it observes that genesis occurs and explains how we can understand and classify this becoming of being without drawing upon any strong opinions about why this creation occurs.

4.3 Aristotle's Theory of Psychical Evolution

The concept of psychical evolution is perhaps the very crux of this work. Even the best anti-reductionist approaches to biology often fail to fully break away from the materialistic and mechanistic limitations or errors of our past thinking. The basic assumption remains that we must somehow work upwards from the material, physical, historical, and bodily to then understand the phenomenal, mental, and spiritual as epiphenomena. Aristotle denies this through his principle that "the

---

the matter (*ikanon to gennōn poiēsai kai tou eidous aition einai en tē hylē*) (*Met.* Z 1033b 33–1034a 5)." Aristotle does not believe that forms are immutable but argues that they can be generated by worldly activity. He does not explicitly explain how man comes from monkey (and quadraped) or ostrich from bird (and quadraped) but he clearly argues that we see an upwards evolution in nature which he calls his *scala naturae*.

Regarding "unnatural occurrences" Aristotle provides a list of known examples including "chicks that are monstrosities (*GA* IV 770a19)", "a snake seen with two heads (*GA* IV 770a24)", "animals…born with too many toes (*GA* IV 770b 31)", "a goat being born with a horn on its leg (*GA* IV 770b 36)", "organs in the wrong places; the liver on the left side and the spleen on the right (*GA* IV 771a 10)", and a hybrid vine called the "smoky vine" (*GA* IV 770b 20). By reference to the "smoky" vine Aristotle explains genetic diversity within a species or form on the basis that: "the alteration is quite small (*ou porrōthen hē metabasis*) and not really contrary to nature (*para physin*), because it is not an alteration to a different nature (*ou gar eis allēn physin*) (*GA* IV 770b 23-24)." He argues regarding "monstrosities" that they result from the process not being able to complete its formal "program" on the basis that: "the "formal" nature has not gained control over the "material" nature (*hotan mē kratēsē tēn kata tēn hylēn ē kata to eidos physis*) (*GA* IV 770b 16-17)." Aristotle also makes some fascinating comments on human genetics, noting that characteristics such as skin colour can skip generations at *GA* I 722a 5-6 10-11.

[15] Aristotle argues that: "We must…inquire in each several case, what is the soul of each individual, for instance of the plant, the man, and the beast. But we must also consider why they are thus arranged in a series (*ephexēs*). For without the nutritive faculty the sensitive does not exist, but in plants the nutritive is divorced from the sensitive faculty…And lastly, and most rarely, living creatures have a power of reasoning and thought (*logismon kai dianoian*) (*De An.* II 414b 33–415a 8)."

whole is more than the sum of the parts" meaning that there is always something meaningful at every level beyond the material, tangible bits of any given whole or organism. This is *not*, however, to say that we should add in to our picture of the world an extra holistic or spiritual bit, as with vitalism. Aristotle's principle is more revolutionary than this. He assumes that the physical follows the spiritual, functional, and meaningful rather than being epiphenomenal upon it as a chance or contingent phenomenon. The psychical is not emergent from the physical but is realised through the physical. This is Aristotle's chicken-and-egg problem which is articulated most clearly in his *Categories*.

Aristotle argues that our "sublunary" world is observably meaning-making because it is mediated and observably passes through a creative "middle." This is the basis for his logic. There is a rich potentiality in nature which is actualised in various pathways. Nature is like a potter which shapes its products with the material it finds before it. These products cannot be explained as automatic chemical interactions such as that sodium and chlorine can combine to make salt. Aristotle hence argues that: "Empedocles was not right when he said that animals have many of the characteristics they do because "it just happened that way" (*dia to symbēnai outōs*) in the process of coming to be (*en tē genesei*)—e.g. even the spine, he says, is such as it is because "it happened to get twisted and thus break". He was unaware, first, that the productive seed (*sperma*) must have a specific potential (*toiautēn dynamin*) and, secondly, that the producing agent is prior, not only in [principle] (*tō logō*) but also in time (*tō chronō*)—for a human begets a human (*genna gar ho anthrōpos anthrōpou*), and therefore it is on account of the parent being of such a sort that the child's coming to be (*genesis*) is such as it is (*PA* I 640a20-28)." He explains this principle further as that: "man builds because he is a builder, and a builder builds in virtue of his art of building. This last cause then is prior: and so generally (*touto toinun proteron to aition, kai outōs epi pantōn*) (*Phys.* II 195b23-25)." The common thread here is that there is a processual chain of function in nature which can only be understood on the basis of a prior organic "whole" which projects its nature into the material world. This whole is not a form as Plato envisaged but an actualised functionality which manifests as the building in the builder, as Aristotle's envisages.

There is, then, an observable creative process in nature which both reproduces and actualises meaning. This reproduction or actualisation is not just automatic (*automaton*) but becomes as an expressed response—as a constructed meaning—towards the world. As regards how the objects of nature are created or evolve or manifest, Aristotle observes regarding "science" that: "one science treats of one class of things, in which complete difference (*teleia diaphora*) is the greatest (*Met.* I 1055a 33-34)." He recognises that "science" is the "special" science of a given object with it being an essential foundation of his metaphysics that science regards the "parts" whereas philosophy regards the "whole." He explains as regards our knowledge of the objects of the world that: "We gain knowledge, commonly speaking, of things that already exist, for in very few cases or none can our knowledge have come into being along with its own proper object. Should the object of knowledge be removed, then the knowledge itself will be cancelled (*Cat.* 7b 24-29)." Knowledge and meaning is dependent upon being which becomes moot if that being is extinguished. That being was, however, a vehicle for its expressed meaning. Beings can be extinguished but they can only come about as an act of sustained transcendent novelty.

As regards "art" Aristotle adds the important parallel that: "after going through many changes tragedy ceased to evolve, since it had achieved its own nature (*epausato, epei esche tēn autēs physin*) (*Poet.* 1449a 13-15)." He hypothesises that "tragedy" achieved *its own nature or completion* through exposure to the world and he explains its development or evolution further as that: "if there had been no Timotheus we should have been without much of our lyric poetry; but if there had been no Phrynis there would have been no Timotheus. The same holds good of those who have expressed views about the truth (*Met.* A 993b 15-17)." He argues here that philosophy also evolves so as to reach a certain completeness on the basis that it is a natural expectation or latent possibility which emerges from humanity. We manifest in ourselves the "learning" that we find in nature itself as an example of, to use a Bergsonian phrase, its "creative evolution." There is a historical development of the various objects of being but this history serves not to achieve something contingent on its inventors or discoverers but realises "its own nature" through them.

Aristotle's conceptualisation of "emergence" is that: "a higher degree of sensation is a property (*idion*) of a higher degree of living thing, a lower degree of sensation would be a property of a lower degree of living thing (*Top.* V 137b 24-26)." He argues in more detail that: "all animals…have an innate faculty of discrimination, which we call sense-perception (*echei gar dynamin symphyton kritikēn, hēn kalousin aisthēsin*). All animals have it, but in some the perception persists, while in others it does not. Where it does not, there is either no cognition at all outside the act of perception, or no cognition of *those* objects of which the perception does not persist. Where perception does persist, after the act of perception is over the percipients can still retain the perception in the soul. If this happens repeatedly, a distinction immediately arises between these animals which derive a coherent impression from the persistence and those which do not (*Post. An.* II 99b 35–100a 4)." It is somehow through sense and mind that our sublunary world or biosphere comes into being.

Aristotle assumes that there are primal cosmic powers of desire and reason that are progressively transformed as they ingress into beings on the basis that: "those [animals] that have more understanding and possess some memory continue the association, and have a more social relationship (*politikōteron*) with their offspring (*HA* VII 589a 1-3)." Physical formation follows this spiritual development in consciousness, a representation of psychical evolution represented in Fig. 4.4 below.

Figure 4.4 shows us that we can move from Aristotle's *Metaphysics* in which he posits the metaphysical need for a possibility of mind and desire to exist in the world, through his *On the Soul* which considers the growing articulation of life in its movement and expression, to his *Ethics* which concludes in accordance with this system by considering "mind" (*nous*) and "friendship" (*philia*).[16] Friendship is the

[16] This structuring of the soul between noetic and orectic sides is explained by Aristotle most thoroughly and convincingly in Chapter VI of his *Nicomachean Ethics*.

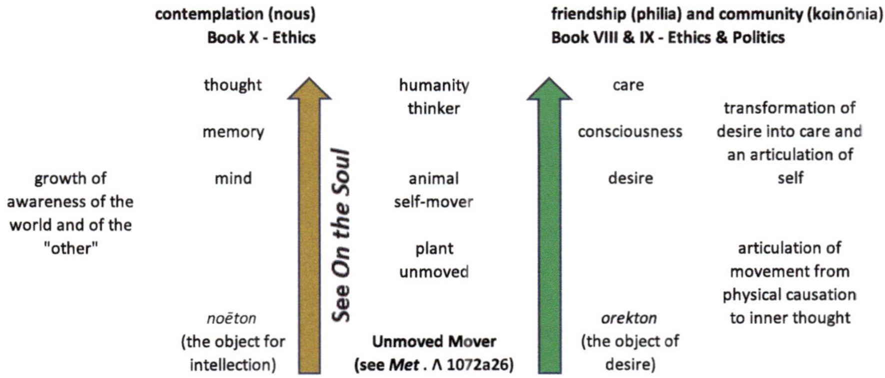

Fig. 4.4 Aristotle's theory of evolution

developed form of dispassionate care as contemplation is the developed form of dispassionate thought. The development of desire into care and friendship ultimately forms the basis of human civilisation and culture as expressed by the political friendship or "*koinōnia*" explained by Aristotle in his *Politics*. For Aristotle, everything exists within the same structure of being and has analogues and pathways across domains.

It is important to stress that Aristotle looks at the world in the full and includes everything in his picture. This holism does not dilute his perspective but makes it full and complete. It is by including plants, chemicals, man, and "God" in the same picture that he can see the commonality between them and the richness of the whole. He can also see that nature cannot be reduced to the play of elements or laws but must fundamentally be understood as "beings" moving through the world, developing, creating, and reproducing. Aristotle's bio-logical picture includes not only what we ringfence as "biology" but also the whole biosphere and semiosphere in which all culture, meaning, and organic interaction takes place. It is this wider philosophical picture that Aristotle brings to the table and which biosemiotics also partially captures but perhaps falls short in coherently and completely philosophically expressing. By bringing Aristotle into the picture we can hopefully develop biosemiotics by helping it fully cast off the reductionistic scientific tradition and to become a true reflection of physical reality as it is expressed in reality as we encounter it in the world in which we live.

Aristotle's theory of evolution is primarily that of a psychic development through a striving for the good passing through generations of beings and creating an uplift of life towards what Aristotle calls "*nous*." As regards this "*nous*," Aristotle explains its principle in his *Metaphysics* Book Λ as that: "the actuality of *nous* is life (*hē gar nou energeia zōē*) (*Met*. Λ 1072b27)" and the process in his *Generation of Animals* as that: "Of the things which are, some are eternal and divine, others admit alike of being and not-being, and the beautiful and the divine acts always, in virtue of its own nature, as a cause which produces that which is *better* in the things which admit of it; while that which is not eternal admits of being <and not being>, and of acquiring a share both in the better and in the worse; also, soul is better than body, and a thing which has soul in it is better than one which has not, in virtue of that soul; and being is better than not-being, and living than not living. These are causes on account of which generation of animals takes place, because since the nature of a class of this sort is unable to be eternal, that which comes into being is eternal in the manner that is open to it (*GA* II 731b 24-34)." The principle that "that which comes into being is eternal in the manner that is open to it" is essential to Aristotle's picturing of the world.

Aristotle explains evolution as the reaching out for meaning and completeness (and eternity) by beings in the world so that "mind" is the "aim of all things, or of all things that possess sensation or reason; or would be, if they could acquire the latter (*ei laboi noun*) (*Rhet.* I 1362a 23-24)." This direction of morphology is manifested in worldly beings both psychically and physically. We could say that Aristotle reintroduces the baggage of values, ideals, goods, and God but this is not a deficiency but a strength since these values do indeed emerge in our world as real aspects of it. For a modern framing of this account of evolution see Theodosius Dobzhansky's essay "Two Contrasting World Views" (1974).

4.4 Aristotle's Developmental Biology

One of the most challenging tasks we face is explaining the value of Aristotle's biology. We perhaps could not expect the *first* biologist to still have something to teach us. In Aristotle's case we can. This is in part because his philosophy, properly understood, is a philosophy of biology. It is a philosophy of "being" (*ousia*). The quality of Aristotle's philosophy can fill a gap in a modern biology that has for various historical reasons eschewed philosophy. It is, however, also important for us to consider how Aristotle presents biology within his philosophy and here we find him amazingly accurate and insightful within this context. This is also a problem faced by biosemiotics. We can easily understand why the biological world should be represented as "signs" which express and exchange meanings between beings but perhaps harder to see how biology can be reformulated on this value-based semiotic basis.

Aristotle's theory of evolution is perhaps best understood as a principle of emergence or of manifestation. It is a niche theory which assumes that the nature of the world is like a forest which will expand and fill itself as an ecosystem if nothing hinders it. He uses the term "*diakosmēsis* (*GA* II 740a 8)" to signify the natural development of a being once an initial seed has been fertilised or become active. He argues that there is a fertility or pregnancy in nature such that: "if anything was on the point of being done, it most probably was done; for it is likely that the one who was on the point of doing something has carried it out…if the sky is cloudy, it will probably rain. And if one thing has been done with a view to another, it is probable that the latter will also be done; for instance, if a foundation has been laid, a house will probably be built (*Rhet.* II 1392b 24–1393a 8)."[17] This is the natural causation from which the final cause or purpose develops as it is expressed in humanity. There is an unconscious propensity for "movement" and "end" in nature, i.e., "nature flees (*pheugei*) the unlimited (*apeiron*), for the unlimited is endless [or incomplete] (*atelēs*), but nature always seeks (*zētei*) an end (*telos*) (*GA* I 715b14-16)," which is ultimately consciously deployed by human beings and other animals in their lives. I note that for Aristotle the human example is special only on the basis that: "The traces of these differentiated characteristics may be found in all animals, so to speak, but they are especially visible where character is the more developed and most of all in man (*HA* VII 608b 5-6)."[18]

Aristotle explains as regards the qualities of human beings that: "in touch he [i.e., a human being] is much more discriminating than the other animals. That is why he is of all living creatures the most intelligent. Proof of this lies in the fact that among the human race men are well or poorly endowed with intelligence in proportion to their sense of touch, and no other sense; for men of hard skin and flesh are poorly, and men of soft flesh well-endowed with intelligence (*De An.* II 421a 22-27)." Aristotle argues that we see even within humanity itself that an organism's increased sensitivity to the world represents an "evolution" of mind and being. As regards the physical aspect of animal being he argues that: "Those [animals] which are endowed with most intelligence show intimacy and attachment towards their offspring even after they have reached their perfect development (human beings and some of the quadrupeds are examples of this); birds show it until they have produced their chicks and brought them up; and on this account hen birds which have laid eggs but omit to sit on them, deteriorate in their condition, as though they were being deprived of one of their natural endowments (*GA* III 753a 12-16)." He argues that nature is expressed through the physical and mental (well-)being of an animal so that if animals are not mentally or spiritually expressing their nature, they will physically deteriorate.

Aristotle argues that the physical structure and behaviour of an animal is based upon their "intelligence"—"semiotic abilities" in biosemiotic terms—and he gives the further evidence elsewhere that "the more perfect and hotter (*teleōtera kai thermotera*) of the animals" produce live births whereas less perfect and colder animals exhibit other "inferior" strategies such as laying eggs, larvae etc. They also produce

[17] Aristotle also comments that: "action follows unless there is some hindrance or compulsion…I ought to create a good, and a house is good, I immediately create a house. Again, I need a covering, and a cloak is a covering, I need a cloak. I ought to make a cloak. And the conclusion "I ought to make a cloak" is an action. The action results from the beginning of the train of thought. If there is to be a cloak, such and such a thing is necessary, if this thing then something else; and one immediately acts accordingly. That the action is the conclusion is quite clear; but the premises which lead to the doing of something are of two kinds, through the good and through the possible (*MA* 701a 15-25)."

[18] Aristotle adds that: "The fact is, the nature of man is the most rounded off and complete, and consequently in man the qualities or capacities referred to above are found in their perfection (*HA* VII 608b 7-8)."

multiple offspring since they invest less energy and attention into each.[19] I represent this situation as regards biological investment as being an expression of development and evolution in Fig. 4.5 below.

We can usefully make some observations as regards Aristotle's representation of "*hylē*" (matter) here, the word originally meaning "wood." Aristotle's "matter" is itself the "subject" of a "being" on the basis that: "what I mean by matter is precisely the ultimate underlying subject (*to prōton hypokeimenon*), common to all the things of nature, presupposed as their substantive, not incidental, constituent (*ex ou ginetai ti enyparchontos mē kata symbekēkos*) (*Phys*. I 192a 32-33)." He adds that: "the matter (*hylē*) is nearly, in a sense is, substance (*ousia*) (*Phys*. I 192a 6-7)." This leads us to Aristotle's concept of "proper matter" which is such that: "the entelechy of each thing is naturally inherent in its potentiality, that is in its own proper matter (*hekastou gar hē entelecheia en tō dynamei hyparchonti kai tē oikeia hylē pephyken enginesthai*) (*De An*. II 414a 26-27)." Matter is not, as Plato and then Descartes suggested, just the inert filling for a being but is the *epigenetic energy* of a species so that: "What is the matter? Not fire or earth, but the matter proper to man (*tis hē hylē; mē pyr ē gēn alla tēn idion*) (*Met*. H 1044b3)" and that: "Fire comes-to-be through the agency of Fire [i.e. fire spreads] and Man through that of Man [i.e man reproduces] (*GC* I 320b 21)."[20] A species manifests when it is a separated flow of "beings" through the channelling of its own "proper matter," this itself being derived from the channelling of the food source (*trophē*) it takes from the world.[21]

---

[19] Aristotle argues that plants and some animals follow a mass reproductive strategy with a "single movement" such that: "even among the animals there are some which generate numerous offspring from one act of coition, a phenomenon which is, indeed, universal with plants; these, as is manifest, produce a whole season's fruit as the result of one single movement (*GA* I 723b 9-12)" with the paradigm of this situation being fish regarding which Aristotle comments that: "Some creatures' eggs are imperfect when laid—e.g. that of the fishes, which become perfected, i.e., grow, outside the creature which produces them. The reason is that these animals are very prolific and this is their function, as it is that of plants (*GA* I 718b 7-10)" (and on larvae see *GA* III 758b 20-22). Aristotle notes further on this situation that: "monstrosities…occur very seldom in animals that produce one offspring only, occur oftener in those that are prolific (*GA* IV 770a 8-9)" and he provides vast further information on these strategies in his fascinating *Generation of Animals*. On biological investment see Jorge Torres "*Philia*: the biological foundations of Aristotle's ethics", History and Philosophy of the Life Sciences 43 (2021).

[20] Aristotle's assertion that the "matter" of a certain "being" is exclusive and "not found apart from the thing itself whose materials they are (*PA* I 645a 37)" is explained by Abraham Bos as that: "A-body-which-receives-soul that has the potentiality only to be a biotic instrument will never produce anything else than a plant or a tree. A-body-which-receives-soul that has a potentiality to be a nutritive and sensitive soul-instrument will in due course manifest animal life ("Why the Soul Needs an Instrumental Body According to Aristotle ([*De An*.] I. 3, 407b13-26)" *Hermes* (2000) p 30)."

[21] Aristotle's own summary is that: "the soul-principle in question is a power of preserving what possesses it as an individual, while food prepares it for work (*hē de trophē paraskeuazei energein*). For this reason it cannot continue to exist while deprived of food…Now all food requires digestion, and that which produces digestion is heat; therefore everything which has a soul has heat (*De An*. II 416b 27-30)." The soul *preserves* the being through the processes of its body, its heat, and its material inputs as food.

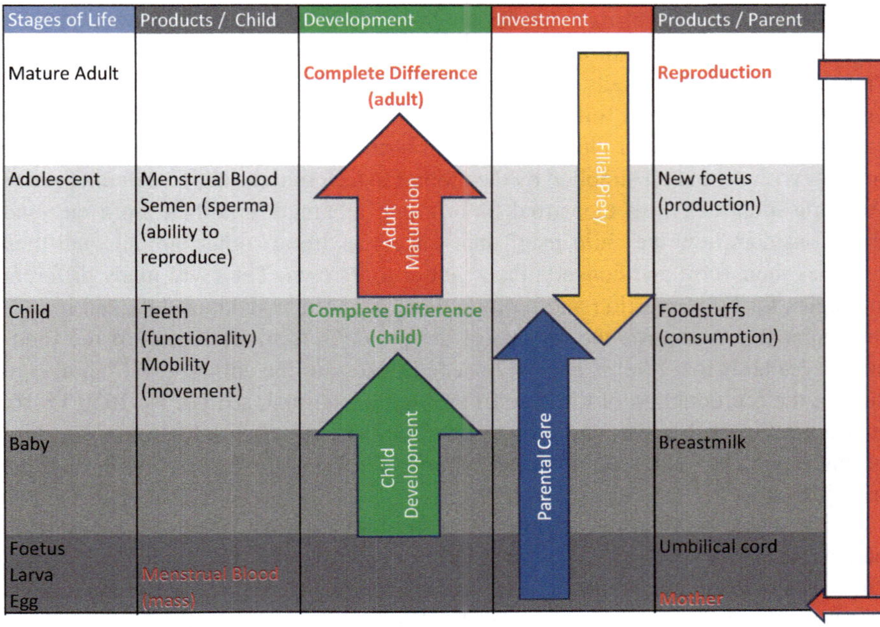

Notes on diagram

Aristotle explains "energy" by reference to the investment that the parent makes in the child (and vice versa). He explains that: "parents love their children as being a part of themselves, and children their parents as being something originating from them (*N.E.* VIII 1161b 18-19)" and that: even animals have the philoprogenitive instinct, which urges them to produce offspring and also to protect the offspring produced. And in fact fathers love their children more than they are loved by them...and these in turn love their children more than their parents, because activity is the greatest good (*dia to tēn energeian einai to ariston*). And mothers love their children more than fathers, because they think that the children are more their work; for people estimate work by its difficulty, and in the production of a child the mother suffers more pain (*E.E.* I 1241a40– 1241b9)"

Aristotle observes regarding the nature of "matter" that any "*ousia*" dependds upon a constant or continuous flow of matter as "food" which will be transformed into the being growing to its maturity and then producing another member of the species. Matter is not for Aristotle something mundane and inert but is a specific energy which is able to take on the various forms of the organism in the world. Matter must be supplied continuously to the organism or it dies. Aristotle meticulously documents this lifeline with one example being the human breastmilk must be "serviceable" as soon as the foetus is not supplied with nutrients by the umbilical cord (see *GA* IV 777a 23-27). On this cut-over of feeding sources he notes that: "In the case of other animals, as there is but a single period of gestation, the concoction of the milk coincides with that; in man, however, as there are more periods than one, the milk must of necessity be available at the earliest of the possible dates (*GA* IV 776a 20-24)."

Fig. 4.5 Aristotle's theory of biological investment

Aristotle also recognises in his *On the Soul* that nutrition and reproduction are a part of the same process of transferring matter between organic beings through the "nutritive soul." He literally tracks matter through this nutritive-reproductive process as it transforms one form or being to become another either through nutrition (as food) or through reproduction (through birth). He locates the beginning of the process with the matter supplied by the mother to the child through the ovum (which Aristotle identifies with menstrual blood), egg albumen, umbilical cord etc. and then considers how the child itself then selects as "food" other animals and then removes their form and converts their matter to its own. There are many different strategies for both nutrition and reproduction. Aristotle adds the subtle and important point that excreta is the residue of the previous form and being of the foodstuff.[22] He adds that food is not only needed to sustain the animal itself but also to sustain the reproduction of the species through the animal; see *GA* IV 765b 15-16. It is hence that both matter and form cycle through beings. Reproduction is a cycling of matter, i.e., as food, sustaining the cycling of a form—or "species"—through its individuals.

What, then, is Aristotle's theory of evolution? Aristotle observes that organic beings are subject to change, are fitted to their environment, and are not "ideal." As "fire" is an element of nature, species and their beings are also emergent elements of nature. All this is evident in fact. Where Aristotle differs from our modern theory is on the basis that he does not believe that life is explicable as a chance material event to be only understood through history. It is of the nature of nature to produce living beings on the basis that: "the actuality of *nous* is life (*hē gar nou energeia zōē*) (*Met.* Λ 1072b27)." Chance and history play roles in the development or coming-to-be of being in the sense that Greek tragedy apparently happened to come to its mature form through Timotheos and Phrynis. Their input to the genre served, however, to help the genre achieve "its own nature" or, in other words, its completion. History, chance, and environment play a role in Aristotle's envisioning of the world but is not reducible to them.

[22] As regards nutrition Aristotle observes that: "As carnivores, the snakes suck dry whatever animal they take and eject them whole with their excrement. It is much the same with the other animals of similar behaviour, for example the spiders; but the spiders suck the juices outside, while the snakes do it within their stomach (*HA* VII 594a12-16)" (and I note that spiders do liquefy their food before ingesting it and that Aristotle discusses other blood-sucking and sap-sucking insects in more detail at *HA* VII 596b). One animal extracts the matter from the body of its food whereas others consume the whole being and expel the old "form" in their excreta with the philosophical principle here being that: "if a waste product must be formed, and a waste product is a contrary; for all change proceeds from a contrary, and the waste product is what remains of the former state (*Long.* 465b17-19)."

4.5 Aristotle and Biosemiotics

Aristotle does, then, have a theory of "evolution" in the sense that his account of the origins of "being" is both developmental and emergent. We cannot, however, claim that he accepts the modern theory of evolution since he would clearly find it insufficient to account for the complexity and variety of life as the product of the chance permutations of agglomerated matter. Instead, his own viewpoint is "structuralist" in the sense that he believes that there are structures in nature which have the possibility of emergence, of what Whitehead calls "ingression."[23] It is also "phenomenological" in the sense that real meaning is conveyed through phenomena. It also suggests the emergent power of "negentropy" through which beings are a defiance of entropy, a positive biological emergence running against physical laws.[24] We also discern a sympathy between Aristotle's philosophy and the "biosemiotics" which finds meaning in *all* worldly being, including the biological. There is, however, some differences between these approaches with Aristotle's, as explained in the previous chapter. We find, for instance, that biosemiotics has developed from out of the Kantian position that meaning is a subjective rather than objective phenomenon which is a destructive error from an Aristotelian perspective.

As regards the above relationships, let us consider, first, that Aristotle is the real founder of semiotics who considers signs (*sēmeia*) and signification directly in his *On Interpretation* and his *Rhetoric*. He argues: "Suppose it were said, "The fact that Socrates was wise and just is a sign (*sēmeion*) that the wise are just." Here we certainly have a sign; but even though the proposition be true, the argument is refutable, since it does not form a syllogism. Suppose, on the other hand, it were said, "The fact that he has a fever is a sign that he is ill," or, "The fact that she is giving milk is a sign that she has lately borne a child." Here we have the infallible kind of sign (*tekmērion*), the only kind that constitutes a complete proof, since it is the only kind that, if the particular statement is true, is irrefutable (*alyton*). The other kind of sign, that which bears to the proposition it supports the relation of universal to particular, might be illustrated by saying, "The fact that he breathes fast is a sign that he has a fever." This argument also is refutable, even if the statement about the fast breathing be true, since a man may breathe hard without having a fever (*Rhet.* I 1357b 12-21)." Aristotle expands upon his "signs" at *Prior Analytics* II 70a where he asks: how can we form facts or data into patterns (*paradeigmata*) which can then be made into logical rules?

Although Aristotle rightly focuses on the example of sickness and what we today call aetiology (from *aitia*, cause) and pathology (from *pathos*, affection) to explain

---

[23] Sydney Hooper explains "ingression" as that: "A colour is something that transcends physical events. It "ingresses" into nature to lend its quality to any event that may require it for a period, but when its function for the time being is over, it disappears, to return when its presence is again relevant. But, when it returns, it is the same colour ("Whitehead's Philosophy: The World as Process" *Philosophy* (1948) p 150)."

[24] See Erwin Schrödinger *What is Life? The Physical Aspect of the Living Cell* (Cambridge, [orig. 1944] 1992).

his position on "signs," he is not just interested in diagnosing sickness. His aim is rather to consider signification and meaning in general. He hence also considers wisdom and child rearing as healthy aspects of life which can be signified or pathologized. His semiotics is also biological and cultural, i.e., it is biosemiotic. His logic follows from his semiotics in the sense that it is primarily empirical, based upon how human beings extract meaning from the living world and then "tie it down." His further scientific studies from his *Physics* to his *Poetics* are concerned with how we approach the various objects of the world, take meaning from them, and then systematise that meaning as human knowledge. The fact that his whole philosophical programme as regards how we interpret the world is based upon biology counters the stubborn idea that meaning is merely a mental phenomenon—a common assumption within semiotics—and insists instead that meaning emerges from the biological world and is only then reflected in and represented by the mind and our culture. Meaning is on this basis a real part of reality and not a human illusion. By contrast, both the existentialist philosophical vision and the neo-Darwinian scientific approach present us only with a random world in which any meaning only emerges through happenstance and struggle.[25]

Both Aristotle and biosemiotics present us with a positive and meaningful world rather than the negative and meaningless world of neo-Darwinism. Both build upon structuralism but do not reduce the world to its structures. They recognise phenomena but do not reduce the world to these mental or cultural representations. We can only appreciate the full picture and its parts by including the whole world—the biosphere, semiosphere, and noosphere—in a single picture. By expanding the picture, the parts come into view as beings with inner movement that externally interact with other beings in the world. To consider this situation let us represent Aristotle and biosemiotics in Fig. 4.6 below.

The breakdown of the world into generic zones—noosphere, semiosphere, biosphere—is important. It shows us that the world is complex and that it is one of complexity within complexity. These complexities are also interlinked and nested. At the same time, they are distinct and separate. The noosphere—the home of ideas, concepts, poetry, and mathematics—is very different than the biosphere—the home of zebras, storks, and cockroaches. The principle stated here is that we must recognise that the world cannot be homogenised as a "world-animal," but must still be appreciated as a single world or "*kosmos*" which possesses a common framing such that we can have an idea (noosphere) of biology (biosphere) and express both through our culture and behaviour (semiosphere).[26] We see here the vitality of the very Aristotelian idea that we can meaningfully understand the world through an analogy between beings and their attributes without subsuming the distinct

[25] Iris Murdoch explains this situation well in her philosophy; see her *The Sovereignty of Good* (London, 1970).

[26] On the world-animal see Dimitri El Murr "Platonic 'Desmology' and the Body of the World Animal (*Tim.*30c-34a)" in Salles, Ricardo (*ed.*) *Cosmology and Biology in Ancient Philosophy: From Thales to Avicenna* (Cambridge, 2021).

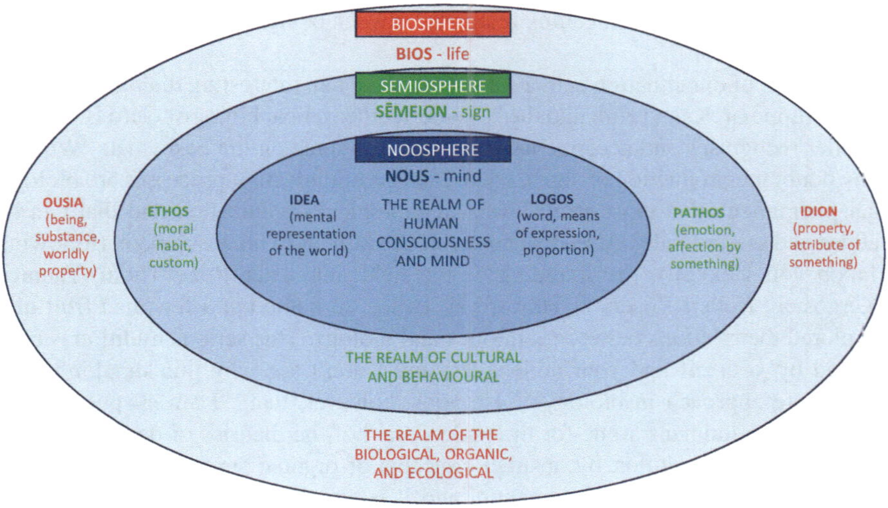

Fig. 4.6 An Aristotelian take on Biosemiotics

signification of beings into a blanket law or abstraction.[27] We see the importance of recognising the centrality of "being" (*ousia*) as the prime actor as regards how we represent the world.

The richness of Aristotle's philosophical thought is of immense importance to this argument as an expression in itself of his anti-reductionism. For consideration of this point see Appendix A which expands upon the terms used in Fig. 4.6 above. Heidegger makes the very interesting point—and explains it in brilliant detail in his work—that the Greek language contains a philosophical pregnancy not similarly enjoyed by the Latin, which has more of a legalistic expression. I would add that the use of English as a *lingua franca* represents a further catastrophic descent into nominalism, into seeing the world through Popper's reduction of it to "bold ideas" and "handy shorthand labels."

4.6 The Meaning of Semiotics

In discussing biosemiotics, it makes sense to go back to its origin as an expansion of semiotics. How can we expand a discipline founded by Ferdinand de Saussure to study "the life of signs within society" and by Charles Peirce to consider signs defined as "something which stands to somebody for something" to make it a new approach for biology? We can surely see that we can, and I would say should,

[27] Hegel makes this point very well in his "The Philosophy of Aristotle" *The Journal of Speculative Philosophy* ([*orig*. 1825–6] 1871).

understand our world as meaning relations between beings, but how can we rethink it through "signs"?

The nub of our situation is in a sense exposed by an interesting dialogue between the semiotician Kalevi Kull and the "evo-devo" theoretical biologist Gerd B. Müller. Müller recognises and accepts the importance of signs on the basis that: "Without any doubt the production of signs, codes, and communication processes are biological phenomena that exist at all levels of biological organization and that can be addressed scientifically. Already Thure von Uexküll and Thomas Sebeok (following Jakob von Uexküll), the founders of the field, and later René Thom, Yannick Kergosien, Robert Rosen, or Howard H. Pattee, to name but a few, had fruitfully explored the relations between semiotics and biology. This serious tradition is continued by yourself and your colleagues, and I don't see why this shouldn't be a legitimate approach in biology."[28] He adds, however, that: "I am skeptical when some biosemioticians argue for the replacement of the theories of development or the theories of evolution by abstract concepts of organic signs and codes. These conceptions usually remain semantic, and it is simply not possible to derive the same kind of testable predictions from them as we do from mechanistically grounded theories. Therefore I can see value in biosemiotics as a particular approach to the biological phenomena of communication, and thus to theoretical biology, but, personally, I do not perceive it as an alternative to natural sciences based theories."[29]

The tension we discern above is in respect to how comfortable we are with challenging or stretching the limits of science and its method. In an essential book of his *Metaphysics*, Aristotle argued that we should accept that a science will necessarily have circumscribed aims and method in accordance with its subject matter and hence that the scientist will "use them [i.e., philosophical concepts] just so far as to satisfy their purposes; that is, so far as the genus to which their demonstrations refer extends (*Met.* Γ 1005a 25-27)." He famously argues in the same book that it is philosophy that studies the "whole," what it means to be a "being" (*ousia*) and what it means to be "one" among "many" or one "whole" among a greater "whole." The sciences are on this basis a different form of investigation with different aims and methods than philosophy. Aristotle would hence say that the scientific study of living beings is very different than the philosophical study of life. We should keep these two approaches to the world separate and value them both in their own spheres of operation. This is the tension we need to address below.

As regards the task immediately before us, let us consider what biosemiotics takes from Peircian semiotics. Donald Favareau explains Peirce's interesting conceptualisation of life expressed through semiotics through his categories of firstness, secondness, and thirdness as per Fig. 4.7 below.

This is what is now known as the "semiotic scaffolding" of reality which Favareau explains as that: ""Reality" writes Peirce, is itself "an affair of Thirdness as Thirdness, that is, in its mediation between [the open possibilities of] Firstness and

[28] Kull, Kalevi "Interview with Gerd B. Müller on Theoretical Biology" *Biosemiotics* (2023).

[29] Kull, Kalevi *Ibid.*

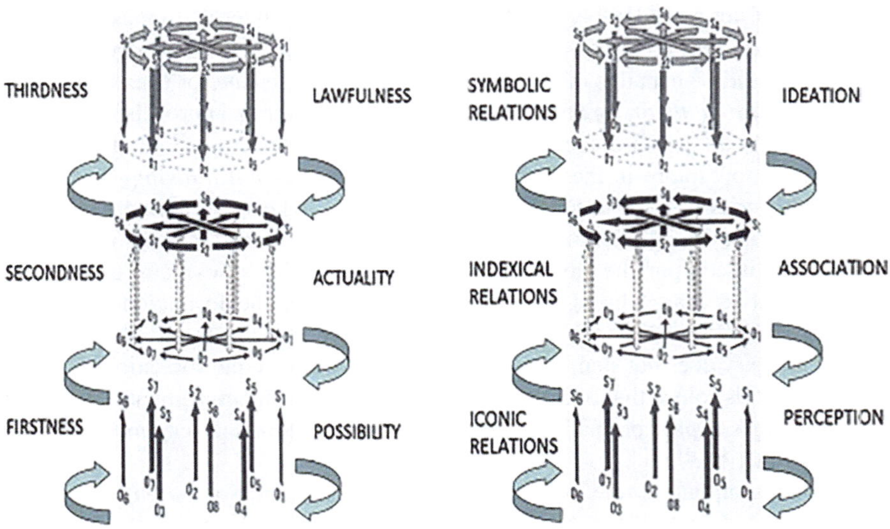

Fig. 4.7 Two representations of Semiotic Scaffolding. (From Favareau "How the Referential Generality of Symbol Scaffolding Grows Minds" (2015) p 241 [modified versions from one appearing in Terrence Deacon's The Symbolic Species (1997) p 79])

the [actualized current brute state of affairs of] Secondness" (EP 2:197, 1903)."[30] He explains this situation further as "an emergent higher-order organizing dynamic (i.e., an emergent thirdness) that allows things to systematically stand for something other than themselves, as a necessary condition for the continuation of the system"[31] and he argues that: "The appearance upon the Earth of such relatively autonomous (in the original sense of "self" (auto) "rule" (nomos) making) systems constitutes a new way of carving out possibility space, as the subset of reactions and relations that is the organism is, at least to some degree, "shielded" from the ongoing larger set of reactions and relations external to it, and is thus free to selectively order and re-order such reactions and relations of secondness under the thirdness of its own emergent organizing dynamic."[32] The inner complexification of the organism gives it an increasing outer self-autonomy.

We could say that this account of semiotic scaffolding bears some comparison with Piaget's account of child development. Josef Perner describes this development as that: "at about 1.5 years infants start to recognise themselves in the mirror. In other words, they become able to *interpret* the mirror image,"[33] "from about 2

[30] Favareau, Donald "Symbols are Grounded not in Things, but in Scaffolded Relations and their Semiotic Constraints (Or How the Referential Generality of Symbol Scaffolding Grows Minds)" *Biosemiotics* (2015) p 242.

[31] Favareau, Donald *Ibid.* p 240.

[32] Favareau, Donald *Ibid.* p 244.

[33] Perner, Josef *Understanding of the Representational Mind* (Cambridge Mass., 1991) p 238.

years on—if not considerably earlier—children use mental terms to make sense of human action. They are—one could say—working on a *mentalistic theory of behaviour*. At around 4 years this original understanding changes into or is extended into a *representational theory of mind*"[34] and, finally: "As children approach school age, then, not only do they become more "logical" in their reasoning about facts—what Piaget tried to capture in the notion of *concrete operational thinking*—but, more importantly, they become rational about their desires."[35] This development of meaning in the being through its nature is paradigmatically explained by Aristotle and his philosophy based upon the model of the foetus. Piaget himself explains this natural development as that: "through developmental study of the formation of mental imagery and its role in thought processes [we] have come to the conclusion that mental imagery does not stem from perception, but from an interiorisation of imitation and that its role is that of a symbol and not of a constituent element of thought; in fact, images appear probably much later than perception and only as part of the semiotic function."[36]

This stress upon the role of "images" (*eidōla*) and "imitation" (*mimēsis*) brings us all the way back to Plato. Aristotle did not reject Plato's position but naturalised it. He retained it as "imagination" (*phantasia*) and "memory" (*mnēmē*).[37] He brought the phenomena down from the ideal into the natural world. Interestingly, we find that Noam Chomsky moves in the opposite direction and re-idealises Jean Piaget's position by arguing that language is a sign of "the existence of an innate kernel" as opposed to Piaget's weaker claim that: "the capacity for learning language can no doubt be explained by the earlier "structuralisation" due to the development of sensory-motor intelligence."[38] Chomsky argues on "creativity" that: "every normal human child immediately carries out this creative act [of language] and they all do it in the same way and without any difficulty, whereas it takes centuries of genius to slowly carry out the creative act of going from evidence to a scientific theory"[39] and ultimately that: "I believe [that]…a fundamental element of human nature is the need for creative work."[40] Creativity is a natural aspect of human existence which is merely raised to heightened levels through such sophisticated mental products as "science." Chomsky adds on "individuality" that: "the individual himself contributes a good deal, an overwhelming part in fact, of the general schematic structure."[41]

---

[34] Perner, Josef *Ibid.* p 239.

[35] Perner, Josef *Ibid.* p 226.

[36] Piaget, Jean & Inhelder, Bärbel "The gaps in empiricism" in Koestler, Arthur & Smythies J. R. (*eds.*) *Beyond Reductionism* (London, 1969) p 132.

[37] On Aristotle on "memory" see his *On Memory and Recollection* and for his "imagination" see especially his *On the Soul*.

[38] Piaget, Jean & Inhelder, Bärbel "The gaps in empiricism" in Koestler, Arthur & Smythies J. R. (*eds.*) *Beyond Reductionism* (London, 1969) p 121.

[39] Chomsky, Noam *Human Nature: Justice Versus Power: The Chomsky-Foucault Debate* (London, [*orig*. 1971] 2011) p 27–8.

[40] Chomsky, Noam *Ibid.* p 45.

[41] Chomsky, Noam *Ibid.* p 4.

Whereas Aristotle reintroduced individuality and creativity by bringing Plato down to earth, Chomsky reclaimed and raised them up again them by insisting that these aspects of humanity are universal and innate. Piaget rejected neo-Darwinian evolution—as does Chomsky—but he does so in favour of a structuralist model of development based upon the real and demonstrable development of the human being in the world. Teresa Houlihan in an interesting comparison of Piaget's thought with Aristotle's explains Piaget's position as that: "From the very outset of his investigations he has maintained that it is only by understanding how concepts are formed in the minds of children that one could explain intellectual activity in the adult. That is to say, there is only one way to understand mental functioning and that is genetically."[42] John Messerly explains this new "genetic" approach to evolution as that: "Though interested from the beginning in reconstructing the evolution of human cognition, Piaget did not have direct access to the psychology of Neanderthal or the intellectual structures of Cro-Magnon. Thus, he was forced to invoke a method useful in biology—replacing phylogenesis with ontogenesis. In other words, the study of the growth of knowledge would focus on children rather than on primitive humans…he hoped that the growth of knowledge would provide a model for biological evolution as well, since knowledge emerged from biology."[43]

Piaget's approach is, I would suggest, a true empirical approach that refutes the dogmas of neo-Darwinism whilst insisting that humanity can be understood as the result of a material, historical, evolution that can be understood scientifically. Chomsky, however, goes further. He dismisses "natural selection" as an ideology and identifies it, rightly, with behaviourism regarding which he argues that the solution should be that of "discarding entirely behaviourism and in fact, in my view, the entire empiricist tradition from which it evolved."[44] We can perhaps usefully address the concerns raised by Gerd Müller above in the context of this dispute between Piaget and Chomsky. Both scientists reject neo-Darwinism but the structuralist Piaget keeps within the bounds of the "empiricist tradition" by keeping meaning as the results of structures and processes that can be understood as a historical, empirical, and mechanistic process.

Chomsky rather finds elemental natural structures in nature that cannot be represented scientifically in the traditional way.[45] The real world has an innate meaningfulness that expresses itself on its own terms and of which we only claim to track its

---

[42] Houlihan, Teresa "A Point of Rapport between Piaget and Aristotle: From Vagueness and Confusion to Clarity and Distinctness in the Acquisition of Human Knowledge" in *Laval théologique et philosophique* (1970) p 233.

[43] Messerly, John G. *Piaget's Conception of Evolution: Beyond Darwin and Lamarck* (Lanham, 1996) p 2.

[44] Chomsky, Noam *Human Nature: Justice Versus Power: The Chomsky-Foucault Debate* (London, [*orig*. 1971] 2011) p 42.

[45] Chomsky tartly comments on "natural selection" that: "It is perfectly safe to attribute this [evolutionary] development to "natural selection" so long as we realise that there is no substance to this assertion; that it amounts to no more than a belief that there is some naturalistic explanation for these phenomena (*Language and Mind* (New York, 1972) p 85)."

measurable mechanisms. This is what Aristotle explained through his contrast between "what is more obscure by nature, but clearer to us" and "what is more clear and knowable by nature (*Phys.* I 184a 20-21)." The contrast here is between the understanding of wholes, of Callias walking towards us, and the parts and elements that make up Callias and enable his existence. Science recognises the parts and their mechanisms in some detail but the wholes only in a derivative and trivial sense. Aristotle's solution, as explained above, was to draw appropriate zones of interest between science and philosophy. In his hands, the science of biology would become a practical science like medicine. We could then hope to obtain knowledge about biological beings through our investigations in whatever way is open to us and most appropriate. We would directly seek testable and replicable truth about the processes of nature. The issue we would then face is, however, whether the biological science possesses the modesty to accept the idea that its "natural sciences based theories" *are usefully limited and necessarily left incomplete.*

4.7 The Meaning of Biosemiotics

So, what is biosemiotics and what does it signify? The sister volume to this work will explore the subject more thoroughly but let us here review what this interdisciplinary approach offers in outline and in principle. And we can consider the problems that come with trying to move beyond disciplinary boundaries slightly later here as well. As the name suggests, biosemiotics finds "semiosis" or "meaning" in biology. This should be uncontroversial but, in fact, it is. The idea has gained traction that the biological world is merely the product of an accumulation of random material interactions. This deflationary approach claims to reduce philosophy to science. There is no why, only what. The inflationary approach—the biosemiotic approach—assumes that the biological world possesses meaning as our cultural and mental worlds built upon it also possess meaning. Its fundamental assumption and assertion is that *the world is meaningful.* This is the essential innovation that we need to expand upon and to defend.

Jakob von Uexküll (1864–1944) is the founder of this semiotic approach to biology, and he provides us with a thorough justification for biosemiotics. His most famous work is his *A stroll through the world of animals and men: A picture book of invisible worlds* (1934) and he also produced works entitled *Theoretical Biology* (1920) and *A Theory of Meaning* (1940). The first title shows us that we are practically and aesthetically engaged with the world of animals and men. Uexküll represents the world in "pictures" as ensembles of beings within the world. The philosophical assertion is made that we picture "invisible worlds." What does Uexküll mean by this? These "worlds" are the subjective "surrounding-worlds" or "*Umwelten*" of each organism, their subjective worldviews which shape their lives. It is also notable that both the *Theory of Meaning* and *Theoretical Biology* refer to "theory." We see in outline that Uexküll's picturing of the biological world is both realistic in the sense that it considers how we must "picture" the world, and also theoretical and philosophical in the sense that he insists upon a meaningful rather than a merely historical world. He argues that the biological world cannot be reduced to what he calls "totally inadequate mathematics"[46] or even by reference to "the [Darwinian] word "evolution" [which] expresses just the opposite of what it is intended to mean."[47]

As regards Uexküll's relationship with or parallelism to Aristotle, we find that he does not himself refer to Aristotle. He takes as his touchstones the rich subjective intellectual and aesthetic worlds of Kant and Goethe. He also refers to his famous predecessor from Tartu, the anti-Darwinian embryologist Karl Ernst von Baer, who had insisted that embryology must be approached through the principle of "effort towards a goal"[48] and whose famous law of embryology "compares to a melody the laws regulating the genesis of an organism…In place of melody, we may also speak of rhythm or of symphony, according to whether we have in mind the rules of the impulses in their simultaneity or in their sequence."[49] By taking "meaning" in life seriously, his philosophical biology inspired philosophers from Heidegger and Cassirer to Merleau-Ponty, Lyotard, and Canguilhem.[50] This philosophical fertility is important in our context; philosophy is now not an add- on ideology or a fig leaf for a discipline simply based upon observation but is part of a real attempt to appreciate and represent how our understanding of biology enables us to make sense of the world around us.

For Uexküll, the impulse from the world is received by the organism and then converted first into sense and then into action. The world of sense can cause a feedback loop changing how the organism addresses and takes stimuli from the world outside. The key qualification made here is that the organism is not merely reactive and mechanical but is active and creative. Uexküll explains in detail that: "The actions directed by the animal towards the outer world (*das Tier*) produce the third world (*dritte Welt*), the world of action (*Wirkwelt*). World of action (*Wirkwelt*) and world-as-sensed (*Merkwelt*) together make a comprehensive whole, which I call the surrounding-world (*Umwelt*). The entire function-circle (*Funktionkreis*) formed from inner world (*Innenwelt*) and surrounding-world (*Umwelt*) (the latter divisible into world of action and world-as-sensed) constitutes a whole which is built in conformity with plan, for each part belongs to the others, and nothing is left over to chance."[51] The world is functional, holistic, ordered, interactive, and creative. Uexküll's model obviously occupies the same philosophical space as that of Aristotle.

---

[46] Uexküll, Jakob von "The Theory of Meaning" *Semiotica* ([*orig*. 1940] 1982) p 67.

[47] Uexküll, Jakob von *Theoretical Biology* (London, [*orig*. 1920] 1926) p 263.

[48] See Jakob von Uexküll *Theoretical Biology* (London, [*orig*. 1920] 1926) p 234.

[49] See Jakob von Uexküll *Ibid*. p 216.

[50] See Tim Elmo Feiten, Kristopher Holland, & Anthony Chemero "Worlds Apart? Reassessing von Uexküll's Umwelt in Embodied Cognition with Canguilhem, Merleau-Ponty, and Deleuze" *Journal of French and Francophone Philosophy* (2020).

[51] Uexküll, Jakob von *Theoretical Biology* (London, [*orig*. 1920] 1926) p 127.

Uexküll's famous paradigm is the life or "*bios*" of the "tick." He *pictures* how the tick *lives* as that it is blind apart from the photosensitivity of its skin which allows it to take its position for predation on top of a twig on which it will wait, possibly for 18 years, for the victim which will provide it with a host for its reproduction. This is the "function-circle" of the tick which pretty much encompasses the whole life-cycle of the tick. Uexküll tabulates this relationship as per Fig. 4.8, as below.

He himself interprets this situation as that: "The meaning of mammals reaches into the tick's *Umwelt*: the melody of the development of mammalian shapes is the theme that also influences the development of the tick's shape. These themes are the smell of butyric acid, the resistance of hair, and the warmth of and ease with which the skin is pierced. That mammals have thousands of other characteristics is of no concern to the tick. Only those characteristics that are common to all mammals appear as themes in the building of the tick; they influence the structure of its perceptual and also its effector organs."[52] We see how biology and semiotics merge into biosemiotics.

How do we understand the phrase "the meaning of mammals reaches into the tick's *Umwelt*"? Uexküll describes how the web of a spider forms a counterpoint with the body of the fly, even before they ever encounter each other. His concern here is to show the counterpoint between inner and outer which he explains as that: "Nothing is left to chance in nature. In every instance a very intimate meaning rule joins the animal and its medium; they are united in a duet, in which the two partners'

***Tick*:** **Meaning-receiver** ***Points*:**	***Any mammal*:** **Meaning-carrier** ***Counterpoints*:**
1. The organ of smell is adapted for one smell, namely, butyric acid.	**1. The only smell common to all mammals is the butyric acid found in their sweat.**
2. A tactile organ is present, ensuring that the tick can exit from the hairs of its prey.	**2. All mammals have hair.**
3. A thermal organ senses a perceptual sign for warmth.	**3. All mammals possess a warm skin.**
4. The tick's stinger is suitable for boring into the skin of each mammal, and at the same time serves as a fluid pump.	**4. All mammals possess soft skin well-supplied with blood.**

***Common meaning rule*:**
Recognition and attack of the prey and extraction of blood on the part of the tick.

Fig. 4.8 "The pursuit of this simple meaning rule constitutes almost the whole of the tick's life". (From Uexküll The Theory of Meaning ([orig. 1940] 1982) p 57)

[52] Uexküll, Jakob von "The Theory of Meaning" *Semiotica* ([*orig*. 1940] 1982) p 72.

properties are contrapuntally made for each other."[53] The animal is united in a "duet" with this environment and responds to it "contrapuntally." Its inner has a living relationship with its outer. He also explains this situation as a "function-cycle" as per Fig. 4.9 below.

Uexküll argues that animals have an encircling ring which both empowers and restricts them on the basis that: "This [function] circle is characterised by the fact that the medium itself exerts no stimulus on the animal; while on the other hand, if the animal forsakes the medium, a stimulus is immediately released, which results in the animal's being guided back into it"[54] and that: "In addition to the function-circle of the medium, we can distinguish the function-circle of food and of the enemy."[55] Underlying all this is the "epigenetics" that Uexküll explains on the basis that there are "three abovementioned actions of protoplasm in the higher animals—construction, running of the machine, and repair. In all cases, something new is achieved; but in accordance with an already existing rule, and always with special relation to the properties of the organs. There is never evolution, but always epigenesis."[56] Uexküll's function circle breaks out of the mechanistic and materialistic cycle *by recognising epigenesis rather than evolution.*

A further scenario or "picture" that Uexküll presents to explain "meaning" is of "the stem of a blooming meadow-flower."[57] He posits in this case the simultaneous existence of *four Umwelts*: of a girl picking flowers, of an ant travelling to its food area in the flower's petals, a cicada which bores into the stem to take the sap needed for construction of its house, and a cow which eats the stem as its fodder. He argues on this multiplicity of attention that: "The same components that are subjected to a certain building-plan (*Bauplan*) in the flower stem are torn asunder into four

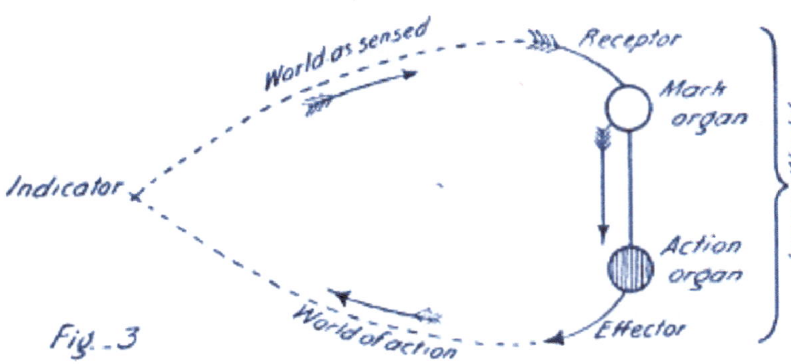

Fig. 4.9 Uexküll's epigenetic loop or function-circle with a "new circle" of response and adaptation to circumstance. (From Uexküll Theoretical Biology ([orig. 1920] 1926) p 157)

[53] Uexküll, Jakob von *Theoretical Biology* (London, [*orig*. 1920] 1926) p 54.

[54] Uexküll, Jakob von *Ibid*. p 127.

[55] Uexküll, Jakob von *Ibid*. p 128.

[56] Uexküll, Jakob von *Ibid*. p 124.

[57] Uexküll, Jakob von "The Theory of Meaning" *Semiotica* ([*orig*. 1940] 1982) p 29.

different *Umwelts* and are integrated, with the same certainty, into various new building-plans (*Baupläne*)."[58] "Attention" is an important principle given some attention.[59] Uexküll argues that meaning is imprinted by subject on object on the basis that: "Every action, therefore, that consists of perception and operation imprints its meaning on the meaningless object and thereby makes it into a subject-related meaning-carrier in the respective *Umwelt*"[60] and then that: "The fact that the body structure is ordered according to a plan (*Planmässigkeit*) seems to contradict the idea that the *Umwelt* structure is also ordered according to a plan (*Planmässigkeit*)."[61]

Uexküll questions whether the structure comes from the environment or "*Umwelt*" or from the being and its "*Bauplan*." This emergent order is clearly similar to Aristotle's dynamic rendering of the concept of "form." Uexküll does, however, take the more distinctly Kantian view that the being creates its own world through a projection of its mind. He ultimately asserts that: "all the properties of objects are actually nothing more than perceptual cues that are imprinted on them by the subject with which they enter into a relationship."[62].

Uexküll concludes generally on the *Umwelt* that: "Once we are finally convinced by an overwhelming number of examples that each *Umwelt* is at base filled only with meaningful symbols, then a second, even more astonishing fact forces itself upon us: Each subject's symbol is at the same time a meaningful theme for the structure of the subject's body."[63] The body is a symbolic manifestation of the being, perhaps as Aristotle suggested as regards courage to a lion. Indeed, Uexküll's semiotic approach argues that we exist in a fundamentally symbolic world in which we are attuned to our own subjective meaning and select impulses from nature on this basis. This is, he argues, the actuality of true *natural selection.* The true nature of "evolution" is in the expression of "form-shaping rules" or "form-shaping melodies" rather than the consequences of "chance."[64] Uexküll's conclusion, again redolent of Aristotle, is that; "Nothing is left to chance in nature. In every instance a very intimate meaning rule joins the animal and its medium; they are united in a duet, in which the two partners' properties are contrapuntally made for each other."[65] He again uses a musical analogy to explain his perspective. The animal is united in a

---

[58] Uexküll, Jakob von *Ibid.* p 30.

[59] Uexküll argues that: "We call this power of noting certain movement-signs more than others, "attention." And since we describe as rhythm any regularly recurring change whatsoever, we may in the last instance refer the power to form numbers to a rhythm of the attention (*Theoretical Biology* (London, [*orig.* 1920] 1926) p 57)." Whitehead takes a similar position using similar language.

[60] Uexküll, Jakob von "The Theory of Meaning" *Semiotica* ([*orig.* 1940] 1982) p 31.

[61] Uexküll, Jakob von *Ibid.* p 30.

[62] Uexküll, Jakob von *Ibid.* p 74.

[63] Uexküll, Jakob von *Ibid.* p 77.

[64] See Jakob von Uexküll *Ibid.* p 72.

[65] Uexküll, Jakob von *Ibid.* p 54.

"duet" with this environment and responds to it "contrapuntally." Its inner has a living relationship with its outer.

It is on this basis that Uexküll directly criticises and refutes Darwin and the philosophical worldview based upon his biology. He rightly criticises the idea that "the species [is] merely a mixture of properties, such as one might find in any mass of fermenting matter"[66] as a manifestation of a poor, materialistic, pragmatic, and empiricist tradition. Uexküll generally asserts that organisms cannot be understood as a product of their material environments but only as an embodied respondent to it and that it is an error on the most fundamental level to conflate cause with effect.[67] The existential fact of being able to drive a car or send a letter should not simply be conflated with the physical facts of a moving car or letter. As regards the alternative problem of idealism, Uexküll argues that: "Perfection is not omnipotence, but merely means the correct and complete exercise of all the means available. Even from the most perfect being conceivable we cannot look for the exercise of resources that it has not got."[68] This is a typically Aristotelian argument: an organism optimises the matter available to it in the best way open to it according to its species niche. Optimisation which is entirely evident in nature from both common-sense and scientific perspectives is not the same as invincibility; "Perfection is not omnipotence."[69]

As does Aristotle, Uexküll stresses that "plan" is not the same as "purpose" but is only the inherited pattern or "melody" (or "habit," "form," or, with Whitehead, "pattern") of the specific being.[70] He addressed this situation though the analogy of the melody which is a sequence of notes standing for the sequences of flows of order having a fittedness that we find expressed in the world as streams of thought, speech, action, and behaviour. Organic existence exists as moving patterns and cycles of meanings. "Species" is on this basis defined as "a rhythmical sequence of acts."[71] "Symphonies" are not just an orderly sequence of acts but a collection of different voices acting in symphony and in concert.[72] The symphony is made to represent an ecosystem of interacting species. This metaphor is picked up by Denis Noble in his brilliant *The Music of Life* which amply shows us that working from music and art

[66] Uexküll, Jakob von *Theoretical Biology* (London, [*orig*. 1920] 1926) p 237.

[67] On this matter see Jakob von Uexküll "The Theory of Meaning" *Semiotica* ([*orig*. 1940] 1982) p 70–1.

[68] Uexküll, Jakob von *Theoretical Biology* (London, [*orig*. 1920] 1926) p 164.

[69] There is substantial evidence of optimisation in nature. See Robert Rosen *Optimality Principles in Biology* (New York, 1967) and John Beatty "Optimal-Design Models and the Strategy of Model Building in Evolutionary Biology" *Philosophy of Science* (1980).

[70] See Jakob von Uexküll *Theoretical Biology* (London, [*orig*. 1920] 1926) p 175.

[71] Jakob von Uexküll argues that: "we participate in the rhythm, and so get the right impression of the species as a rhythmical sequence of acts (*Ibid.*p 245)."

[72] Uexküll argues that: "In music, we distinguish between melody and symphony. By "melody" we understand the orderly sounding of musical notes one *after* the other; by symphony, their sounding *together* (*Ibid.* p 29)."

is not a trivial or quixotic approach to the world but is of the essence of the reality of human and living existence that we want to explain.

4.8 Some Thoughts on the Problems Faced by Biosemiotics

One of the most evident aspects of the discipline of biosemiotics and of the biosemiotics community itself is its broad and interdisciplinary nature. This fact and its importance will be drawn out and explained in the sister volume to this work which will consider the history of biosemiotics and the various interdisciplinary conferences that have enriched and typified its tradition.

The significance of this interdisciplinary approach is that it offers us a width of insight to provide a counter to the reductionist approach to the world by showing us the richness of the world through a range of perspectives. This width reveals a common meaningfulness of being whilst also forcing us to appreciate the fact and principle that these various dimensions of being cannot be reduced to a single law or a single perspective. This richness of being—best exemplified by its individuality and its creativity—cannot be reduced to mathematics or language or to any other medium of thought but must be held up, wondered at and appreciated, and its manifestation understood. We also find, however, that the reductionistic project emerges even here. The point of taking an interdisciplinary spectrum from this perspective is to find the reduction. The only difference to previous reductions is that whereas in the past this reductionism has been a reduction to the discipline of mathematics, physics, or chemistry, the new reductionism of *our* age is to computers, codes, information, etc. Modern biosemiotics certainly takes pains to refute and overcome this reduction and its success in doing so will be considered in the sister volume to this work.

As regards how we overcome reductionism generally, we have seen that sciences overcome the enthusiasm of their founding once their rich complexities have been revealed. This is in a sense the story of physics in our own age. The enlightened brilliance of the "machine age"—and it was brilliant—gave us hope that we could understand and master the whole universe as a "clockwork universe." This thought project has even been broadly successful in its objectives and few movements can claim this feat. We see, however, its error by considering its dream or claim that all the sciences could ultimately combine to form one single *mathesis universalis*. It assumes that we can add up all the science to achieve one big science that will understand everything. The monstrous optimism of the seventeenth century demonstrated here supposed that constructing a big tower of Babel would effectively allow us to become "God." From one perspective, we did indeed come close. A more cynical position argues, however, that like many conventions of that age, we have merely invented a mythology or ideology which usefully simplifies in the short term, but which ultimately presents us with a catastrophic obstruction. We could say that the discipline of biology has gone through these struggles for identity and place with one result being the moderated science of biosemiotics which (at its best) both gives meaning to nature and explains it in human terms.

The problem we face today—perhaps as ever—is that reductionism seeks to reduce the world to the Parmenidean "One." It also reduces the world to Anaxagorean chance and necessity. It makes the world flat and formal or a meaningless flux. It distorts reality to fit an ideology. Paul Feyerabend expands upon *our* problem in this context as that: "A scientific approach, says [Peter] Medawar, "annihilates" them [i.e., individuals] and "relieves" science of their "burden"…I now add that [Nicolae] Ceauşescu did in practice what the hypothesis (and Medawar) do in theory."[73] He explains this situation through his final work, aptly titled *Conquest of Abundance: A Tale of Abstraction versus the Richness of Being*. This brilliant work is centred upon a rich and eloquent defence of Aristotle which has, entirely typically, been completely ignored by Aristotle scholars. When Aristotle challenged our modern preconceptions, his modern guardians worked tirelessly to silence him. As regards biosemiotics, we also find that some of its proponents have similarly sought to rein it in, to make it acceptable in accordance with what they take to be modern sensibilities and, we could say, with the modern power structure. Doing this would reduce biosemiotics to a biology of codes or information. By offering it relevance—and, ultimately, power—by incorporating it within the reductionist paradigm, they would turn biosemiotics into another obfuscating smokescreen.

What are the reductionistic risks for biosemiotics? The first is that its multidisciplinary approach can be co-opted as part of an attempt to see the world through a "unity of science." The exact opposite is the case. We need an ever richer and multiplying family of sciences to unpack and represent the world before us. The idea that we can join up all the sciences to produce one science that explains everything is a nonsense, though an understandable one. This is not to say that this idea is not useful for encouraging a dialogue between the sciences and a collegiate approach between scholars used to living in caves. Its drawback is simply that it reduces and annihilates all it touches. The other reductions we face are similarly tired products of "seventeenth century" mythologies and ideologies. The idea that the world is built upwards from out of bits of matter and through historical chance—as a sort of "at-at ontology"—just repeats interminably. The sweep of this ideology is put forward by Richard Rorty as: "the conviction that seventeenth-century science discovered that everything could be explained by atoms and the void, and that philosophy has a moral duty to preserve this insight…[and] if there was supposed to be answers to such general questions as "How is abstraction (recognition, constancy) possible?" [we] can say, there is *no* answer, except the pointless remark that nature has evolved suitable hardware to get the job done."[74] We see a mixture of vagueness, pragmatism, and obstruction here.

---

[73] Feyerabend, Paul "Ethics as a Measure of Scientific Truth" in *Conquest of Abundance: A Tale of Abstraction versus the Richness of Being* (Chicago, 1999) p 250.

[74] Rorty, Richard *Philosophy and the Mirror of Nature* (Princeton, 1979) p 217 and 235.

The risk that biosemiotics faces is being pulled into this miasma. One way this has been achieved has been to coopt the ability of biosemiotics to restate the world in terms of "information theory." Don Favareau explains the problem here as that: "while great strides have subsequently been made in our understanding of the purely material relations underlying "the genetic code"—conceptually unclear still is the absolutely scientifically legitimate question of just in what sense information—defined not just as the inanimate sequence of nucleotides themselves but as the functional relation of those nucleotides to a system for which they serve as "sequences of code"—can be thought to be an intrinsic property of things. For Francis Crick, articulating his "central dogma" of genetic inheritance, "information" was synonymous with "the sequence of amino acid residues" per se (1985: 1); and for Claude Shannon and Warren Weaver, "information" was the diminution of uncertainty in a system absolutely without regard to cognitive or semantic considerations (1949: 8). Both Crick's notion of "information flow" from gene to protein and Shannon and Weaver's mathematical theory of "communication across a channel" thus explicitly deny that the "information" that they are talking about "means" anything in the sense that we associate with the word "meaning.""[75] Having spent decades trying to strip all "meaning" from the world, when "meaning" was correctly reintroduced it was so defined that this new "meaning" is utterly vague and meaningless. We could reasonably harbour the suspicion that it is meaningfully left meaningless.

The argument that the world can be reduced to information, information flows, codes etc. has largely been an attempt to reduce the world to computers in the "computer age," as a previous age sought to reduce the world to machines in the "machine age," and another age sought to reduce it exclusively to the mind of "God." In all cases, we find little attempt to contextualise these core ideas or frames. A certain language is adopted and a medium or model is taken and conflated with reality, an ideology is formed and applied. This is Aristotle's problem with Plato's reduction of the world to forms. The "third man" is not only a distraction or error but is a systemic obfuscation. To interpose a reduction to matter or mechanism is similarly obfuscating. To reduce the world to information without recognising that it is the information of beings is perhaps even more absurd than these as what Aristotle would call "*atopon*," i.e., something without a place. The end result of this absurdity is the idea that the world can be understood as competing codes that is "code biology."[76] Biosemiotics is thereby fused with neo-Darwinism in a very literal sense since biology is reduced to the genetic code and its evolution through competition.

---

[75] Favareau, Donald "Introduction: An Evolutionary History of Biosemiotics" in *Essential Readings in Biosemiotics: Anthology and Commentary* (Dordrecht, 2010) p 30–1.

[76] Code biology is an offshoot of biosemiotics for which see the foundational text of Marcello Barbieri *The Semantic Theory of Evolution* (Chur, 1985) and for a development of this thought see his "The Paradigms of Biology" *Biosemiotics* (2013). For a critique of this development from within the biosemiotics community see Arran Gare "Code biology and the problem of emergence" *Biosystems* (2021) and Kalevi Kull "Codes: Necessary, but Not Sufficient for Meaning-making" *Constructivist Foundations* (2020).

The goal here is not just to bring biosemiotics into line with neo-Darwinism, a model that it was brought into being to refute, but to repurpose it as a defence of this very ideology.

The problem is ultimately even deeper than this revanchism and is a fundamental problem with the role of science itself. We find that biosemioticians with a biological background seek to find a philosophical framing that can develop and evolve the reductionist philosophy through which their biological tradition has emerged. Howard Pattee hence gives us thoughts on "How Does a Molecule Become a Message?" (1969) and Terrence Deacon seeks to explain "How Molecules Became Signs" (2021). The present author from a very different perspective would say that the very idea that "molecules" evolve to become the beings we encounter today is a derivative of the seventeenth-century atomistic fallacy which should be laid to rest rather than humoured. The same applies to the various brave and brilliant attempts to either reduce the world to chemical matter such as Alan Turing's "The Chemical Basis of Morphogenesis" (1952) or to define what it means to be a self-replicating system such as expressed by John von Neumann in his "The general and logical theory of automata" (1956). These perspectives may have once been reasonable hypotheses and speculations and they may indeed still be perfectly useful as scientific tools and hypotheses to frame scientific research, but they are also philosophically useless or harmful. *Their error is only harmless if philosophy is not conflated with science.*

The issue we face is, simply put, the limits of science and of the scientific method as we understand it. Paul Feyerabend, a good Aristotelian, perhaps comes closest to exposing the problems incumbent upon this method, for instance in his dialogue with Imre Lakatos in their *For and Against Method.* It is, of course, immensely useful to be able to reverse engineer nature and to break it down to its component parts. Nature itself is, however, a matter of wholes, of beings, and not of competing bits. We can't truly get It from Bit. The whole is fundamentally different to the parts, the atoms. This is a philosophical truth and it opens up a different world of beings, of reality, not reducible to human engineering. This is not, of course, to denigrate science itself or its standard methods—just as we would not denigrate logical forms or poetic meters—but it is only to identify the problem of confusing the medium with the message. I suggest that we need to reject empirical, analytic, or scientistic philosophy and to return to what I would call true philosophy. The full importance of this move is to release our lives from the grip of a misapplied technology. Noam Chomsky expresses the philospohico-political nature of this movement when he calls for the paradigm shift of "discarding entirely behaviourism and in fact, in my view, the entire empiricist tradition from which it evolved."[77] Biosemiotics can implement this transformation by understanding and representing the word as meaningful worldly existence by providing the necessary philosophical framework for the valuable performance of biological science to take place.

---

[77] Chomsky, Noam *Human Nature: Justice Versus Power: The Chomsky-Foucault Debate* (London, [*orig*. 1971] 2011) p 42.

4.9 Conclusions

Some monsters are hard to slay. One of these is scientific reductionism. Even most non-reductionist scientific movements are thoroughly reductionistic. For instance, cybernetics, complexity theory, and general systems theory (GST) recognise that we cannot reduce the world to matter, form, etc. whilst reducing the world to their own mechanical worldviews. They are simply more sophisticated forms of reductionism based upon more sophisticated machine metaphors.

What we need is not only more sophisticated science but also a reformulated, revitalised philosophy. We need to reinvent and recentre a philosophy that has become subordinate, derivative, and utilitarian. The development of this new philosophy should go hand in hand with the development of a confident and nuanced science that understands its limits as well as its frontiers. The role of biosemiotics in this context should not be to replace the reductionism to genes with one to memes (following Richard Dawkins) but to demonstrate the importance of framing science with philosophy. Its introduction of signs and semiosis should be transformational not as a technical improvement but because it properly reintroduces qualities and values to our view of the world. It reintroduces the "being" as an actor having an inside world and an outside presence. Its actions have "meaning." It provides us with a philosophical "framing" as an additional surround which allows us to find meaning through science. This is a true complexity theory in which we find a range of nested perspectives reflecting a matrix of meaning in the world itself.

Aristotle helps us to transform our view so as to represent the qualitative world. He therefore provides us with an alternative to the seventeenth century ideological framing which strips meaning from the world and removes it to "mind" or "God." He shows us that meaning is "in" the world and conveyed through the meaning-making of semiosis. Aristotle achieves this meaning-making through his paradigmatically interdisciplinary approach which seeks to not to reduce our vision of the world to a unity of science but to represent our complex, value-laden reality.

Chapter 5
Aristotle and the Problem of Abstraction

Abstract This chapter considers the meaning of meaning. Far from being archaic and superseded, Aristotle's metaphysics—as also his logic—is a necessary framework for allowing us to structure the world as an architecture of meaning. The world is, on this basis, comprised of beings interacting through the "means" of natural and human processes. We consider a range of these processes as examples, including that of building a wall. The "four causes" are a related intellectual tool for understanding the movement of beings through the world. This framing reduces the world to its basic components but avoids reducing it to a single idea or ideology. Reality is on this basis necessarily and meaningfully complex. We are drawn back to Aristotle's philosophy of realism as a correction to Plato's philosophy of idealism. The ultimate comparison we face is between Aristotle's world of beings with their senses of meaning and a world represented through mathematics as abstract laws to which we can (somehow) reduce meaning.

Keywords Aristotle and abstraction · Aristotle and process · The "four causes" and irreducible complexity · Aristotle's realism

Aristotle was an excellent scientist, but he is perhaps an even better philosopher. His conflict with Plato, his master, was monumental and has shaped the Western philosophical tradition ever since. His realism and Plato's idealism is a fundamental duality in philosophical thought.[1] The great classical scholar W.D. Ross could, however, comment: "Depend upon it; whatever Plato or Aristotle may say, nine times out of ten they mean the same thing…And where they do not [agree], the advantage is by

[1] See on this matter David Newsome *Two Classes of Men: Platonism & English Romantic Thought* (London, 1972).

P. N. Jackson, *Aristotle on Meaning in the Living World*, Biosemiotics 30,
https://doi.org/10.1007/978-3-032-00602-8_5

no means always with Aristotle."[2] As we have seen already, Aristotle accepted all of Plato's key concepts, even his forms, but he *transformed* them. He gave them only a different slant and oriented them towards explaining reality rather than being merely an idealistic or ideological map. Aristotle's brilliant innovation—or one of them at least—was appreciating the problem of abstraction.

Let us begin in considering this matter with Aristotle's observation that: "a boy may become a mathematician, but not a philosopher or a physicist …because the objects of mathematics exist by abstraction (*di' aphaireseōs*), whilst the first principles (*hai archai*) of these other subjects come from experience (*ex empeirias*) (*N.E.* VI. 1142a 16–19)." Aristotle asks not only what mathematics, physics, and philosophy require but also what it means to be a mathematician, physicist, *or* philosopher. He explains the key difference as being that mathematical idealism does not deal with reality since: "the mathematical sciences are concerned with forms; they do not confine their demonstrations to a particular substrate (*ta gar mathēmata peri eidē estin, ou gar kath' hypokeimenou tinos*) (*Post. An.* I 79a 6–7)." Mathematics is purely formal, and this is not a virtue but a deficiency. It is deficient because it *takes away* the substance, substratum, and matter to leave only a mathematical residue with the term for abstraction being *di' aphaireseōs* which suggests *sub*-traction, a stripping away of reality. Mathematics on this account does not address reality but only a simplified and homogenised model of the world. In arithmetic, for instance, we represent a certain aspect of reality—its being(s)—through our mental language of number.

Aristotle argues that theoretical reductions of reality are necessarily simplifications of that reality. His logic, his metaphysics, and all his sciences are knowingly intellectual simplifications or mental representations of the world which together provide us with a fuller perspective upon the whole, the whole reality and its objects and beings. It is by recognising the gaps in our knowledge—what Niels Bohr would call a "principle of complementarity" (1927) and Kurt Gödel would explain through his "incompleteness theory" (1931)—that we can reasonably approach reality. It is this "bottom up" standpoint of degrees of knowability that leads Aristotle to formulate his concept of the "Unmoved Mover." This explicit deliberate "gap" demonstrates that we need to recognise gaps in our knowledge so as to conceptualise and give place to that which is concretely beyond our capacity. It is also critical to recognise here that Aristotle *completely overturns* the approach to the world put forward by Pythagoras and Plato. The language of mathematics and even the language of language are merely artificial means or technologies for representing or tokenising the world and not a presentation of reality itself. We should recognise the problem of conflating the medium with the message or the means with the ends. This problem of the nature of images (*eidōla*) and representations of the world goes back at least to Plato.

I note as regards this "problem" with mathematics that the modern philosopher Henri Bergson expressed the situation identified by Aristotle two millennia

[2] W.D. Ross "Platonic Philosophy and Aristotelian Metaphysics" in *Philosophy and Metaphysics; Proceedings of the Aristotelian Society* (1925) p 156–7.

previously through his philosophy of "creative evolution." He argued that: "[t]he aspect of life that is accessible to our intellect—as indeed to our senses, of which our intellect is the extension—is that which offers a hold to our action. Now, to modify an object, we have to perceive it as divisible and discontinuous"[3] so that: "We are at ease only in the discontinuous, in the immobile, in the dead. The intellect is characterised by a natural inability to comprehend life."[4] His reference to continuity and discontinuity is strikingly Aristotelian, as we will see below. Alfred North Whitehead, the great philosopher and mathematician, similarly argued that: "The nature of any type of existence can only be explained by reference to its implication in creative activity, essentially involving three factors: namely, data, process with its form relevant to these data, and issue into datum for further process—data, process, issue. The alternative is the reduction of the universe to a barren tautological absolute, with a dream of life and motion. The discovery of mathematics, like all discoveries, both advanced human understanding, and also produced novel modes of error. Its error was the introduction of the doctrine of form, devoid of life and motion."[5]

5.1 Adding or Subtracting Meaning

What does it mean to add meaning to or subtract meaning from the world? Subtracting meaning is, as above, a selection of meaning by a conscious being from a rich and fuller reality. Peirce's concept of "abduction" possibly represents a similar train of thought. The idea of adding meaning, on the other hand, normally suggests that either God or man somehow imposes meaning upon the world through their mental intervention. Whereas subtraction is a synthetic and aporetic approach to meaning, addition is an analytic approach through which the analysis is the added meaning. Let us inquire further: Aristotle assumes subtraction, but what happens if we assume addition?

William James (1842–1910) takes this alternative position. He explains his opposition to Aristotle's position of "subtraction" as that: "We operate…by physical subtraction…[and t]his supposes that the consciousness is one element, moment, factor—call it what you like—of an experience of essentially dualistic inner constitution, from which, if you abstract the content, the consciousness will remain revealed to its own eye. Experience, at this rate, would be much like a paint of which the world pictures were made…my contention is exactly the reverse of this. Experience, I believe, has no such inner duplicity; and the separation of it into consciousness and content comes, not by way of subtraction, but by way of addition."[6]

---

[3] Bergson, Henri *Creative Evolution* (New York, [*orig*. 1907] 1998) p 36.

[4] Bergson, Henri *Ibid.* p 143.

[5] Whitehead, A.N. "Forms of Process" in *Modes of Thought* (New York, 1938) p 93.

[6] James, William "Does "Consciousness" Exist?" *The Journal of Philosophy* (1904) p 480. For a critique see Arthur J. Deikman "I = Awareness" *Journal of Consciousness Studies* (1996).

Since we would expect "ego" to be fundamental to the "addition" of meaning, let us ask how and why James thinks he can even dispense with the idea of an "inner" consciousness which provides the addition? We find that the "addition" he commends is that of "relation" which exists on the basis that: "Consciousness connotes a kind of external relation, and does not denote a special stuff or way of being. The peculiarity of our experiences, that they not only are, but are known, which their "conscious" quality is invoked to explain, is better explained by their relations—these relations themselves being experiences—to one another."[7] James' theory of mind is an explanation of the world through relativism.

James adds that our thought and language does not reflect upon the world but *is* the material world in the sense that the thing and its representation are the same on the basis that: "Why…do we call a fire hot, and water wet, and yet refuse to say that our mental state, when it is "of" these objects, is either wet or hot? "Intentionally," at any rate, and when the mental state is a vivid image, hotness and wetness are in it just as much as they are in the physical experience."[8] James, rather ironically since he is a darling of modern philosophy, argues for the position attributed to Aristotle (without basis) to denigrate him that: "the organ literally and physiologically undergoes an alteration: the eye-jelly turns red, the nose becomes smelly."[9] What are the philosophical implications of James' position? We see, first, a materialistic reductionism. The inner mind is conflated with the outer material world. "Matter" is just a blanket concept and not, as with Aristotle, a conceptualisation of "energy" according to which there are as many expressions of energy as there are beings, e.g. there are both sensible matter and noetic matter. James rather depends upon the reduction of the world to "material objects" on the basis that: "There is…no aboriginal stuff or quality of being, contrasted with that of which material objects are made, out of which our thoughts of them are made."[10]

The human "subject" is also ultimately reduced to a "material object" on the basis that: "I believe that "consciousness," when once it has evaporated to this estate of pure diaphaneity, is on the point of disappearing altogether. It is the name of a nonentity, and has no right to a place among first principles. Those who still cling to it are clinging to a mere echo, the faint rumour left behind by the disappearing "soul" upon the air of philosophy."[11] James' ultimate assertion is that the concepts of "consciousness" and "soul" should "disappear" from our philosophical account of the world.

The contrast here is between a world in which mental action is a subtraction through which a being *actively* selects a dimension of worldly being and a world in which it is an addition as a *passive* effect or appearance caused by the relations

[7] James, William "Does "Consciousness" Exist?" *The Journal of Philosophy* (1904) p 486.

[8] James, William *Ibid.* p 488–89.

[9] Burnyeat Myles "Is an Aristotelian philosophy of mind still credible? (A Draft)" in Nussbaum, Martha & Rorty, Amelie O. (*eds.*) *Essays on Aristotle's* De Anima (Oxford, 1992) p 19.

[10] James, William "Does "Consciousness" Exist?" *The Journal of Philosophy* (1904) p 478.

[11] James, William *Ibid.* p 477.

between different material objects. The first option obviously represents the reality that we experience and live. We are human beings evidently living in the world with our own mental perception of the objects of the world filtered and represented through our beings, minds, souls, and consciousness. How, though, could William James come to the strange position that he takes, and can he be dismissed as a mere radical outlier of modern thought? James is by no means a radical outlier. He is the notable founder of the James-Lange Theory of emotion and of the idea of the "stream of consciousness." As regards the radicalness of James' position, Karl Popper explains that: "The idea of what I call "objective truth"—that is to say, a statement is true, or a story is true, if it corresponds to the facts—is a very old idea. It is explicitly formulated by Aristotle, and it has since been very much under attack by philosophers—and, since William James, especially by pragmatists"[12] with his own conclusion being that: "the attacks against truth are just bad philosophical eyewash."[13] Even though we can add "truth" to the list of essential human facts attacked by James, he is still by no means the worst offender. Indeed, he is one of the more reasonable voices in modern philosophy. This shows us the problem we face in countering the problem of reductionism.

We see that the lure of technology pervades even the best modern thought. Popper argues that the underlying problem behind this modern situation (and perhaps even more fundamentally) is that: "most philosophers mix up the idea of truth with the idea of a criterion of truth. They think that if there is an idea of truth, there has to be a criterion of truth attached to it. In other words, they are operationalists. There has to be an operation by which we find out whether or not a thing is true. Now it is quite clear that such an operation does not exist."[14] The idea we face is that there is a sort of layer of truth or operation in the world whereas there is rather the truth or operation of a being in the world. Our modern error is to replace reality with an artificial construction for operating that reality. It is hence that Gödel explained that mathematics is incomplete because it interposes its own artificial structures over our vision of natural reality, a problem that Aristotle had surveyed two millennia previously as that: "number qua number has peculiar attributes (*hē arithmos idia pathē*), such as oddness and evenness, commensurability and equality, excess and defect (*Met.* Γ 1004b11–13)." Mathematics is its own world and its own language which relates to the world in symbolic form but only captures a very circumscribed dimension of its being and its meaning.

Aristotle also observes that both number and word are only reflective constructions of the world which are not living or real so that: "two fives will make ten…there is no common limit (*koinon horon*) whatever at which these two fives coalesce. And the same with the parts three and seven… The same may be said about speech…No common limit (*koinon horon*) exists, where those parts—that is, syllables—join

[12] Popper, Karl "Description, Argument, and Imagination" in *Knowledge and the Body-Mind Problem: In defence of interaction* (London, [*orig.* 1969]1994) p 96.

[13] Popper, Karl *Ibid.* p 96.

[14] Popper, Karl *Ibid.* p 96. On operationalism see Alain Boutot "Les théories de la morphogenèse et le dilemme de la substance" *Revue de Métaphysique et de Morale* (1998).

(*Cat.* 4b 28–39)." The living and natural world is fundamentally different than our representations of it. Aristotle draws out the fundamental point that an understanding of the world *as* abstractions rather than aporetically *through* abstractions is erroneous and dangerous.

5.2 Aristotle and the "Place" of Mathematics

This is not, however, to say that Aristotle has no place for mathematics. By considering the "*periodos*" we see that Aristotle builds outwards from mathematics to seek to represent the physical and then biological worlds. He unpacks the "*periodos*" to understand it as a point (*stigmē*), a straight line or path (*grammē*), and a circle (*kyklos*) as being the bases for the principles (*archai*), movements (*kinēseis*), and cycles (*periodoi*) of natural beings. We could, for instance, embody these principles as event, life, and lifecycle. Aristotle himself explains this *embodied mathematics* through the case study of the "snub nose" which considers how the "concave" is a mathematizable shape that routinely emerges, or as Whitehead or Plato world say "ingresses" or "participates," in nature as a generic possibility.[15] We further find that it is not only fundamental categories of being such as time and space but also shapes, structures, and patterns of being that exist in nature as a sort of "natural law." The metaphysical approach to the world presumes a form of structuralism.

Aristotle is interested in how the shapes of nature are mathematically morphological and this Aristotelian theme was famously picked up by the Aristotelian scholar and mathematician D'Arcy Wentworth Thompson in his magnum opus *Growth and Form* (1917). I provide an example of his "theory of transformations" in Fig. 5.1 below.

I comment that this focus upon scaling, proportion, and symmetry behind natural form and morphology goes directly back to Aristotle who wrote a treatise on the scaling and size of living beings, *On the Longness and Shortness of Life*, in respect to how they grow from a seed or foetus to maturity and fill out their form through growth in space and over time. Aristotle also considers the structuring through which animals move in his *Movement of Animals* and for a scientific review of the "points" (*sēmeia*) through which animals move, see his *Progression of Animals*. He provides for specific case studies the fact that the hand grows in proportion to its form and the maturing skeleton, i.e., the hand and the bones grow proportionately as, say, a crystal also grows in nature; see *PA* I 640b 18–23 and *GA* II 745a 4–9.[16] The concept of symmetry also notably runs throughout Aristotle's work.

[15] On the "snub nose" see Suzanne Mansion "Τὸ σιμόν et la définition physique" in *Naturphilosophie bei Aristoteles und Theophrast* (Heidelberg, 1969).

[16] On the scaling of life see J. B. S. Haldane "On Being the Right Size" *Harper's Magazine* (1926), Alexander W. Shingleton "Allometry: The Study of Biological Scaling" *Nature* (2010), and Andrew Clarke and Peter Rothery "Scaling of body temperature in mammals and birds" *Functional*

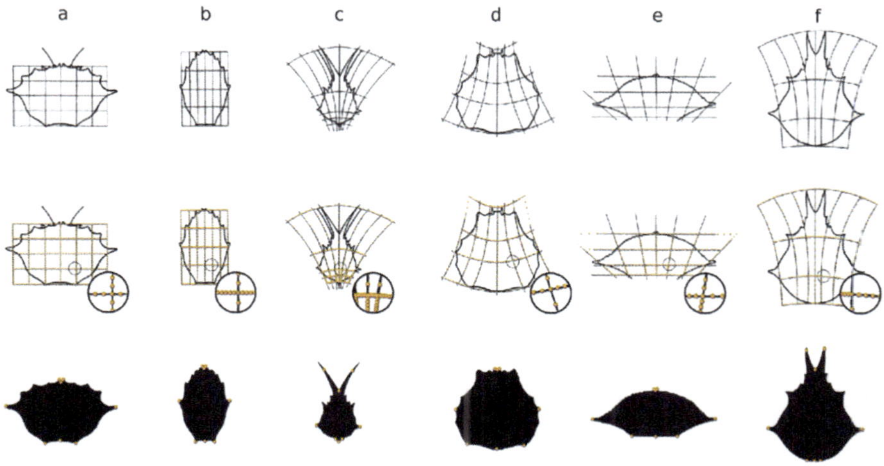

Fig. 5.1 D'Arcy Thompson's morphology of crab carapaces. Carapaces of various crabs: (**a**) *Geryon*, (**b**) *Corystes*, (**c**) *Scryamathia*, (**d**) *Paralomis*, (**e**) *Lupa*, (**f**) *Charinus*. (Taken from Scholz et al. "D'Arcy W. Thompson's Cartesian transformations: a critical evaluation" *Zoomorphology* (2020))

Orogee Dolsenhe nicely observes that Aristotle is, in all his work—on biology, psychology, ethics, sense perception etc.—fundamentally a structuralist who regards the world through the prism of analogous qualitative structures by means of a "square of opposition…showing the interrelationship between the four types of predicates [through] predictive modality."[17] As regards Aristotle's own position, *his* structuralism, we find an explicit case study for the commensurability of being at *N.E.* V 1133a 8 *ff.* where Aristotle explains how human beings relate to the objects and beings of the world through the example of the trade between a builder and a shoemaker with Aristotle again using craft as a metaphor for human action generally. Aristotle also considers how vision, memory, and mind relate the subject to the object in the world at *Mem.* 452b 9–19. Aristotle's *Movement of Animals* considers the symmetry and structure needed for organised movement. I note that a non-reductionistic structuralist position does not *reduce* the world to structure but rather accepts that the world and its beings is such as to produce structures, often replicable, which can logically be captured and represented. Aristotle explains these structures as "forms." By retaining an objective point of view describing the relationships between beings and objects, Aristotle avoids falling into the traps of subjectivism and relativism.

For an example of the mathematical aspect of Aristotle's thought we find that his *Movement of Animals* interestingly considers the relation between the "one" and

---

Ecology (2017). On growth in general see Claudia Zatta's Chapter 3 in *Aristotle and the Animals: The Logos of Life Itself* (Abingdon, 2022).

[17] Dolsenhe, Orogee "The Revival of Structuralism: The Periodic Table of Mental Science" *Journal of Scientific Exploration* (2022) p 318.

"two"—an important question in Greek thought after Pythagoras and Parmenides—on the basis that: "if any one of their parts moves, another part must necessarily be at rest; it is on this account that animals have joints. For they use their joints as a centre, and the whole part in which the joints is situated is both one and two, both straight and bent, changing actually and potentially because of the joint (*MA* 698a 17–21)." This principle also applies to animals themselves which exist over and above their parts (and hence the whole is greater than the sum of the parts) in the sense that: "every animal…must have in itself something that is at rest, in order to provide that which is moved with the origin of its movement, supported upon which it will move both as an integral whole and in its several parts (*MA* 698b 5–7)." I comment that the reference to rest/movement, one/two, straight/bent, and actual/potential is significant in addressing the duality and possibility of "being" and is a development of Pythagoras' Table of Opposites.

Aristotle argues further as regards how potentiality/actuality relates to one/two that: "what is capable of desiring and what is capable of fleeing are not different, either from one another or from what is capable of perceiving, but their being is different (*De An.* III 431a 12–14)" with it being the case that: "the actualisation must be one (*mia estai hē energeia*) (*Phys.* III 202b 37)." Potentiality is inherently a duality of possible action, e.g. of either being attracted and repulsed, which concludes in line with René Thom's "catastrophe theory" with *one* actual expression or "catastrophe."[18] This is the true significance of Aristotle's principle of non-contradiction; something *actually* will only be one thing, but it *potentially* can be other than this.[19] Aristotle describes this potentiality as that there is an innate contingency in nature according to which cases which "for the most part and commonly, tend in a certain direction (*hōs epi to polu*)…may issue at times in the other or rarer direction (*De Int.* 19a 21–22)" (and see also *GC* II 336b 1–4). I stress that Aristotle's thought is defined by its consideration of this "becoming" through "potentiality" (*dynamis*) as opposed to the simplification of Plato and the Megarians in Aristotle's day and those thinkers who do not recognise the power of "nature" in ours.

I note generally that the *Movement of Animals* is a remarkable scientific study of the mechanics and implications of animal movement based on the principle of symmetrical structure, even for that of the snake. It is also a study both of physiology and psychology. His *Peri Psychēs* (or *On the Soul*) on psychology or, in Aristotelian terms, the "soul," explains the structures behind the movement of animals on the basis that: "The sentient subject (*to aisthanomenon*) must be extended, but

[18] "Catastrophe" here simply means a sudden turning from one state to another, e.g. a startled dog either fleeing or standing its ground; see E.C. Zeeman "Catastrophe Theory" *Scientific American* (1976).

[19] Aristotle argues that: "A single motion must be caused by a single mover (*Met.* Λ 1073a 28)" and that: "that which is capable of receiving the object of thought, i.e., the substance, is thought. And it is active when it possesses the object (*to gar dektikon tou noētou kai tēs ousias nous, energei de echōn*) (*Met.* Λ 1072b 22–23)." A selection must be made by being activated from multiple possibilities.

sensitivity and sense cannot be extended; they are a kind of ratio and potentiality (*logos tis kai dynamis*) of the said subject (*De An*. II 424a 27–28)." Aristotle argues that there is a scaling and structuring of the soul as well as of the body and he expands upon these inner (though also outward facing) structures of character in his *Ethics*. The "self" is based upon "sense" (*aisthēsis*) with its placement in and representation of the world being formulated on this basis. The being will move and sense in outside space based upon the inner calculations and reflections of its *psychē*. Aristotle explains that: "the soul is analogous to the hand; for as the hand is a tool of tools (*organon estin organōn*), so the mind is the form of forms (*ho nous eidos eidōn*) (*De An*. III 432a 1–2)." Both the hands and the soul which is between them as their mental centre are tools of the being, one physical and the other spiritual.[20] We find in nature a fittedness and scaling of both "soul" (*psychē*) and "body" (*sōma*).

5.3 The Movement from Mathematics to Biology

Let us consider how Aristotle moves from mathematical points all the way to biological human beings. I re-summarise Aristotle's position as a movement from a point, a monad, to a biological being (*ousia*) in Fig. 5.2 below.

It is, of course, simply true that Aristotle did not have access to Bohr's atomic model with its three shells. He did, however, clearly understand the power of the circle as the natural structuring of a chemical entity, notably the planets. He also

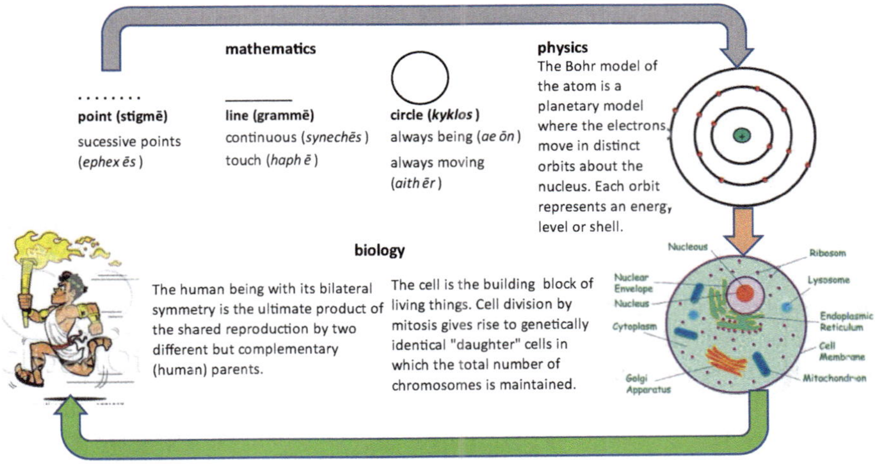

Fig. 5.2 Aristotle's theory of evolution (again)

[20] Aristotle stresses that the "soul" is not abstract or separate from the being but is a mental or spiritual projection of that being from self to body to world on the basis that: "to say that the soul is like fire is like saying that the craftsman, or his craft, is the saw or the augur which he uses (*PA* II 652b 13–15)."

understood how the bilateral symmetry of an organism could evolve from this fundamental structure, as from an ovum. It is hence that we find in Aristotle what we could call an "evolution of shape" or of form, literally morphology. René Thom comes closest to explaining this transformational morphology in his "semiophysics."

Thom explains this situation in principle as that: "For hundreds, maybe thousands of years, Aristotle has been *the only one to think in terms of the continuous*. To my mind, that is his chief merit."[21] As regards Aristotle's own argument from "continuity" (*synechēs*) we find that: "…"continuous" [*synechēs*] as the very name indicates [means] "held together" (*synechētai*) (*Phys.* V 227a13)" whereas "…"next in succession" (*ephexēs*) is a property of things of a higher order of abstraction (*en proterois tō logō*), such as numbers, where there is no question of contact (*haphē*) (*Phys.* V 227a20–21)." Aristotle constantly refers to "*synechēs*" as signifying an organic movement of a formed something and he contrasts this with "*ephexēs*" which is not consolidated but patterned. He uses "*ephexēs*," for instance, to denote a sort of generic framing of possibility for figure and for the hierarchy of the types of soul; see *De An.* II 414b 20–27. Thom explains this ontological principle of being as that: "at most it [i.e. space] is the place of some entity (*ousia*), for it is never empty. This decision to relegate space to a kind of total ostracism [e.g. by Plato] led him [i.e. Aristotle], by a singular rebound, to multiply the kinds of matter. Each type of change (μεταβολή), each genus (γένος), requires a specific matter. But all these matters have one thing in common: they are continua (συνεχές); in the sense they have the character of spatial extension."[22] Aristotle is considering here the fundamental physics of shapes, powers, and resultant movements and states of being.

Thom explains Aristotle's "semiophysics" in detail and profoundly in his work as a real alternative *today*. We should all ponder upon the fact that it is the active and vital challenge that Thom presents to modern thought that is likely the reason why his "semiophysics" has been so studiously ignored by it. Let us hence ourselves move to consider Aristotle and Thom's very careful and systematic conceptualisation of continuity as a holding together of a pathway of events or structures of what can then become an organically connected being with an internalised "end" (*telos*). Aristotle provides us with a full and clear account of this "continuity" in "nature" as that: "natural things are exactly those which do move continuously (*synechōs*), in virtue of a principle inherent in themselves (*apo tinos en autois archēs*), towards a determined goal (*eis ti telos*); and the final development which results from any one such principle is not identical for any two species, nor yet is it any random result (*to tychon*); but in each there is always a tendency towards an identical result, if nothing interferes with the process (*Phys.* II 199b16–18)." Continuity is a speciate principle which seeks an end according with its form unless something "hinders." It is not and

[21] Thom, René *Semiophysics: A Sketch: Aristotelian Physics and Catastrophe Theory* (Redwood City, 1990) p viii.

[22] Thom, René *Ibid.* p 166.

cannot be “random.” The key differentiating principle for Aristotle as for Plato is the “species” or “form” (*eidos*), though expressed on a very different basis.

Aristotle argues that a given being is precisely a given being (*ousia*) that possesses a certain unity (at a point in time) which will “move” in accordance with *its* nature on the basis that: “the real cause of the motion of a ball rebounding from a wall is not the wall but the thrower. So it is clear that in all these cases the thing does not move itself, but it contains within itself the source of motion (*kinēseōs archēn echei*)—not of moving something or of causing motion (*ou tou kinein oude tou poiein*), but of suffering it (*tou paschein*) (*Phys.* VIII 255b 27–31).” Aristotle places emphasis upon the potentiality or power of “*dynamis*” which actualises as the movement of a being in the world. Deborah Achtenberg explains this situation further as regards the different potentialities and pathways of beings as that: “the same movement on a mover’s part may cause different movements in different cases: if, using a certain arm movement, I throw a piece of chalk on a table, the chalk will roll across the table and fall off on to the floor, while if I use the same arm movement to throw some water on a table, it will splatter and sink in. The mover, then, appears to be a cause but not the primary cause of movement. The material of that which I throw is different in the two cases.”[23] Aristotle makes this point through his concept of the “*automaton*” which concerns how a material being naturally “falls”.[24]

According to Aristotle, we cannot understand the world in full if we abstract away its material and its being and the potentialities that reside in them. He adds as regards the nature of abstraction that there is a hierarchy of abstraction at increasing removes from physical being with the salient principle being that: “the mathematician… treats of these things [but] does not treat of them as the limits of a physical body; nor does he consider the attributes indicated as attributes of such bodies. That is why he separates them; for in thought [*viz.* though not the world itself] they are separable from motion (*Phys.* II 193b32–35)”; and see also *N.E.* VI. 1142a 16–19.[25]

---

[23] Achtenberg, Deborah *Cognition of Value in Aristotle’s* Ethics (New York, 2002) p 149.

[24] Aristotle clearly argues that the “middle” presents a space for meaning which is not automated but is in some sense contingently selected. In his *Physics* he argues that there is a role for such random events as a stone falling from a roof and killing someone *by chance* but that “chance” is primarily “contingency” in the sense that: ““*Automaton*” …may be used to describe the behaviour of brute beasts and even of many inanimate things. For instance, we attribute it to *automaton*, if a horse escapes a danger by coming accidentally to a place of safety. Or again, if a tripod chances to fall on its feet for a man to sit down upon, this is due to *automaton*, for though a man would put it on its feet *with a view to* its being a seat, the forces of nature that controlled its fall had no such aim (*Phys.* II 197b14–18)” and that: “neither inanimate things nor brute beasts nor infants can ever accomplish anything by *tychē*, since they exercise no deliberate choice (*Phys.* II 197b 7–8).” “Chance” always takes place in the realm of meaningful beings interacting in the world according to their natures.

[25] On Aristotle’s remarkable theory of abstraction see Michael J. White “The Metaphysical Location of Aristotle’s Μαθηματικά” *Phronesis* (1993), Stephen Gaukroger “The One and the Many: Aristotle on the Individuation of Numbers” *The Classical Quarterly* (1982), Daniel P. Maher “Aristotle on Mathematical and Eidetic Number” *Hermathena* (2011), and W.A. Hammond “The Significance of the Creative Reason in Aristotle’s Philosophy” *The Philosophical Review* (1902).

Aristotle emphasises "the limits of a physical body" as the "contact" between being and world, and he prioritises this physical engagement over any mathematical abstraction or other subtraction from this ontology. Reality is prioritised over abstraction. Whereas we reduce the world to the homogenous and isotropic forces of physics, Aristotle regards a power of continuity, of the line, and of circularity, of the circle or cycle, pervading the world. Whereas our basic unit is the atom, Aristotle's is the cell. Whereas the paradigm of the modern approach is chemical, that of the Aristotelian approach is biological. Whereas the modern mind reduces the biological to the physical, the Aristotelian approach assumes that the biological is based upon and perhaps evolved from the physical.

5.4 Aristotle and the Construction of Being

There are sympathies and similarities between the accounts of being and life by Aristotle and modern philosophers such as Henri Bergson. This does not, however, mean that Aristotle has been superseded by Bergson along with other thinkers such as A.N. Whitehead. Aristotle gives us a distinct and excellent account which peculiarly manages to integrate science, notably biology, into it.

We find that Aristotle's *Posterior Analytics* is concerned with the touching, sticking, standing, and cohesion of being(s) and with the question: "what is the bond of continuity (*ti to synechon*) that makes a present process follow the completion of a past event (*meta to gegonenai to gignesthai hyparchein*) (*Post. An.* II 95b 3–4)." He similarly considers how various things are so constructed at *Met.* H 1042b 16–25 and at length in his *Meteorology* Book IV. We see from this that Aristotle's "physics" is not just a study of movement but also of the nature of physicality-in-the-world and of thingliness. Aristotle argues as regards how this sticking leads to the formation of a "being" (*ousia*) that: "…"to be one" means "to be indivisible" (being essentially a particular thing, distinct and separate in place or form or thought (*hoper tode onti kai idia chōristō ē topō ē eidei ē dianoia*)), or "to be whole and indivisible" (*Met.* I 1052b 16–18)." Being is therefore something that is achieved as a completion of becoming rather than just as the beginning of a being. The creation or becoming of a being is as significant and as ineffable as its full maturation or being.

This formation of a "being" leads us to consider the mathematical principle of "contact" (*haphē*) which can be used to explain the biological fact of "touch." Aristotle explains "contact" as that: "when bodies come into contact or are separated, their boundaries instantaneously become one at one time—when they touch, and two at another time—when they are separated (*Met.* B 1102b 1–3)." Touch is hence a connection between beings, a transfer of information between them which in a sense unites them over an interval. Aristotle's ultimate wider suggestion is that "being" is built up from empirical instances as a sort of channelling, accumulation, or actualisation of being through epigenetic memory and the imprinting of being. Sense in this sense becomes being. Aristotle hence argues that: "Organic unity differs from contact (*diapherei de symphysis haphēs*); for in the latter case there need not be anything besides the contact, but in organic unities there is something identical in both parts, which makes them grow together (*sympephykenai*) instead of merely touching (*Met.* Δ 1014b 23–25)." We take information from the world through "touch" or "contact" (*haphē*) but a difference importantly remains between the toucher which must be an entity and the touched which is merely a surface. Through "organic contact" differences diminish and the sides "grow together" to become a being (*ousia*).

This hypothesis on eidetic emergence is most famously recounted in the conclusion of the *Posterior Analytics* which regards how particulars "take a stand"; see *Post. An.* II 100a 15–100b 2. As regards how the instances produce the being, Aristotle observes that to be "one" is to be "something" determinate which becomes "active" or "actual" and hence *loses* its "prior" and "posterior" in any meaningful sense since "we go on packing the space between until the intervals are indivisible or unitary (*Post. An.* I 84b35–36)." Aristotle is concerned with the fact that a being *forms* or emerges from the world and becomes a being from a process in time of becoming. Aristotle also adds that this "growing together" can become a malformation—as with Siamese twins—when the principles of growth and combination are confused or pressurised; see *Met.* Z 1040b 13–16. His point here, clearly drawn from his study of foetuses but applicable to the processes of the animal world in general, is that there are natural processes which create and develop beings when placed in the right place and given the right time. None of this emergence is attributable to chance since the process has and reaches for its "end" which in the case of a human foetus is a human being.

How does Aristotle account for "creation" and "formation"? For Aristotle, the space of the "middle" (*to meson*) and the contingency of "chance" (*tychē*) only exist due to the nature of the "sublunary" world. The world is meaningful in an entangled rather than mechanical way. The "middle" exposes to Aristotle the fact that the world operates on the principle of quasi-necessity in which, once initiated, things in the world will *normally* but not *necessarily* complete according to their nature "if nothing interferes (*an mē ti empodisē*)." It is, however, only with the emergence of the "final cause" that we find a real shift from quantity to quality on the basis that: "The final cause produces motion as being loved, but all other things move by being moved (*Met.* Λ 1072b 3–4)." We see that an active discretionary power of love and thought emerges in the biological world which is increasingly distinct from the purely material cycle of the physical world. Aristotle adds the complexity that one cause of man is the sun's "ecliptic" (*ho loxos kyklos*) which man then surpasses on the basis that: "We must, then, think of the action of the lower stars [i.e., of the sun and moon] as similar to that of animals and plants. For on our earth it is man that has the greatest variety of actions (*De Cael.* II 292b 1–3)."[26] Aristotle again arrives

[26] Aristotle argues in full regarding the "cause" of man that: "the cause of a man is (1) his elements: fire and earth as matter, and the particular form (*ta stoicheia [kai] to idion eidos*); (2) some external formal cause, *viz.* his father (*ti allo exō*); and besides these (3) the sun and the ecliptic, which are neither matter nor form nor privation nor identical in form with him, but cause motion (*ho hēlios*

again at the concept of the "circle" or "cycle" as an ideal and also regards an evolution of complex movement.

We can add here that we do factually know *that* but not *why* our life is measured by the circadian and circalunar cycles of the cosmos. Aristotle interprets this situation as that the cycling of the planets does contain a cosmic significance that he describes as a pure existence without a "middle" (*amesa*) of pure actuality which is "*aiōn*" ("always being") and "*aithēr*" ("always running").[27] Aristotle therefore provides us with two variants, a cosmic view and a human view. On the one hand, the cosmos represents a massive system of energy and of mathematical perfection which represents something "divine." On the other hand, humanity occupies the highest expression of creation and creativity that is derived from the cyclical generativity of nature which the animal world imitates but needs to struggle to achieve through a "middle." It is hence that humanity can manifest both "art" (*technē*) which imitates the process of nature "*peri genesin*" and "science" (*epistēmē*) which understands being "*peri to on*"; see *Post. An.* II 100a 6–9. As regards how human beings stand in relation to other animals we find that: "other animals mostly live under the guidance of nature alone, although some are guided a little by habit (*ethos*) (*Pol.* VII 1332b 4–5)" (and see also *Poet.* 1456b 23–24) with our full situation being that animals "make things neither by art nor after inquiry or deliberation…[and] people discuss whether it is by intelligence (*nous*) or by some other faculty that these creatures work (*Phys.* II 199a20–22)."

We ultimately have two hard problems in respect to "movement"; what is the meaning of the necessity of nature, and what is the meaning of the creative expression of man? We again return to the question of the nature of "chance" and "necessity, but Aristotle's response is not vague and general but detailed and realistic. Aristotle argues as regards human "mind" that: "human beings live under the guidance of reason as well [as desire], since they alone have reason (*logos*). Consequently, all three of these factors [i.e., nature, habit, and reason] need to be harmonised (*symphōnein*) with one another (*Pol.* VII 1332b5–6)." The human being has acquired elevated powers in all three directions and a peculiar self or ego from the need to "harmonise" these diverse aspects of human being. It is the *lack* of automation and the *need* for humanity to "know thyself," to integrate a sense of our being and our direction and end in our minds, that defines the human being. The human mental representation of the world is in a sense an expression of the world itself and its objects, but it is also a "pausing" upon the world which allows the development of our relationship with the world through reflection.[28] It is hence that: "human

kai ho loxos kyklos, oute hylē onta out' eidos oute sterēsis oute homoeides alla kinounta) (*Met.* Λ 1071a 14–18)."

[27] These etymologies are Aristotle's; see *De Cael.* I 270b 23 & *De Cael.* I 279a 23.

[28] Aristotle explains thought as a "pause" as that: "thinking seems to be more like a state of rest or a halting than a movement and the same thing is true of the syllogism (*De An.* I 407a 34–35)." Gavin Lawrence explains this principle as that: "…"*Episteme*" is a coming to a stop, a *stand-still*: *eph-istemai* (an etymology Aristotle explicitly plays with in *Post. An.* II 100a3-b5). Taking episteme here to be a matter of understanding, of having *the why* as well as *the that* (for example *Meta.*

thought…does not possess the good at this moment or that, but its best, being something different from it, is attained only in a whole period of time (*Met.* Λ 1075a7–10)." We transcend nature by passing through it, internalising its possibilities, and deepening and transforming its meaningfulness within ourselves.

We again see how humanity is defined by its discontinuities. It is both elevated and integrated (and we will consider the importance of "integration" later) but it is also divided, with this division somehow allowing its self-reflection. Aristotle's theory of "formation" is explained further and well by John Peter Anton as that: "The fulfilment of intelligence will become *form*. Prior to its operations on the content of sensibility, intelligence is but a power. Experience is the necessary condition for the actualisation of intelligence; it is its material cause, as it were…Intelligence as an accomplishment, as active *nous*, becomes knowledge distilled, communicable, independent, universal; it is a good of the rarest kind, in fact it is the good. It is the intelligibility of the universe stated in the medium of human *logos*. Intelligence as a power is individual, distinct, a given capacity in individual human beings; it is mortal. But intelligence fulfilled becomes immortal as a vision of the principles of nature and logic; it is immortal in its communications of the permanence of nature and the eternality of its structures. Here man changes his psychological ego into philosophical universality. Not that man becomes eternal in reality; he comes to understand the meaning of eternality by making his intelligence what the universe is in actuality."[29]

5.5 The Architecture of Meaning

As so often, the excesses and simplifications of our modern age has muddied our thought on even basic points of reality. Aristotle "craft metaphor" has often been conflated with the "watchmaker argument" which posits that mechanistic function

A.1), the idea is that when one understands, one has come to a stop: enquiry is now ended, and the soul is stilled and settled, properly stable—not (so) capable of being misled or overturned by a false theory as when someone only has *the that's* ("Acquiring Character: Becoming Grown Up" in Pakaluk, Michael and Pearson, Giles (*eds.*) *Moral Psychology and Human Action in Aristotle* (Oxford, 2012) p 274)."

[29]Anton, John Peter *Aristotle's Theory of Contrariety* (London, 1957) p 163 and 167. Markus Gabriel adds that: "our intellect is divine by thinking the thought that there must be a principle, the substance of which is its activity. In this manner, at the peak of its activity, it touches (θιγγάνων) God who is nothing but a transcendent pure activity of being, an *actus purus*. It does so by engaging in the active thought that God is a pure activity. Thus, it achieves the state of a pure self-awareness, reflecting upon its very activity of thinking. As a consequence, it has no other object than itself without thereby ever becoming itself a transcendent pure activity. Our intellect's pure self-awareness comes as close to the divine as possible by achieving a state of thinking the thought that God is a principle, the substance of which is its activity and nothing else ("God's Transcendent Activity: Ontotheology in "Metaphysics" 12" *Review of Metaphysics* (2009) p 401–2)." See also the various work of Aryeh Kosman on this situation.

must infer a "maker," an argument explicitly denied by Aristotle.[30] Rather, when Aristotle considers the philosophical implications of mundane and banausic reality, he is setting forth a worked example of the depth of meaning which evidently infuses our world in all its aspects. He sets forth his "ergon argument" in his *Ethics* which considers the structures of human being. He uses "*technē*" here as a parallel by means of which to explain the nature of "virtue" and human habit (ethics) and he also, as we will see below, produces a penetrating parallel between the ergon of craft and the ergon of biological form in his biological works. The inability of modern thought to self-critique has led to such weak arguments as Thomas Nagel's supposed refutation of Aristotle's "ergon argument" by reference to a "combination corkscrew," as discussed earlier.

How does Aristotle represent craft? He routinely refers to the example of a house, the art of construction, and the builder who possesses the art and produces the houses. He also routinely refers to carpentry, shoemaking, medicine, and sculpting. He makes it clear that his detailed consideration of art and of the production of tables by artists is intended as a correction to Plato's conclusion that a "table" should be explained as a copy of an ideal "Table." We do not, he suggests, "remember" the table or the knowledge of how to make it but this knowledge rather arises in us through experience and through indoctrination. I give Aristotle's representation of the situation in Fig. 5.3 below.

We see in an elevated sense Karl Popper's three "worlds"—of the physical, the mental, and the product of the two.[31] We see the human being who possesses the mental skill and has the physical tools and produces products, e.g. tables and chairs, which are the product of physical and mental energy combined. In more detail, Aristotle's "master-craftsman" (*architectōn*) is not a mere technician but is one who understands the art of which he is a master, the tools which are both the physical and ideal vehicles of his craft, the styles and nuances of the "form" produced according to its designated "function," and the final milieu of both the product and the craftsman.[32] In this case this is the "*oikos*" (home) with this being the root of "ecology" as a world created by (biological) beings.

Let us observe in Fig. 5.4 below how very different beings are transformed through craft into something very different.

We see the importance of "process" (*kinēsis*) as well as of "being" (*ousia*). We see the hylomorphic combination of "form" (*eidos*) with "matter" (*hylē*) and also the significance of Aristotle's four causes at play; the formal, moving, material, and

[30] Aristotle argues that: "It is absurd to suppose that purpose (*heneka*) is not present because we do not observe an agent deliberating (*Phys.* II 199b26–28)."

[31] See Karl Popper "World 3 and Emergent Evolution" in *Knowledge and the Body-Mind Problem* (London, [*orig.* 1969] 1994) and the other essays in this volume.

[32] For further consideration of the *architectonic* "master craftsman" see Claudia Baracchi *Aristotle's Ethics as First Philosophy* (Cambridge, 2008), David Charles "Wittgenstein's Builders and Aristotle's Craftsmen" in Sharples, R.W. (*ed.*) *Perspectives on Greek Philosophy* (Aldershot, 2003), and Monte Ransome Johnson "Aristotle's architectonic sciences" in Ebrey, David (*ed.*) *Theory and Practice in Aristotle's Natural Science* (Cambridge, 2015).

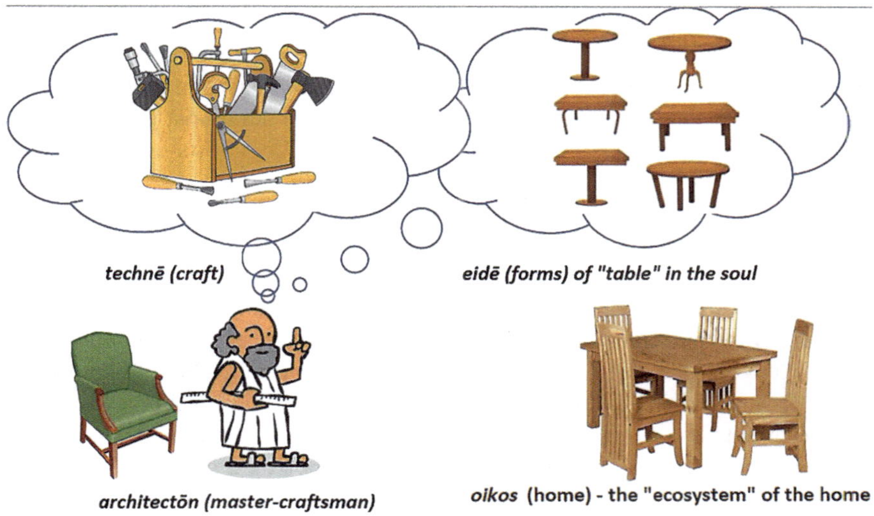

Fig. 5.3 The craftsman and his "world"

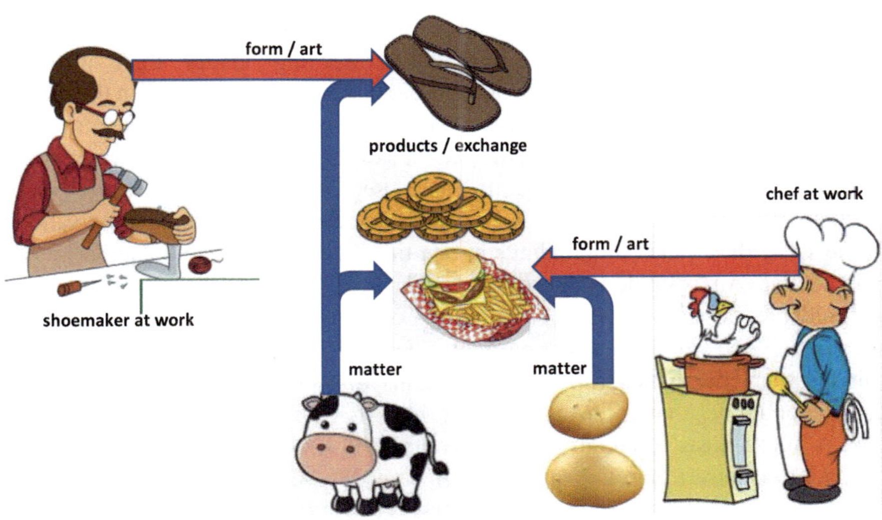

Fig. 5.4 The "entangled" world of craft

final causes. At the centre of the picture, literally and symbolically, we find money (*nomisma*) which is the symbolic "means" by which "goods" (aptly named as signifying products of thought and labour) are exchanged.

Aristotle explains the craft metaphor in relation to biological reproduction as that: "it is obvious that only one table can be made from one piece of timber, and yet he who imposes the form upon it, although he is but one, can make many tables. Such too is the relationship of male to female: the female is impregnated in one coition, but one male can impregnate many females. And these relations are analogues of the principles referred to (*Met.* A 988a3–8)." Aristotle moves beyond abstract theory to consider the implications of the real divisions of nature. He adds regarding this entanglement that: "the house and the builder do not perish together (*Met.* Δ 1014a24–25)," i.e., the artefact takes on a "being" of its own after it has been enformed and produced by its creator.

He explains the situation in detail in his *Generation of Animals* as that: "nothing passes from the carpenter into the pieces of timber, which are *his* material, and there is no part of the art of carpentry present in the object which is being fashioned: it is the shape and the form which passes from the carpenter, and they come into being by means of the movement in the material. It is his soul, wherein is the "form," and his knowledge which cause his hands (or some other part of his body) to move in a particular way (different ways for different products, and always the same way for any one product); his hands move his tools and his tools move the material. In a similar way to this, Nature acting in the male of semen-emitting animals uses the semen as a tool, as something that has movement in actuality; just as when objects are being produced by any art the tools are in movement, because the movement which belongs to the art is, in a way, situated in them (*en ekeinois gar pōs hē kinēsis tēs technēs*) (*GA* I 730b11–25)." Aristotle considers what is entailed in the biological act of creation by reference to the creation of products as a craft analogy.

Aristotle comments further in principle (1) regarding "matter" that: "different things can be generated by the moving cause (*tēn kinousan aitian*) when the matter is one and the same, e.g. a chest and a bed from wood. But some different things must necessarily have different matter; e.g. a saw cannot be generated from wood, nor does this lie in the power of the moving cause, for it cannot make a saw of wool or wood (*Met.* H 1044a25–30)," (2) regarding "motion" that it covers "motivation" in the sense that: "each thing admits at one time of being active, at another of not being active. An example is the buildable. The being at work of the buildable, just as buildable, is building…But building is a certain motion. And surely the same account will exactly fit the other motions (*Phys.* III 201b10–13)" with this "practical syllogism" being most fully explained in his *Movement of Animals*,[33] (3) regarding "potentiality" that nature is pregnant with *transferable* meaning in the sense that: "a house exists potentially if there is nothing in X, the matter, to prevent it from becoming a house…similarly in all other cases where the generative is external (*exōthen hē archē tēs geneseōs*) (*Met.* Θ 1049a9–13)," (4) regarding "being" that the product itself *must always be something* in the sense that "if it comes to be, it is its concrete unity that comes to be (*Met.* Z 1034b12)" and (5) regarding "mind" that:

[33]Aristotle's interest in nutrition and animal movement and instinct leads him to what has been considered as an attempt to formulate a "practical syllogism" along the lines: "My appetite says, I must drink; this is drink, says sensation or imagination or thought, and one immediately drinks (*euthus pinei*). It is in this manner that animals are impelled to move and act, the final cause of their movement being [appetite] (*MA* 701a34–5)"; see Michel Crubellier "Le "syllogisme pratique" ou Comment la pensée meut le corps" in Laks, André & Rashed, Marwen (*eds.*) *Aristote et le mouvement des animaux* (Villeneuve-d'Ascq, 2004).

"artefacts like houses and statues which arise "from thought" *never* arise from necessity (*Post. An.* II 95a4–5)."[34]

Aristotle uses the analogy of craft for a very different purpose than the "watchmaker argument." He uses this analogy to demonstrate (1) that matter is a potentiality for various forms, (2) that the creative principle exists as a principle which can be applied in various ways, (3) that the creative principle most properly works through an external principle acting upon something else, (4) that the world exists as whole beings so that matter is one thing *or* another, e.g. carbon monoxide *or* carbon dioxide, so that change occurs through formal catastrophes rather than through material gradualism, and (5) that thought is a creative principle which must work through the world and hence is never merely a material mechanism. The watchmaker argument is not as philosophically risible as it is often taken to be, adn nor is it Aristotle's position.

5.6 Aristotle and the Meaning of "Means"

That the world is a world of "beings" is clearly at the centre of Aristotle's thought along with the issue as regards how and why these "beings" come-to-be. The "mean" (*to meson*) is on this basis a core of Aristotle's philosophy not just as one of his core theses about the construction of our world but as regards the very purpose of his philosophy. Our sublunary world is itself defined by contrast with the cosmos by its possession of a "means." Aristotle also corrects Plato on the basis that meaning in the world emerges *from* the "means" and not *through* ideas which descend into the world and "participate" in it. The "moving cause" is on this basis Aristotle's key innovation.

We can, drawing on the passage *GA* I 730b11–25 as quoted above and Aristotle's work generally, represent some of the key "media" involved in production (from "*pro-ducere*"; to bring forth) as per Fig. 5.5 below.

The above figure unpacks Aristotle's "middle" world of movement. We find, first, the "medium" in which we exist which can be air (for land animals) or water (for marine animals) and we also live in a categorical world of space and time and also live through the light through which we find direction. Second, we have the "medium" of our mental and spiritual world which is driven by the images and conceptions with in us. We also have a body which is a centre between our eyes and

[34] Rèmi Brague explains this last point brilliantly as that: "Our gaze sees what is buildable (the materials, that is) because it appears to us in a determinate and graspable form; but the buildable as such—what in the materials, makes them apt to be assembled into a building—is not something we could factually ascertain. It is something that we "say." It belongs to the possible and not to the real. We grasp the possible as such, and we ascribe to it something like "being" by means of a logos…The paradigm of "seeing" is superseded…in favour of something like "saying"…That a block of marble is a possible David may only be discovered by a faculty which Aristotle here implicitly calls "*legein*" (to say) ("Aristotle's Definition of Motion and its Ontological Implications" *Graduate Faculty Philosophy Journal* (1990) p 17–18)."

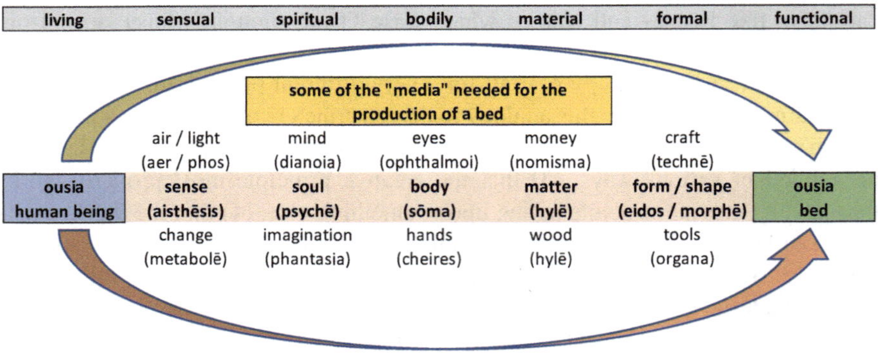

Fig. 5.5 The meaning of the "means"

hands. Third, we have "matter" (*hylē*) which is a technical term taken from "wood" (*hylē*) as "*ousia*" originally meant "possessions, property." "Money" (*nomisma*) is an archetypal "means" based upon the word "*nomos*" (law, convention). "Law" (*nomos*) and "speech" (*logos*) are other human tools which act as "media" shaping our milieu. Fourth, we have the "media" peculiar to the craft itself, its tools and shapes. The final outcome is that one *living* "*ousia*" produces an *artificial* "*pragma*." The bed is constructed for the form of the living being as also shoes or medicine.

It is hopefully clear how useful and fertile Aristotle's craft analogy actually is for showing how one being can implant forms (*eidē*) and principles (*archai*) in another on the basis of the principle that: "there is nothing paradoxical in the actualising of the potentiality of one thing having its effective seat in another (for the potentiality of an agent—the "teacher" for instance—is actualised not *in vacuo* or in isolation, but in the "taught": it is the actualising *of* this *in* that) (*Phys*. III 202b6–8)." I represent this situation in Fig. 5.6 below.

We can return to Socrates as a "midwife" for the ideas of students. The teacher can teach about a subject area about which he has knowledge based on his experience, having synthesis based on analysis.[35] This transfer of an "object" of knowledge from one person to another represents the transfer of a spiritual or mental rather than a physical good. Aristotle argues that the world is irreducibly complex as an evolution and cycling of forms (or species or ideas) in matter.

Aristotle adds that the world is a "middle" though which principles or seeds of being—called "*archai*"—pass through the world to their ends or "*telē*." Nor are these movements all in the same direction but some move from one mind to another,

---

[35] Aristotle's position is "synoptic" in the sense coined by Plato to describe his own dialectical position at *Rep*. 537C as a "seeing the whole together" and "taking a comprehensive view." Plato also employs the apt simile that we are "walking around and examining from all sides" the subject of our inquiry (see Jon Moline's *Plato's Theory of Understanding* (Wisconsin, 1981) p ix and xi). Aristotle also argues that we should be able to "walk around" a city, i.e., a society should retain a human scale, and that we must build up a knowledge of the properties of something in order to understand its essential being; see *De An*. I 402b 22–25.

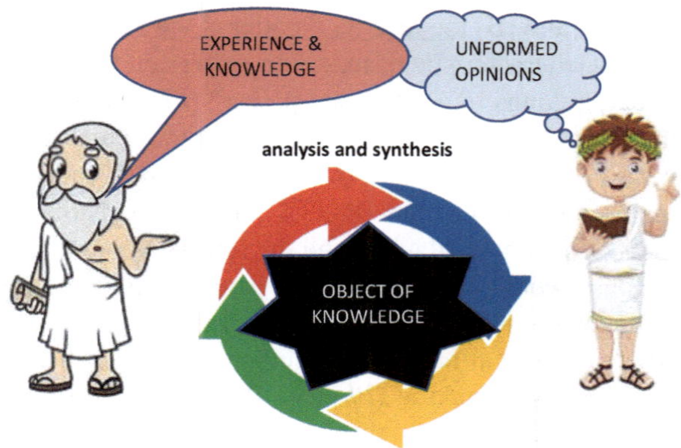

Fig. 5.6 The cycle of knowledge

Primary Means	Inner	Outer	Movement	Nature
Mother	Child	Sperm of Father	Body to Body	Nature / Nurture
Builder	Form of House	House	Mind to World	Craft / Virtue
Poet	Poem	Experience of World	World to Mind	Art / Imagination
Doctor	Health	Body of Patient	Mind to Body	Knowledge / Treatment
Teacher	Knowledge	Mind of Student	Mind to Mind	Knowledge / Indoctrination

Fig. 5.7 The pathways of principles in nature

some from one body to another, some are internal representations and others external expressions as set out in Fig. 5.7.

The first and key "cycle" is the reproductive cycle of nature through which the principles of male and female come together to produce and then raise a child. The second is the craftsman through which a human being physically projects himself in the world by replicating a form which exists in his soul to the world outside. The third is the poet who is literally a "maker" who creates a world through himself but does so by "imitating" the world, as we will see in more detail below. The doctor creates through his understanding nature and its principles "health" in another body and the teacher imparts knowledge in another mind. It is perhaps in this sense that Aristotle does regard the world as a circulation of forms and ideas but with his ideas being embodied or "living" ideas and not ossified abstractions.

Aristotle himself observes as regards science and medicine, as other systematised transfers of knowledge, that their very existence require us to reject both: "[A] [the idea that Forms alone explain worldly being since] if the Forms are causes, why is their generating activity intermittent instead of perpetual and continuous (*synechōs*)...Besides, in some instances we *see* that the cause is other than the Form. For it is the doctor who implants health and the man of science who implants science [and [B] the idea that matter alone explains worldly being on the basis that] it is the characteristic of matter to suffer action, i.e., to be moved: but to move, i.e., to act, belongs to a different "power" (*heteras dynameōs*). This is obvious both in the things that come-to-be by art and in those that come-to-be by nature. Water does not of itself produce out of itself an animal; and it is the art, not the wood, that makes a bed (*GC* 335b 18–33)." Aristotle adds the interesting thought experiment that: "if you planted a bed and the rotting wood acquired the power of sending up a shoot, it would not be a bed that would come up, but wood—which shows that the arrangement in accordance with the rules of the art is merely an incidental attribute, whereas the real nature is the other, which, further, persists continuously through the process of making (*Phys.* II 193a 12–17)." His point of emphasis here in this thought experiment is that we impose our own artistic form upon the matter of nature which is the primary nature of being and will reassert itself as such. He similarly observes regarding living beings that they revert to an elemental form once the body does not possess the integral principle of life, this being an early statement of negentropy.

The craft analogy is useful for Aristotle owing to the open clarity it provides as a parallel for natural processes although he adds that the full picture is that: "virtue is more exact and better than any art, as nature also is (*N.E.* II 1106b 14–15)." Both nature and virtue (good human nature) are more complex phenomena than crafts which are in themselves a more peripheral example of being. Craft is, however, useful as a parallel which exposes the principles which also exist behind nature and behind human nature and its "goods," "ends," and "virtues."

5.7 Building a Wall on Aristotelian Principles

It is perhaps truer to call Aristotle's philosophy an "archology" (from *archē*) but we can also explain it through a term that has been passed down to us: "aetiology" (from *aitia*). The modern etiological approach chiefly studies and diagnoses the causes for movements away from the health of an organic being. Aristotle's aetiology is rather a *positive* science of living being as opposed to a reactive science of diseased being. The basis for this perspective is that Aristotle regards the positive principles or "causes" of the real world and does not merely pragmatically assume chains of causation or idealistically assume that they are merely an absolute expression of "divine" or other "higher" principles. This is Aristotle's realism in a nutshell. It is an insistence that meaning and being emerges from the reality of the world, i.e., what we see is what we get, but with this meaning being assumed to be a positive expression of the nature of being rather than some sort of chance.

It is hence that a man building a wall becomes significant. Aristotle's opponents seek to explain it on a religious or pragmatic basis, as per Fig. 5.8 below.

This Platonic-Cartesian worldview has meaning flowing down from above (as a "descent of Ideas") to the world as a concentrated (divine and/or human) "will" exercised on matter (*res cogitans* applied to *res extensa*). Plato's account admirably explains

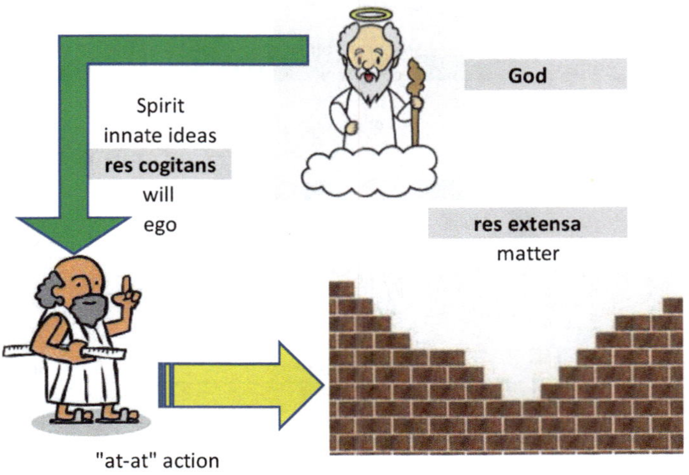

Fig. 5.8 The Platonic-Cartesian model for being and building

the fact of "Being" that there is a wall and a word for a wall and an idea of a wall but does not so satisfactorily explain the becoming of the wall and the building which he attributes to a "reminiscence" of divine Ideas. As always, Plato's position is both built upon and refuted by Aristotle, in this case in his *On Memory and Reminiscence*. It is because Descartes reduces all meaning in the world to the mind, will, and ego of human beings that he must argue the seemingly strange point that animals have no souls or minds (or, in a sense, bodies) but are merely inanimate machines.

Aristotle's philosophical explanation for the building of the wall is represented in Fig. 5.9 below.

The Aristotelian ontological and metaphysical worldview is centred upon "beings" and recognises a "nature" (*physis*) through which all organic beings act (or behave) on a speciate basis in the world. The full complexity of this situation is that a human being becomes a builder once he has the requisite "form[s] in his soul" of the building craft or *technē* and he must then go out and find the matter so as to match it with his form and hence produce an end-product, in our case a wall. Both the craft-form and the artifact-form itself, e.g., the wall, only exist in the human soul and not in nature.

Aristotle is interested in the emergence of "function" (*ergon*) and of an "organ/tool" (*organon*) in nature. He considers the human situation because of what it tells us about this situation. Everything peculiarly human has its "traces and seeds" (*ichnē kai spermata*)" (see *HA* VIII 588a 34–588b 3) in the animal world. He observes on this situation that: "it is thought that every animal has its own special pleasure (*hēdonē oikeia*), just as it has its own special function (*ergon*): namely, the pleasure of exercising that function (*kata tēn energeian*). This will also appear if we consider the different animals one by one: the horse, the dog, man, have different pleasures—as Heraclitus says, an ass would prefer chaff to gold, since to asses food

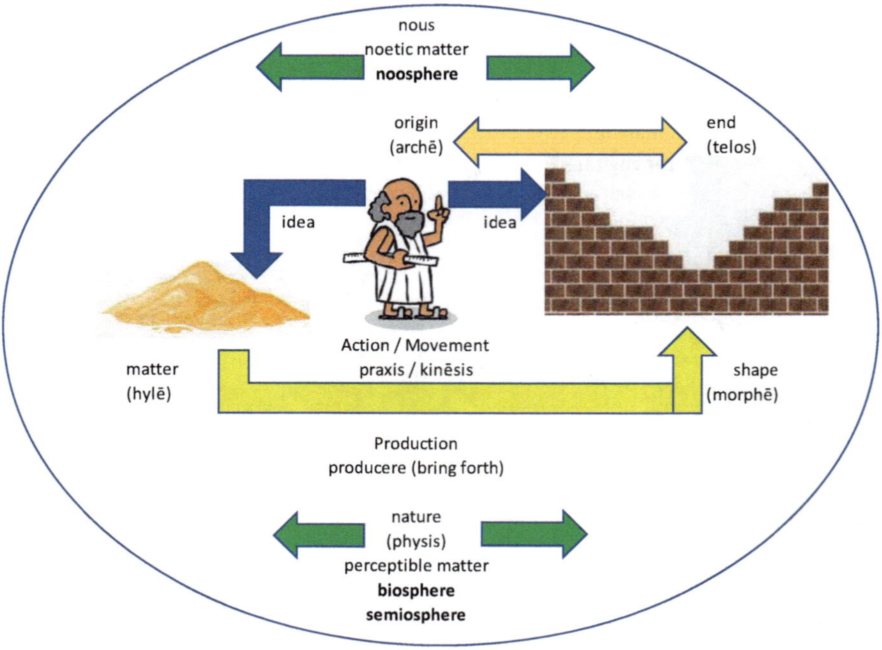

Fig. 5.9 The Aristotelian cycling of being and action

gives more pleasure than gold. Different species therefore have different kinds of pleasures (*N.E.* X 1176a 3–9)." Aristotle seems to argue that "being" is in some sense a product of the "pleasure" (as well as the form, matter etc.) which is "proper" to a being. As regards how pleasure informs and "completes" the lives of human beings he argues that: "one cannot get the pleasure of the just man without being just, nor that of the musical man without being musical and so on (*N.E.* X 1173b29–31)" and that: "pleasure (*hē hēdonē*) completes the activities (*teleioi tas energeias*), and therefore life (*kai to zēn*), which they [i.e., people] desire (*N.E.* X. 1175a15–16)."[36] Aristotle typically does not regard pleasure merely as a superficial addition to action but as a meaningful dimension of nature driving the expression of action.

[36] Aristotle's clear and subtle explanation of the nature of "pleasure" is that: "It is not right to say that pleasure is perceptible process, but it should rather be called activity of the natural state (*energeian tēs kata physin hexeōs*), and instead of "perceptible" "unimpeded" (*anti de tou aisthētēn anempodiston*) (*N.E.* VIII 1153a 12–15)." He argues that pleasure is a factor of sense which "completes" the life of animals on the basis that: "all animals have one sense at least, *viz.* touch, and whatever has a sense has the capacity for pleasure and pain and therefore has pleasant and painful objects present to it, and wherever these are present, there is desire, for desire is just appetition of what is pleasant (*De An.* II 414b 3–7)" with there being a scale of pleasures such that: "As their activities are diverse, so also are their pleasures. Sight excels touch in purity, and hearing and smell excel taste; and similarly the pleasures of the intellect excel in purity the pleasures of sensation, while the pleasures of either class differ among themselves in purity (*N.E.* X 1175b 37–1176a 3)."

The significance of the human being to Aristotle is that we reveal the nature of nature itself by studying its—and our—nature. Aristotle puts forward the remarkable hypothesis that the human being as a naked and strangely unformed animal must find and take on its functions through the peculiar plasticity of its hand and mind rather than, as with normal animals, being fitted and formed to a function by nature.[37] The lack of innate form and function as a "blank slate" requires the use of "mind" for reflective and creative thought and also alienates us and detaches us from nature in a peculiar manner. Aristotle is, of course, famously aware of the problem of Idealism that arises from a philosophy of abstraction. To deal with this problem he concentrates upon proving the existence of a creative pathway of nature through his study of the human example in nature.

It is this pathway, this "middle," which requires Aristotle to formulate his own new "moving cause" and "final cause" alongside the "formal cause" and the "material cause" of his predecessors. His new causes are *explicitly* manifested in human beings and through the "craft analogy" but are *implicitly* manifested in nature itself.

5.8 Aristotle and the Meaning of Man

The instructive peculiarity of the human being has several aspects. First, we see that human beings represent an explosion of "ends" (*telē*) with each individual—whether tinker, tailor, or candle-stick maker—having different "ends" and different "forms" in their souls. As regards these forms in general, Aristotle argues that they are a canalisation of being in the sense that: "The form of which we have spoken is a kind of power immersed in matter—a duct, as it were (*Touto de to eidos aneu hylēs, oion aulos, dynamis tis en hylē estin*) (*GC* I 322a 28)." This is how a being moves to its ends. As regards the forms "in the soul" of human beings we find that they are peculiar as an imitation of formal nature in the sense that: "Things are generated artificially whose form is contained in the soul (*apo technēs de gignetai hosōn to eidos en tē psyche*) (by "form" I mean the essence of each thing, and its primary substance (*eidos de legō to ti ēn einai hekastou kai tēn prōtēn ousian*)) (*Met.* Z 1032b 1–3)." Such human expressions as "art" (and, we can add, "science") are similarly explained as human *formations* on the basis that: "generally art either completes

[37] Aristotle observes on the nature of the hand: "Now it would be wrong to say, as some do, that the structure of man is not good, in fact, that it is worse than any other animal. Their grounds are: that man is barefoot, unclothed, and void of any weapon of force. Against this we may say that all the other animals have just one method of defence and cannot change it for another: they are forced to sleep and perform all their actions with their shoes on the whole time, one might say…For man, on the other hand, many means of defence are available, and he can change them at any time…Take the hand, this is as good as a talon, or a claw, or a horn, or again, a spear or a sword, or any other weapon or tool: it can be all of these, because it can seize and hold them all. And nature has admirably contrived the actual shape of the hand so as to fit in with this arrangement…One finger is placed sideways: this is short and thick, not long like the others. It would be impossible to get a hold if this were not placed sideways as if no hand were there at all (*PA* IV 687a 23–687b 14)."

(*epitelei*) that which nature is unable to make work, or imitates (*mimeitai*) it. So if that which is in accordance with art (*kata technēn*) is for the sake of something (*heneka tou*), it is clear that so too are things in accordance with nature (*kata physin*) (*Phys*. II 199a 15–18)."[38] Aristotle stresses both that human beings create *within nature* and also that they have the power *to use nature* for their own ends.

Second, we find that the possession of artificial forms in the human soul engenders a creative power in human beings which is not present in the other animals. If we see the "form" as being like a seed which will channel a form of being, we see that the human ability to create "forms" also enables human beings to create their own paths and lives through these creative "*archai*" and their "*telē*. Third, Aristotle argues that man achieves his "ends" and his "pleasure" *beyond* the direct remit of his own physical "form" or "species." Whereas other species have their ends *as* themselves so that "perhaps the latter two [causes i.e., final and formal] are the same (*Met*. H 1044b 2)" human beings also express finality *through* our lives as human beings. This sense of finality is consequential to the detachment of man from nature which results from his development of an independent reasoning power separate from the instinctual pleasures of his species, this being the effect of "*nous*" rather than "*dianoia*."" It is hence that Aristotle concludes his *Ethics* by consideration of the *sophos*, the philosophical man, who can interpret the world in such a way as to surpass his human being as a whole. The *phronimos* and the *spoudaios* also paradigmatically transcend their biological human being in other ways.

We finally see that human beings imagine a whole "world" on a basis which seems to mirror but go far beyond the "*Innenwelt*" of other animals. How bio-logy becomes bio-semiotics is a mysterious though important transformation.

5.9 Conclusions

We ultimately see that Aristotle does address reality *through* mathematics. He considers how the world is geometrical in the sense of expressing itself *through* geometrical shapes or forms. He also considers how the world is material in the sense of expressing itself *through* matter. In respect to the "beings" which move *through* the world he argues that they have beginnings (*archai*) and ends (*telē*), matter and form, parts and whole, and becoming and being. The organic beings of the world are beings with an identity moving towards their ends. Aristotle's physics—his representation of nature (*physis*)—is centred around the meaning that arises from the contact (*haphē*), continuity (*synechēs*), and separation (*chōristos*) of being(s) in the world. Aristotle's "nature" is a positive emergent meaningfulness with its paradigms being the growth of a foetus into a mature adult which reproduces and dies and an

[38] Aristotle elsewhere questions whether these images in our soul are genuine "forms," by noting that: "many things are generated (e.g. house, ring) of which we hold that there are no Forms (*Met*. A 991b 7–8)."

ecosystem such as a forest which is rich with life because this is nature's norm. This is his basic scientific worldview. It is based upon the concept of emergent meaning rather than mechanical rules.

Aristotle's physics ultimately provides us with a reasonable means for understanding the chemical, biological, mental, and cosmic which also saves the phenomena. The relationship of each "sphere" within the whole is explained through a reasonable working hypothesis. The "habits" (*hexeis*) of beings are taken to express a natural striving to replicate and build upon the order of nature.[39] In all cases, Aristotle avoids an argument premised upon divine intervention, rejecting an explanation from divine intervention in his *On Divination in Sleep*; see *Div*. 463b 12–15. It is hence that he places such emphasis upon the "craft analogy" that I have represented above as the building of a wall. The full significance of this process and this act of creation is that "craft" is a circular process through which the plane of the noosphere interacts with that of the biosphere and of the semiosphere to realise meaning in the world. We see that we have not merely one world of "meaning" but many, all nested and interacting. The pivot of our lived reality is, however, the biosphere which acts as a platform for the noosphere and exists within the semiosphere of the world. I stress here the pedagogical and persuasive power of Aristotle's argument from blunt and unavoidable reality. This can be discerned in the clear and penetrating argument that: "man is a moving principle (*archē*) or begetter of his actions as of children (*N.E.* III 1113b 17–19)" which *forces us* to recognise the parallelism between our biological and our mental/ethical/habitual/acting being.

For Aristotle's basic theological worldview, he presents us with the picture that: "Coming-to-be and passing-away will, as we have said, always be continuous (*aei...synechēs*), and will never fail...And this continuity has a sufficient reason. For in all things, as we affirm, nature always (*aei*) strikes after the better. Now being (we have explained elsewhere the variety of meanings we recognise in this term) is better than not-being; but not all things can possess being, since there are too far removed from the principle (*archē*). God therefore adopted the remaining alternative, and fulfilled the perfection of the universe, by making coming-to-be uninterrupted; for the greatest possible coherence would thus be secured to existence, because that coming-to-be should itself come-to-be perpetually is the closest approximation to eternal being. The cause of this as we have often said, is circular motion; for that is the only motion which is continuous (*synēches*) (*GC* II 336b 25–337a 2)." Aristotle breaks out of the mechanistic paradox that "something cannot come from nothing" and observes that being something is a fundamental rule of nature, whether as chemical matter or as animal being. His theological principle is to explain the "evolution" from chemical to biological as movements back towards the innate meaningfulness of nature and of the universe.

---

[39] Aristotle argues that: "that which has become habitual becomes as it were natural; in fact, habit is something like nature (*homoion gar ti to ethos tē physei*) for the distance between "often" and "always" is not great, and nature belongs to the idea of "always", "habit" to that of "often" (*estin d' hē men physis tou aei, to de ethos tou pollakis*) (*Rhet.* I 1370a 5–8)."

We see that our habitual being is based upon but is not reductive to our biological being as our intellectual being is also based upon whilst not being reductive to our behavioural being. We also find a hierarchy of principles in which biological processes such as reproduction, healing, and digestion run parallel to spiritual processes such as production, contemplation, and action. This striving to act, create, and complete represents an Aristotelian "theory of evolution" as a growth of consciousness and of "mind" in the world as expressed by the *scala naturae*.

Chapter 6
Aristotle and His Philosophy of "*Ousia*"

Abstract This chapter considers why "*ousia*" is the first and most important of Aristotle's "categories" and is a key concept for his philosophy. It signifies a "being" or "substance," with "substance" being usefully broad as a term so as to allow a consideration of what any something is. The sorites problem is unpacked to explore this situation. Aristotle compares a heap of sand (*sōros*), a human being (*ousia*), and the *Iliad* as a structured mass of words (and ideas, images) to consider the fundamental nature of "*ousia*" or "being." As entities with identity they each in some sense possess "movement" (*kinēsis*). Aristotle typically possesses a much wider concept of "movement" than ours. It is the becoming of a being, the movement of a being to its *telos,* whether this be getting a glass of water or writing the *Iliad*. For Aristotle, movement is more than just mass in motion. He does not conflate chemical matter with organic mass as does the modern Newtonian position. He both differentiates the organic and non-organic, as we clearly must, but his physics also relates to his biology and explains how the biological is an evolved or derived form of the physical.

Keywords Aristotle on ouisa · Ousiology · Sorites problem · Aristotle and Newton

The emphasis that for a being to be alive we must imagine it as being capable of swinging between opposites and "making turns" (*Phaedo* 72bc) goes back to Plato. Typically, Aristotle expanded upon and grounded Plato's general principle by showing in detail how each living being possesses such an individual identity and pathway. This theme is most obvious in his *Ethics* but applies to each of the subjects treated in his corpus. His *Generation of Animals* is based upon the primal fact that the "form" cycles through beings so that a man generates a man and a pig a pig but with the process allowing a certain small amount of independence, variation, and "self-movement" for the individual. His observation that: "Though it was generated by another thing bearing the same name (e.g., a man is generated by a man), it [i.e., the child] grows by means of itself (*GA* II 735a 20)" shows us that he is concerned both with the epigenetic stability of form and also with the separation, autonomy, and variation of individual life. His *On the Soul* argues for the principle that the

P. N. Jackson, *Aristotle on Meaning in the Living World*, Biosemiotics 30,
https://doi.org/10.1007/978-3-032-00602-8_6

"soul" (or innate energy) of a body and of a being must be specific to that being; see *De An*. I 407b 20–27.

Aristotle explains his concept of "*ousia*" in principle most intensively in his *Metaphysics* Book Z. He argues elsewhere that matter *is* substance at *Met*. H 1042a 32–34 and that it is "nearly substance" at *Phys*. I 192a 6–7. The primary question addressed is as regards the "evolution" of matter as regards how elemental energy (or "prime matter") is and becomes material being (or "sensible matter") and then spiritual being (or "noetic matter"). Matter becomes being by becoming individual matter ("proper matter") and enformed being ("proximate matter").[1] This is Aristotle's central question of "emergence." Aristotle's wider position is that the "emergence" or "becoming" of a "being" must be explained by considering how "matter" (*hylē*) becomes "determinate" (*hōrismenon*) as "substance" (*ousia*) and gains "continuity" (*synechēs*), "boundaries" or "limits" (*peras* or *horos*), "separation" (*chōristos*), and, ultimately, "total difference" (*teleia diaphora*). This is how all things, including even such things as "lyric poetry," come about. We find that everyday beings, whether (natural) biological or (artificial) artisan, come about on parallel principles even if the principle is in one case the being replicating itself and, in another case, a being replicating or producing something outside of itself as another being.

Aristotle does not conflate biological being with or derive it from physical matter. He rather accepts that living beings have a structure of life—what we would call "negentropy"—which provides a unifying principle for the organism, holds it together, and channels its energy as that of an "open system." This is the increasing concentration, formation, and integrity of an organism. This organism as an open system is self-replicating and flows as the species or "form" through the "matter" of reproduced instances. Aristotle seeks to understand the mechanism and meaning for why these instantiations of the species each have an end as the biological *telos* of continuing the species as well as individual death when the "*periodos*" of any individual being ends as its initial energy fully expires and the heart fails.[2] This is Aristotle's philosophy of "*ousia*," his ousiology, which is also the basic framing for his scientific inquiries. The closest modern fit with Aristotle's thought is the school of biosemiotics and I will consider this relationship further in the sister volume to this work.

[1] Aristotle argues that: "the proximate matter and the shape are one and the same (*hē eschatē hylē kai hē morphē tauto kai hen*) (*Met*. H 1045b18–19)" regarding which Joseph Owens explains that: "The various types of form require different kinds of matter…Properly speaking, the matter and the potency of a thing is the *proximate* matter, and not the "primary matter," unless the two happen to coincide (*The Doctrine of Being in the Aristotelian* Metaphysics 2nd ed (Toronto, 1963) p 344)."

[2] Aristotle argues that: "An animal is defined by the fact that it possesses sensation: and the part of the body to have sensation first is the part that has blood [and hence nutrition and the principle of material growth] in it first—in other words, the heart, which is the source (*archē*) of the blood and the first part to have it (*PA* III 666a 34–666b 1)" and that: "the heart is the place where life fails last of all (*GA* II 741b 19–20)." He also argues from another angle that: "of necessity, life must be coincident with the maintenance of heat (*thermou toutou sōtērian*), and what we call death is its destruction (*Juv*. 469b 18–20)" and that an organism is not just an open system in the sense of being a "fire" or "flame" (*phlox*) but is as energy which needs to be managed and (using the fire metaphor) "banked up" over its life; see *Juv*. 470a 13.

Here I note only that we should *not* simply see Aristotle as an inspired forerunner of what is likely to prove to be the full and most "true" scientific picturing of the world. I will rather contend below that Aristotle's philosophical insight and intervention can lead us to understand our world on a much deeper and more precise philosophical level than even our most sophisticated scientific thought today.

6.1 The "Sorites Problem"

In respect to this development from elemental matter to biological organism and then to the functions and possibilities of this organism we find Aristotle's key "sorites problem." The problem refers to what a "heap" (*sōros*) is and how it becomes a "being" (*ousia*) with this being Aristotle's own consideration of evolution. The statement of this problem that we will consider is given in Fig. 6.1 below based upon the following text: "In the case of all things which have several parts (*pleiō merē echei*) and in which the totality (*to pan*) is not, as it were, a mere heap (*mē estin oion sōros*), but the whole is something besides the parts (*esti ti to holon para ta moria*), there is a cause (*esti ti aition*); for even in bodies (*en tois sōmasi*) contact (*haphē*) is the cause of unity (*aitia tou hen einai*) in some cases, and in others viscosity (*glischrotēs*) or some other such quality (*ē ti pathos*). And a definition (*ho horismos*) is a set of words (*logos*) which is one by being connected together (*syndesmō*), like the *Iliad*, but by dealing with one object (*henos einai*). What, then, is it that makes man one; why is he one and not many (*dia ti hen all' ou polla*), e.g. animal + biped (*oion to te zōon kai to dipoun*) (*Met.* H 1045a 8–15)."

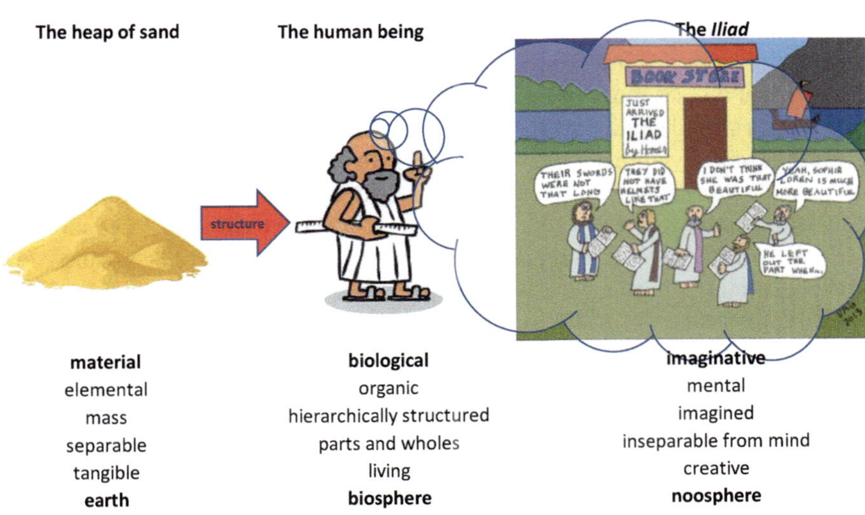

Fig. 6.1 Aristotle's "sorites problem"

We see in outline that Aristotle deliberately combines and compares a range of types of being: (1) a "heap" e.g. of sand, (2) an organism e.g. a man, and (3) a mental construction, e.g. the *Iliad*. The subtext of this situation is that Aristotle presents us with a range of everyday objects in order to explain how they correspond as "beings" and hence to philosophically illuminate the nature of "being" itself. I note the very careful and technical precision and conciseness of the above text in considering how a being is "connected together" as "one."

Aristotle's "heap" represents the nature of "mass" as opposed to "organism" (man) and "mental creation" (*Iliad*). We can therefore begin here by considering Aristotle's basic position on mass. He argues that: "any water is said to be specifically (*tō eidei*) the same as any other water because it has a certain similarity to it, and water from the same fountain differs in no other respect than in its more striking degree of similarity (*Top*. I 103a 19–23)." He draws the further distinction between determinate "solid" and indeterminate "liquid/gas" as that: "wine is said to be one, and water one, as being something formally indistinguishable. And all liquids are said to be one (e.g. oil and wine), and melted things; because the ultimate substrate of all of them is the same, for all those things are water [i.e., liquid] or vapour [i.e., gas] (*Met*. Δ 1016a 21–24)." He also observes, however, that: "a drop of wine does not mix with ten thousand measures of water, for its form is dissolved (*luetai gar to eidos*) and it changes so as to become part of the total volume of water (*GC* I 328a 27–29)." Aristotle's basic position on the elements is therefore (a) that they find their own "places" in the sense that water becomes a sea and sand becomes a sandbank, and (b) that elements may mix and become a composite but only in due proportion and that without this proportion the wine and its "form" are lost in the sea.

What is the further significance of the heap of sand or sandbank? Aristotle is clearly considering why elements move as and separate out as "masses" moving as *heaps* of sand, *patches* of colour, and then become such things as foetuses or ideas. Regarding the colour of things, Aristotle argues that: "when bodies are thus mixed, their resultant colour presents itself as one and the same at all distances alike (*pantothen*); not varying as it is seen nearer or farther away (*Sens*. 440b 17–18)"; i.e., he observes that masses move in accordance with a certain entropic Brownian motion to produce a shared equilibrial state which produces shared characteristics. I note as regards the validity of this interpretation of reality that Johann Rafelski and Berndt Müller argue from a modern scientific vantage point that there is a natural minimum of mass required for matter to *phenomenally* exist on the basis that: "if you think of a crystal, you must have an extended crystalline structure, and if you speak of a droplet of water, it must have a sufficient size such that the properties of the liquid dominate the properties of the surface."[3] We see that elements exist and move *as*

[3] Rafelski, Johann and Müller, Berndt *The Structured Vacuum: Thinking About Nothing* (Deutsch, 1985) p 100–101. We find in respect to how chemical substances exist in nature and move as a mass or body of matter that water is not *just* a mass or body of "water molecules" but is, rather, an ionic mass which cannot be understood on the basis of its atomic components and for discussion of this point see Holly VandeWall "Why Water is not H_2O, and Other Critiques of Essentialist Ontology from the Philosophy of Chemistry" *Philosophy of Science* (2007) and also Mae-Wan Ho

masses and, as we will see later, that they operate in accordance with their surfaces and their boundary conditions, i.e., on the basis of *their forms*.[4]

Having considered masses in outline, let us move on to consider Aristotle's "organism," i.e., his question: "What, then, is it that makes man one; why is he one and not many, e.g. animal + biped." Aristotle moves from the heap to consider the "whole" and the "parts" (and "the one and the many") which are in a human context the various organs which constitute the "parts" of the human being. Aristotle works up to this structure through consideration of the intermediating nature of "flesh" and "blood." He argues that "flesh" is a fluid mass (which is itself irrigated by blood) with the "organs" being en-formed "solids" derived from it, his paradigm for an irrigated organ being the "hand"; see e.g. *PA* I 640b 18–23. He adds that food is elemental matter broken down by water (liquid) and food (solid)—taken along with air (gas) and fire (sun/heat)—and with the "solid" being consumed by breaking down *its* forms and converting them into *our* forms; on this mastering see *Mete*. IV 379b 33–380a 4. Aristotle considers here the states of being and how solids possess "form" and "shape" whereas liquids and gases do not. He argues that in a physical sense beings become concrete through a process of solidification and he interprets the physical states of nature *philosophically* through his observation that the (fluid) matter *becomes* the (solid) substance.

As already seen, Aristotle supposes that such organism and integrity of being emerges from out of the power of touch or sense on the basis that: "Organic unity differs from contact (*diapherei de symphysis haphēs*); for in the latter case there need not be anything besides the contact, but in organic unities there is something identical in both parts, which makes them grow together (*sympephykenai*) instead of merely touching (*Met*. Δ 1014b 23–25)." The elements of a being ultimately "grow together." As regards the end result of this organisation, Aristotle explains this modality further as that: "substance means a "one" and a "this" (*hē ousia hen ti kai tode ti sēmainei*) (*Met*. Z 1037b 27)" and that: "you predicate "man" of *a* man (*ton gar anthrōpon tou tinos anthrōpou katēgorēthēseis*) (*Cat*. 2a 23)." We understand the nature of natural processes by considering that the processes of becoming in nature are not chaotic or chance but normally end in a definite species of organism or material. Aristotle argues that there is both form and process (being and becoming) in a sense inherent in matter and naturally arising from out of it so that, although we find teratomas in nature we also find that beings normally actualise as beings,

---

"Quantum Jazz: Liquid Crystalline Water Music of the Organism" *Subtle Energies & Energy Medicine* (2009). Leibniz also observes that: "as we see liquids collect according to the will of nature in spherical drops, so in the nature of the universe the series of maximal capacity came to be ("The Twenty-Four Statements" in Martin Heidegger *The End of Philosophy* (New York, [*orig. c*. 1690] 1971) p 51)."

[4] On Aristotle's "heaps" see Paul A. Bogaard "After Substance: How Aristotle's Question Still Bears on the Philosophy of Chemistry" *Philosophy of Science* (2006), Richard Sharvy "Aristotle on Mixtures" in *The Journal of Philosophy* (1983), and Paul A. Bogaard "Heaps or Wholes: Aristotle's explanation of Compound Bodies" *Isis* (1979).

e.g. as foetuses or houses, switching from one form to another rather than existing in a chaotic mixture.

Having considered Aristotle's sorites problem in respect to (1) the heap of sand and (2) the organic being and its parts we finally arrive at (3) the creative work of art, the *Iliad*. As an organic being is built upon chemical elements, an imaginative world is similarly built upon an organic being. As regards the human ability to create an imaginary world, Aristotle comments in his *Poetics* that "epic" possesses a "beginning, middle, and end (*echousan archēn kai mesa kai telos*), so that epic, like a single and whole animal, may produce the pleasure proper to it (*hōsper zōon hen holon poiē tēn oikeian hēdonēn*) (*Poet.* 1459a 18–20)." He also notes that although an epic is a single something "like a single and whole animal," the *Iliad* can be broken down into self-contained smaller tragedies (see *Poet.* 1459b 1) and (explaining *syndesmō*) that: "the *Iliad* is unitary by combination, but the definition of "man" by signifying one thing (*hē Ilias men syndesmō eis, ho de tou anthrōpou tō hen sēmainein*) (*Poet.* 1457a 29–30)." He therefore argues (a) that "epic" is a distinct form of artistic production, (b) that "epic" has its own particular "end" or "pleasure," (c) that although the *Iliad* is a complete whole or "cycle" it can be broken down into "parts," and (d) that it is not a "whole" in the same sense as "man."

I conclude that from Aristotle's perspective our world is obviously and unavoidably based upon "beings" and hence upon ontology. Aristotle explains on this situation, returning to his core *Metaphysics* Z, that: "Nothing, then, which is not a species of a genus will have an essence (*mē genous eidōn hyparchon to ti ēn einai*)—only species will have it, for in these the subject is not thought to participate in the attribute and to have it as an affection, nor to have it by accident (*tauta gar dokei ou kata metochēn legesthai kai pathos oud' hōs symbebēkos*) (*Met.* Z 1030a 11–15)." He concludes that only a species and not an individual or a genus will have an "essence" (*to ti ēn einai*). Hence only "epic" ("man") and not "entertainment" ("animal") or "*Iliad*" ("Socrates") will be something we can properly define "like a single and whole animal." His emphasis upon a sandbank, man, and the *Iliad* provides us with important and complex paradigms which make us think and do not allow us to rest satisfied with shallow, disconnected generalisations.[5]

6.2 Aristotle on the Matter of Matter (*hylē*)

Aristotle disputes the Newtonian idea that biological being can be reduced to physical forces and treated on the same basis as chemical matter. He argues that living matter and being is evidently different in kind to chemical matter even if they do in some sense possess a clear relationship as an artefact is an analogue for a biological

[5] On the gradation of matter see Joseph Owens *Aristotle's Gradations of Being in* Metaphysics *E-Z* (South Bend, 2007), Jonathan Beere "Potentiality and the Matter of Composite Substance" *Phronesis* (2006), and Enrico Berti's "Multiplicity and Unity of Being in Aristotle" *Proceedings of the Aristotelian Society* (2001).

being. The issue we face is how to define and to understand the differences and continuities between the inorganic and the organic and to understand how and why one is transformed into the other. If we set aside the dogma of our modern ideology which reduces the world to the blanket matter of Descartes' *res extensa*, it is obvious that biological matter—and energy—is not just some sort of inert physical stuff. Biology simply cannot be reduced to chemistry or to physics. The sister volume of this project will show beyond doubt that *modern science* does itself reject the reduction of the organic to the inorganic, the moving to the inert, the living to the dead. Here we need to consider why and how Aristotle himself unpacks and philosophises upon the principle of matter (*hylē*).

Aristotle is the founder and greatest thinker of ontology, *viz*. of the consideration of "being" (*to on*) itself. He also contemplates the meaning of "nonbeing" as well as "being," a discussion which again goes back to Plato. He explains regarding "being" that: "…"being" is used in many senses, but all of these are related to one principle (*all' apan pros mian archēn*), for some are called "being" in view of the fact that they are substances, others by being attributes of substances (*pathē ousias*), others by being on the way to becoming substances (*hodos eis ousian*), or else by being destructions or privations (*sterēseis*) or qualities of substances, or productive or generative either of substances or of whatever is related to substances, or negations of any of these or of substances. On account of this, we even say that nonbeing *is* nonbeing (*dio kai to mē on einai mē on phamen*) (*Met*. Γ 1003b 5–10)." The strong sense of the existential "is" insists that even absence or non-being is something in the context of being. Aristotle adds as regards how "being" comes from "non-being" that there is a "path to substance (*hodos eis ousian*) (*Met*. Γ 1003b 7)." His assumption is that "becoming" is a rising up of energy and structure to the already existing formal pattern of a new (and old) "being." Non-being is the context and pregnancy through which being becomes. It is hence that we find the fascinating concept of deprivation or "*sterēsis*" which itself represents a "kind of form (*eidos pōs estin*) (*Phys*. II 193b 20)" which can be understood as a pregnant possibility and as a creative space from which something determinate may emerge to fill the gap.

As regards this creative space, Aristotle's realism simply recognises the evident "negentropy" of biological being(s) and frames this fundamental fact of nature in scientific, philosophical, and theological terms. He explains "negentropy"—a term coined by the physicist Erwin Schrödinger in his seminal *What Is Life?* (1943)—as that: "loss of power is contrary to nature (*hē d' adynamia para physin*) (Instances of loss of power in animals are all contrary to nature, e.g. old age and decay, and the reason for them is probably that the whole structure of an animal is composed of elements whose proper places are different (*diapherei tois oikeiois topois*); none of its parts is occupying its own place (*tēn autou chōran*)) (*De Cael*. II 288b 17–19)." Aristotle recognises that the biological is an ordering of matter that does not derive *from* the chemical but is imposed *upon* it. This is the difference between entropy and negentropy. The composite, organic, articulated, formal animal is a superposition on chemical matter which is perishable on this basis. Aristotle expands upon this situation as that: "perishable things, [have] the potentiality for the opposite, which makes the continuity of the motion distressing (*epiponon einai tēn synecheian tēs kinēseōs*); this results when the substance is matter and potentiality; not actuality (*hē gar ousia hylē kai dynamis ousa, ouk energeia*) (*Met.* Θ 1050b 26–28)." Worldly beings, he argues, are built to expire, to suffer, to tire themselves out. The perishability of beings is presented as a striving for meaning arising from a presumed limitation of the cosmos to project its order leading to the natural tendency of an organism to wear down. It is hence that Aristotle ultimately argues that "desire," a striving for being, underlies our world and even, at a remove, the "Unmoved Mover" itself.

The role of matter, of the material dimension or cause, is therefore at the epicentre of Aristotle's account of "being." It is not assumed but taken as being a creative principle of *worldly* being. Aristotle argues that being in the world is dependent upon the existence of beings in the world on the basis of the principle that: "everything comes out of that which actually is (*esti gar ex entelecheia ontos panta ta gignomena*) (*De An.* III 431a 3–4)." This is his realistic approach.[6] The main question he asks is not "what is reality?" but "why is reality as it is?" How do the cyclical drivers of the world manifest as different species in the way they do? Why, also, is there an apparent evolution of being in nature from the inanimate object to the human being that instantiates the greatest level of consciousness of the animals. I contend that Aristotle seeks to understand *how meaning is emergent on an epigenetic basis as a suspended reality based upon the contingent actuality of being in the world.*

Mariska Leunissen explains on this situation that Aristotle's *scala naturae* is based upon "matter" as well as "form," as per Fig. 6.2 below.

Aristotle argues as regards the material evolution of beings that this evolved structure is reflected in the generative methods of animals (as per Fig. 6.2 above) and in their diet on the basis that: "Hunger and thirst are desire, the former for what is dry and hot, the latter for what is cold and wet; flavour is a kind of seasoning of these (*De An.* II 414b 12–14)." He argues that the quality of blood, which he takes to be a kind of liquified food, determines the quality of the consciousness of the animal; see *PA* II 650b 19–651a 15. In respect to the constitution of the being, he speaks of the "*krasis*" or blend of the body and this is a theme of his biology, especially of his *History of Animals* in which he considers the relationship between an organism and its physical environment.[7] Aristotle's biology moves from his

[6]Aristotle argues that knowledge and being is derivative from beings, as follows: "If all animals ceased to exist, there would be no knowledge at all, though there might in that case, notwithstanding, be still many objects of knowledge…If the animal itself is destroyed, then perception is also destroyed. But perceptibles yet will remain, such as body, heat, sweetness and bitterness and everything else that is sensible (*Cat.* 7b 33–8a 6)."

[7]Aristotle explains "*krasis*" on the basis that: "one country differs from another in these respects, and one water from another, on account of the same causes, for the quality of the nourishment especially and of the bodily condition of a person depends upon the blend (*krasis*) of the surrounding air and of the foods which the body takes up…[and] hard, cold water in some females causes barrenness, in others the birth of females (*GA* IV 767a 28–35)." In respect to the effects of environmental factors he argues that: "the formation and destruction of habits may involve the modification of certain factors, (say) the heat or cold or dryness or moisture of the physical elements, or the

	Animal species	Type of material nature
	Live-bearing (blooded)	***Optimal: hot, moist, and pure***
1a	Internal: humans, mammals, dolphins	Hottest, moistest, almost no earth
1b	Internal: eggs: selachia	Hot, moist, very little earth
	Egg-laying (blooded)	***Lacks moisture***
2a	Perfect eggs: birds, snakes, reptiles	cold, dry, earthy
2b	Imperfect eggs: fish, crustaceans, cephalopods	colder, dryer, earthier
	Grub-producing (bloodless)	***Lacks moisture and heat***
3	Insects	Coldest, driest, uses external earth

Fig. 6.2 Aristotle's scale of perfection as based on differences in material natures in GA 2.1. (Redrawn from Leunissen 2018 p 60)

Generation of Animals, through his *Progression of Animals* and *Parts of Animals*, to his *History of Animals* in order to unpack and exhibit this evolution of beings.

Let us consider in this context that Aristotle explains "evolution" as a path running from the inanimate through plants and animals to man (see *N.E.* I 1098a1*ff*; *Resp.* 477a 15*ff*) through insects which are an intermediate form (see *Long.* 467a20–21; *Juv.* 468b 2–3). Aristotle also treats the sponge as an intermediate form between plant and animal commenting that "it has sensation of a sort (*HA* I 487b 9)" even if "the whole thing is similar to a lung (*HA* V 549a 8)." His *scala naturae* is hence primarily a growth of consciousness and of the complexity associated with it. Through his scale of being Aristotle argues regarding the "testacea," e.g. snails, molluscs etc., that they "stand midway between animals and plants (*GA* I 731b 8)" and that: "they take shape and are generated out of a certain earthy and fluid coagulation (*synistatai kai gennatai ek tinos systaseōs geoeidous kai hygras*) (*GA* I 731b 14)." He contends that these testacea are produced from the putrefaction which releases a "frothy bubble" which contains heat and pneuma; see *GA* II 762a 19 *ff*.

Aristotle's position on "*hylē*" in biology is therefore that beings must be materialised and that their nature (*physis*) must be either derived from or reflected in their material basis. Life begins with a sort of hot foamy aerated substance, something like semen. From this initial material, he assumes that there is an evolution and refinement of energy (matter or *hylē*) which is expressed in the different forms (*eidē*) of different beings (*ousiai*). We could interpret this primal matter as protoplasm.

Aristotle ultimately, and famously, argues that the most basic forms of life likely emerge naturally and asexually from matter as "spontaneous generation." We could perhaps say that Aristotle expects to see something like the Miller-Urey experiment which hypothesised the emergence of organic compounds from inorganic matter but with Aristotle's hypothesis being that this emergence of life is not just a hypothetical one-off event but a pregnancy of matter which is itself a part of a constant

proper seats of the habits, whatever they may be. For excellence and defect are in every case concerned with influences whereby their possessor is, according to its natural constitution, liable to be modified (*Phys.* VII 246b17–19)."

cycle of nature. As regards the difference between organic and inorganic matter, he refers to the action of rennet to make cheese regarding which "the milk is the body, and the fig-juice or rennet contains the principle which causes it to set (*GA* I 729a 13–14)." He uses this organic action to hypothesise that the male sperm causes the female matter to "set" in a parallel manner; see *GA* I 739a 6–739b 20. He thereby identifies and considers the difference between organic and inorganic chemistry and between beings generally. He argues on this situation that we see an evolution in complexity and consciousness in nature which ranges from the primitive and microbial life which he judges to be "spontaneous" all the way to the human complex at the other end of the scale. He makes the point that living matter is not merely an inert mass but something with definite powers that act upon the world. If anything, modern science suggests not only that living matter is not inert, but no matter is.[8]

In all cases, Aristotle shows himself to be an empirical "materialist" in the sense that he assumes that all things emerge from "matter." He seeks to understand "matter" and "form" as biological constructs with a philosophical depth (allied with scientific observation) that we do not approach today. He seeks to represent the world rather than to reduce it to abstract laws. He assumes that matter is simply the manifestation of being(s) that will in some way, maybe not open to us, represent the qualities of what it is. Aristotle even assumes that thought and consciousness must possess their own "noetic matter" and Russell Winslow explains the possible meaning of this hypothesis well as that: "[the] Aristotelian intellectual perception discovers the being under inquiry in its singularity by becoming impregnated by the *ousia* of the other being…[and this]…*nous* does not have attributes, it does not have an organ, it does not have shape, *except* as the form in potency of what it perceives. *Nous* is the most primordially *open* part of the soul. It can become any intelligible thing that works upon it. If there is such a thing as *primary matter* in Aristotle, from this description it would seem that, rather than some sort of lowly material substrate, *nous*—the highest potency in the cosmos—is a kind of *primary matter*."[9] Aristotle assumes that body and mind are part of the same cosmos and informed by its processes on different but parallel levels. Meaning is taken as being *in* beings rather than being a general law applied to them.

---

[8] David Bohm argues, for instance, that: "the quantum theory, which is now basic, implies that the particles of physics have certain primitive mind-like qualities which are not possible in terms of Newtonian concepts (though, of course, they do not have consciousness). This means that on the basis of modern physics even inanimate matter cannot be fully understood in terms of Descartes' notion that it is nothing but a substance occupying space and constituted of separate objects. Vice versa, it will be argued that mind can be seen to have always a physical aspect, though this may be very subtle ("A new theory of the relationship of mind and matter" *Philosophical Psychology* (1990) p 272)." See also Harald Atmanspacher "Dual-Aspect Monism à la Pauli and Jung" *Journal of Consciousness Studies* (2012).

[9] Winslow, Russell "On the Life of Thinking in Aristotle's *De Anima*" *Epoché* (2009) p 309–310.

6.3 Aristotle's Concept of Ontological Movement (*kinēsis*)

Aristotle's conceptualisation of movement is ontological rather than mechanistic, energic rather than mechanistic, real rather than abstract. It is concerned with beings moving through their milieu rather than with objects moving in space. It assumes that something will move as that something in accordance with the nature of its being. This being and its movement will also become articulated as its *psychē* develops and it moves from a stone, through plant, through animal to man. As we will explore below, we see first an outward expression of movement so that a chemical interacts, a plant grows, an insect moves, an animal feels, and a human being acts. We then find an inward expression of movement, or consciousness, which ultimately allows a human being to develop principles (*archai* and *eidē*) *in* itself and to then implant these principles *in* other things. The paradigmatic activities here are technological and craft reproduction. It is clearly these principles as regards the evolution and exchange of principles between beings that stands behind Aristotle's core distinction between potentiality (*dynamis*) and actuality (*energeia*). These principles only seem to prevail in our material, "sublunary" world which manifests milieux through which beings need to become.

The philosophical framework for this world of being and becoming is represented primarily in the categories and the four causes. As regards the categories, Aristotle argues that: "If, then, the categories are enumerated as substantive, quality, whereness, whenness, relation, quantity, action, and being-acted-on, it follows that there are three kinds of movement—qualitive (*poiou*), quantitive (*posou*), and local (*kata topon*) (*Phys*. V 225b 6–9)." He expands upon this position as that: "the common feature that characterises them all [i.e., natural things] seems to be that they have within themselves a principle of movement (or change) and rest (*en heautō archēn echei kinēseōs kai staseōs*)—in some cases local (*kata topon*) only, in others quantitative, as in growth and shrinkage (*kat' auxēsin kai phthisin*), and in others again qualitative, in the way of modification (*kat' alloiōsin*) (*Phys*. II 192b 13–16)." Aristotle insists, contrary to the modern position, that movement is always *the movement of something*. It is also qualitative and can be growth and other forms of change and not just change of place. We see here the significance of the term "*alloiōsis*" meaning "modification" or "othering." This is the mechanism for variation in the world through the "*pathē*" and other such structures, for more detail of which see Appendix One.

Aristotle explains further as regards change of place that: "local movement is the final characteristic that manifests itself in generated creatures (*teleutaion de phora pasin hyparchei tois en genesei*)…of all kinds of change local movement is the one which takes its subject the least away from its essential nature, for it is the only one that affects no intimate characteristic, as modification affects quality and growth or shrinkage quantity (*Phys*. VIII 261a 18–24)." Local movement is a real projection of the being itself and is in a sense a choice of that being, as Aristotle will continue to argue in the following paragraph. Typically, what *we* take to be the whole picture, i.e., movement as movement in space, is only the final step of Aristotle's picture with its further significance having been severed. This final step is, however, not trivial for Aristotle but is given new importance as the exteriorisation and articulation of meaning from the being into the world through its external expression. Aristotle thereby lends our own mundane sense of meaning a deeper consideration and significance.

Let us represent this new meaning of movement as Fig. 6.3, as below.

Aristotle provides us with an "evolutionary" account of movement as an inert quantitative growth moving to a qualitative growth as the "being" (*ousia*) develops more tools (*erga*) to support a more sophisticated existence; see *Phys.* VIII 261a 15–17, *Top.* V 137b 24–26, *Rhet.* I 1362a 23–24 and *GA* I 731b 1–8. It is hence that independent movement is for Aristotle an evolution of movement as an expression of increasingly complex being. This is the meaning of Aristotle's famous maxim that: "nature does nothing in vain (*hē physis outhen poiei matēn*) (*IA* 704b 15)," i.e., that nature will only give an organism an organ, e.g. a head or eyes or legs, if that being is able to use it.[10] The gradation of movement as an evolution of being is understood as a being increasingly projecting itself into the world, first through physical movement and then through spiritual and mental awareness and motivation. How we act or behave in the world is an expression of both mental and physical powers.

Let us on this basis compare Aristotle's interpretation of "motion" with our own "at-at ontology" put forward by Bertrand Russell as that: "Motion consists merely in the fact that bodies are sometimes in one place and sometimes in another, and that they are at intermediate places at intermediate times" through Fig. 6.4 as below.

The Russellian approach is a simple observation that a material body has a force which creates an effect in time and place. The "motion" of the cause is simply the time during which the force has an effect. The Aristotelian position rather observes

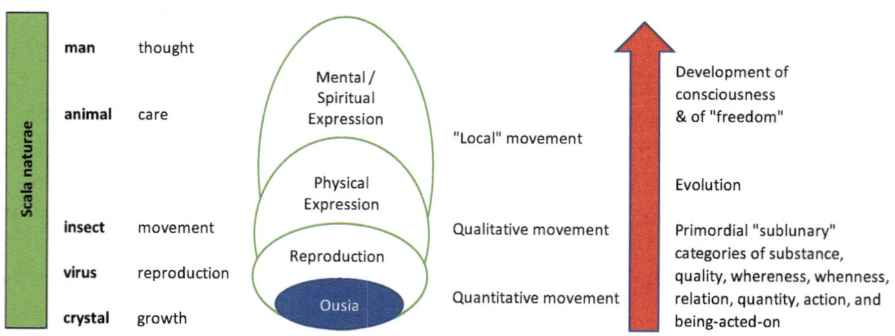

Fig. 6.3 Aristotle's evolution of "movement"

[10] Aristotle observes that: "Animals, simply as animals, do not have heads of necessity (*Cat.* 7a 18)" and (using his favourite example of sight as a gradated power) that: "what is not yet at the stage when it naturally ought to have sight is not called either seeing or sightless (*Cat.* 13a5–6)." Aristotle deduces that when "life" moves from being unconscious and inanimate to being conscious and animate it needs to develop an appropriate "head" and "sight."

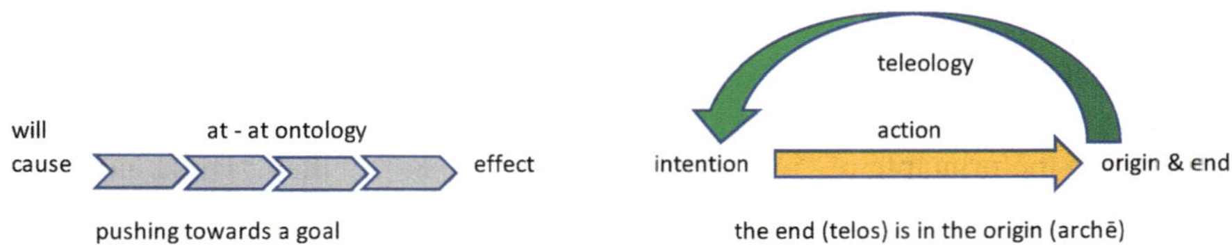

Fig. 6.4 Causal and teleological movements

that the effect (or *telos*) is included in the intention (*archē*) so that there is both a mental and a bodily component to any action which expresses the nature of a being acting within the nature of nature itself. Aristotle explains this situation through a lion seeking its prey. Hendrik Lorenz explains Aristotle's approach as that: ""What pleases the lion," he [i.e., Aristotle] insists, "is not the sight of "a stag or wild goat," but that he is going to get a meal (*N.E.* III 1118a18–23)." The lion's pleasure, Aristotle thinks, is a pleasure of anticipation, and so he must take it to involve apprehending the prospect of having a meal."[11] In the animal world, "movement" is not just a material object having an effect but a being acting to get results. This is the "practical syllogism" put forward in Aristotle's *Movement of Animals*.[12] Carl Jung usefully explains this situation as a contrast between "mechanistic" and "energic" standpoints.[13]

Aristotle gives as examples of these sought ends or "final causes" a person seeking a drink of water and a person seeking the philosophical life. It could be claimed in line with modern thinking that Aristotle's supposed belief that nature itself exhibits "final causes" discredits his philosophy. This is a false and disingenuous position. The reality is that Aristotle goes out of his way *not* to impute "final causes" to

[11] Lorenz, Hendrik *The Brute Within: Appetitive Desire in Plato and Aristotle* (Oxford, 2006) p 149.

[12] See Michel Crubellier "Le "syllogisme pratique" ou Comment la pensée meut le corps" in Laks, André & Rashed, Marwen (*eds.*) *Aristote et le mouvement des animaux* (Villeneuve-d'Ascq, 2004).

[13] Jung argues that: "It is a generally recognised truth that physical events can be looked at in two ways: from the mechanistic and from the energic standpoint. The mechanistic view is purely causal; it conceives an event as the effect of a cause, in the sense that unchanging substances change their relations to one another according to fixed laws...The energic point of view on the other hand is in essence final; the event is traced back from effect to cause on the assumption that some kind of energy underlies the changes in phenomena, that it maintains itself as a constant throughout these changes and finally leads to entropy, a condition of general equilibrium. The flow of energy has a definite direction (goal) in that it follows the gradient of potential in a way that cannot be reversed. The idea of energy is not that of a substance moved in space; it is a concept abstracted from relations of movement. The concept, therefore, is founded not on the substances themselves but on their relations, whereas the moving substance itself is the basis of the mechanistic view...the principle of finality is the logical reverse of the principle of causality. Finality is not only logically possible, it is also an indispensable explanatory principle, since no explanation of nature can be mechanistic only. If indeed our concepts were exclusively those of moving bodies in space, there would be only causal explanation; but we have also to deal conceptually with relations of movement, which require the energic standpoint. If this were not so, there would have been no need to invent the concept of energy ("On Psychic Energy" in *The Structure and Dynamics of the Psyche* (London, [*orig.* 1928] 1960) p 3–5)."

material nature and to argue that elemental and cosmic matter does *not* operate on the same basis as living matter. Chemicals and planets do not manifest "final causes" because they exist in such manner as not to need them. Aristotle does, however, rehabilitate the principle of "imitation" (*mimēsis*) by saying that the chemical elements "imitate" the cosmic cycles and remain vital when they are moving within these ecological cycles rather than when they are disconnected and stagnating. Animals also in some sense imitate or follow these cosmic cycles.[14] They must be understood as emerging *from* inanimate matter in the sense of following its cyclic patterning and systematisation of movement, but they transcend this mechanism even further.

In respect to how Aristotle systematically considers "movement," we have not only his *Physics* but also his *Movement of Animals* and his *Progression of Animals*. I suggest that these latter two works are eminently scientific exercises which have not been adequately recognised or taken seriously as such because they do not fit into our current mould for scientific inquiry. Aristotle considers in these works the mechanisms and principles through which an animal moves with there being some useful comparison to be made with Leibniz's idea that an animal is a master monad built from out of monads.[15] Aristotle pursues this line of thought not as a purely philosophical exercise but as a scientific exercise by considering the "centre" of an animal—what we could call its centre of gravity—and also the points and structures that enable and define its movement.

Aristotle also considers elsewhere in great detail—in the *Parva Naturalia* and his biological works—all of the fundamental structures of life and how they are expressed in different life forms according to their environmental niche. As regards their fundamental structures, he approaches animal being on the basis that it possesses a metabolism covering: "those [functions] which belong to both body and soul, such as sensation, memory, passion, desire, and appetite generally, and in addition to these pleasure and pain (*Sens*. 436a 8–11)" and "waking and sleep, youth and age, inhalation and exhalation, life and death (*Sens*. 436a 15–16)." As regards the assumed principle behind the remarkable variability of being upon this common

[14]Aristotle argues that the elements "imitate" the cycles of the cosmos at *GC* II 337a 1–8 and he argues regarding the nature of "imitation" generally that: "generally art either completes that which nature is unable to make work, or imitates it. So if that which is in accordance with art is for the sake of something, it is clear that so too are things in accordance with nature (*Phys*. II 199a 15–18)."

[15]Leibniz's full picture is that: "there are also infinite degrees of life in the monads, some dominating more or less over others. But when a monad has organs that are adjusted in such a way that, through them, there is contrast and distinction among the impressions they receive, and consequently contrast and distinction in the perceptions that represent them [in the monad] (as, for example, when the rays of light are concentrated and act with greater force because of the shape of the eye's humours), then this may amount to sensation, that is, to a perception accompanied by memory—a perception of which there remains an echo long enough to make itself heard on occasion. Such a living thing is called an animal, as its monad is called a soul. And when this soul is raised to the level of reason, it is something more sublime, and it is counted among the minds ("Principles of Nature and Grace Based on Reason" in *Philosophical Essays* (Indianapolis, [*orig*. 1714] 1989) p 208)."

basis of living being, Aristotle posits that: "nature will often quite spontaneously take some part that is common to all animals and press it into service for some specialized purpose (*PA* III 662a 19–20)."[16] Aristotle's approach to living being is as morphology, homology, and archology (aetiology). It is a systematic study of being and becoming as is peculiar to the living world and is a valid scientific *and* philosophical project on this basis.

As regards the ultimate source of the fundamental structuredness of being itself, Aristotle brilliantly hypothesises the "Unmoved Mover." The ontological basis for being and meaning is that which makes the cosmos one and that which any "*ousia*" must imitate and seek in moving towards its own focussed singularity of being, its own status as "*ousia*." Without it we are, as Aristotle comments, as lost "as mice walk(ing) upon loose soil." He argues that: "Any quality of rest…in an animal is of no effect unless there is something outside it (*exōthen*) which is absolutely at rest and immovable. And it is worthwhile to stop and consider this dictum; for the reflection which it involves applies not merely to animals, but also to the motion and progression of the universe (*pros tēn tou pantos kinēsin kai phoran*). For just as in the animal there must be something which is immovable if it is to have any motion, so *a fortiori* there must be something which is immovable outside the animal, supported upon which that which is moved moves. For if that which supports the animal is to be always giving way (as it does when mice walk upon loose soil and when persons walk on sand) there will be no progress, that is, no walking, unless the ground were to stand still, and no flying or swimming unless the air or sea were to offer resistance (*MA* 698b 8–18)." We must find an inner of the being and an outer of the world which in some sense relate and coordinate.

Aristotle posits that the cosmos must itself enable the principles of the inner and outer, motion and rest, and, ultimately, being and becoming. He thereby concludes his philosophy with a typically subtle, sane, modest, necessary, and inspired cosmology of divinity.[17]

[16] Aristotle explains his niche theory as that nature has always limited matter but uses it in the best way for the being at hand, like a potter with his clay; see *GA* II 730b. He explains "*kata physin*" on the basis that: "if it is better to do a thing in a particular manner, it is also in accordance with nature (*kata physin*) (IA 704b 17)."

[17] Jan Patočka gives the excellent assessment that: "It is clear that the first unmoved mover is divinity and, more precisely, the Greek divinity. Like the gods of Olympus, it has the property of endowing everything with meaning and preserving everything (*sōtēria*) while remaining intrinsically without participation in all inferior things, a detachment carried here to the point of absolute separation (*Aristote, Ses Devanciers, Ses Successeurs* (Paris, [*orig.* 1964] 2011) p 228)." On Aristotle's theology see also Markus Gabriel "God's Transcendent Activity: Ontotheology in "Metaphysics" 12" *Review of Metaphysics* (2009), Stephen Menn "Aristotle and Plato on God as *Nous* and as the Good" *Review of Metaphysics* (1992), Fabienne Baghdassarian *La question du divin dans la philosophie aristotélicienne* (HAL Thesis, 2011), Jonathan Beere "Thinking Thinking Thinking: On God's Self-thinking in Aristotle's Metaphysics Λ.9" (online 2010), Charles H. Kahn "The Place of the Prime Mover in Aristotle's Teleology" in Gotthelf, Allan (*ed.*) *Aristotle on Nature and Living Things* (Pittsburg, 1985), Philip Merlan "Aristotle's Unmoved Movers" *Traditio* (1946), and Susan Sauvé (Meyer) "Unmoved Movers, Form, and Matter" *Philosophical Topics* (1987).

6.4 Aristotle and the Degrees of Being

It should, of course, be obvious that it is both impossible and ridiculous to reduce the organic to the inorganic world, biology and organism to chemistry and elemental matter. This fact has not, however, prevented the high priests of our civilisation from systematically and dogmatically seeking to force our world and our worldview into precisely this reduction. Aristotle recognised the allure of reductionism from the off and countered it by insisting upon life, being, matter, and individuality. Aristotle's principle is not just an academic argument, however valid and necessary. Rather, Aristotle shows us how being emerges from, is naturally created *from*, matter every moment of every day. The concretion and metabolism of being is simply a fact of nature that we must face up to rather than merely assume.[18] The historical argument commonly associated with evolution is not only partial but also leads to a toxic philosophical contingency which reduces nature and meaning to the play of chance.

Aristotle explains the significance of the "face" of beings, i.e., their functional parts or what Aristotle terms their "non-uniform parts" (also referred to as "anhomoeomerous parts" in the literature), as being the meaningful "boundary conditions" of our world. He explains this situation as that: "[the first degree of synthesis is the] fluid and solid, hot and cold, [which] form the material of all composite bodies; and all other differences are secondary to these…The second degree of synthesis (*systasis*) is that by which the homogenous [uniform] parts (*tōn homoiomerōn*) of animals, such as bone, flesh, and the like, are constituted out of the primary substances. The third and last stage is the synthesis which forms the heterogeneous [non-uniform] parts (*tōn anhomoiomerōn*), such as face, hand, and the rest. Now the order of actual development and the order of logical existence are always the inverse of each other. For that which is posterior in the order of development is antecedent in the order of nature, and that is genetically last which in nature is first (*PA* II 646a 17–27)." We see another careful representation of Aristotle's distinction between "what is more obscure by nature, but clearer to us" to "what is more clear and knowable by nature (*Phys*. I 184a 20–21)." The person of Socrates may be immediately apparent to us, but he is also the very last product of a long natural process of becoming.

I represent the above stages of the material becoming of animal being as Fig. 6.5 below.

We see that food is taken by an animal and the original form removed, i.e., that of the unfortunate chicken, and converted into energy that can be used by the

[18] A.N. Whitehead stresses and develops this point in his work. He argues that: "Even in sight, we enjoy our vision because there is no eyestrain. Also we enjoy our general state of life, because we have no stomachache…the functioning of our body has a much wider influence than the mere production of sense experience. We find ourselves in a healthy enjoyment of life by reason of the healthy functionings of our internal organs—heart, lungs, bowels, kidneys etc. The emotional state arises just because they are not providing any sensa directly associated with themselves ("Nature Alive" in *Modes of Thought* (New York, 1938) p 159)."

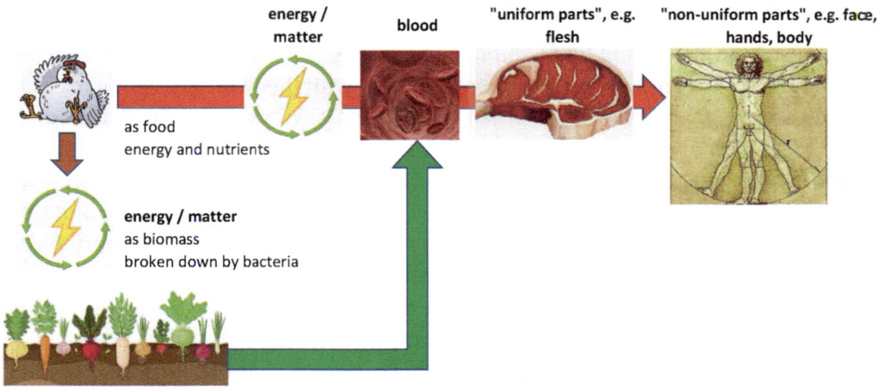

Fig. 6.5 The emergence of a formed being from matter

receiving organism. The remains of the chicken can be broken down and consumed by vegetables which can then also be consumed by the receiving organism on the same terms as previously. This matter or energy is then channelled by the receiving organism through its blood, which Aristotle regards as a flow of liquified food, to the uniform parts of the body, e.g. the flesh. The non-uniform parts are boundaries of the uniform parts which serve a functional purpose. The uniform parts support the non-uniform parts which represent the distinct "being" (*ousia*) of the organism.

Aristotle explains this situation further in his *Meteorology* which considers how—and to what extent—we can really understand the cycles and structures of the natural world. He argues in respect to great meteorological processes such as the salinity of the sea that it "occurs according to a certain order (*taxin*) to the extent that things here [in the sublunary realm] can have a share in (*metechein*) order (*Mete*. II 358a 25–6)." He adds that things exist in this sublunary realm only on the basis that: "as long as the determining proportion holds a thing's nature is maintained (*heōs gar an en autē ho logos, physis tout'estin*) (*Mete*. IV 380a 1)."[19] We see the use here of "*logos*" as a "proportion," as a meaningfulness which gives a being nature, form, function, direction, and movement. Each entity, organism, or ecosystem has its own inner logic, its order or "*logos*," to which it converts any incoming matter. We see here the important observation that the ordering or "*logos*" of animal beings is itself a further development or, we could say, evolution of the *logos* of the natural world as expressed in its physical and chemical productions. Aristotle retains

[19] Aristotle explains further that: "simple generation and natural change are the result of these properties (*hē haplē genesis kai hē physikē metabolē toutōn tōn dynameōn estin ergon*), as well as the corresponding natural destruction: and these processes occur both in plants and in animals and their constituent parts. Simple, natural generation is a change effected by these properties, when present in the right proportions (*hotan echōsi logon*), in the matter underlying a particular natural thing (*ek tēs hypokeimenēs hylēs hekastē physei*), this matter being the passive properties of which we have spoken (*Mete*. IV 378b29–35)."

but transform Plato's idealism when he argues that the "sublunary" world is seeking to express higher ideals in itself and its movements.

As regards our knowledge of and engagement with this world Aristotle argues in the same work that: "the end (*hou heneka*) is least obvious where matter predominates most (*pleiston tēs hylēs*). If you take the extremes (*ta eschata*), matter is pure matter (*hylē ouden allo par' autēn*) and the essence is pure definition (*ousia ouden allo ē logos*); but the bodies intermediate between the two are matter or definition in proportion as they are near to either (*ta de metaxy analogon tō engus einai hekaston*). For each of those elements has an end (*heneka tou*) and is not water or fire in any and every condition of itself, just as flesh is not flesh[20] nor viscera, viscera, and the same is true in a higher degree with face and hand. What a thing is is always determined by its function (*apanta d' estin hōrismena tō ergō*): a thing really is itself when it can perform its function (*ta men gar dynamena poiein to autōn ergon alēthōs estin hekaston*); an eye, for instance, when it can see (*Mete*. IV 390a 4–12)." His point here is that we simply see and understand function and form in our lives on a mesoscopic but not on a cosmic or microscopic scale. We must bank the meaning that we can see and touch whilst recognising that there is implied higher or further meaning that is not open to us.

As regards the nature of matter itself, Aristotle regards the remarkable solidification and formation of matter as a basic principle of animal metabolism. The flow of matter is the flow of life lived on the surface of the being—its form or shape—on the basis that: "that which surrounds is on the side of the form, that which is surrounded is on the side of the matter (*to men periechon tou eidous einai, to de periechomenon tēs hylēs*) (*De Cael*. IV 312a 12–13)." The matter even *becomes* the form on the basis that: "the proximate matter and the shape are one and the same (*hē eschatē hylē kai hē morphē tauto kai hen*) (*Met*. H 1045b18–19)." We see how quantity and mass *become* qualitative and articulated as a general rule in parallel with von Baer's laws which rule in reference to a more limited focus upon the development of the foetus. Aristotle provides us with a remarkable variant of von Baer's theory that matter begins as energy and then builds up form—through the structures of the bodily structure—until it becomes the boundary which *is* form.[21]

Aristotle's revealed hypothesis is that: "the action of heat is to make things more compact, thicker, and drier [and] concoction is an imperfect state due to lack of proper heat (*apepsia de ateleia di' endeian tēs oikeias thermotētos*) (*Mete*. IV 380a

[20] Aristotle also argues that: "if man and the animals and their parts are products of nature, then account must be taken of flesh, bone, blood, in fact all of the "uniform parts" (*homoiomerōn*) and indeed of the "non-uniform parts" (*anhomoiomerōn*) too, *viz*. face, hand, foot; and it must be explained how it comes to pass that each of these is characterised as it is, and by what force (*kata poian dynamin*) this is effected. It is not enough to state simply the substances out of which (*ek tinōn*) they are made, as "out of fire," or "out of earth" (*PA* I 640b 18–23)."

[21] In relation to foetal development Aristotle recapitulates von Baer's law through his argument that the immediate driving force of a being's identity is found in the particular which is based upon the species and then the genus; see *GA* IV 767b 30–36. On this subject, see René Thom "Homéomères et Anhoméomères en Théorie Biologique d'Aristote à Aujourd'hui" in Devereux, Daniel & Pellegrin, Pierre (*eds.*) *Biologie, Logique et Métaphysique chez Aristote* (Paris, 1990).

4–6)." He argues that the applied energy or "heat" of a being should actively "master" what it is applied to and concretise, transform, and complete it. He moves us from "form" (*eidos*) to "proportion" (*logos*) but he does not therefore reduce being to constitution or to process. He rather argues that there is a creative pregnancy (*dynamis*) or process in nature which enables the realised actuality (*energeia*) of being. His outline conclusion here is on this basis that: "the actuality of each thing is naturally inherent in its potentiality, that is in its own proper matter (*hekastou gar hē entelecheia en tō dynamei hyparchonti kai tē oikeia hylē pephuken enginesthai*) (*De An*. II 414a26–27)" which shows us that the "actuality" (*entelecheia*) of a being is "inherent in" (*hyparchonti*) its "potentiality" (*dynamis*) which is, namely, its "proper matter" (*oikeia hylē*). Aristotle's realism pulls out the requirements of "reality" and shows us a suspended reality of beings, qualities, and goods which naturally form a meaningful milieu or "world." Whereas Plato's driving meaning is found in the imitation of ideal forms, Aristotle's is found in the development of the innate potentiality of beings in the world.

Returning to the sorites problem with which we began, we now see how there must be an implicit directional structure inherent in being which creates beings. It is this inherent ordering which creates patches of colour, sandbanks, animals, human beings, and works of art such as the *Iliad*. Aristotle is here building upon Plato (and Pythagoras) who discerned mathematical ratios underlying the world like a harmony. Eric Sanday correctly interprets this situation as that there is a right "proportion" of a determinate body on the basis that: "the soul demands a certain bodily ratio (*logos*), capable of supporting a mean, in order to perform its functions."[22] The soul demands a *logos* which supports a mean and its functionality. Aristotle uses "*logos*" to signify order through "concoction" (*pepsis*) on the basis that: "In some cases of concoction the end of the process is the nature of the thing—nature, that is, in the sense of the form and essence (*eidos kai ousian*). In other cases it leads to some latent form (*eis hypokeimenēn tina morphēn*) which is attained when the moisture has acquired certain proportions or a certain magnitude in the process of being broiled or boiled or of putrefying, or however else it is being heated (*Mete*. IV 379b 25–29)." Aristotle's "realistic" position fully satisfies the existence of body and soul in one being. Some processes of nature lead to some sort of material form (in the substratum) and other lead to the full creation of beings having a "form and essence." This is indeed our experienced reality in which some processes are creative and others merely supportive. Principles (*archai*) hence form a hierarchy of meaning such as that required by horse-riding in which the riding art is premised upon many other subsidiary crafts; see *N.E.* I 1094a1–15.[23]

Aristotle's ultimate subtle position is that: "the potential is still on the way to actuality (movement being an actualising, though not a complete one), whereas the mover must already be actualised (*to de dynamei eis entelecheian badizei, estin d'*

[22] Sanday, Eric "*Phantasia* in *De Anima*" in Claudia Baracchi (*ed.*) *The Bloomsbury Companion to Aristotle* (London, 2013) p 111.

[23] On this subject see Monte Ransome Johnson "Aristotle's architectonic sciences" in David Ebrey (*ed.*) *Theory and Practice in Aristotle's Natural Science* (Cambridge, 2015).

hē kinēsis entelecheia kinētou atelēs. To de kinoun ēdē energeia estin) (*Phys.* VIII 257b8–9)." The egg, we find, must always be produced by a chicken. There is no vagueness about this situation which is that of a living event rather than of a chemical process. We must factor into our picture *an act of creation*. This creation of a being is the internalisation of a principle or mass of a self, regarding which George Blair explains that: "motion is an incomplete *internalisation* of an end [from the mover], if it is precisely what the being is doing to get the end inside it, and it has so far only partially succeeded.[24]" Neither the striving after being and the good nor their achievement as completion are automatic or mechanistic. Beings must be achieved to the initial point as to have mere viability as autonomous beings but should also conclude in their full maturity and the actualisation of their potentialities and goods.

6.5 Modern Views on Aristotle's Physics

In outline, Aristotle's *Metaphysics* considers the nature of being (*to on*) whereas his *Physics* considers the nature of nature (*physis*). His physics explains the same subject as our own, i.e., physical movement, but from a very different but not necessarily inferior perspective. The core conclusions of modern physics—such as "F = ma" (force = mass x acceleration) and "$E = mc^2$" (energy = mass x speed squared)—show us that the science of physics is fundamentally concerned with mass, energy, and movement. Aristotle's physics are also concerned with these fundamental concepts, though his physical concepts are not mathematical but biological, for better or worse. Aristotle's change does *always* have a direction, but this is a richer sense of direction than mere trajectory. His study of nature (*physis*), his "physics," does not primarily consider "movement" through the lens of our mechanics, kinematics, or ballistics but as qualitative change and growth as well as change of place. In all cases, Aristotle's physics is qualitative rather than quantitative. It is not a *mathematical* representation of movement but an *ontological* one. I note that the modern dismissal of ontology itself is actually a dismissal of the idea that there are beings as opposed to things; it is properly understood as a war on nature.

Werner Heisenberg, one of the great founders of quantum physics, argues that the aim of modern physics should be formulated on the basis that: "The term "complete understanding" would mean that the forms of matter in the sense of Aristotelian philosophy would appear as results, as solutions of a closed mathematical scheme representing the natural laws of matter."[25] He adds that: "we can say that the matter of Aristotle, which is mere "potential," should be compared to our concept of energy, which gets into actuality by means of the form, when the elementary particle is

---

[24] Blair, George A. "Aristotle on Entelexeia: A Reply to Daniel Graham" *The American Journal of Philology* (1993) p 96.

[25] Heisenberg, Werner *Physics and Philosophy* (London, 1958) p 144.

created."[26] Heisenberg takes Aristotle seriously and recognises that the culmination of physics could be Aristotelian concepts represented mathematically by modern physics. This is, indeed, precisely what René Thom seeks to achieve through his "semiophysics." It is notable that Heisenberg in an important lecture agreed with the morphological *field* theory of Goethe as opposed to the mathematical *force* theory of Newton.[27] As we will consider in the sister volume to this work, Heisenberg, along with many of his peers, sought to naturalise physics and to transform it from a technology into a science. On this point see especially Kurt Riezler's *Physics and Reality: Lectures of Aristotle on Modern Physics at an International Congress of Science* (1940).

René Thom draws out this situation from a mathematical perspective as that: "If one wished to mark the crucial spot where modern science parted company with Aristotle, one would find it in the stone thrown up into the air and falling back to earth. In Aristotelian physics, there are two continuous movements, consecutive but distinct: the forced upward movement (βίαιος) followed by the natural downward movement. At the highest point, where the movement changes direction, there is *discontinuity*. For us, after Galileo, there is one single movement described by a single parabolic law of the form $z = z_0 - [1]/[2]\, gt^2$ (where the z is the height and t the time). The rising motion is analytically continued in the falling motion. So between Aristotle and Galileo we have *changed our criterion of individuation*. Instead of considering the instantaneous target, we have adapted the purely mathematical definition of the analytical curve."[28] The action of throwing a stone at something is replaced by the calculation of the trajectory of the object thrown. Thom concludes on this matter that: "The miracle of physics is that this should be successful. But there should be no extrapolation of the miracle outside its proper domain (that of fundamental laws describing the infinitely great and the infinitely small brought somewhat curiously together)."[29] Thom suggests that mathematics "curiously" married the micro- and macroscopic whereas Aristotle best addresses the mesoscopic and human/living zone of existence.

Carlo Rovelli explains further as regards this situation that: "Aristotle's physics is the correct approximation of Newtonian physics in a particular domain, which happens to be the domain where we, humanity, conduct our business. This domain is formed by *objects in a spherically symmetric gravitational field (that of the Earth)*

---

[26] Heisenberg, Werner *Ibid.* p 139.

[27] See Werner Heisenberg "The Teachings of Goethe and Newton on Colour in the Light of Modern Physics" in *Philosophical Problems of Modern Physics* (Woodbridge, [*orig.* 1941] 1979).

[28] Thom, René *Semiophysics: A Sketch: Aristotelian Physics and Catastrophe Theory* (Redwood City, 1990) p 214. Thom adds that: "Instead of considering the instantaneous target, we have adopted the purely mathematical definition of the analytical curve. It is, of course, important to consider the direction of the movement when one wishes to describe the process qualitatively. But we really want to define the individuality of a process—as only mathematics can do. Our quantitative physics rests in its entirety on the use of analytical objects (Lie groups, for example) and of analytic continuation (*Ibid.* p 214)."

[29] Thom, René *Ibid.* p 214.

immersed in a fluid (air or water) and the main celestial bodies visible from Earth…For a student who has learned physics in a modern world it may sound strange to start physics by studying objects in a fluid. But for somebody who hasn't it may sound strange not to: everything around us is immersed in a fluid. Aristotle's physics is a highly nontrivial correct description of *these* phenomena, without mistakes, and consistent with Newtonian physics, in the same manner in which Newtonian physics is consistent with Einstein's physics in its domain of validity."[30] He adds that: "The bad reputation of Aristotle's physics is undeserved, and leads to widespread ignorance: think for a moment, do you really believe that bodies of different weight fall at the same speed? Why don't you just try: take a coin and piece of paper and let them fall. Do they fall at the same speed? Aristotle never claimed that bodies fall at different speed "if we take away the air." He was interested in the speed of real bodies falling in our real world, where air or water is present. It is curious to read everywhere "Why didn't Aristotle do the actual experiment?" I would retort: "Those writing this, why don't *they* do the actual experiment?" They would find Aristotle right."[31]

The philosopher of movement Jan Patočka also expands further on this situation through his deep engagement with Aristotle. He comments on the importance of the "medium" that: "The stone can only be moved by the medium if the latter receives not only the motion but also the capacity to move—a capacity to move that the medium must retain for a certain time after it has itself ceased to be moved."[32] The medium is itself something ontologically significant in enabling the movement of any object or being within it. It is the substratum within which a being exists, its milieu. Patočka adds as regards how we move from Aristotle to modernity that: "The first germ of the modern concepts of force and energy would need to be sought in Aristotle not in the couple *dynamis-energeia*, but rather in *bia* [i.e., violence]. The independence of this force (*bia*) is all the more manifest because force cannot produce an internal transformation since it does not create any quality in the thing itself—this would amount to mixing a forced movement with a natural movement."[33] Modern "billiard ball" physics is on this basis really a study of force and violence, the imposition of force and order, rather than the study of the natural physical progression or process of beings.

Patočka reiterates Aristotle with his observation that: "all true motion (*kinēsis*) is from something to something. This is the case wherever we say that something

---

[30] Rovelli, Carlo "Aristotle's Physics: A Physicist's look" *Journal of the American Philosophical Association* (2015).

[31] Rovelli, Carlo *Ibid.* Although all objects (regardless of their mass) free fall with the same acceleration—9.8 m/s/s which is the acceleration of gravity—when we factor in air resistance we find that a heavier object will accelerate to higher speeds than a lighter object before reaching a terminal velocity. More massive objects therefore fall faster than less massive objects because they are acted upon by a larger force of gravity; they accelerate to higher speeds until the air resistance force equals the gravity force.

[32] Patočka, Jan *Aristote, Ses Devanciers, Ses Successeurs* (Paris, [*orig*. 1964] 2011) p 334.

[33] Patočka, Jan *Ibid.* p 335.

moves (the hand carrying the spoon goes from the plate to the mouth, the man comes home from work, the rock rolls down the mountain into the valley) or changes (the green apple becomes red when it ripens)."[34] His further assessment of Aristotle's "physics" is that it is not a primitive form of physics but the "most ambitious philosophy of movement known to the history of thought which aims to reunite the lowest and the highest, to provide a unitary explanation of all being, from the falling stone to philosophical thought and the historical deeds and gestures of the community."[35] As regards the problem of this qualitative approach, he adds that: "Aristotle creates a systematic science of being in motion, but it is, fundamentally, without mathematics. In the depths of its essence, the Aristotelian movement-process is not mathematizable. Mathematization, in relation to this movement, can only mean a fictitious procedure, entirely external, which facilitates description but does not grasp its own nature."[36]

6.6 Modern Views on Aristotle's Kinematics

We can perhaps appreciate Aristotle's "kinematics" further with Richard Hassing, in the unexpected context of our "spending some time in a garden, or with pets." Hassing explains movement in this context as that: "cats act and move in an obvious, characteristic way called feline. Dogs act in a canine way. Birds fly with movements and flight paths characteristic of their kind. These ways of moving *specific* or particular to the natural living kinds have never been derived from the *common* or general laws of physics. Rather, these specific activities seem to be irreducible to the motions and properties of common elementary parts. The source of a bird's characteristic way of moving seems to be internal to the bird in an essential way. Of course airplanes have internal sources of motion, too—engines, fuel, a pilot. For airplanes, however, something other must produce them and start them up. In contrast, birds have babies and do not need pilots."[37] We see that the animal moves according to its *internal* nature according to its form (species), medium, and ends.

We see that "physics" as it is commonly understood today only tells us about very proscribed materialistic properties of any being—as Hassing puts it in another article concentrating on cats: "To make gravitational force fully govern the local motion of the cat, we must throw the cat out of the window, that is, compel it to be moved like a falling body or projectile."[38] Hassing explains the problem further as that: "Normally, the way two bodies interact depends on what kind or species each

[34] Patočka, Jan *Ibid.* p 177.

[35] Patočka, Jan *Ibid.* p 213.

[36] Patočka, Jan *Ibid.* p 26.

[37] Hassing, Richard "Animals Versus the Laws of Inertia" *Review of Metaphysics* (1992) p 31.

[38] Hassing, Richard "Laws versus Teleology" in Kaldis, Byron (*ed.*) *Encyclopaedia of Philosophy and the Social Sciences* (SAGE, 2013) p 56.

is. A dog or cat, for example, interact in a certain way, a cat and mouse in another. In fundamental contrast, the law of gravity expresses a principle of motion in nature that is independent of the kind, size, shape, internal structure and function, i.e., the nature, of the two interacting bodies."[39] This is the fundamental difference of including the "being" (*ousia*) which differentiates Aristotle's physics from Newton's. I note that J. B. S. Haldane also considers the remarkably different results between a mouse and a horse falling down a mineshaft in his "On Being the Right Size" (1926) and see also Justin Broackes' consideration of the cat "Tibbles" in his "Substance" (2005–6).

Modern physics is precisely useful *within the domain of external physical forces* but with the problem we face being its claim to possess a wider, universal scope. This problem arises only when a materialistic, mathematical representation of an aspect of reality is conflated with reality itself. This formalism is precisely the objection that Aristotle presses against Plato and his idealism. To reiterate, none of this is to say that mathematics or its abstractions are not useful or even philosophically revealing, only to say that it and they have limits and come with problems when misapplied. Even if, however, we note that quantum physics is actually compatible with Aristotelian philosophy (as we will consider in the sister volume of this work), we are not able or required to replace our modern physics with Aristotle's. We must rather see that Aristotle's physics has a different objective than our own and one that shows us that our own mathematical representation of natural forces has limits even if it is eminently successful on its own very narrow terms. This is precisely as René Thom argued above.

Patočka's basic position as regards the essential concept of the "movement" of "continuity" is that: "Continuity in general, and that of movement in particular, shows that movement is something simple, not composed of parts, and that it is like an organic whole because it is always the realisation or actualisation of something. And insofar as this something is inserted into opposites, movement is always either an attempt to break through to a higher realization, a more perfect *energeia*, or a decline. It testifies in any case to a lack in the being that moves."[40] Movement is on this basis *the movement of something for a reason to satisfy a lack of completeness in some sense*. Patočka adds further that: "continuity is a necessary and sufficient condition of the unity of movement…But the unity of movement has another higher meaning as designating the complete movement which has reached its goal. Just as completeness is the realisation of unity, irregularity is a weakening of it."[41] The image presented here is that of the perfection of a circle or straight line in the perfectly functioning being or the enformed being in the world. The full evolution of a

---

[39] Hassing, Richard *Ibid.* He adds that: "For *all* bodies—celestial and terrestrial, natural and artificial, living and non-human, human and non-human—possess mass and relative position; also velocity, momentum, acceleration, kinetic energy etc. These measurable and mathematically representable terms of classical physics are common to all bodies and their parts, unlike Aristotelian forms. As such they are form- or species-neutral."

[40] Patočka, Jan *Aristote, Ses Devanciers, Ses Successeurs* (Paris, [*orig.* 1964] 2011) p 120.

[41] Patočka, Jan *Ibid.* p 180.

being is either achieved in the effortless regularity of its movement or is directly embodied in its form.

Patočka adds as regards the nature and role of "time" in respect to "being" that: "It is clear that the "now" makes this movement both continuous and divided,"[42] i.e., the "now" represents both a creative break and a continuity along an ontologically and modally possible pathway. He explains further that: "time is the measure of motion; as rest is a deficient mode of motion as the deprivation of motion in that which is capable of it so that time incidentally measures rest also, i.e., one body remains in the same place while another moves, and the same number which enumerates the stages of the motion of the latter indicates the length of the abode of the former in the same place since that which is at rest can "be in the same number" as that which is in motion."[43] He explains here how Aristotle's concept of time and movement are related in a relatedness that is mathematically representable through time and space. Patočka concludes on this situation that: "At this point, we also see what it means to be in time. This does not mean "being at the same time as time" considered as a natural medium, a sort of space of movement. To be in time is to be measured by time, to be in it as in a number. It means that each thing, each process in time, is encompassed by it because there is always something else still—there is a time that exceeds their limits."[44] For Aristotle, the world *is* representable by number but is *not* reducible to number, measure, or time.

6.7 Conclusions

We previously observed that even such moderate modern thinkers as William James sought to dispense with such concepts as "consciousness" and "soul." The positivistic prejudice he displayed extends to all modern thought of *all* its schools and even to such a concept as "being" itself. The simple fact is that modern thought has sought to dispense with all non-tangible dimensions of being, including even the concept of "being." As regards how modernity displaced Aristotelian thought, Jan Patočka observed that the forces of modernity needed to actively struggle to overcome Aristotle's deliberate anti-reductionism. It required a real battle to strip such things as "*ousia*" and, infamously, "final causes" from our representation of the world.

The hard problem presented by Aristotle is his anti-reductionism, encapsulated by his idea that the whole is greater than the sum of the parts. The paradigm of the "whole" is the "organism" which stands at the base of all his thought. His first and preeminent "category" is hence "substance" (*ousia*) with the full picture being that: "the categories are enumerated as substantive (*ousia*), quality, whereness,

[42] Patočka, Jan *Ibid.* p 205.

[43] Patočka, Jan *Ibid.* p 207.

[44] Patočka, Jan *Ibid.* p 207.

whenness, relation, quantity, action, and being-acted-on…[and] there are three kinds of movement—qualitive (*poiou*), quantitive (*posou*), and local (*kata topon*) (*Phys*. V 225b 6–9)." This is a map for a being moving through the world according to its "being" within its milieu. We can also add the "four causes" on the same basis. These provide another matrix of four completely different dimensions of being—matter, form, finality, and movement—which each act on that being and constitute one dimension of its worldly existence. This is Aristotle's "realism" and as regards his critique of "idealism" Patočka explains that: "Without rejecting on principle the geometrization of movement, Aristotle holds it to be a fruit of abstraction which does not penetrate its nature. The true source of the movement is, despite all the change, the process and with the process being indissolubly linked to the nature of the beings engaged in it. The process is nothing other than the history of these beings, their "life." One can, from this history, explain the mathematizable movement by geometric idealisation, but it is not possible, conversely, to explain the processes in their diversity within the living and ever-growing universe from the complication of mechanical movements."[45]

We return to the Aristotelian conceptualisation of "creation" as a cosmic field of potentiality. The world is filled up with being seeking its good on the basis that: "Coming-to-be and passing-away will, as we have said, always be continuous (*aei…synechēs*), and will never fail…And this continuity has a sufficient reason. For in all things, as we affirm, nature always (*aei*) strikes after the better. Now being (we have explained elsewhere the variety of meanings we recognise in this term) is better than not-being; but not all things can possess being, since they are too far removed from the principle (*archē*). God therefore adopted the remaining alternative, and fulfilled the perfection of the universe, by making coming-to-be uninterrupted; for the greatest possible coherence would thus be secured to existence, because that coming-to-be should itself come-to-be perpetually is the closest approximation to eternal being. The cause of this as we have often said, is circular motion; for that is the only motion which is continuous (*synēches*) (*GC* II 336b 25–337a 2)." This picturing of the world "saves the phenomenon" in the sense that it represents reality as we experience it, even if it speculates on matters beyond our direct knowledge in order to fill in gaps in our picture. The alternative is to ignore these intangibles and these gaps and to reduce our picture of the world only to what can be quantifiably measured.

[45] Patočka, Jan *Ibid.* p 211.

Chapter 7
Can We Learn from Aristotle's Science Today?

Abstract This chapter considers Aristotle's scientific method as opposed to the modern Baconian empirical/pragmatic and Cartesian mathematical/rationalistic approaches. Aristotle is the father of systematic science as well as the founder of logic and metaphysics. He provides us with an integrated and deep philosophical picturing of the world and then applies this philosophical framing to produce meaningful individual sciences. I show that Aristotle's science provides us with a topological view of the world with each of its objects analysed, considered on its own terms, and then considered in context. I take his "Poetics" and his "Rhetoric" as examples of Aristotle's science to show how he takes a subject area and treats it not only technically but also metaphysically. Following his emphasis upon the "*ousia*," Aristotle's special sciences work from the nature of their subject matters. They represent how we engage with each type of being with poetics, rhetoric, logic, biology, physics, chemistry etc. all being shaped by how we can engage with each respective subject matter or field. Aristotle's observations on science are evidently true, but are they valid? Can something be true and invalid? The issue at stake is the role of science in our lives. Do we want the knowledge of science or the power of technology?

Keywords A defence of Aristotle's science · Aristotle's scientific method

Let us ask: how good is Aristotle's science? Does it stand up today and, if so, in what sense? In a previous age, Dante Alighieri was famously able to describe Aristotle as "the master of those who know," to which proposition Descartes dissented three centuries later when he put forward his famous "method" along with his *Meditations* seeking to "destroy the principles of Aristotle."[1] The common modern mythology that has arisen from this history is that Aristotle reawakened Europe

[1] Descartes, René "Letter to Mersenne 28th Jan 1641" § 298 *CSM* III p 171.

P. N. Jackson, *Aristotle on Meaning in the Living World*, Biosemiotics 30,
https://doi.org/10.1007/978-3-032-00602-8_7

during the Renaissance but then became a burden preventing the advance of knowledge in the modern age.

As usual, there is some lateral truth behind the myth. Scholasticism did indeed represent an obstacle for the modern philosophy of progress seeking to channel the effort of mankind into pragmatic approaches and material results. It was with some justification thought that the positive benefits of unfettered science outweighed the benefits of scholastic speculation. This does not, however, mean that the modern "Aristotelians" were necessarily philosophically wrong. Thomas M. Lennon rather concludes on the scholastics that it was "Not that the Aristotelians were wrong or that they were espousing a doomed cause; they were just not very good."[2] Nor do we find that the moderns are necessarily philosophically right or even very good. The modern position is riddled with the confusion of game-playing and apologetics leading to such odd results as that stated by Jorge Secada that: "I believe, in fact, that Descartes would have preferred to remain an Aristotelian rather than end up a "new" empiricist."[3] The modern concentration on method—whether Cartesian or Baconian—is, on close inspection, a double-edged sword. Leibniz himself commented on the real quality of Descartes' method that it was "like the precepts of some chemist; take what you need and do what you should, and you will get what you want."[4] What we gain in reduction, we lose in subtlety.

I add as regards the relationship between the neo-Aristotelians and Aristotle himself that scholastic Aristotelianism centred upon religious apologetics is clearly radically different in intention and result from the thought and work of Aristotle himself. Whatever we conclude about the positions taken by his medieval or early modern followers, it is simply false to say that Aristotle *himself* sought to reduce or subordinate science to philosophy or religion. Not only was Aristotle the real practising founder of many of our core sciences—physics, biology, meteorology, poetics, ethics, psychology, logic, politics, rhetoric etc.—but he also carefully considers the value and limits of both science and philosophy taken per se. Aristotle argues in principle on this point that each science "cut[s] off a part of being and investigate[s] the attribute of this part; this is what the mathematical sciences for instance do (*Met.* Γ 1003a25–6)" whereas regarding philosophy "There is a science which investigates being as being and the attributes which belong to this in virtue of its own nature (*estin epistēmē tis hē theōrei to on hē on kai ta toutō hyparchonta kath' auto*). Now this is not the same as any of the other so-called special sciences (*Met.* Γ 1003a21–2)." Aristotle recognises that philosophy is fundamentally different than science.

Aristotle simply argues in principle that we can understand the world by considering its "parts" as distinct domains of investigation through science and its "whole"

[2] Lennon, Thomas M. *The Battle of the Gods and Giants: The Legacies of Descartes and Gassendi, 1655–1715* (Princeton, 1993) p 52.

[3] Secada, Jorge *Cartesian Metaphysics: The Late Scholastic Origins of Modern Philosophy* (Cambridge, 2000) p 185.

[4] Leibniz, Gottfried "Untitled Essay" in *Die philosophischen Schriften* (*ed.*) C.I. Gerhardt IV (Berlin, 1875–90) 329. On the problem of the idea of a scientific method per se see Paul Feyerabend *For and Against Method* (Chicago, 1999).

as the domain proper to philosophy. He illustrates the philosophical "whole" further as that: "It is evident…that it belongs to one science [i.e., philosophy] to be able to give an account of these concepts [i.e., opposites, plurality, unity, negation, privation etc.] as well as of substance (*ousia*)…and that it is the function of the philosopher to be able to investigate all things (*Met.* Γ 1004a32–1004b1)." It is notable that he places the concept of "substance" (*ousia*) under the remit of philosophy along with the situational or cosmic placing of substance. It is by establishing the proper boundaries and proper ambit of both science and philosophy that we can give fullness and substance to them both. Let us assess the value and validity of this hypothesis below.

7.1 Aristotle on the Limits of Science

What, then, *is* science? Aristotle argues that: "being immediately falls into genera; for which reason the sciences too will correspond to these genera (*Met.* Γ 1004a4–6)." Each "special" science will have its own genus, its own "object," but with philosophy, also explained as a science, being an exception as a general science, or knowledge, of the conceptual whole in respect to being and being. As regards the limits of "science" in general, why it is ordinarily limited to its specific object(s) and hence cannot supplant philosophy, Aristotle argues precisely that a scientist should not be expected to perform subtle metaphysics in order to engage with his own "special" science. He should simply draw up a rough-and-ready method to approach any worldly object according to the best avenue of attack for doing so.[5] He should also, however, not conflate his specialised part with the general whole or assume that adding all the sciences up produces a full picture of the whole.

Aristotle observes further as regards the problem of conflating science with philosophy that scientists (1) apply philosophical concepts in a partial and utilitarian manner because they: "use them [i.e., philosophical concepts] just so far as to satisfy their purposes; that is, so far as the genus to which their demonstrations refer extends (*Met.* Γ 1005a 25–27)" and (2) misunderstand the key philosophical concept of "substance" (*ousia*) by: "forgetting that substance, of which they have no correct idea, is prior to these other things. For number *qua* number has peculiar attributes, such as oddness and evenness, commensurability and equality, excess and defect (*Met.* Γ 1004b9–12)." Aristotle stresses that scientists are prone to reduce the world to their own scientific remit and to conflate the "part" of the world with which they are occupied with the "whole" of worldly reality. The emphasis upon "*ousia*" is vital here. Aristotle observes that *science "forgets" the substance (ousia) because it is interested only in the special dimension of reality that it recognises.*

---

[5] Aristotle argues that: "it is the mark of an educated man to look for precision in each class of things just so far as the nature of the subject (*hylē*) admits; it is evidently equally foolish to accept probable reasoning from a mathematician and to demand from a rhetorician scientific proofs (*N.E.* I 1094b23–27)."

The problem breaks down further, according to Aristotle, as that scientists fall into (1) the *ontological error* of empiricism by supposing that: "all attributes are accidental (*panta kata symbebēkos*) (*Met.* Γ 1007a30–31)" because they cannot themselves philosophically (or though their scientific perspective) make sense of the architectonic structure of the world and into (2) the *epistemological error* of relativism or of supposing that the world revolves around the scientist and that: "everything [is] relative—relative to opinion and perception (*pros ti poiein apanta kai pros doxan kai aisthēsin*), so that nothing either has come to be or will be without some one's first thinking so. But if things *have* come to be or will be, evidently not all things are relative to opinion (*Met.* Γ 1011b5–7)." This stress upon the need for ontology before epistemology is a critical point of Aristotle's philosophy as also Hegel's contra Kant. Aristotle rejects, in effect, the reduction of the world to ideologies which "bring everything into line with hard-and-fast theories (*De Cael.* III 306a 8–9)." The problem we ultimately face is in respect to people, even intelligent scientists, who do philosophy without proper education or without a mind suited to the pursuit (with scientific and philosophical minds not necessarily being interchangeable) leading to the erroneous belief that the world is accidental and relative when it is actually *primarily* substantive and formal.

The conflation of philosophy, which needs to see the whole, with science, which needs to focus upon the part and which takes from philosophy only what it needs in order to do so, is *the fundamental problem of reductionism.* Aristotle, by point of contrast, shows himself to be a model of considered pragmatism and empiricism who examines the meaning of this world because he seeks to preserve the fullness of nature. It is on this basis that the attack on the Aristotelian approach is often of a political and theological character—as typified by Martin Luther[6]—framing his philosophy as a denial of the assertion that the "higher principles" of the world must be approached intuitively and theologically rather than empirically and speculatively. This is the true significance of Newton's "*hypotheses non fingo.*" The modern approach, for various reasons, demonstrates a strange collapse of meaning which is the opposite of the Aristotelian approach which is a building up of meaning, a "saving the phenomena."

7.2 Aristotle on "Special Sciences"

Aristotle argues in respect to the "special sciences" that: "We are seeking the principles and causes (*hai archai kai ta aitia*) of things, but clearly, qua things (*hē onta*). For there is a cause of health and of good physical condition, and there are

[6] Martin Luther ominously complained that: "The universities also need a good, thorough reformation…[since] everything which the papacy has instituted and ordered is directed only towards the increasing of sin and error…[and] the Holy Scriptures and the Christian faith are little taught, and the blind, heathen master Aristotle rules alone, even more than Christ (*An Open Letter to The Christian Nobility of the German Nation Concerning the Reform of the Christian Estate* (1520))."

principles and elements and causes of mathematical objects, and in general, every science which proceeds by *thinking* or which participates in *thought* (*dianoētikē ē metechousa ti dianoias*) to some extent is concerned with causes and principles, whether these are very accurate or rather simplified. But all these sciences, marking off some being or some genus, conduct their investigations into this part of being qua being, and they say nothing concerning whatness; but starting from the whatness of their subject, which [whatness] in some sciences is made clear by sensation but in others is laid down by hypothesis, they thus proceed to demonstrate more or less rigorously the essential attributes of their genus. Consequently, it is evident by such induction from these sciences that there is no demonstration of *substance* or of whatness but that these are made known in some other way (*dioper phaneron hoti ouk estin apodeixis ousias oude tou ti estin ek tēs toiautēs epagōgēs, alla tis allos tropos tēs dēlōseōs*) (*Met.* E 1025b 3–16)." Scientific induction does not reveal the "substance" but takes it as given. This is what Aristotle means when he begins his categories with "substance" (*ousia*) and argues that all other dimensions of being are subsidiary to this.

Aristotle draws out above the fact that mathematics and biology (as medicine) have different objects and hence are qualitatively different as sciences. He implies that although there is an important principle of analogy and of common reference in the sense that all objects are objects within the world, there is no single "scientific method" for approaching them.[7] We can approach many objects in the world and do so in many ways, sometimes through "*technē*" and at other times "*epistēmē*." It is hence interesting that Aristotle's own paradigmatic science is, interestingly and importantly, the *art* of boxing, on the basis that: "By a science we mean a disposition (*kata diathesin*); each science, too, has its own name, such as boxing, for instance, or wrestling. And those who are that way disposed get their name from the name of the science (*Cat.* 10b 4–6)." From our anticipations of modern physics or chemistry as a model science, Aristotle's approach seems immediately primitive. In reality, however, we see that how a human being is trained to become, and then is, a boxer is quite as good an example of a science as the study of astronomical objects. The difference is that one is a passive and inorganic science which observes the movement of physical objects whereas the other is an active, engaged, and organic science which trains and shapes movement in beings. It is notable that Aristotle emphasises the essential "naming" of a science, its concentration upon a certain named objective.

I will try to show below that Aristotle's thought becomes far more original and penetrating when it is considered on his own terms. Aristotle does not approach "movement" in the same way as we do today, but his contrary interpretation of movement is quite as meaningful as our own. Aristotle's own concept of movement is not as a generic force but as the *inherent expression of something* on the principle

[7] Aristotle argues that we must approach a subject scientifically as the matter itself requires, as follows: "it is the mark of an educated man to look for precision in each class of things just so far as the nature of the subject (*hylē*) admits; it is evidently equally foolish to accept probable reasoning from a mathematician and to demand from a rhetorician scientific proofs (*N.E.* I 1094b23–27)."

that: "there are as many types of motion or change as there are of being (*hōste kinēseōs kai metabolēs estin eidē tosauta hosa tou ontos*) (*Phys.* III 201a 9)." Movement is also determined by form in the sense that: "the many movements are incomplete and different in kind (*all' hai pollai ateleis kai diapherousai tō eidei*), since the whence and the whither give them their form (*eiper to pothen poi eidopoion*) (*N.E.* X 1174b 4–5)." It is hence that Aristotle's science *follows* the movement of being on the basis that: "Most of the principles…which are connected with a particular science are peculiar to it. Hence to convey to us the principles connected with each particular science is the task of experience. I mean, e.g. that it is for astronomical experience to convey to us the principles of astronomy (for it was not until the phenomena had been thoroughly apprehended that the demonstrations of astronomy were discovered); and the same applies to any other art or science (*Pr. An.* I 46a 18–22)."

We ultimately see that Aristotle understands the importance of sustained and specialised empirical observation for the production of knowledge. His consideration regarding how a given subject matter is articulated and organised as "movement" informs *his* whole scientific and philosophical research programme and produces not only his *Physics* but also his *Movement of Animals*, *Progression of Animals*, *Generation of Animals*, and the *Ethics*. From a distance, Aristotle's *On the Heavens* reflects upon the nature of what we can take from our human engagement with the planets, *viz.* that the heavens demonstrate beyond doubt the mechanical principles of inert matter. His *Topics* consider how we approach different subject matters in principle, and his other logical works—collected as the *Organon*—consider and categorise how we think and talk about the world around us. He questions the sense in which our mathematical concepts relate to and represent the outside world in his *Metaphysics*.

Aristotle not only invents the sciences but also defends them against a mathematical, idealistic, or ideological position. He argues that beings have their own space—what we could call from biosemiotics their "*Umwelt*"—to exist. He also argues that we expect each being to have its own "matter" as well as its own "form" and its own "becoming" as well as its own "being." Aristotle is also focussed upon how biological matter emerges from chemical matter not in the sense of some chance act of "creation" but as an everyday creative possibility of matter. This explains his concentration upon the nature of "food" and metabolism (from "*metabolē*" meaning "change") which I will consider below. I will show that Aristotle's philosophical consideration of matter and being possesses an immediate precision and close concentration upon the nature of reality which is difficult but possible for us to rise to.

7.3 Aristotle on Abstraction

Aristotle's approach to the world is conceptual without being abstract. Whereas "theory" ("*theōria*") and "theoretical" signify scientific "reduction" to the modern mind, it meant expansive philosophical "contemplation" about being(s) to the

Greeks.[8] This distinction translates into logic in the sense that Aristotle's logic of beings and of human beings, i.e., of how human beings as beings think about and engage with the world, is different from our mathematical logic which does not recognise beings, human or otherwise. This mathematical worldview is a product of the principle of Galileo (*contra* Aristotle) that the being of a thrown object is irrelevant to the path of its thrownness. The important difference in character—in richness, intent, and objective—between our and Aristotle's terminology and thought is exhibited in the glossary to this work and in Appendix 1.

Aristotle himself argues regarding this "mathematical" worldview that: "a man *qua* man is one indivisible thing (*adiaireton ho anthrōpos hē anthrōpos*); and the arithmetician supposes one indivisible thing, and then considers whether any attribute belongs to a man *qua* indivisible. But the geometer treats him neither *qua* man nor *qua* indivisible, but as a solid (*Met.* M 1078a 23–27)." He recognises the artificiality of mathematical abstraction in each of its forms. Daniel Maher explains this principle as regards how mathematicians *abstract or subtract* from the world as that: "While the arithmetician approaches man from the side of form, the geometer does so from the side of matter."[9]

Aristotle considers the peculiarity of mathematics as a science itself and identifies different beings, objects, and aspects within it with his argument being that: "the mathematical sciences are concerned with forms; they do not confine their demonstrations to a particular substrate (*ta gar mathēmata peri eidē estin· ou gar kath' hypokeimenou tinos*). Even if geometrical problems refer to a particular substrate, they do so only incidentally. As optics is related to geometry, so is another science to optics, namely, the study of the rainbow (*Post. An.* I 79a 6–9)." Monte Ransome Johnson explains this structuring of being and of the knowledge that arises from it through the table set out in Fig. 7.1 below.

Johnson explains this representation or structuring or principles (or forms) or sciences as that: "an intermediate or mixed "mathematical-physical" science (e.g. optics) supplies a middle term that allows the application of the axioms of an autonomous mathematical science (e.g. geometry) to the explanation of a phenomenon described by empirical science (e.g. meteorology). Thus the first, independent

[8] See Nicholas Lobkowicz *Theory and Practice: History of a Concept from Aristotle to Marx* (Notre Dame, 1967) p 48–9.

[9] Maher, Daniel P. "Aristotle on Mathematical and Eidetic Number" *Hermathena* (2011) p 38. Stephen Gaukroger explains this situation as that: "Mathematical abstraction is distinctive in that it is a twofold process: we must abstract the mathematical properties of the object or collection by disregarding what it is that has those properties (i.e., the matter), but there is also a second part to the abstraction in which we disregard the *properties* of sensible objects so that what has these properties becomes the object of investigation. These two parts of the abstraction taken together yield mathematical properties and a noetic matter of which these are the properties. In abstracting numbers we "detach" them from sensible things, but it is an essential characteristic of numbers (and geometrical figures) that they be properties, so we must therefore "attach" them to something else; otherwise they would be "free floating" properties, so to speak, and this is as impossible in the case of numbers as it is in the case of kinds ("The One and the Many: Aristotle on the Individuation of Numbers" *The Classical Quarterly* (1982) p 318)."

	Example 1:	Example 2:	Example 3:
Mathematical science:	1. Arithmetic	1. Stereometry	Geometry
Physical science:	2. Harmonics	2. Mechanics	Optics
Empirical science:	3. Acoustics	3. Levers	Meteorology

Fig. 7.1 Johnson's "architectonic" examples of science(s)

science stands in a relation of architectonic superiority to the two subordinate sciences."[10] Aristotle presents us with a structured and hierarchical world of principles, parts, and wholes.

It is significant and interesting that Aristotle's rejection of formalism and reductionism is most fully demonstrated in what is to us the unlikely place of the *Ethics*. His *Ethics* is typically not just a work on ethics but a study of *what ethical structures mean*. He provides in this work his most thorough proof for the ousiological structuring of meaning itself—as actions, goods, inquiries etc. His introductory statement hence reads: "Every art and every inquiry, and similarly every action and pursuit, is thought to aim at some good; and for this reason the good has rightly been declared to be that at which all things aim. But a certain difference is found among ends; some are activities, others are products apart from the activities that produce them. Where there are ends apart from the actions, it is the nature of the products to be better than the activities. Now, as there are many actions, arts, and sciences, their ends also are many; the end of the medical art is health, that of shipbuilding a vessel, that of strategy victory, that of economics wealth. But where such arts fall under a single capacity—as bridle-making and the other arts concerned with the equipment of horses fall under the art of riding…in all of these the ends of the master arts are to be preferred to all the subordinate ends (*en hapasais de ta tōn architektonikōn telē pantōn estin hairetōtera tōn hyp'auta*) (*N.E.* I 1094a1–15)." I speculate that we can represent Aristotle's position here in Fig. 7.2 below.

I suggest that it is easy to see how human action and life is structured generally on this basis with distinct patterns of knowledge and action realising themselves in very different physical structures and outcomes. It would be interesting to see how such a definite and evident representation of the structuring of human being could be refuted.

It is important for us to consider on what basis this principle of structure and hierarchy is built into the very foundation of Aristotle's study of ethics. His position in the *Ethics*, as elsewhere, is that our world is composed of organised beings with subordinate physical organs (*organa*) with psychic functions (*erga*) which subserve the being. This organisation and structuring of being and their complementary relationship with the world are fundamental facts of existence which need to be included in our interpretation of that being. We additionally find that even in respect to mathematics the

[10] Johnson, Monte Ransome "Aristotle's architectonic sciences" in Ebrey, David (*ed.*) *Theory and Practice in Aristotle's Natural Science* (Cambridge, 2015) p 177.

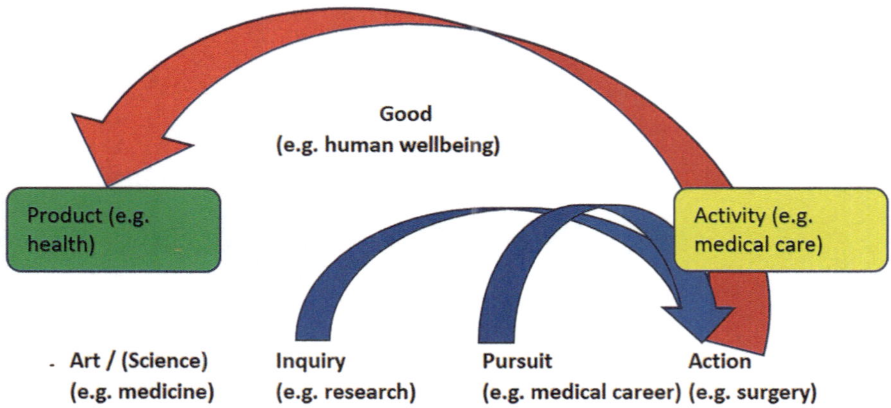

Fig. 7.2 Aristotle's model structuring of human action

"being" is ultimately unavoidable whether as light, meteorological phenomena, planets etc. Aristotle observes several millennia before Gödel that even at the very extremes of reductionism, i.e., mathematics, a *full* reduction to abstraction is lacking. As regards his logic and metaphysics, Aristotle does not reduce the world to abstractions but he does understand that it possesses elemental, foundational, categorical structures that help us to understand it. Aristotle does not simply apply abstractions to the world but understands that the world is axiomatic, *built upon* axioms. As regards his full picture of reality, we find that Aristotle presents us with rich, meaningful principles that provide the basis for the richness of being(s). The world is a plenum of being, as Leibniz argued, like "lake full of fish" and not just "atoms and the void."

7.4 Aristotle on Symbolism

Aristotle rejects abstraction and idealism, but his worldview is fundamentally symbolic and realistic. What do we mean by this? For Aristotle, there are always *two* sides of meaning—a mental and a physical—which come together over a shared interface of reality, as a *symbolē*. It is this "meaning," this nexus of being, that Aristotle refers to as "mind" or "*nous*." *Nous* is the "order" and "good" which infuses the world that we encounter and are able to engage with.

Aristotle observes as regards how human "*nous*" engages with the "*nous*" of the world that: "evidently knowledge and perception and opinion and understanding have always something else as their object, and themselves only by the way (*autēs en parergō*) (*Met.* Λ 1074b 35–36)." He adds that: "the soul is analogous to the hand (*hōste hē psychē hōsper hē cheir estin*); for as the hand is a tool of tools (*kai gar hē cheir organon estin organōn*), so the mind is the form of forms (*kai ho nous eidos eidōn*) (*De An.* III 432a 1–2)." I represent this mental and physical reaching out and grasping of being in Fig. 7.3 below.

Fig. 7.3 The concentration of reaching and grasping

Aristotle argues that we can grasp the world with our mind through mental concentration in a parallel way as we grasp the world with our hands by handling its objects. This idea that we are shaped by the objects with which we are engaged, i.e., our habits, is the basic premise of his *Ethics*. We become carpenters, musicians, philosophers, politicians etc. in large part through the problems we need to face in life which inform the habits, talents, and functions we come to embody. We also find that both the end-states of Aristotle's ethics, *sophia* and *phronēsis*, are dimensions of seeing the world and grasping the meaning of its beings, as "virtue" is an excellence in navigating through the world and its beings, as a knife or axe cuts.[11] As regards how we reflect upon the world through "*nous*" see *N.E.* VI 1143a 35–1143b 3 and the *Posterior Analytics*.

As previously considered, Aristotle builds upon Plato by taking his concepts and repurposing them from idealism to realism. One of these concepts is that of "ratio" which underlies Plato's ontology in a strictly mathematical sense. Plato mythologises ratio in his *Timaeus* on the basis that there is "one that really and truly makes

[11] Aristotle argues that we see objects *en parergō* or "by the way" or "on the side" on the basis that: "we have already a common faculty (*aisthēsin koinēn*) which apprehends common sensibles (*tōn koinōn*) directly. Therefore there is no special sense for them…The senses perceive each other's proper objects incidentally, not in their own identity, but acting together as one, when sensation occurs simultaneously in the case of the same object, as for instance of bile, that it is bitter and yellow (*De An.* III 425a27–425b2)." We have the capacity to develop a general "common sense" understanding of the world by bringing together and integrating the various sensory inputs that we experience automatically and naturally. Our mind or "*nous*" is also in some sense formed upon the basis of this "common sense." On "*en parergō*" see John Edward Russon "Self-Consciousness and the Tradition in Aristotle's Psychology" *Foi et Raison* (1996).

a unity of itself together with the things bonded by it, and this in the nature of things is best accomplished by proportion (*touto de pephyken analogia kallista apotelein*) (*Timaeus* 31c)."[12] We find that Plato ascribes the existence of the universe and of all things to the workings of *mathematical* ratios. Aristotle, by contrast, takes this mathematical concept and naturalises it. He makes an ideology into a symbology. His own contrary concept of "*analogia*" is exemplified by the statement that: "a single definition can be given of soul only in the same sense as one can be given of figure (*De An*. II 414b20–21)." For Aristotle, shapes, souls, forms, and mathematics are useful shorthand representations for the primordial common aspects of beings. The meanings of the world and its symbolic relations are drawn out by Aristotle into a symbolic topology understood through words, concepts, and analogies.

Aristotle explains his concept and use of "*analogia*" well through his *Ethics*. He explains here as regards friendship that: "In all friendships between dissimilars it is, as we have said, proportion that equalises the parties and preserves the friendship (*to analogon isazei kai sōzei tēn philian*); e.g. in the political form of friendship the shoemaker gets a return for his shoes in proportion to his worth, and the weaver and all other craftsmen do the same. Now here a common measure (*koinon metron*) has been provided in the form of money (*N.E.* IX 1163b32–1164a1)." Let us consider further what Aristotle describes as friendships of utility as between "farmer A, food C, shoemaker B, shoemakers product equalised D (*N.E.* V 1133b 5–6)" with money being a measure (*to metron*) on the basis that: "money serves as a measure which makes things commensurable and so reduces them to equality (*to dē nomisma hōsper metron symmetra poiēsan*) (*N.E.* V 1133b 17) in Fig. 7.4 below:

We see here that Aristotle is not just making an ethical, practical, or political point in his *Ethics* but is rather drawing out a philosophical position using human action and behaviour as a case study.

Aristotle gives the further example of "justice" in this work as that: "The just...is a species of the proportionate (*estin ara to dikaion analogon ti*) (proportion being not a property only of the kind of number which consists in abstract units (*to gar analogon ou monon esti monadikou arithmou idion*), but of number in general). For proportion is equality of ratios (*hē gar analogia isotēs esti logōn*), and involves four terms at least (*N.E.* V 1131a29–32)." Regarding these "four terms" we find that: "for

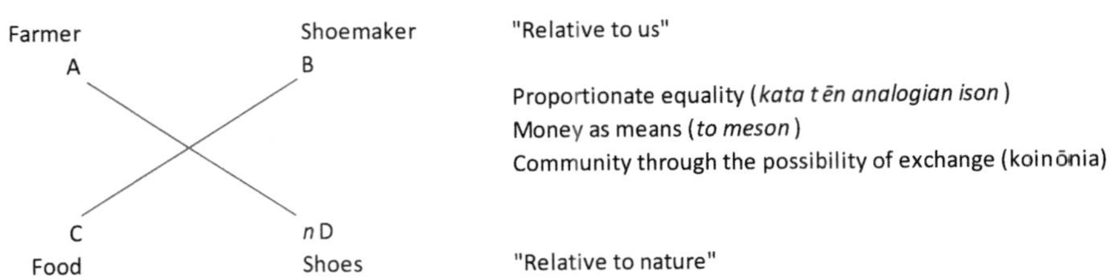

Fig. 7.4 Commensurability of the farmer and the shoemaker

[12] See Coleen Zoller "Plato on Hypothesis, Proportion, and the Education of Philosophers" *Auslegung* (2007).

the persons for whom it is in fact just are two, and the things in which it is manifested, the objects distributed, are two. And the same equality will exist between the persons and between the things concerned (*N.E.* V 1131a19–21)." Aristotle's expanded position on justice is that there are various "substantive" objects—i.e., the judge, the plaintiff, the claimant, the (current) written law, and the (historical) customary law—at play and that these "substances" have a generic relationship and patterning in which they are constantly changing and interacting. Aristotle also refers to business partnerships at *N.E.* V 1131b30 and see also *E.E.* VII 1242b. In all cases, the world is symbolic in the sense of being the coming together of two sides in a "symbolic" (and "semiotic") zone of meaning.

7.5 Aristotle and Homeostatic Proportion

Aristotle argues that "*analogia*" is "an equality of ratios (*isotēs esti logōn*)" and that this is *not* merely a mathematical relation. We also see the use of "*logos*," also meaning word or sentence, to denote "ratio." Other terms used by Aristotle (that are clearly within the language and culture to some degree) as part of this framing of the world as beings possessing pathways are "*symmetria*" (symmetry), notable as a biological principle, "*kairos*," meaning a sort of fittedness in one's destiny, and "*metrios*," as denoting measuredness. Aristotle also speaks of the "*prepon*" as that: "due proportion (*to prepon*) consists in contraction and amplification as the subject requires (*Rhet.* III 1404b 15–16)."

Concentrating on Aristotle's use of "*logos*" we find that: "each being is produced according to a definitive ratio (*logō tini hekston ginetai*) (*GA* II 740b 33)." He explains in another context that: "If harmony (*symphōnia*) is a species of voice, and voice and hearing are in one sense one and the same, and if harmony is a ratio (*logos d' hē symphōnia*), then it follows that hearing must be in some sense a ratio (*tēn akoēn logon tina einai*). That is why both high and low pitch, if excessive, destroy hearing; in the same way flavours in excess destroy taste, and in colours the over-brilliant or over-dark destroys vision, and in smelling the strong scent, whether sweet or bitter, destroys smell; which implies that sense is some kind of ratio (*hē d' aisthēsis ho logos*) (*De An.* III 426a27–426b4)." "Meaning" in the world is, on this basis, precisely that which operates in the "mean" (*to meson*) accessible by a given "being" (*ousia*).[13] I suggest that this is a good representation of homeostasis which leads on to a proper philosophical consideration of what the constitution of a being represents and signifies. Hegel also puts forward this philosophical position on homeostasis, as will be explained in more detail later, when he observes that "the

[13] Aristotle argues in principle that: "we must observe that in every divisible continuum there exists excess, deficiency, and a mean, which may be either relative to each other or relative to us (*E.E.* II 1220b21–24)." The mean is always "relative" to the being itself and its body and soul.

cause as the "original thing" that continues to exist outside of its effect...haunts representational thinking."[14]

We can also consider Aristotle's emphasis in respect to "*pepsis*" or digestion in this context. Aristotle argues that "*pepsis*" is a "mastering" and regulation of matter operating on the principle that: "as long as the determining proportion holds a thing's nature is maintained (*heōs gar an en autē ho logos, physis tout' estin*) (*Mete.* IV 380a 1)." We can add further as regards how a "*logos*" holds the body together and "saves" or preserves it that: "if the excitement of the sense organ is too strong, the ratio of its adjustment (which, as we say, constitutes the sense) is destroyed (*luetai ho logos*) (*De An.* II 424a31–32)." The sense is itself a "ratio" and yet Aristotle contextualises it as that: "The sentient subject must be extended, but sensitivity and sense cannot be extended; they are a kind of ratio and potentiality of the said subject (*megethos men gar an ti eiē to aisthanomenon, ou mēn to ge aisthētikō einai oud' hē aisthēsis megethos estin, alla logos tis kai dynamis ekeinou*) (*De An.* II 424a27–28)." We cannot reduce the world to symbols and proportions since the proportion of "sense" only exists as part of a sensing being and is the sensing faculty of that being, as a kind of controlled and channelled energy. We also find that just as the senses along with the "common sense" are a dimension of the body and of the being, the mind is also.

Aristotle explains that "sense" is in some sense a "harmony" (*symphōnia*) or a "*logos*" which exists within the milieu of the body as an energic channel and representation of that body. He applies the same principle to "mind" on the basis that: "the mind (*dianoia*) does not think of large things at a distance by stretching out to them, as some think that vision operates (for the mind will think of them equally if they are not there), but one thinks of them by a proportionate mental impulse (*alla tē analogon kinēsei*); for there are similar figures and movements in the mind (*esti gar en autē ta homoia schēmata kai kinēseis*)...all internal things are smaller, and as it were, proportionate to those outside (*panta gar ta entos elattō, kai analogon* [*kai ta ektos*]). Perhaps, just as we may suppose that there is something in man proportionate to the forms (*esti d' isōs hōsper kai tois eidesin analogon labein allo en autō*), we may assume that there is something similarly proportionate to their distances. E.g. if one experiences the impulses AB, BE, he can imagine CD; for AC and CD are in the same ratio as AB: BE (*Mem.* 452b 9–19)." The mind is also a "measure" of being with its word (or *logos*) being derived from the Latin word "*mens*" meaning "measure" and interestingly linked with "*mederi*" or "to become whole" from which we derive "medicine."

Aristotle also explains in his dedicated work on the subject of sleep that it is a channelling of being in the sense that: "a potentiality and its actuality reside in the same subject; and what we call sensation, as actuality, is a movement of the soul through the agency of the body (*aisthēsis hōs energeia kinēsis tis dia tou sōmatos tēs psychēs esti*)...[and] If...some part has sensation as its function, this too, if it exceeds the due period of its capacity for continuous sensation, will become

[14] Hegel, G.W.F. *Lectures on the Proofs of the Existence of God,* (1829) §313.

exhausted, and cease to perform its function…[and hence] The animal is defined by the possession of sensation, and we hold that sleep is in some way the immobilisation or fettering of sensation, and that the release or relaxation of this is waking…it follows that the object of sleep is to preserve animal life. But the end (*telos*) of an animal is the state of waking (*Somn.* 454a 8–455b 23)" (and see also *Somn.* 455a 21–27). Let us ask: what does the above passage (as an example) signify about Aristotle's objectives and his approach vis-a-vis our own? What does the phenomenon of sleep tell us, if anything, about our being? Does sleep have any *philosophical* significance?

Aristotle asks what sleep is *per se*. He defines sleep as the resting state of the whole organism, its soul and body, and he hypothesises that it is the passion needed to counter its action. Aristotle uses sleep—as also sight, sensation, activity etc.—to understand the meaning and significance of active as opposed to passive consciousness. There are states and degrees of being and of activity and actualisation. As regards life and being, the "end" (*telos*) is metaphorically in the active state of wakefulness and not just in the potentiality for wakefulness. The *telos* of "life" is not "death" but the full expression of maturity. Being is on this basis a meaningfully suspended reality and not just a chemical state. We see here how Aristotle blends science with philosophy. He asks: what is the meaning for worldly phenomenon as we observe them? Aristotle hence considers not just how sleep manifests but also why (and also the why of the why). As regards his investigation of the phenomenon of sleep we also find, as generally, that the same observable phenomena of reality are used in each of his works—whether biological, metaphysical, logical etc.—but with a different emphasis. One work will consider what the mechanism of sleep teaches us about the life of different animals, another will consider why sleep should per se be needed by animals.[15] Aristotle considers the phenomena of nature from a "synoptic" (or "peripatetic") perspective which lends both clarity and richness to the real world by considering it from multiple perspectives and dimensions.

Aristotle's philosophy treats reality itself through a philosophical lens. He even makes some interesting comments on mirrors in the context of the symbolic and homeostatic world presented above at *De Insomn.* 459b. By considering the mirror as a boundary or interface (as well as a "means") between a subject and an object he discusses the relationship between objective, subjective, and means. We see how the subjective is shaped through consideration of the objective (and of the subject as

[15]Aristotle explains "sleep" in a critical passage of his *On the Soul* as that: "both sleep and waking depend upon the presence of the soul, and waking is analogous to the exercise of knowledge, sleep to its possession but not its exercise. Now in one and the same person the possession of knowledge comes first. The soul may therefore be defined as the first actuality of a natural body potentially possessing life; and such will be any body which possesses organs (*De An.* II 412a 23–29)." Aristotle uses "sleep" as a critical part of his explanation of what the soul *is* in his treatise on this subject matter. Alternatively, in his biological works he details the various ways in which animals—dolphins, elephants etc.—do actually sleep within their environments as an understanding of their "*bioi*." On this subject see Marcia Sá Cavalcante-Schuback "The Hermeneutic Slumber: Aristotle's Reflection on Sleep" in Baracchi, Claudia (*ed.*) *The Bloomsbury Companion to Aristotle* (London, 2013).

object).[16] The idea that we engage with the world as through a mirror where we are reflected back to us by nature is essential to Aristotle's approach and is diametrically opposed to Richard Rorty's denial of this "mirror of nature."[17]

7.6 Aristotle on Standards and Relations

Aristotle reformulates Plato's concept of "participation" as the principle that the dimensions within a being relate to the dimensions outside the being. The main disconnect between Plato and Aristotle on these matters is in respect to the place and importance of the "being" (*ousia*). Aristotle states this principle in his *Ethics* as that: "This distributive proportion is not continuous (*esti d' ou synechēs autē hē analogia*); for we cannot get a single term standing for a person and a thing (*N.E.* V 1131b 16–17)." Aristotle conceptualises this "order" (*taxis*) as being mathematically representable in its structuring and scaling but *not* reducible to this mathematics. The point he makes here is that we cannot simply remove the beings and understand the world as laws and processes with this situation being most obvious when it comes to human behaviour and action.

Aristotle's general assessment, exemplified by his representation of ethics, is that nature is something possessing a working principle around the "*horos*" or "standard" (as a metaphorical boundary stone) which can give rise to an "*orthos logos*" or "right reason" in human beings. He hence explains further that human ethical behaviour is achieved on the basis that: "the intermediate is determined by the dictates of the right *logos* (*to de meson estin hōs ho logos ho orthos*)…In all the states of character we have mentioned, as in all other matters, there is a mark to which the man who has the *logos* looks (*esti tis skopos pros hon apoblepōn ho ton logon echōn*), and heightens or relaxes his activity accordingly, and there is a standard which determines the mean states (*kai tis estin horos tōn mesotētōn*) which we say are intermediate between excess and defect, being in accordance with the right *logos* (*kata ton orthon logon*) (*N.E.* VI 1138b19–25)." Underlying this situation is an observable, dynamic space or "middle" through which beings are able to *meaningfully* interact. It is notable that Aristotle's picturing of the world is not that of an abstraction but of a framing or explication of *how we do physically see and engage with the world and with other beings within it.*

---

[16] René Lefebvre discusses this passage in detail and comments that: "The passage must be placed in its context: the study of the dreamlike imagination. It serves a double purpose; establishing the existence of an initiative coming from the site of sensation, Aristotle intends to show how the passage from sensation to imagination takes place, especially in dreams. Highlighting the impressionability of certain surfaces, including that of the eye, he aims to establish the conditions for the receptivity of the image ("Le Miroir de L'Âme" (Aristote, *De Insomnis*, 459 b 23–460 a 26)" *Les Études philosophiques* (1988) p 200)."

[17] See Richard Rorty *Philosophy and the Mirror of Nature* (Princeton, 1979).

As ethics are a proper calibration of human being to the world, so is human mind and its products. Aristotle hence argues that: "when mind thinks the highly intelligible, it is not less able to think of slighter things, but even more able; for the faculty of sense is not apart from the body, whereas the mind is separable (*to men gar aisthētikon ouk aneu sōmatos, ho de chōristos*). But when mind has become the several groups of its objects, as the learned man when active (*kat' energeian*) is said to do (and this happens, when he can exercise his function by himself (*hotan dynētai energein di' autou*)), even then the mind is in a sense potential (*esti men kai tote dynamei pōs*), though not quite in the same way as before it learned and discovered; moreover the mind is then capable of thinking itself (*kai autos di'autou tote dynatai noein*) (*De An.* III 429b3–10)." We find the key aspects of Aristotle's physics and metaphysics directly explained and contextualised through his representation of psychology. Aristotle hence asks: what does it mean for the mind to be "separable"? What does it mean for a mind to "become" its objects? How does the "potential" become "actual"? How does the mind complete itself by closing the circle and thinking about thinking?

Mind is not an abstraction for Aristotle but a real physical and spiritual organ of engagement with the world. It can hence be understood as homeostatic on the basis (a) that the "mind" comes into being by means of its content but does so according to its essential formal nature which is to be "the form of forms (*kai ho nous eidos eidōn*) (*De An.* III 432a 2)," (b) that the "mind" is like an organ in the sense that its power expands in proportion to its use and hence it becomes "more able…to think slighter things," (c) that the human "mind" is different from other human faculties such as sense only on the basis that it is detached from the world and therefore actively works upon the world, and (d) that it is precisely this detachment which properly and causally explains our power for reflection, and for self-reflection. The same account of *nous* "thinking itself" by "sharing in the objects" of the world is famously given from another angle at *Met.* Λ 1072b 19–24.Aristotle's overall position on our power to abstract from objects (*di' aphaireseōs*) and to become ourselves through objectification (*en parergō*) explains how and why he argues for the principle that a human being is formed or habituated from an initial "blank slate" position. He argues that a person understands the world by seeing *through* its objects to the forms beneath and that a person comes into being by internalising these objects and becoming their mirror image as a philosopher, scientist, musician, soldier, carpenter, farmer etc. He also argues that we are shaped as a whole in this process and somehow become a whole character or archetype based upon our occupation. We mature into a human role as something and hence become something in accordance with our place in society and in the world. Aristotle also argues in the concluding section of his *Ethics* that friendship is a completion of human being through the emulation, comparison, and reflection between peers.[18] I note

[18] Aryeh Kosman adds the context that: "In none of these teachings [of Aristotle] does the individual subject disappear; whether in friendship, polity, or contemplation, the self is enhanced by

that Hegel's position on "shapes" (*Gestalten*) and on the power of Ideas to shape the world bears relation to Aristotle's whilst exceeding his naturalistic approach to the situation.

On the possible objection that Aristotle's "education of forms" in nature does not fully explain human being and individuality, I comment that Aristotle is not concerned primarily with the "*Sturm und Drang*" of human existence but with the structure of being within which a human being makes sense. Aristotle's method is to stop where the science ends and to leave a space for the individual and personal to manifest their own being and their own sense.

7.7 The Logos of Aristotle's Logic

Aristotle's conceptualisation of "*logos*" as the meaning that expresses being leads to his logic as representing the mental structures that allow us to "tie down" this meaning. Aristotle's logic is always about something else whilst *also* being *en parergō* about the human mind itself. His logic is designed to allow us to understand both the powerful reach of our human logic and the limits of that logic by reference to the subject matter we are studying. Aristotle's logic is ultimately an "Organon," a tool which allows us to engage with the world, whereas our own mathematical, computational logic is a blunter tool unconcerned with beings and ultimately with mind as well.

From an Aristotelian perspective, it is by recognising the "gaps" which naturally emerge in our framework of thought that we understand ourselves better and our thinking better and also understand the nature of the world better by understanding how and in what sense it is knowable. We also gain self-awareness from this self-reflection and this self-awareness prevents us from being overwhelmed or occupied by the subject matter with which we are engaged. Aristotle explains this maturation of being and of thought as that: "human thought…does not possess the good at this moment or that, but its best, being something *different* from it, is attained only in a whole period of time (*Met.* Λ 1075a 7–10)." Our human being exhibits an independence of mind which allows us to engage in genuinely creative acts. Aryeh Kosman explains this situation well in his *The Activity of Being: An Essay on Aristotle's Ontology* as that: "A basic tenet of Aristotle's ontology…concerns this relation between determinacy and determinability, that is, receptivity to further determination. It is because a substance is the determinate being it is that it is able to be the subject of attribution without being overwhelmed by it."[19] It is by maintaining an

incorporation, not diminished ("Aristotle on the Desirability of Friends" in *Virtues of Thought: Essays on Plato and Aristotle* (Cambridge Mass., [*orig*. 2004] 2014) p 182)."

[19] Kosman, Aryeh *The Activity of Being: An Essay on Aristotle's Ontology* (Cambridge Mass., 2013) p 233. He explains further that: "to think of some particular thing is to surrender the author-

intellectual detachment and not being overwhelmed by the subject of our work that we can be a "master craftsman" who can intellectually contextualise and meaningfully utilise our knowledge. This is perhaps the grounds for Aristotle's concern with the banausic mind and for what Socrates meant when he argued that an unreflective life was not worth living.

The idea of "*nous*" is critically important to Aristotle's interpretation of the world since it is through this meaning-making force that all things are formed and through this that we think about the world. Aristotle argues that: "intuitive reason (*ho nous*) is concerned with the ultimates (*tōn eschatōn*) in both directions (*ep' amphotera*); for both the first terms (*prōtōn horōn*) and the last (*eschatōn*) are objects of intuitive reason (*nous*) and not of argument (*logos*), and the intuitive reason which is presupposed by demonstration (*kata tas apodeixeis*) grasps the unchangeable first terms (*akinētōn horōn kai prōtōn*), while the intuitive reason involved in practical reasonings (*ho d' en tais praktikais*) grasps the last and variable fact (*tou eschatou kai endechomenou*) (*N.E.* VI 1143a 35–1143b 3)." We see that our thought as well as our being is formed by its whence and whither. The form is the actuality in both directions as the source and the end, and the movement is the material potentiality. It is interesting that Aristotle's most careful explicit unpacking of the relationship between "*nous*" and "*logos*" occurs in his *Ethics*.

Aristotle's above account includes not only the differentiation between *nous* and *logos* but also between *a priori* and *a posteriori* logic. Aristotle explains that we can both engage with the concrete individual and with our abstractions and can determine how the two can be brought into alignment through logic and through contemplation upon them both. Aristotle's logic makes sense on this basis as a form of "seeing" with *a priori* reasoning being "demonstration" from principles and his *a posteriori* reasoning being "induction" from facts.[20]

I represent Aristotle's logical approach in Fig. 7.5 below, giving what I take to be its main themes and dimensions. Aristotle's review of logic clearly reflects how we do engage with the world as human beings in reality.

---

ity of thought to that which one thinks of, and it is then still the case that thinking is not what thought is but only what it happens to be doing given that one is thinking of that object (*Ibid.* p 220)…Aristotle appears to offer reflexivity as a way to preserve thought's intentionality without allowing its value to bleed out of an object other than it (*Ibid.* p 224)."

[20] Aristotle explains that: "the expression "to know that every triangle has angles equal to two right angles" is ambiguous, meaning to have knowledge either of the universal or of the particulars…For it never happens that a man starts with a foreknowledge of the particular, but along the process of being led to see the general principle he receives a knowledge of the particulars, by an act (as it were) of recognition (*Pr. An.* II 67a 17–25)" which approach Aryeh Kosman explains as that: "The explanatory activities of science involve at once learning the principles of explanation and coming to understand the phenomena that are explained: as Aristotle puts it, "revealing the general by making clear the particular"—δεικνύντες τὸ καθόλου διὰ τοῦ δῆλον εἶναι τὸ καθ' ἕκαστον (*Posterior Analytics* 71a8). This is the single activity that may be thought of alternatively as ἀπόδειξις or as ἐπαγωγή, as proceeding from or to the principles: grasping the phenomena clearly by explaining them, and in the process, grasping the principles of explanation ("Saving the Phenomena: Realism and Instrumentalism in Aristotle's Theory of Science" in *Virtues of Thought: Essays on Plato and Aristotle* (Cambridge, Mass., [orig. 2000] 2014) p 148–9)."

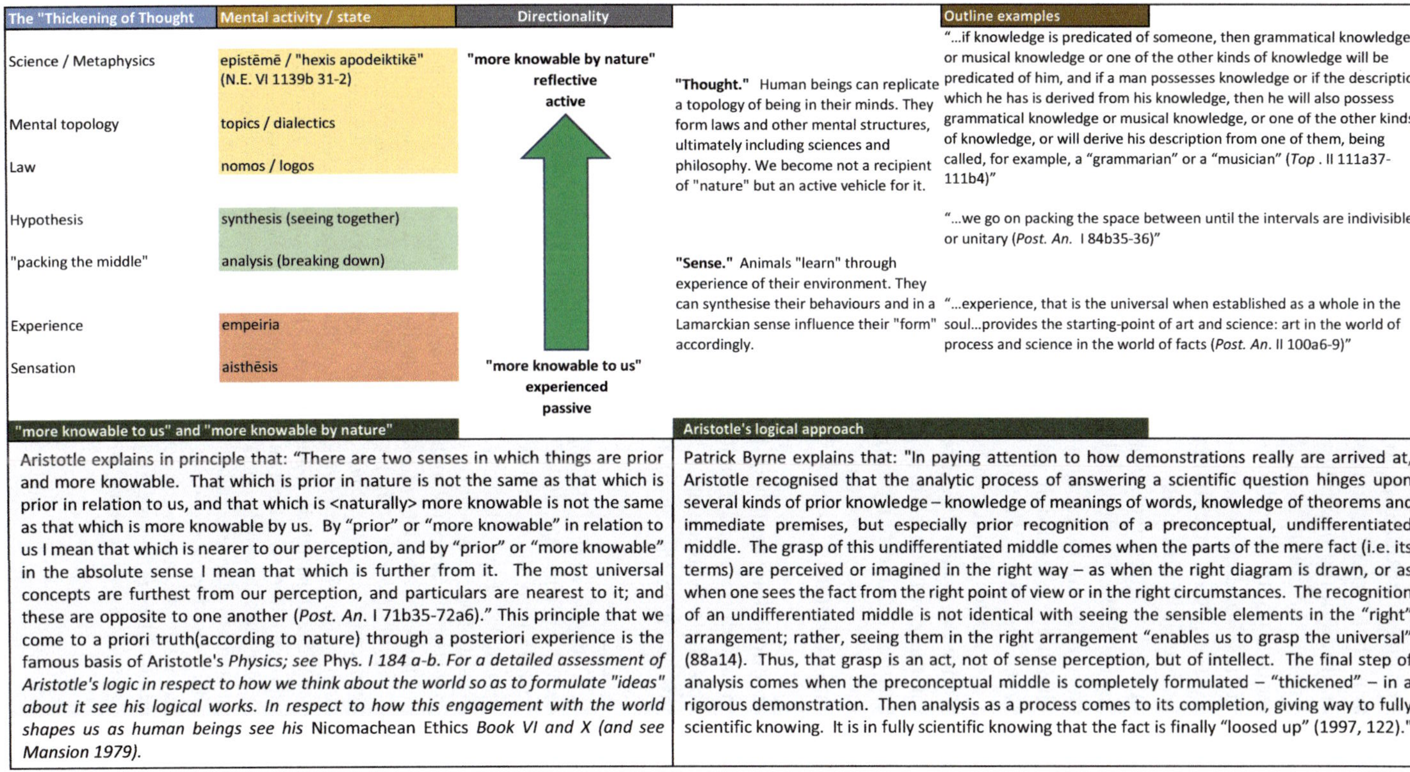

Fig. 7.5 Aristotle's logical structuring of the human mind

Imre Lakatos explains, from a contemporary perspective, the situation as regards how the products of our own mind, *our* logic, relate to the *logos* of the world. He explains that "truth" emerges from the process of carefully "injecting" our attention through the positing of "occult hypotheses" on the basis that: "The main point about the classical analysis-synthesis is that it links known and unknown together by a chain of deduction, or, by a circuit of truth and/or falsity. We inject truth or falsehood at some point and it will be carried to every part by the circulatory system. Now circulatory troubles arise both near the reasoned facts and near the occult hypotheses."[21] The introduction of the "occult" or irrational or superrational is important. The very act of doing so rejects reduction, chance, and necessity. We infer that reality is explicable neither as "chance" nor as "necessity" but only as an interaction between a creative existence and a natural law. We see that there is and that we primarily exist within a middle ground of meaning between the two. We find a necessary strangeness, richness, and entanglement in the world which can (strangely enough) be understood by the human mind but only ever partially, through reduced abstractions. Lakatos ultimately draws out the complementarity of being which enables us to see the world from one aspect or another but not both together and in both cases precisely but deficiently.[22] This is perhaps a philosophical version of Niels Bohr's "principle of complementarity" which we will consider in the sister volume of this project.

As regards his "circulatory troubles," Lakatos explains that the outcome of this lack of circulation (and of deep understanding) is the modern mathematical logic on which he comments that: "Mathematics has been trivialised, derived from indubitable, trivial axioms in which only absolutely clear and trivial terms figure, and from

---

[21] Lakatos, Imre "The Method of Analysis-Synthesis" in *Mathematics, science and epistemology: Philosophical Papers Vol 2* (Cambridge, 1978) p 76. On Aristotle and "occult hypotheses" see Peter T. Manicas "Aristotle, Dispositions, and Occult Powers" *Review of Metaphysics* (1965).

[22] Imre Lakatos argues that: "(1) I call a deductive system a "Euclidean theory" if the propositions at the top (*axioms*) consist of perfectly well known terms (*primitive terms*), and if there are *infallible truth-value-injections* at the top of the truth-value *True*, which flow downwards through the deductive channels of truth-transmission (*proofs*) and inundate the whole system…(2) I call a deductive system an "*empiricist theory*" if the propositions at the bottom (*basic statements*) consist of perfectly well known terms (*empirical terms*) and there is a possibility of *infallible truth-value-injection* at this bottom which, if the truth-value is *False*, flows upward through the deductive channels (*explanations*) and inundates the whole system." His conclusions are (1) that: "Both programmes offer truths, which are trivial and uninteresting if taken in isolation, but *because of its location*, trivial truth inundates the whole Euclidean theory but not the empiricist one" (2) that: "When in science one wants to prove everything from below, one has first to redefine, reconstruct, everything in the perfectly well known terms of the bottom" and (3) that: "It makes an immense epistemological difference at what level we inject truth-value in the theory; for this determines the pattern of flow of truth and falsity in the system ("Infinite Regress and Foundations of Mathematics" *Proceedings of the Aristotelian Society* (1962) p 158–63)."

which truth pours down in clear channels. Concepts like "continuity," "limit," etc. gave way to concepts like "natural number," "class," "and," "or" etc. The "arithmeticisation of mathematics" was a most wonderful Euclidean achievement."[23] Concepts such as "continuity," "limit" etc. are explicitly and notably Aristotelian. According to Lakatos it is their sacrifice that produces a "triviality" regarding which he wonders: "One is amazed how sophisticated triviality can be."[24]

I represent this problem as described by Lakatos as Fig. 7.6 below.

We see on both sides a *simple and clear method* working from fact *or* theory. As Aristotle has explained above, though, this either/or position is a false dichotomy. Even the idea that there is a single approach for dealing with the various objects of the world is false. We need analysis *and* synthesis in all cases. The idea that the world can or should be understood as chance and/or necessity is also clearly erroneous, and I represent this error through Figs. 7.7 and 7.8 below.

The two diagrams above show us (1) that science when taken as a full technological explanation of "being," must regard any product of nature as a monster or as a machine and (2) that it is only by having a fuller vision of the world—in this case an Aristotelian approach—using a philosophical (or theological) framing which takes into account the cycle of being, as "analysis" and "synthesis," that we have the methodological and intellectual resources to represent the true reality of being rather than an abstract construct produced by science.

Jaakko Hintikka also explains Aristotle's position vis-à-vis the alternative on the basis, oddly enough, that he recognises existence in his logic. He argues that: "According to this theory of Aristotle's, in chains of scientific syllogisms existence

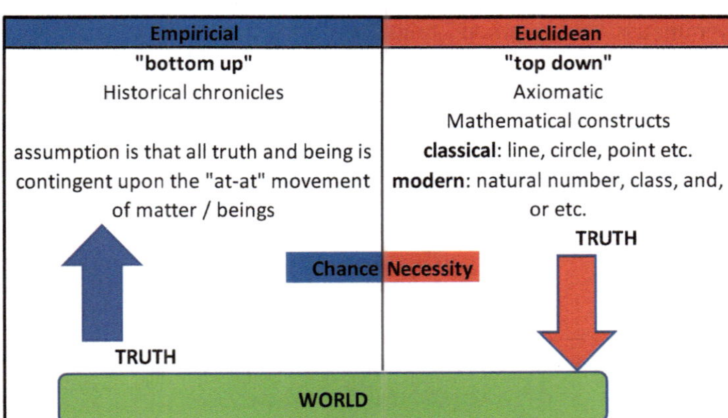

Fig. 7.6 Lakatos' Two "infallible truth-value-injections"

[23] Lakatos, Imre *Ibid.* p 166.

[24] Lakatos, Imre *Ibid.* p 182.

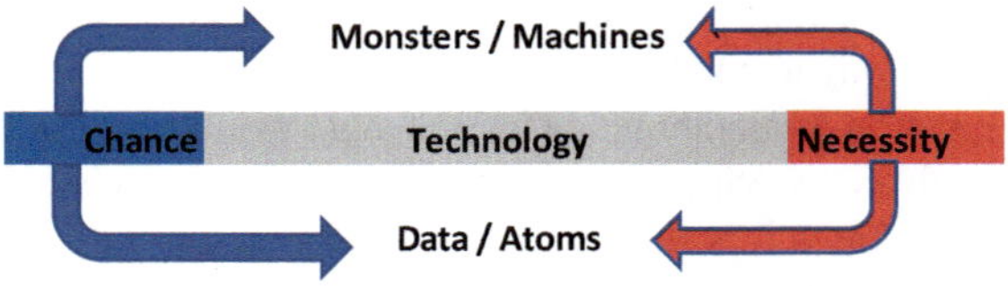

Fig. 7.7 The analytic dualism of chance and necessity

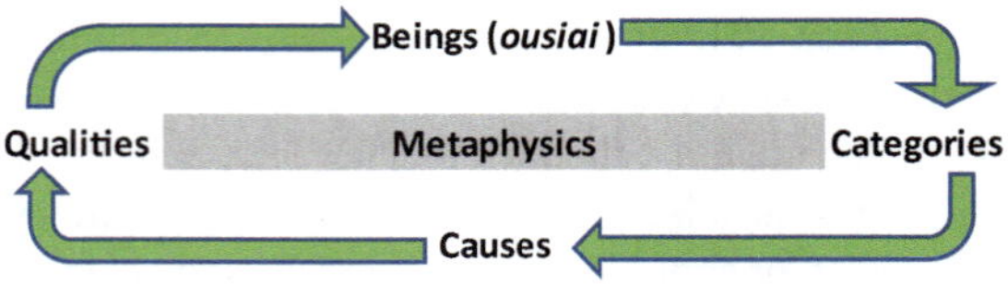

Fig. 7.8 The synthetic cycle of qualitative being(s)

is transferred from top down, from more general premises to less general ones. Thus in any one science, it suffices to assume existential force for the major term. As a part of the syllogism, this existential force is transferred to the minor term mediated by the middle term."[25] Behind this interpretation of being we find an "asymmetry" of beings which, according to Hintikka, is "rampant in Aristotle's modal logic." Hintikka explains: "what are the sources of this asymmetry? One place where it shows up is in epistemic logic…Thus according to epistemic logic knowledge (being known) trickles down in a chain of syllogisms from the major premise to the conclusion in each syllogism. This asymmetry matches the asymmetries of Aristotelian modal syllogisms. For according to him only the major premise needs to be necessary in a barbara-type syllogism in order for the conclusion to be necessary."[26] Aristotle's thought is not only concerned with "symmetry" but also with "asymmetry" and symmetry breaking.

Hintikka usefully concludes as regards the import of this situation that we *can and should* employ Aristotle to challenge even the "mostly firmly entrenched habits of twentieth-century philosophers."[27] This is no subtle shift. What it signifies is a paradigm shift making logic and philosophy concern *values* rather than *mechanisms* with this shift of emphasis utterly transforming how we look at and philosophically approach the world.[28]

---

[25] Hintikka, Jaakko "On Aristotle's Notion of Existence" in *Analyses of Aristotle* (New York, [*orig.* 1999] 2004) p 11.

[26] Hintikka, Jaakko *Ibid.* p 9–10.

[27] Hintikka, Jaakko *Ibid.* p 5.

[28] On Aristotle's logic see also Russell Winslow's *Aristotle and Rational Discovery* (London, 2007) and Walter Elsasser "A Form of Logic Suited for Biology" in Rosen, Robert (*ed.*) *Progress in Theoretical Biology, Vol. 6* (Academic Press, 1981).

I add that Patrick Byrne also explains the value of Aristotle's logic and science well—in a substantive rather than trivial way—as that: "Aristotle's criteria for genuinely *epistemic* premises require that the major and minor premises could *not* be knowledge of *mere* facts. Rather, the premises themselves must be formulations of reasoned facts, facts whose reasons are known. Hence, the analysis of a compact, mere fact via demonstrative episteme resolves mere knowledge of the merely given but obscure association of terms into a set of connections that is thoroughly illuminated, whose reasons are fully manifest. In light of all this, one would be missing the point to say that Aristotle regarded *the earth* as the cause of the moon's deprivation of light. It is, rather, the cause as expressed in the middle term (not "the earth" but "the earth *as* optical obstruction"), in the intelligible interconnections fashioned by the demonstrative syllogism, which is that cause."[29] The reasoned position that Byrne imputes to Aristotle fixes the problem of Lakatos' triviality by connecting and circulating the meaning of the world.

We can also return to consider G.W.F. Hegel to explain this situation and, in part, to understand why something so obvious should also be controversial. Hegel—who rightly points us to Aristotle as the authority on the importance of recognising and navigating a "system" of nature[30]—argues that: "When the world is taken as this aggregate, even only of natural things, it is not represented as *nature*, by which we understand something that is inwardly a systematic whole, a system of arrangements and levels, and especially of laws. The term "world" merely expresses the aggregate in a way that, resting as it does simply on the existing mass of things, it has no advantage, at least no qualitative advantage, over worldly things."[31] A.N. Whitehead similarly explains the problem with science as that: "Science can find no creativity in nature; it finds mere rules of succession."[32] The difference we find here is between a mathematical logic and a natural logic as regards how being becomes.

7.8 The Full Meaning of the *Poetics*

We see that Aristotle's logic is qualitatively different than ours. Is it inferior? It does not reduce the world to mathematical calculation but provides an assessment of meaning in the world. It subtracts but does so knowingly and judiciously. It

[29] Byrne, Patrick *Analysis and Science in Aristotle* (New York, 1997) p 91.

[30] See G.W.F. Hegel "The Philosophy of Aristotle" *The Journal of Speculative Philosophy* ([orig. 1825–6] 1871).

[31] Hegel, G.W.F. *Lectures on the Proofs of the Existence of God* (1829) §274.

[32] Whitehead, A.N. "Nature Alive" in *Modes of Thought* (New York, 1938) p 154.

provides a scaffolding for explaining our understanding and engagement with the world without conflating that scaffolding with the world itself. I contend that Aristotle's scientific studies operate on a similar basis. He thereby seeks to understand the "special" meaning of any subject area or science and also to appreciate the meaning of this meaning "*en parergō*." I take Aristotle's *Poetics* as an example of this situation. His exercise is not just a study of literary form and style but is a study of the significance we can take from cultural and creative activity. It is a philosophical representation of what "*poiēsis*" or the power of making *means*. We can take this as a paradigm for Aristotle's science, as his own paradigm was the art of boxing.

Aristotle argues on the act of *poiēsis* that: "Now, epic and tragic poetry, as well as comedy, dithyramb, and most music for aulos and lyre, are all, taken as a whole, kinds of *mimesis* [imitation]. But they differ from one another in three respects: namely, by producing *mimesis* in different media, of different objects, or in different modes. Just as people (some by formal skill, others by a knack) use colours and shapes to render mimetic images of many things [i.e., as artists], while others again use the voice [i.e., as actors], so too all the poetic arts mentioned produce *mimesis* in rhythm [i.e., dance], language [i.e., poetry], and melody [i.e., music], whether separately or in combinations (*Poet.* 1447a13–22)." Nathan Spiegel explains the situation that Aristotle is addressing in the above passage as that: "Every work of art uses the means appropriate and essential to it. The means in dancing are rhythmic gestures, in sculpture-forms, in painting-colours, in poetry and music-rhythm, words and melody. These means can be used either separately or in certain combinations. Thus music uses both rhythm and melody, whereas dancers use rhythm alone, and by the cadence of their movements express feelings and actions and create personalities. Some arts, like comedy, tragedy and dithyrambic poetry use all the three means together."[33]

Francis Wolff explains the situation we face here further by assessing the commonality between Aristotle's *Poetics* and his *Physics* and the relation between "*technē*" and "*mimesis*" as that: "a *technē mimētikē* work has two different kinds of matter, depending on whether it is considered the product of a *technē* (as theorised in the *Physics*) or the product of *mimēsis* (as theorised in the *Poetics*). The matter of a work of art is for example bronze, wood, and even sound, which is shaped by the artist. The matter of mimetic works varies in relation to the types of representation and designates the medium "in which" (1447a17) the representation is made: each imitation has its own medium. This question, dealt with by Aristotle in chapter 1 of the *Poetics*, leads him to an initial classification of the fine arts in relation to the medium: sculpture, the medium of which is shapes (*schēmata*, 1447a19), painting

[33] Spiegel, Nathan "The Significance of "Mimesis" in the Light of Aristotle's Doctrine of the Four Ontological Causes" *Revue Belge de philologie et d'histoire* (1975) p 18.

the medium of which is shapes and colours, and so forth for literature (art and language), instrumental music (art of rhythm and melody), dancing (art of rhythm and shapes), and singing (art of melody and language)."[34] Aristotle, as Wolff shows, scientifically unpacks the analogy between the activities and objects of physics and those of poetics. He presents us with two worlds; a physical world of objects in space and a mental world of imaginary beings.

Aristotle considers on what basis both worlds can simultaneously exist in our human mind and being. He establishes here at the outset of his work the principle that the artistic *medium* enables the relationship between a subject (e.g. an artist's "vision"), a medium (e.g. shapes and colours), and an object (e.g. an artwork) to be expressed. Each of these expressions and causes—the matter, form, movement, and end—is different and distinct even if they have *analogous* goods and ends. The act of creation is not a process but an event, a nexus of combined energies and completed significance. I note that the passage considered above is, once again, an introductory passage which raises the philosophical problem to be scientifically analysed and explored further in the research project. I find the conciseness alongside the depth and perceptiveness of its perception to be breathtaking.

We find that Aristotle's *Poetics* is not just a work concerned with poetic style and structure, as we would approach the subject today, but it also and ultimately covers the very fact of mental constructions. The poetic work imitates nature in the sense that it must have a "beginning, middle, and end (*echousan archēn kai mesa kai telos*), so that epic, like a single and whole animal (*hōsper zōon hen holon*), may produce the pleasure proper to it (*Poet.* 1459a 18–20)." The idea that a poem is like an animal and that it develops and makes demands according to *its* ends and function defines Aristotle's theory of being(s). Aristotle's *Poetics* considers the chain through which we form an imaginative and representative world and also how our art itself mimics nature. *Nomos* ultimately copies or fills in for *physis*. Human beings construct a world of beings either as sciences or as arts. These disciplines take over us and direct us according to *their* needs so that: "the genre's own nature teaches poets to choose what is apt for it (*autē hē physin didaskei to harmotton autē haireisthai*) (*Poet.* 1460a 3–4)." Aristotle considers poetry to understand how it is a natural product and how it relates to nature.

We can add that this difference between the scientific world of intellectual abstractions and the artistic world of imaginative and worldly projections is explicitly addressed in Aristotle's *Ethics* which divides human intellect between the practical reason (*phronēsis*) and the intellectual reason (*sophia*). He also here explains

[34] Wolff, Francis "The Three Pleasures of Mimēsis According to Aristotle's *Poetics*" in Bensaude-Vincent, Bernadette & Newman, William (*eds.*) *The Artificial and the Natural: An Evolving Polarity* (Cambridge Mass., 2007) p 55. He adds that: "The form of the representation is thus a reality coming from the artist's mind. What the *mimēsis* represents—this will be its form. So generally, every *mimēsis* is the representation of at least a possible if not an actual reality (see ch. 25, 1460b7–11) (*Ibid.* p 55)." The question is raised regarding artistic imagination and how far it must imitate the outside world in its creations.

the relation between the human being and the objects of the world as that: "every art [or form of understanding] (*pasa epistēmē*) does its work well (*to ergon eu epitelei*)—by looking to the intermediate (*pros to meson*) and judging its works by this standard (so that we often say of works of art that it is not possible either to take away or to add anything, implying that excess and defect destroy the goodness of works of art, while the mean preserves it; and good artists, as we say, look to this in their work) (*N.E.* II 1106b8–14)." Aristotle argues that the work of art has its own full expression which is shaped and produced by the artist, as an engineer could similarly design a bridge. The creative actualisation of thought is on this basis better than the thought and intention that precedes it, if it is realised. This consideration of "products" brings us back to the introduction of the *Ethics* where Aristotle argued, as we have seen, that: "Where there are ends apart from the actions, it is the nature of the products to be better than the activities. Now, as there are many actions, arts, and sciences, their ends also are many…"; see *N.E.* I 1094a1–15.

We ultimately see that for Aristotle his *Physics*, his *Poetics*, and his *Ethics* are not merely discrete or "special" studies of their respective subject matters but are rather studies of their subject matters within an overarching research programme in respect to being, becoming, and movement. Aristotle's approach from the method of analogy enables a true ontology on the basis that it asks how the various genera of being—chemical, physical, animal, human, scientific, artistic, poetical etc.—can *all* analogically be explained as *shaped* and *moving* things-in-the-world.[35] Whatever the value of its successors, the all-encompassing nature of Aristotle's project is not a failing preventing quality specialisation but a spectacular success in drawing out the shared and parallel values and drivers of beings within being; as Hegel correctly observed in his study of Aristotle.

7.9 The Full Meaning of the *Rhetoric*

Let us also consider Aristotle's *Rhetoric*. We would be wrong if we treated Aristotle's *Rhetoric* merely on the basis of *our* definition of rhetoric as a study of "the art of effective or persuasive speaking or writing, especially the exploitation of figures of speech and other compositional techniques." Clearly, the *Rhetoric* is an excellent example of precisely such a study (as the *Poetics* and *Ethics* etc. are *also* excellent studies of their subject matters) but it is also much more than this. It is a philosophical reflection as well as a scientific textbook. Aristotle himself explains this wider

[35] Aryeh Kosman explains Aristotle's synthetic approach that: "The *Poetics* may be seen as the sequel to the *Ethics* and the *Politics* in a deeper sense than that in which it continues philosophical interests central to the theory of human action in these works. For, more importantly, the *Poetics* continues the vision that is developed in these companion volumes: a vision of the *polis*, or more generally of organised social and communal activity, as the source of our capacity to live satisfying and fulfilling human lives ("Acting: Drama as the *Mimesis* of *Praxis*" in *Virtues of Thought: Essays on Plato and Aristotle* (Cambridge Mass., [*orig*. 1992] 2014) p 117)."

context as that: "it appears that rhetoric is as it were an offshoot of dialectic and of the science of ethics, which may be reasonably called politics (*Rhet.* I 1356a 25–6)" and he also invokes his *Topics* on the basis that: "By specific topics I mean the propositions peculiar to each class of things, by universal those common to all alike. Let us then first speak of the specific topics, but before doing so let us ascertain the different kinds of rhetoric (*Rhet.* I 1358a 31–33)." We see that there is a way of logically approaching each "class of things" and of mentally assimilating and rhetorically representing them.[36]

We find, in other words, that Aristotle is engaged in a complex comparative or analogical study of various "spheres" of being and beings for which each "sphere" has its own science; see *Met.* E 1026a *passim.* Hegel rather brilliantly explains this situation as that: "Each of the parts of philosophy is a philosophical whole, a circle rounded and complete in itself. In each of these parts, however, the philosophical Idea is found in a particular specificality or medium. The single circle, because it is a real totality, bursts through the limits imposed by its special medium, and gives rise to a wider circle. The whole of philosophy in this way resembles a circle of circles. The Idea appears in each single circle, but, at the same time, the whole Idea is constituted by the system of these peculiar phases, and each is a necessary member of the organisation…In this manner philosophy exhibits the appearance of a circle which closes with itself, and has no beginning in the same way as the other sciences have."[37] Hegel shows us how the various sciences can as parallel spheres of meaning produce a virtuous circle of general understanding. This higher understanding manifests not only as philosophy but also as rhetoric, as the representation of what is generally reasonable in a culture back to its members.

Aristotle hence relates the *Rhetoric* not only to the *Ethics*, *Politics*, and Dialectics (*Topics*) but also to the *Poetics* to which he refers in his concluding section when he is seeking to understand the philosophical application of metaphor.[38] Aristotle explains that: "metaphor is a kind of enigma, so that it is clear that the transference is clever (*Rhet.* II 1405b 5–6)," i.e., that "metaphor" ("*meta-phora*") is a way of transferring meaning between domains. Aristotle again refers to the *Topics* here; see *Top.* VI 139b 33–35.

Let us on this basis consider the fact that Aristotle's *Rhetoric* is concerned with the "bringing-before-the-eyes"[39] of a subject matter and that his *Topics* is a

[36] Aristotle argues that rhetoric and dialectic are, as also ethics and philosophy, additional levels of expression and articulation which must be added to the scientific facts in order to give them full expression and allow them to be communicated. He explains the lower qualitative bar for dialectic as compared with logic as that: "scientific discourse is concerned with instruction, but in the case of such persons instruction is impossible; our proofs and arguments must rest on generally accepted principles, as we said in the *Topics*, when speaking of converse with the multitude (*Rhet.* I 1355a 26–29)."

[37] Hegel, G.W.F. *Hegel's* Logic (1830) § 15 & 17.

[38] See Fran O'Rourke "Aristotle and the Metaphysics of Metaphor" *The Boston Area Colloquium in Ancient Philosophy* (2005).

[39] See Sara Newman "Aristotle's Notion of "Bringing-Before-The-Eyes": Its Contributions to Aristotelian and Contemporary Conceptualizations of Metaphor, Style, and Audience" *Rhetorica:*

complementary consideration of how we should approach and represent these "spheres" or "topics" of being. I represent this situation in Fig. 7.9 as below.

Whereas the *Rhetoric* is concerned with the rhetoric through which we present a case, the *Topics* is concerned with the dialectic through which we formulate and argue a case. We find that the basic assumption upon which dialectic stands is that: "things which create or preserve something are co-ordinate with that of which they are creative or preservative, for example "healthy things" are co-ordinate with "health," and "things which produce a good condition" are co-ordinate with "good condition," and so with the other cases (*Top*. II 114a 29–32)." As with his logic, physics, ethics, and metaphysics, Aristotle is concerned to consider how qualities and values map on to quantities or beings. Aristotle adds as regards how epistemology corresponds with ontology that: "the bases of arguments are equal in number and identical with the subjects of reasonings. For arguments arise from "propositions" while the subjects of reasonings are "problems." Now every proposition and every problem indicates either a genus or a peculiarity or an accident (*ē genos ē idiom ē symbebēkos*); for the differentia also, being generic in character, should be ranged with the genus (*Top*. I 101b 14–19)." He seeks to represent how the mind and its arguments/problems maps to the world through a proportionate *logos* between them.

The *Topics* is a magisterial but little studied treatise which we can only touch upon here for the purpose of showing how Aristotle builds his argument through the various exercises which make up his corpus. Aristotle shows us in this work how knowledge relates to sense as that: "if knowledge is a conceiving, then the knowable is conceivable; and if sight is a sensation, then the visible is sensible…it is through the sensation connected with sight that we recognise the form which is in each thing (*Top*. II 113a32–114a 20)." He concludes the work with a consideration of "judgement" such that: "For such a process [i.e., to have taken in at a glance the results of each of the two hypotheses] one must possess a certain natural ability (*euphya*), and real natural ability consists in being able correctly to choose the true and avoid the false. Men of natural ability (*hoi pephykotes*) can do this; for they judge correctly what is best by a correct feeling of love or hatred for what is set before them (*Top*. VIII 163b 13–17)." Aristotle shows us how we see the form through our senses and

Inductive Logic	Deductive Logic	Dialectic	Rhetoric	Sophistry
Posterior Analytics	Prior Analytics	Topics	Rhetoric	Sophistical Refutations
reasoning using patterns to arrive at a conclusion (conjecture)	syllogistic reasoning using facts along with definitions, accepted proportions, and the laws of logic	the art or process of finding the truth via an exchange of ideas about the world	the art of persuasing and of "bringing-before-the-eyes" a subject matter	the use of words to convince without regard for truth and substance

Fig. 7.9 Aristotelian expressions of human logic

A Journal of the History of Rhetoric (2002).

how we discern truth in these forms and situations. He considers here the human dimension in respect to how we assimilate the world in order to make true judgements upon it, and care or love it, whereas his *Rhetoric* is concerned with how we assimilate the facts so as to be able to properly present them. This conclusion from his *Topics* makes explicit the assumption of his *Ethics* that good ethical behaviour is largely a natural characteristic rather than something that can be taught, though it can be trained and enforced.

In all cases, Aristotle remains in the human sphere. The last tableau of our thought-world presented in Fig. 7.9 refers to his *Sophistical Refutations*, sometimes considered part of his *Topics*, which treats how we can define arguments without falling into category errors or errors of logical typing. It is by avoiding these errors that we can "very nearly approach to the truth (*Soph. El.* 169a 33)" since our truth is primarily a right "seeing" of the world. The main sophistic error that Aristotle treats in his *Sophistical Refutations* is that of nominalism regarding which he argues that: "a man is apt to be deceived when he carries on his inquiry by means of words (*epi tou logou*) (*Soph. El.* 169b 1–3)" and, in more detail, that: "[when] we use names as symbols in the place of things (*alla tois onomasin anti tōn pragmatōn chrōmetha symbolois*), we think that what happens also in the case of the names happens also in the case of the things…those who are unacquainted with the power of names are the victims of false reasoning (*Soph. El.* 165a 7–17)." Aristotle argues that we must understand that the medium through which we engage with the world is not the same as the world itself or its "substance" (*ousia*) but only an interpretation of it. He argues that we are ourselves a potentially corrupting medium for knowledge and another of Aristotle's criticisms of sophism is that it is a cult of personality.

The picture Aristotle ultimately presents is of a landscape of symbolic structures between our sense of the world and the world itself which moves from the structures of the *Topics*, namely that: "A philosopheme is a demonstrative inference, an epichireme is a dialectical inference, a sophism is a contentious inference, and an aporeme is a contentious inference of contradiction (*Top.* VIII 162a 16–18)." He also considers the "enthymeme" in the *Rhetoric.* We see how Aristotle's consideration of a topic is constantly put forward in the context of other topics, with there being a clear and explicit analogical method employed.[40] Through this semiotic scaffolding Aristotle philosophically represents the world as a *meaningful* landscape of beings in which there is a real and variegated engagement between the insides of a being with its outside. This conversation with the world leads us to see the world in semiotic terms as a transfer and capturing of information and of memory. The critical aspect of Aristotle's analysis is that we can explain and quantify the language of reality and of its beings and yet we will always find that we are operating within a

[40] Analogy stands at the very beginning of Aristotle's thought; see *Cat.* 1a. For Aristotle's observation regarding the shortcomings of analogy see *GC* II 333a 30ff. He argues in principle that: "The underlying nature can be known by analogy. For as the bronze is to the statue, the wood to the bed, or the matter and the formless before receiving form to any thing which has form, so is the underlying nature to substance, i.e., the "this" or existent (*Phys.* I 191a 8–12)."

middle bookended by concrete facts and worldly beings; see *N.E.* VI 1143a 35–1143b 3. It is both liberating and limiting to accept that we can never fully know the whole but only represent it partially through our thinking about its middle.

Aristotle treats in his *Rhetoric* and *Topics* how we fully engage with the world through the dialectical rather than purely logical expression of language. Martin Heidegger explains this exercise as that: "Aristotle investigates the "*pathē*" [affects or affections] in the second book of his *Rhetoric*. Contrary to the traditional orientation, according to which rhetoric is conceived as the kind of thing we "learn in school," this work of Aristotle must be taken as the first systematic hermeneutic of the everydayness of Being with one another."[41] Heidegger correctly understands, here and throughout, how Aristotle's work goes beyond *our* boundaries, *our* categories, and *our* preconceptions.

7.10 Conclusions

Aristotle's approach to science is reasonable and penetrating. It is not loose or "unscientific" but methodological and empirical in nature. It builds up from the facts and does so in accordance with a methodology appropriate to the subject matter under investigation and the questions being asked of it. Paul Feyerabend, the philosopher of science, well explains that Aristotle provides us with "the only reasonable empiricism that ever existed [whereas] Bacon…was an anti-empiricist plus an anti-rationalist ("rationalism" in the sense of Descartes) and in this respect the father of [modern] science."[42]

The rigour employed may be comparable but there is a vast philosophical difference between Aristotle's science and ours. This difference is, in a nutshell, between a purely analytic approach and the classical pursual of analysis followed by synthesis. The problem we face today, as Niels Bohr observed in his, is that theoretical explanation has been replaced by mathematical formulae. Science has become technology. Method has replaced thought. The problems we face here are philosophical or spiritual rather than practical. They do not regard the practical limitations we must accept to pursue science but our ability and willingness to accept limitations. Can we accept that our research and its conclusions are only ever a subtraction from and representation of reality and that we must supplement this practical inquiry with a philosophical framing which brings us back to the rich reality before us? This is what Aristotle's realist programme asks of us. His great achievements are precisely to formulate evidence-based science which works from below and a speculative philosophy which fills in the gaps left and explains how the sciences are a part of the whole of nature.

---

[41] Heidegger, Martin *Being and Time* (Oxford, [*orig*. 1926] 2001) I 5 § 138.

[42] Feyerabend, Paul *For and Against Method* (Chicago, 1999) p 197. For Feyerabend's thorough defence of Aristotle, see his *Conquest of Abundance: A Tale of Abstraction versus the Richness of Being* (Chicago, 1999).

Aristotle's science is, as I have shown above generally and will show below in respect to his biology, of a quite exceptional character. It does not force meaning but it allows the meaning of the various subject matters of nature—including in respect to human being and culture—to be exposed. His method allows this meaning to be "tied down" and also to be understood as part of the whole. His philosophy explicitly explores this metaphysics, this underlying framing of being and of thinking which explains how our experience of the world is not "as mice walk(ing) upon loose soil." The genius of Aristotle's philosophy goes hand in hand with the genius of his science. He does not, however, argue that all scientists must also be philosophers. Science is a restricted activity. Aristotle clearly argues that scientists must be specialists who are occupied by their research program. We should not expect them to also be philosophers. This is an example of Aristotle's pragmatism which is always concerned with the "gaps" in our knowledge and our approach, and the fact that we are always subtracting from reality in various ways.

Science is, as Aristotle observes, correctly wary of recognising values outside its direct scientific remit. It is, however, a conceit that science can be conflated with reality or that our full engagement with reality should be value-free and hence, supposedly, scientific. This positivistic approach is rightly opposed by such modern approaches as ecology and virtue ethics. Although many of the holistic, "*Gestalt*," and homeostatic approaches of the modern world are clearly Aristotelian in character, perhaps the best surrogate is biosemiotics. Its direct emphasis upon meaning is important, and also its emphasis that the whole biological world is implicitly meaningful. It represents a scientific approach but one which also recognises values and meaning. It recognises "beings" as meaning-making actors in the world. This framing does not in principle water down or disrupt science but merely frames it so that it addresses our reality holistically and in the full.

Chapter 8
Can We Learn from Aristotle's Biology Today?

Abstract This chapter considers a selection of approaches from Aristotle's biological work in order to show the quality, seriousness, and *philosophical* power of Aristotle's science of biology. It seeks to show the remarkable insight of Aristotle's scientific work in this field. His observations are becoming increasingly pertinent in an age of epigenetics and evo-devo, not less. Even more significant, however, is the fact that the great philosopher Aristotle produces a sophisticated philosophy based upon biology and hence provides us with a philosophical basis for biology. Aristotle's philosophy seeks to explain the biological fact that a human foetus grows into a human being and the matures into an ethical human being. The ultimate subject of Aristotle's whole project is not the meaning of matter or God, but life as ultimately expressed in the human being. Aristotle's approach is not anthropocentric but merely argues that we must surely include our own existence within any explanation of the world.

Keywords A defence of Aristotle's biology · Philosophical holism

We have seen above that Aristotle's work on rhetoric, logic, poetics, and ethics are far more than special studies of specific, technical areas of human activity. They are practical and technical studies which are *successfully* framed by metaphysics. By understanding the mechanics of any given part of the whole, we can appreciate the whole better. This principle applies not only to the various dimensions of human thought and behaviour but also, as we will see below, to more objective dimensions of being such as biology. Aristotle's philosophy and his logic is even itself fundamentally *bio*-logical in nature. It is through close and philosophical consideration of the foetus and the organism that we can grasp the central truths of being and becoming in the world on an observable, empirical, and demonstrable basis. It is through these metaphysics that Aristotle shows us that approaching the world as a whole is not a distraction or an error but is, rather, an important truth. His fundamental challenge to our standard philosophical—atomistic and relativistic—assumptions explains why such exertions have been made to dismiss, distort, or ignore his thought and its implications. The most famous and egregious attempt to achieve this

P. N. Jackson, *Aristotle on Meaning in the Living World*, Biosemiotics 30,
https://doi.org/10.1007/978-3-032-00602-8_8

destruction or dismissal was performed by Bertrand Russell who moved from Aristotle's error of fact as regards the number of women's teeth to the conclusion that Aristotle's thought and work is in general "intellectual rubbish" and "stupidity."

It is useful to consider the sleight of hand that Russell employs to come to these conclusions. Russell's contention as regards Aristotle's error is that: "Aristotle could have avoided the mistake of thinking that women have fewer teeth than men, by the simple device of asking Mrs. Aristotle to keep her mouth open while he counted. He did not do so because he thought he knew. Thinking that you know when in fact you don't is a fatal mistake, to which we are all prone. I believe myself that hedgehogs eat black beetles, because I have been told that they do; but if I were writing a book on the habits of hedgehogs, I should not commit myself until I had seen one enjoying this unappetising diet. Aristotle, however, was less cautious. Ancient and medieval authors knew all about unicorns and salamanders; not one of them thought it necessary to avoid dogmatic statements about them because he had never seen one of them."[1] Russell concluded that: "Aristotle, in spite of his reputation, is full of absurdities. He says that…[giving a few other supposed examples which we cannot treat here] an elephant suffering from insomnia should have its shoulders rubbed with salt, olive-oil, and warm water; that women have fewer teeth than men, and so on. Nevertheless, he is considered by the great majority of philosophers a paragon of wisdom. Superstitions about lucky and unlucky days are almost universal."[2]

Since Russell uses Aristotle's biology to dismiss his corpus of work and his whole point of view, it is useful for us to consider Aristotle's biology in detail in this context. To achieve this, let us consider (a) the validity and quality of Russell's critique, (b) the relative quality of Aristotle's work, and (c) the subtext of Russell's critique. I will show that Russell's critique of Aristotle is an insistence upon modern dogmas amounting to fraud to a similar or greater degree as pertained in respect to previous critiques from the perspective of fundamentalist religion. The problem we face is that Russell's dogmatic dismissal of Aristotle's biology (and to some degree of his philosophy) has been widely accepted in the literature as merely a statement of fact and an expected price of progress. I note that Russell is not the exception but is a paradigm as regards how Aristotle's biology, and his worldview, has been approached in our age through a mixture of fear, mendacity, and contempt.

[1] Russell, Bertrand *An outline of intellectual rubbish: a hilarious catalogue of organized and individual stupidity* (Girard, 1943) p 22. Other examples given of Aristotle's errors are "He says that children should be conceived in the Winter, when the wind is in the North, and that if people marry too young the children will be female. He tells us that the blood of females is blacker than that of males; that the pig is the only animal liable to measles." I cannot address these matters in detail—whether they are true or false, and accurately represent Aristotle's opinion—though a quick internet search proffers the following: ""Measles" and "gid" are terms used to designate certain diseases in our domestic animals caused by the young or bladderworm stage of certain tapeworms; thus we find measles in cattle and pigs, and gid amongst sheep and goats." As regards blood: "men generally have higher levels of haemoglobin than women, which gives their blood a redder appearance." I note that an underlying subtext of the examples that Russell gives is Aristotle's supposed misogyny.

[2] Russell, Bertrand *Ibid.* p 19.

8.1 A Critique of Russell's Critique of Aristotle's Biology

What is the problem with Russell's assessment of Aristotle's biology? We see from the off a clear sleight of hand since Aristotle does not do anything remotely like making claims about unicorns (though he does invent the "goat-stag" to consider the status of imaginative constructs) or indulging in superstitions about lucky or unlucky days. Aristotle is careful to stick to observable reality and refuses on principle to work from the claims of numerology, mythology, and religion.[3] He is the father of realism and of empiricism who develops the most fundamental critique of ideology and idealism and almost singlehandedly invents science in order to address concrete reality. He does not make dogmatic comments about such empirical things as the food taken by hedgehogs and in all cases relies upon close observation. His treatment of hedgehogs, as regards how they are behaviourally and physically adapted to mate, is demonstrably empirically sourced.

Some of Russell's few selected examples of Aristotle's supposed dogmatism and superstition are themselves in error. For example, the full passage from Aristotle that Russell refers to in his critique as regards elephants reads as follows: "Elephants suffer from the flatulent ailments through which they are unable to discharge either the fluid excrement or that of the belly. And if it eats earth it becomes weakly, unless it eats it continually: if it does so continually, it is not harmed. It also swallows stones sometimes. It is attacked by diarrhoea too; when attacked, they treat it by giving it hot water to drink, and they dip the fodder in honey when giving it to eat, and each of these settles the ailment. When it has been exhausted by not lying down to sleep, they are made sound by having their shoulders rubbed with salt and olive oil and hot water (*HA* VII 605a 23–605b1)." Considering the matter empirically, we find that Aristotle's account is perfectly respectable since elephants *do* have a problem with diarrhoea, they *do* eat stones and earth (actually as part of their diet), and they *do* experience problems with lowering themselves for sleep. Aristotle's account is not only broadly correct, it is also clearly based upon observations given to him by working specialists.[4]

Do elephant bowels matter? Does Russell's hit job on Aristotle matter? What exactly does matter? Aristotle cared about elephant nutrition and digestion because he was interested to understand the rules and requirements of animal metabolism. His method to achieve this was to consider each genus and species or organic life open to him and to compare them and find patterns. This detail is provided in the

[3] Not only does Aristotle explicitly dispense with mythology in his research but he comments on "divination" in dreams that: "since some of the lower animals also dream, dreams cannot be sent by God, nor is this [i.e., divination] the purpose of their appearance, but they have a divine origin; for nature is divinely ordained, though not itself divine (*Div*. 463b 12–15)." As always, we find a balanced and reasonable approach to worldly phenomena.

[4] Aristotle provides an account of how elephants are captured and tamed at *HA* VII 610a 25–33. See J.M. Bigwood "Aristotle and the Elephant Again" *The American Journal of Philology* (1993). See also Rose Kivi "How Do Elephants Behave?" *Sciencing* (2019) & Joan Mayol "Why do Elephants Eat Soil?" *Metode* (2013).

Parva Naturalia, *Movement of Animals*, *Parts of Animals*, *History of Animals* etc. His interest in elephant movement is also clearly linked to his extensive work on the articulation of animals explored in his *Movement of Animals* and *Progression of Animals*.[5] We ultimately see that Aristotle's approach to the elephant's bowels is also more empirical and more accurate than Russell's account of it. Russell ironically exhibits the unreflective dogmatism that he claims to critique. That Russell's smear about Aristotle's laxness with the facts has been routinely recycled in the literature is also troubling since a cursory review of the source material should have flagged up Russell's own misrepresentation. The critical thing here is not that Russell is wrong or even lax but that he is dogmatically pushing a metaphysical position and misrepresenting facts to do so. We ultimately find that moderns have no monopoly on honesty or facticity. They also have their own dogmas and mythologies to promote and defend.

What is the subtext for Russell's misrepresentation of Aristotle? Let us remember that Russell sought to refute Aristotle's position on "substance" (*ousia*) on the basis that "Socrates" exists only as a "bundle" of events and we find here that the above lecture is designed to conclude that: "Perhaps the world would lose some of its interest and variety if such beliefs were wholly replaced by cold science."[6] Russell's conviction is to defend scientific reductionism and technocracy on the basis that we could achieve great efficiencies by tidying up the messy facts of the world. I suggest that in truth this reductionism does not merely tidy but that it misrepresents reality and seeks to reduce it *in its image*. "Socrates" is hence reduced to a meaningless bundle of events. What Russell is arguing for is precisely (and curiously) what his former collaborator A.N. Whitehead described as the "Hume-Newton situation…[which is]…the primary presupposition for all modern philosophic thought. Any endeavour to go behind it is, in philosophic discussion, almost angrily rejected as unintelligible."[7] So desperate is his desire to defend this unholy narrative that Russell creates a straw man from out of Aristotle's error on the subject of the number of women's teeth and lathers this up into an accusation of misogyny in order to discredit Aristotle on this basis.[8]

---

[5] Aristotle explains the descent of elephants as that: "The elephant does not behave as some used to allege, but settles down and bends its legs, though it cannot on account of its weight settle down on both sides simultaneously, but reclines either on to the left or on the right, and in that posture goes to sleep. Its hind legs it bends just as a human being does (*HA* II 498a 8–12)."

[6] Russell, Bertrand *An outline of intellectual rubbish: a hilarious catalogue of organized and individual stupidity* (Girard, 1943) p 26.

[7] Whitehead, A.N. "Nature Lifeless" in *Modes of Thought* (New York, 1938) p 135.

[8] See Robert Mayhew *The Female in Aristotle's Biology: Reason or Rationalization* (Chicago, 2004) and Jessica Gelber "Females in Aristotle's Embryology" in Falcon, Andrea & Lefebvre, David (*eds.*) *Aristotle's* Generation of Animals: *A Critical Guide* (Cambridge, 2028).

8.2 A Review of Aristotle's Biology

Let us on the basis of Russell's critique perform a review Aristotle's biology below and consider (a) the nature of Aristotle's biology, (b) its value on its own terms, and (c) its value to us today.

8.2.1 "If You Cut the Chick Open, There Will Be a Small Piece of Yolk Still Remaining Attached to the Gut"

It is not generally appreciated that Aristotle is the father of biology even though fully half his extant work deals with zoology, anatomy, embryology etc. and with his biology clearly underpinning the rest. Aristotle is also the pioneer of dissection as a method although his eight-book volume of dissections is sadly not extant today. As regards his own "value-added" we see that Aristotle did not simply invent and perform dissection for purely scientific research purposes but also as a method to understand the "parts" of a whole. Aristotle's whole research programme is based upon the method of analogy and homology which he draws from his biological research which gives him such taxonomical observations as that: "the land-tortoise, the turtle, the lizard, both types of crocodile, and generally all animals of this sort: they have a simple stomach, simple in formation, resembling in some cases the pig's, in other's the dog (*HA* II 508a 5–9)." Aristotle looks both for rules, models, and commonalities and for such *exceptions* as that the camel does not possess horns but still has no bottom set of teeth and four stomachs in common with horned ruminants (see *PA* III 674a 30*ff.*).[9] His emphasis upon morphology is, importantly, not to be confused with formalism.

In Aristotle's mapping of life we even find that he is happy to record the *commonality* between the digestion of both man and lobster (!) so that: "The nature of the internal parts of both [i.e., testacea and cephalopod] is as if one were to conceive it in a straight line as in fact it is in the four-footed animals and the human beings: first, at the extreme upper point of this straight line is the mouth designated as A, next the gullet B, then the stomach, C; and from the intestine to the residual outlet, D (*PA* IV 684b 21–26)." Aristotle considers the common *Bauplan* of a range of organisms—even happily including plants along with animals—and considers the common process for extraction and digestion and also the different methods or strategies through which this is achieved in different beings. He considers the channelling and canalisation of both food (matter) and being (form) through the world most explicitly as movement in his *Movement of Animals* but with his conceptualisation of movement (*kinēsis*) not generally being merely simple ambulation—movement

[9] On Aristotle's study of the camel see Allan Gotthelf & James G. Lennox (*eds.*) *Philosophical Issues in Aristotle's Biology* (Cambridge, 1987) p 184–5.

in space—but the expression of energy, i.e., kinetic energy, in the world, including for processes such as digestion.

As regards the process of nutrition, Aristotle argues (a) that the embryo must have a source of nutriment *before* "a shoot and a root are thrown out from it (*GA* II 739b 37–38)" and that: "In the seeds of plants there is some nutritive matter, which at first has a milky appearance (*galaktōdes*); and it may be that in the same way, in the material of the animal, the residue left over from its construction is present as nourishment for it from the outset (*GA* II 740b 6–8)" and (b) that: "Some animals have their primary matter (*prōtēn* [*hylēn*]) within themselves, having derived it from the female parent, e.g., those animals which are produced not viviparously but out of larvae or eggs. Others derive it from the mother for a considerable time by being suckled (*GA* II 733b27–29)." We *then* see (c) that: "When the nourishment that passes through the umbilical cord is no longer sufficient for the foetus, owing to its size, at that same time, the milk is becoming serviceable, and when no nourishment is entering by way of the umbilical cord, then the blood-vessels to which the cord acts as a sheath collapse; and for these reasons and at that time the exit of the foetus takes place (*GA* IV 777a 23–27)" and *then* (d) that: "if an animal is going to get any nourishment after the period of its suckling is over, it must of necessity possess instruments [i.e., teeth] with which to deal with its nourishment (*echein organa pros tēn ergasian tēs trophēs*) (*GA* V 788b 23–25)." Aristotle finally provides us with the remarkable observation (e) that teeth are sufficient to serve for a life, are replaced only once, and would be insufficient "if life lasted for a thousand or ten thousand years (*GA* II 745a 34)." The tools given to a being are adapted to that being—this especially evident as regards teeth, as Aristotle routinely observes—and are fitted to serve it over its full life.

What is the purpose of Aristotle's study of nutrition in animals? Aristotle seeks to understand the "principle" (*archē*) of being not just as an abstraction but as a "seed" of matter having meaning. He hence argues that: "The "principle" (*archē*) of the chick is derived from the white [in the egg but] its nourishment comes through the umbilical cord from the yolk (*HA* VI 561a24–25)," and that: "if you cut the chick open, there will be a small piece of yolk still remaining attached to the gut (*HA* VI 562a15–16)." He similarly comments on the development of fish embryos that: "Nourishment is present inside the small fishes' stomachs similar to that in the stomach of chicks; some of it is white and some yellow (*HA* VI 565a 10–13)." In respect to cuttlefish Aristotle professes bemusement about the operation of the umbilical cord but still concludes that "as the young cuttlefish grows, the white portion continually diminishes, and ultimately disappears, just as the yolk does in birds' eggs (*HA* V 550a 22–23)." In respect to quadrapeds Aristotle notes that: "Quadraped embryos contain excrements too, when their development is complete, both liquid and solids, the latter in the last part of the intestines and urine in the bladder (*HA* IX 586b 8–11)." Finally, as regards human beings, he discusses the residue of "poppy" that a human baby produces after birth; see *HA* IX 587a 25–33.

Aristotle seeks to understand the "form" of "being" and the "process" of "becoming" on the basis that: "there is a receptacle (*esti ti dektikon*) for the food at each of its stages, and also for the residues that are produced; and as the blood-vessels are a sort of container for the blood, it is plain that the blood (or its counterpart) is the final form of that food in living creatures. This explains why the blood diminishes in quantity when no food is taken and increases when it is; and why, when the food is good, the blood is healthy, when bad, poor (*PA* II 650a33–650b2)."[10] Aristotle interestingly simplifies the flow of *matter* as a flow of milk and then blood through the organism delivering "food" to its parts.[11] He also observes that each of these stages has its own "tools," one of which is teeth on the basis that: "it did not rain on purpose to spoil the crop, but the result was merely incidental to the raining…[whereas] natural organs like the teeth…seem to embody a purpose (*dokei hyparchein to heneka tou*) (*Phys.* II 198b 23–28)."

Aristotle's approach to the sublunary world is not abstract or mathematical but is fundamentally chemical in nature since it assumes that all dimensions of a physical being, including mind with its "noetic matter," have a material basis. On top of this chemical and physical basis, we find an additional bio-logical layer of form and function. We are reminded of the heap of sand (chemical) upon which the human being (biological) and then its mental product of the *Iliad* (mental) is based. In respect to the biological, Aristotle is concerned with the provision and extraction of matter in a continuous stream to a being from inception to death and also with the parallel process of digestion and excretion of by-products. The essential point for us to note here is that the "principle" (*archē*) is not for Aristotle a formal abstraction but something substantive and hence enmattered like a seed. And not only is the flow of matter through form essential but also the by-products of life to which Aristotle fascinatingly pays great attention. He is interested in how living being is made up of both solids—food, organs, etc.—and liquids—water, bile, blood, urine, etc.—with the philosophical hypothesis he puts forward being that life is a synthesis of solid (earth), liquid (water), air (gas), and fire (energy).

I represent Aristotle's theory of biological investment of energy in Fig. 8.1 below.

Not only does the organism reach maturity, however, but it also reaches the end of its lifespan or "*periodos*." Aristotle regards the organism as a "form" which degrades over time and hence becomes in respect to its material constitution increasingly degraded, losing memory, getting grey hair, cancers etc. The meaning of death

---

[10] Aristotle also argues that: "The food from which the parts of animals directly derive their growth is the natural substance blood. But the blood and the veins must have the same source; for the latter exist for the sake of the former, as its vessel and receptacle (*hōs angeion kai dektikon*) (*Resp.* 474b 4–7)."

[11] Aristotle regards the movement of being through the world and also a flow of matter through the being. The liquids feed the solids of being on the basis that: "bodily excellences such as health and fitness we ascribe to the mingling of the warm and cold humours in due proportion, in relation either to each other or to the environment (*Phys.* VII 246b5–6)." This is the basis for the Aristotelian system of elements based on "fire" as "warm and dry," "earth" as "cold and dry," "air" as "warm and moist," and "water" as "cold and moist." The further neo-Aristotelian use of this system is as "four humours," namely "blood/warm and moist/spring/air/(heart)/sanguine," "yellow bile/warm and dry/summer/fire/(liver)/choleric," "black bile/cold and dry/autumn/earth/(spleen)/melancholic," and "phlegm/cold and moist/winter/water/(brain)/phlegmatic" (and for the original Aristotelian approach see the *Physiognomics*).

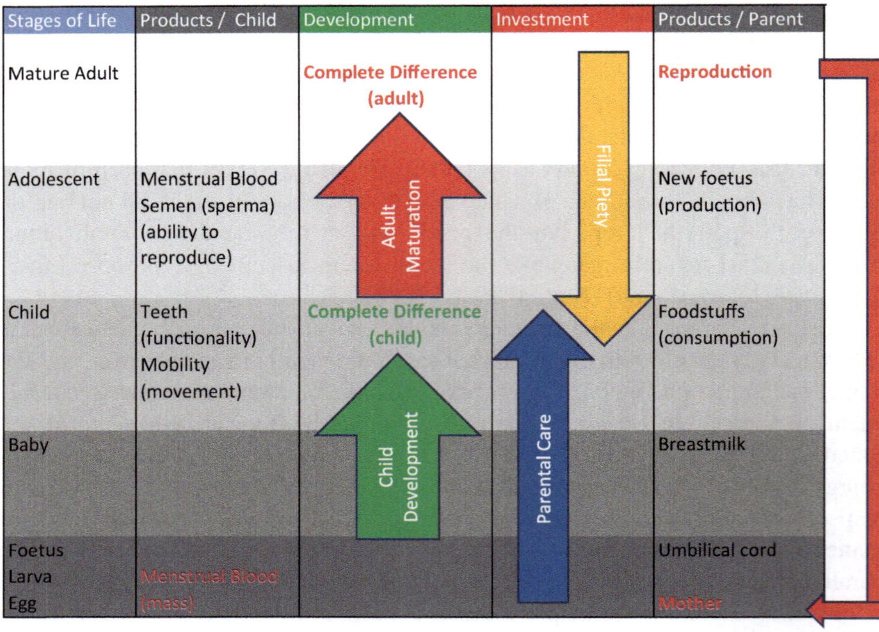

Notes on diagram

Aristotle explains "energy" by reference to the investment that the parent makes in the child (and vice versa). He explains that: "parents love their children as being a part of themselves, and children their parents as being something originating from them (*N.E.* VIII 1161b 18-19)" and that: even animals have the philoprogenitive instinct, which urges them to produce offspring and also to protect the offspring produced. And in fact fathers love their children more than they are loved by them...and these in turn love their children more than their parents, because activity is the greatest good (*dia to tēn energeian einai to ariston*). And mothers love their children more than fathers, because they think that the children are more their work; for people estimate work by its difficulty, and in the production of a child the mother suffers more pain (*E.E.* I 1241a40– 1241b9)"

Aristotle observes regarding the nature of "matter" that any "*ousia*" dependds upon a constant or continuous flow of matter as "food" which will be transformed into the being growing to its maturity and then producing another member of the species. Matter is not for Aristotle something mundane and inert but is a specific energy which is able to take on the various forms of the organism in the world. Matter must be supplied continuously to the organism or it dies. Aristotle meticulously documents this lifeline with one example being the human breastmilk must be "serviceable" as soon as the foetus is not supplied with nutrients by the umbilical cord (see *GA* IV 777a 23-27). On this cut-over of feeding sources he notes that: "In the case of other animals, as there is but a single period of gestation, the concoction of the milk coincides with that; in man, however, as there are more periods than one, the milk must of necessity be available at the earliest of the possible dates (*GA* IV 776a 20-24)."

Fig. 8.1 Aristotle's theory of biological investment

is that the being will cease to live once its life cycle is complete and as its core energy has dissipated. As regards the energy of life Aristotle significantly speaks of plants, insects, animals, and man interchangeably in his *On the Longness and Shortness of Life* and argues that they all "dry up" and lose their sap (water) and energy (fire) over their lives; see *Len.* 466a 18–23. Aristotle's explanation for ageing and for the need for reproduction is based upon the idea that these chemical elements must be artificially held together in the organism through a process of "negentropy."[12] This holding together (*synechēs* and *symphysis*) is not merely a chemical action—indeed, it runs against mere chemical action—but clearly occurs to achieve a function (*ergon*), an action (*praxis*), an actuality (*energeia*), a being (*ousia*), and an end (*telos*). Organic being is observably framed by both the "form" and finality of life and the "matter" of its efficient existence.

Hopefully, we can see the quality and sophistication of Aristotle's biology even from this slight sampling of it. Even this brief account of Aristotle's basic approach refutes Russell's accusation that: "Thinking that you know when in fact you don't is a fatal mistake, to which we are all prone. I believe myself that hedgehogs eat black beetles, because I have been told that they do; but if I were writing a book on the habits of hedgehogs, I should not commit myself until I had seen one enjoying this unappetizing diet. Aristotle, however, was less cautious."[13] I rather contend that there is *nothing* in Aristotle's work that is not reasonable and that there are surprisingly few errors of fact. It would be interesting to know whether Russell *or anyone quoting him since* had bothered to consult any of Aristotle's biological work and, if not, what this situation truly signifies.

8.2.2 "Nature Does Nothing in Vain"

Perhaps the most controversial aspect of Aristotle's philosophy is the maxim that "nature does nothing in vain." Leibniz also argued that "nothing is without reason" (and "no effect is without reason") with Hegel also positing that "every grain of dust is an organization."[14] Martin Heidegger explained Leibniz's principle of reason in outline as that: "In all that surrounds and concerns it, human cognition seeks reasons, often only the most proximate ones, sometimes even the more remote reasons, but in the end it seeks the first and last reasons…We are, in our conduct and cognition, on the way to reason as though this came about on its own. We constantly have, as it were, the principle of reason in view: *nihil est sine ratione*. Nothing is without reason. Our conduct everywhere takes into account what the principle of reason

[12] Aristotle argues that chemical elements which have their *own* movements must be "held together" in and by the complex and composite organism with a further pressure upon the integrity and actuality of something being provided by the environment; see *Phys.* VIII 259b 11–14. The being is itself a *suspended reality* on the basis that it is not a circular movement but that: "with everything its corruption and perversion are not in any chance direction, but leads to the contrary and intermediate states (*E.E.* II 1227a33–34)."

[13] Russell, Bertrand *An outline of intellectual rubbish: a hilarious catalogue of organized and individual stupidity* (Girard, 1943) p 22.

[14] Quoted in Ivan Il'in *The Philosophy of Hegel as a Doctrine of the Concreteness of God and Humanity Vol. I* (Evanston, [*orig.* 1918] 2010) p 158.

says."[15] Heidegger's statement applies to Aristotle who, as we saw above, considered the human process of reasoning by reference to "the first and last reasons."

Heidegger goes on to unpack and explore this principle in detail and with immeasurable subtlety, though it would be a distraction to explore it too far here. I simply note that there is nothing fanciful or Panglossian about Heidegger's position. As regards where he leads us, we find that Heidegger continues with an analysis of the poetic fragment of Angelus Silesius that: "The rose is without why, it blooms because it blooms/It pays no attention to itself, asks not whether it is seen." Nature is different from our questioning of it. Ground and meaning is different than cause and effect. Heidegger explains as regards the philosophical consequences of this situation that: "Aristotle already knew what belongs within the compass of axioms. But even today we lack an adequate illumination of the deeper insights Aristotle developed, though not immediately but mediately, concerning the essence of axioms"[16] and also that: "Today nothing takes root any more. Why? Because the possibility of a thoughtful conversation with a tradition that invigorates and nurtures us is lacking, because we instead consign our speaking to electronic thinking and calculating machines, an occurrence that will lead modern technology and science to completely new procedures and unforeseeable results that probably will push reflective thinking aside as something useless and hence superfluous."[17]

Heidegger stresses that Aristotle can teach us something about the nature and meaning of meaning and its principles or axioms. What *is* a principle or axiom? Let us answer this question by showing how Aristotle digs through to principles (*archai*) *through the research programme of his science*. It is essential to recognise that Aristotle's logic and his philosophy is based upon his reflections upon the biological world, a point that Walter Elsasser explains as that: "it is certainly more than a coincidence that Aristotle, the founding father of scientific biology, was also the founding father of logic. It should then not be very surprising that an inquiry into matters biological turns into a discussion of logic."[18] Underneath all this we find a clash between "classical" or Aristotelian logic based upon how human beings address and interact with the world and the "mathematical" and computational logic of Bertrand Russell (and A.N. Whitehead) based upon what we could call "electronic thinking and calculating machines." This is the distinction between different logics drawn out, as we have seen already, by Imre Lakatos.

What, then, does Aristotle express through his core principle or conclusion that "nature does nothing in vain"? To take an example of his method and of his proof, Aristotle argues regarding the suckers of various types of octopus that: "All these creatures have two rows of suckers, except a certain kind of octopus, and these have only one, because owing to their length and sliminess they are so narrow that they

---

[15] Heidegger, Martin *The Principle of Reason* (Indiana, [*orig*. 1957] 1996) p 3.

[16] Heidegger, Martin *Ibid.* p 15.

[17] Heidegger, Martin *Ibid.* p 15.

[18] Elsasser, Walter "A Form of Logic Suited for Biology" in Rosen, Robert (*ed.*) *Progress in Theoretical Biology* (Academic Press, 1981) [online Vordenker version] p 2–3.

cannot possibly have another. Thus they have the one row only, not because this arrangement is the *best* (*beltiston*), but because it is *necessitated* (*anagkaion*) by the particular and specific character of their being (*dia ton idion logon tēs ousias*) (*PA* IV 685b 13–16)." Aristotle speaks of an optimisation of form in matter. The octopus has a "form" (*eidos*) and a "way of life" (*bios*) which necessarily has tentacles and suckers but with the material resources and scaling of the octopus determining whether the octopus has one or two rows of suckers.[19] We see here the principles of fittedness and optimisation and also what Aristotle calls the "correction" of nature which we will consider in more detail below. I note that the octopus is a perfect example of an Aristotelian organism having a head as a centre of control and tentacles as canals of operational matter. Aristotle goes into considerable detail on the life (*bios*) and environment of the octopus, as he does for all the animals summarily considered below.

Returning to consider (human) teeth, we find the same operating principles. We not only have milk teeth and then adult teeth but also find a patterning of teeth themselves which Aristotle posits as a general principle of function and optimisation, as follows: "Why then should it not be the same with the parts in nature, e.g. that our teeth should come up of necessity—the front teeth sharp, fitted for tearing, the molars broad and useful for grinding down the food (*Phys.* II 198b24–26)." He adds by reference to hair, another peripheral and detachable addition to the body, that: "Man has the hairiest head of all animals…the hair affords shelter both from excessive cold and from excessive heat. The human brain is the biggest and most fluid of all brains; therefore it needs the greatest amount of protection (*PA* II 658b 1–9)" and in principle that: "nature everywhere gives to another part what it takes from elsewhere. And in those cases where nature has made the body quite hairy, the hair around the tail is wanting, as happens with the bears (*PA* II 658a 36—658b 2)."[20] This idea that nature uses matter in the optimal way is what Aristotle means when he says that nature is like a housekeeper (and a potter) which does nothing "in vain."[21] He explains this principle of "the more and the less" in the context of teeth and tusks etc. as that: "Nature allots defensive and offensive organs only to those creatures which can make use of them "in a greater degree" (*PA* III 661b 29–31)." Aristotle's is evidently a science of analogous qualities and values rather than of isolated quantities and facts.[22]

---

[19] See Michael Chase "Teleology and Final Causation in Aristotle and in Contemporary Science" *Dialogue* (2011).

[20] Aristotle adds regarding the elephant that: "The elephant is the least hairy of the quadrupeds. In those animals which have considerable size, the tail tends to correspond with the rest of the body in respect of the thickness and thinness of its hair. This does not apply to those which have very small tails (*HA* II 499a 9–12)."

[21] See Mariska Leunissen "Nature as a Good Housekeeper: Secondary Teleology and Material Necessity in Aristotle's Biology" *Apeiron* (2010).

[22] Aristotle explains his taxonomy on the basis that: "it is practically by resemblance of the shapes of their parts, or of their whole body, that the groups are marked off from each other: as e.g. the groups Birds, Fishes, Cephalopods, Testacea. Within each of these groups, the parts do not differ so far that they correspond only by analogy (as a man's bone and a fish's spine); that is, they differ

This fittedness resulting from the history of a being finding and using a function also applies on a macro-level as an explanation for species (*eidos*). Aristotle argues that the ape "dualises" on the basis that it spends *some* of its time upright and hence "the ape spends most of its time on all fours rather than upright (*HA* II 502b 20–21)" and that: "Apes are hairy on the back, in virtue of being quadrapeds, and hairy on their fronts in virtue of being man-like (*anthrōpoeideis*) (*HA* II 502a 23–24)." He puts forward two distinct evolutionary arguments here (1) that a being's morphology shifts to meet its functional requirements according to the innate possibilities of its generic and speciate pathway, and (2) that nature is such that it will tend to optimise the matter available to it and will hence "place" hair in the optimal places on the animal's body. He adds regarding the ostrich that it is the only one of the birds (and Aristotle also comments that there are no known fishes) which possesses eyelashes; see *PA* II 658a 13–14 and that: "as being a quadraped, it has upper eyelashes, and it is bald in the head and upper part of the neck, as a result of which its eyelashes are hairier than they would otherwise be; as being a bird, it is feathered on its lower parts (*PA* IV 697b 18–21)."

We generally find an interplay between cross-cutting explanations on animal morphology and evolution, namely (1) that animals are characterised by the "type" to which they belong but they can also "dualise" between types giving these animals the possibility for taking features from several types, e.g. the ape is man/quadraped and the ostrich is bird/qudraped but the ostrich cannot take a feature from "man" which is on another pathway, and (2) that the physical materials of an animal are as a natural point of principle "used" in the manner most beneficial to an animal. It is hence that an ape has fine eyelashes and an ostrich has bushy eyelashes and this is part of Aristotle's view that an animal is created by a flow of matter finding its form. The "form" can on this basis be understood as "boundary conditions."[23]

Aristotle considers how the matter and form meet at the boundary and create boundary conditions. He argues that there can be both a material and a causal purpose for a biological phenomenon with one taking advantage of the other. He hence argues *both* that: "the eyelashes are placed at the ends of small blood-vessels, which have to stop where the skin itself comes to an end. Thus, owing to the fact that the moisture which comes off is corporeal in composition, hair must be formed at these places even on account of a necessary cause such as this, unless some function of nature impedes by diverting the moisture to another use (*PA* II 658b 21–24)" and *also* that: "The purpose of hair is to give protection; and as the quadrupeds go on all fours, they need more protection on their backs (*PA* II 658a 15–20)."As regards

not structurally, but only in respect to bodily qualities, e.g. by being larger and smaller, softer or harder, smoother or rougher, or so forth, or, to put it more generally, they differ "by the more and less" (*mallon kai ētton*) (*PA* I 644b 8–15)." He comments that: "Such projecting parts quickly get broken if they are brittle. Cartilage and bone are the same in kind and differ only by "the more and less" (*mallon kai ētton*); so neither of them continues to grow when it has been cut out of the living organism (*PA* II 655a 32–34)." Aristotle gives us an example here of how and why he uses the tool of dissection.

[23] Regarding the idea that the "form" of something is expressed at the "limit" of matter we find that: "that which surrounds is on the side of the form, that which is surrounded is on the side of the matter (*De Cael.* IV 312a 12–13)."

(human) nails, Aristotle comments that: "There are nails on the toes for the same reason that there are nails on the fingers: the extremities have but little strength and therefore need to be specially protected (*PA* IV 690b 8–10)." He interestingly argues as regards these material attachments to the body that they are *both* the products of a chemical process resulting from their being at the interface of a being with the world *and* also growths directed to their best use for protection.

This, then, is how Aristotle treats the (in a sense) detachable outgrowths of hair, teeth, nails etc. Moving inwards, he comments that a finger is a "part" in the sense that: "the finger is defined by means of the whole body; for a finger is a particular kind of part of a man (*Met.* Z 1035b11–12)" which he explains in detail as that: "the parts of the soul are prior, either all or some of them, to the concrete "animal" and so too with each individual animal; and the body and its parts are posterior to this, the essential substance, and it is not the substance but the concrete thing that is divided into these parts as its matter: this being so, to the concrete thing these are in a sense prior, but in a sense they are not. For they cannot even exist if severed from the whole; for it is not a finger in any and every state that is the finger of a living thing, but a dead finger is a finger only in name. Some parts are neither prior nor posterior to the whole, i.e., those which are dominant (*kyria*) and in which the formula (*logos*), i.e., the essential substance (*ousia*), is immediately present (*hama*), e.g. perhaps the heart or the brain (*Met.* Z 1035b18–27)." Aristotle argues that the finger may not be an essential part of a human being but that it is a quintessential and non-contingent part of human being.

8.2.3 "The Whole Is more than the Sum of the Parts"

As we see from the above section, Aristotle draws out the function of each of the parts in his anatomy. We also see that he considers the whole and how it relates to the parts.

As regards the organ of skin, Aristotle considers this philosophically and scientifically not only as the boundary of a being but also as the sense organ of touch (*haphē*) or contact. The skin is on this basis the core function of the identity and self of a being. Enfleshed as form built upon the matter of flesh, the skin is an archetypal "non-uniform" part as a definite solid for the whole being as a hand or face is a definite solid "non-uniform" part of a whole. Figure 8.2 below is René Thom's diagrammatic representation of the lungs as one of Aristotle's "non-uniform" or "anhomoeomerous" parts.

Thom puts forward the lungs as a "non-uniform" part on the basis that the lungs are "a surface where gas transfers take place between air and blood" located within a cycle of hysteresis and he concludes that this "supporting [heterogenous] organ is the "edge" of the homogenous [uniform] matter which leads to it."[24] The lungs are

[24] Thom, René "Homéomères et Anhoméomères en Théorie Biologique d'Aristote à Aujourd'hui" in Devereux, Daniel & Pellegrin, Pierre (*eds.*) *Biologie, Logique et Métaphysique chez Aristote* (Paris, 1990) p 502.

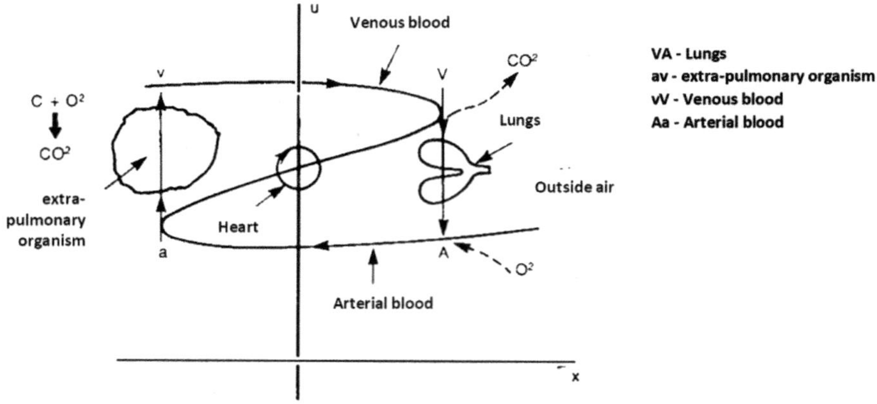

Fig. 8.2 The lungs as a "non-uniform" part. (Redrawn from Thom "Homéomères et Anhoméomères" p 501)

a solid interface between the being and the world in a comparable way as the skin is. Thom arrives at the axiom that: "the act is the edge of power."[25] He shows us how an organ—in the case the lungs—can be understood as a functional interface or membrane between self and world. We can even consider the mind or "*mens*" as a membrane between being and world with Aristotle interestingly representing the brain as a regulating mechanism rather than a purely mental instrument.[26]

We see above how gas, liquid, and solid interact in a body and on this principle of setting see *GA* I 729a 13–14 and on the uniform and non-uniform, see Sect. 6.4 and Fig. 8.3 below.

We find that Aristotle does not draw a sharp line between the organic body itself, its organic processes, and the fuel, products, and residues of those processes. He rather seeks to unpack and account for *everything naturally or actually found in the body* including phlegm, bile, and food as well as the blood which is more obviously an integrated "part" of the body rather than being something external to it; see *HA* I 520b 3 *ff*, i.e., it is a proprium.[27] He hence breaks human beings (and other organ-

[25] Thom, René *Ibid.* p 502.

[26] Aristotle argues that the brain is a regulator of heat or energy on the basis that: "Of all the animals, man has the largest brain for his size…the largeness is due to there being a great deal of heat and blood in the region around the heart and the lung. This too explains why man is the only animal that stands upright (*PA* II 653a 27–31)." I note, as per Aquinas, that "*mens*" or "mind" is a term derived from the Latin word "*mensurare*" which means "to measure."

[27] Aristotle argues that: "in one respect blood is essentially hot, and in another respect it is not. Heat will be included in the *logos* of blood, just as fairness is included in the *logos* of a fair man, and in this way blood is essentially hot; but in so far as it is hot owing to external influence, blood is not essentially hot (*PA* II 649b 24–29)," i.e., warmth is a "proprium" of blood since blood can only be (human) blood in any meaningful sense if it is warm, i.e., or it congeals and is "useless," though

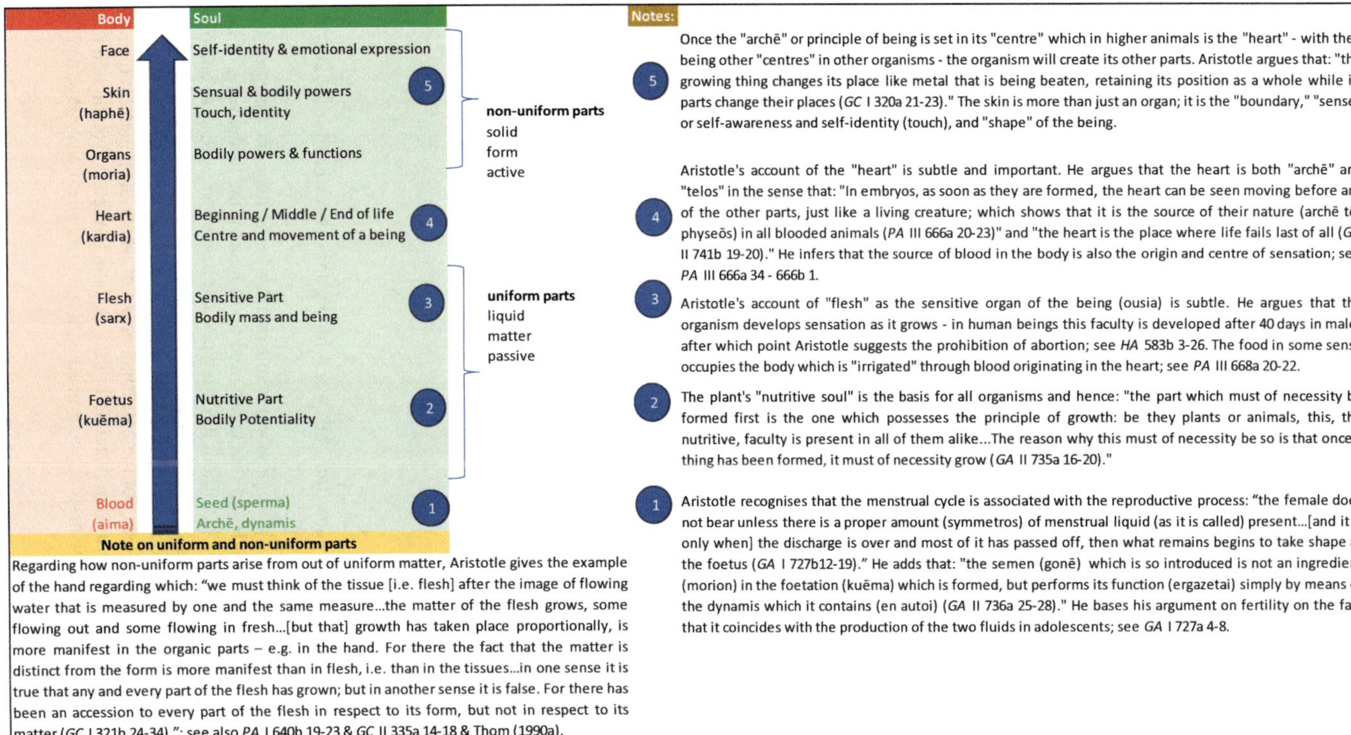

Fig. 8.3 Aristotle's theory of uniform and non-uniform part

isms) down into "solids" and "liquids" on the basis that: "Now every substance, whatever it may be, that we find in the body, must of necessity be one of the following: (1) one of the parts which are there in accordance with nature, in which case it will be one of the uniform or non-uniform parts; (2) one which is there contrary to nature (*para physin*), e.g. a tumour; (3) residue; (4) colliquescence; (5) nourishment. By residue I mean that which is left over as surplus from the nourishment; by colliquescence that which is given off as an abscession from the material that supplies growth, as a result of decomposition proceeding contrary to nature (*para physin*) (*GA* I 724b 23–28)." I suggest that we see the breadth of Aristotle's mind in his ability to structure the body and what is "in it" with this level of wide clarity. At the same time as seeing the whole and the parts of the whole he employs a hypothesis and a definite conceptual framework centred upon the flow of "matter" as food into "bodies" and "beings."

Nourishment and excreta are natural inputs and outputs for the life process. Excreta is the necessary residue of the food produced on the basis that: "food is acted upon by what is nourished by it, not the other way round (*De An*. II 416b1–2)" and that: "the waste product is what remains of the former state (*Long*. 465b 19)." We see a clear purposive structure unless we strive not to see it. Not everything, however, has a purpose. Aristotle includes in his list not only the tumour which is "contrary to nature" (*para physin*) but also "colliquescences" which he explains as that: "it seems probable that, just as the bile elsewhere in the body is a residue or colliquescence, so this bile around the liver is a residue and serves no *purpose* (*ouch heneka tinos*)—like the sediment produced in the stomach and the intestines. I agree that occasionally nature turns even residues to use and advantage, but that is not reason for trying to discover a purpose in all of them (*PA* IV 677a 13–18)." Aristotle argues that the useful is in accordance with nature ("*kata physin*") when it is in its "own proper place (*en te tois oikeiois topois*) (*GA* II 739a 3)"[28] and that *some such things which are out of place are* indeed "in vain."

not all blood is warm as not all men are "fair" but warm-bloodedness goes hand-in-hand with warm blood.

[28] The "useless" is contrary to nature ("*para physin*") with examples being tumours, phlegm, and the fat which is formed from "surplus blood that is not used up to nourish the fleshy parts of the animal (*PA* II 651a 22–24)" and which is without its own place and hence causes disease and disability (see *GA* I 726a 11 *ff* and *PA* II 651b 2 *ff*). Aristotle adds that semen, menstrual fluid (this being Aristotle's equivalent to the ovum), and milk are obviously useful residues and also the marrow which is taken as being "the enclosing of the nourishment out of which the bones are formed (*PA* II 652a 5)." Blood is for Aristotle a source and part of the body in the sense even though it is "not continuous (*ou syneches*) with the flesh nor conjoined with it organically (*sympephykos*) (*PA* II 650b 7–8)."

8.2.4 *"Tasting Would Be Like Smelling, If We Lived in Water, and Perceived at a Distance Before Making Contact"*

Aristotle ultimately argues that the "good" exists in the world situationally in the sense that it emerges in or through a complex and "entangled" world. He takes the epiglottis as an example of this ability of being(s) to self-correct on the basis that: "nature has remedied (*iatreuken hē physis*) the inefficiency of the position of the windpipe by constructing the part called the epiglottis (*PA* III 665a7–9)." He explains further (a) that "birds and oviparous quadrupeds have no epiglottis, but they achieve the same result by contraction of the windpipe; for when taking in food the latter closes the epiglottis. As the food continues on its way, the latter class contracts the windpipe, the former closes the epiglottis (*Resp.* 476b 1–3)" and (b) that "Creatures which have gills receive food through the mouth as they discharge water through the gills; for they have no windpipe, so that they cannot suffer harm from the water's falling into it, but only from entering the stomach. This is why they make the discharge and the reception of food rapidly, and have sharp teeth, set like a saw in most cases; for it is impossible for them to chew their food (*Resp.* 476b 5–12)." He adds that the gills serve the same purpose as the dolphin's blowhole. I note Aristotle's reference to the optimal fittedness of fish teeth to their way of life.

In all cases, Aristotle seeks to understand how nutrition and respiration can share a common channel by means of a remedy achieved in various ways depending on the being and its history. His hypothesis on this matter is that: "the time which water animals have for perceiving tastes is short; hence, since the use of this sense [i.e., taste] is short, so is the articulation of its organ [i.e., the tongue]. The reason why their food passes very quickly into the stomach is because they cannot spend much time sucking out its juices, otherwise the water would get in as well. So unless you pull the mouth well open, you will not be able to see that the tongue is a separate projection (*PA* II 680b 18–23)." To demonstrate this point, he turns to the example of the crocodile which "dualises" on the basis that: "the river crocodile…apparently has none [i.e., no tongue], but only a space for it; and the reason is that in a way he is both a land-animal and a water-animal. In virtue of being a land-animal, he has a space for a tongue; as a water-animal, he is tongueless…The reason for this is that these creatures have not much need for a tongue because they cannot chew their food or taste it before they eat it: they can perceive the pleasantness of it only while they are swallowing it. This is because the perception of juices is effected by the tongue (*PA* IV 690b 21–30)."[29] Aristotle not only makes observations about the crocodile's lack of tongue but also makes a hypothesis about why it may have become residual due to its physiognomy and way of life.

[29] Aristotle identifies a different adaptation in the dolphin and the other cetaceans which "take in sea water, and discharge it by the blow hole (*HA* VII 589b 2)" and also: "The crustacea, such as crabs and lobsters, [which] discharge the water through the folds by the hairy parts, but the cuttle-fish and polypus through the hollow about the so-called head (*Resp.* 477a 3–6)."

Aristotle additionally observes regarding the seal—another animal existing between land and sea—that: "The seal, sure enough, has passages, clearly visible, through which it hears; whereas the dolphin has the power of hearing, but no ears (*HA* I 492a 28–29)" with his resolution being that: "The reason is that it spends its life in a fluid medium. The ear is a part of the body which is an addition made to the passages in order to safeguard the movement of the air which comes from a distance, and therefore it is of no use to the seal (*GA* V 781b 24–28)." Aristotle also concentrates upon the seal's "stunted feet (*HA* I 487b 23)" as parts of a *Bauplan* which have owing to an animal's way of life (*bios*) become useless. He adds as further examples the mole's eyes; see *De An.* III 425a10 and a bird, probably the swift, which is "footless" since it has atrophied feet since it spends most of its life in the air; *HA* I 487b 24–26. Aristotle observes that an animal evolves to fit its niche but can only do so in accordance with the *Bauplan* it has inherited.

Aristotle interestingly also hypothesises that ants and other insects can sense food, i.e., smell, without breathing. With his usual judicious boldness, he is willing to ask whether there is perhaps another sense than the usual five; see *Sens.* 444b 7–20. He moves us from a prosaic view of sense with such observations as that: "Tasting would be like smelling, if we lived in water, and perceived at a distance before making contact (*Sens.* 447a7–9)." He looks past the physical sense to the underlying function of the senses which form a "common sense" (*aisthēsis koinē*) and then serve the "mind" (*dianoia*). We could observe on this situation that various animals such as lizards do smell with their tongues, i.e., smell and in a sense see through their sense of taste. I will sadly need to forgo discussion of Aristotle's detailed philosophical account of the senses. I stress here, however, that his treatment of the senses is a philosophical consideration as regards how a being relates to the world in its "way of life" (*bios*) as his biology is also, as outlined above.

8.2.5 "if One Way Is Better Than Another That Is the Way of Nature"

Aristotle's approach to the world is fundamentally comparative, as seeking the essential meaning that lies behind the multiplicity and diversity of being. This is both the empirical approach and the principle of taxonomy. We see here the philosophical basis for Aristotle's occupation with comparative anatomy as briefly considered above. The further underlying philosophical hypothesis is that nature abhors a vacuum so that its possibilities populate all its niches. It is hence that Aristotle regards the world as a plenum rather than as "atoms and the void." From this perspective the evident emergence of meaning in the world itself defeats the idea of materialism through which the products of the world are simply the result of inert matter shaped by collision and mixture. We cannot merely see the world as a primordial soup of chemicals out of which different permutations of clump could randomly appear. We must rather assume that there is a primordial meaningfulness that emerges in the world through its beings. This is what Aristotle captures under the heading of "*nous*."

Aristotle approaches the world by recognising such *aporiai* as that "something cannot come from nothing." For this he takes the idea of chaos and observes alongside it a world of "beings" existing on the basis that: "everything that changes is something and is changed by something and into something (*Met.* Λ 1069b 36–1070a 1)." Beings don't just become or happen but transform from one thing to another. Aristotle expands upon this principle of being that something is always something on the basis that: "natural things are exactly those which do move continuously, in virtue of a principle inherent in themselves, towards a determined goal; and the final development which results from any one such principle is not identical for any two species, nor yet is it any random result (*to tychon*); but in each there is always a tendency towards an identical result, if nothing interferes with the process (*Phys.* II 199b 16–18)." This is a beautiful and precise statement of Aristotle's philosophy of the contingency of "beings" (*ousiai*) such that behind the seeming chaos of the world there is a natural structuring in time and place of being based upon "form" or "species" (*eidos*). We can look through the surface difference and see that the individuals of a species are the same in the sense that they embody the *same* basic set of drivers and ends.

Aristotle draws out the further nuances of the role of "chance" in being(s) as that: "Our first presupposition must be that in nature nothing acts on, or is acted on by, any other thing at random (*to tychon*), nor may anything come from anything else, unless we mean that it does so in virtue of a concomitant attribute (*symbebēkos*). For how would "white" come from "musical," unless "musical" happened to be an attribute of the not-white or of the black? No "white" comes from "not-white"—and not from *any* "not-white," but from black or some intermediate colour. Similarly, "musical" comes to be from "not-musical," but not from *any* thing other than musical, but from "unmusical" or any intermediate state there may be (*Phys.* I 188a31–188b3)." As there are grades of being (see *Met.* Γ 1003b 5–10) there are also grades of contingency ranging from *tychē* (chance, as above), *symbebēkos* (accident; see *Met.* Δ 1025a 14–32), and *automatos* (natural propensity; see *Phys.* II 197b 14–18). Aristotle argues that there is a categorical structuring of being into different genera of being which possess their own scales, contraries, and dynamics. Behind Aristotle's division of being into genus, species, and individual, we find a further division into essence and contingent attributes or characteristics. Chance is only ever a contingency, a form of "*kairos*," within this ontological framework.[30]

[30]Aristotle's situationism is based upon the principle of potentiality that: "to feel them ["right" feelings] at the right times, with reference to the right objects, towards the right people, with the right aim, and in the right way, is what is both intermediate and best, and this is characteristic of excellence [virtue] (*meson te kai ariston, hoper esti tēs aretēs*). Similarly with regard to actions also there is excess, defect, and the intermediate (*N.E.* II 1106a21–24)." On this "kairology" see Pierre Aubenque "Kairos et Prudence de la Détermination Temporelle de L'Acte Morale Selon Aristote" in *Problèmes Aristotéliciens: Philosophie Pratique* (Paris, 2011).

We also see that not only are beings not based upon randomness but they also exhibit a paradigmatic health such that: "to discover what is natural we must study it preferably in things that are in a natural state, and not in specimens that are degenerate. Hence in studying man we must consider a man that is in the best possible condition in regard to both body and soul, and in him the principle will clearly appear—since in those that are bad or in a bad condition it might be thought that the body often rules the soul because of its vicious and unnatural condition (*Pol.* I 1254a36–1254b3)." Aristotle's aetiology asserts that the operating principles of a being are seen best in a healthy organism rather than a diseased, confused, primitive, or unhealthy one. His assumption is that "nature" is a force or potentiality which produces beings which it observably seeks to fit to the world in line with some evident "good" or meaningfulness. This hypothesis is supported by the action of every foetus, action, and chemical action in the world, so it is reasonable and important as an explanation of the reality we experience even if it is unprovable.

Perhaps the doctor's son appreciated the paradigmatic importance of *good* health as an expression of the natural "good" for every "being." He hence argues on the medical standard that: "The doctor has a standard (*tis horos*) by reference to which he distinguishes a healthy from an unhealthy body, and the degree to which each activity is advisable and healthy, and beyond which in either direction it ceases to be so. Similarly, in regard to actions and choices of things that by nature are good but not praiseworthy, the good man should have a standard of possession, choice, and avoidance concerning abundance and scarcity of wealth and other gifts of fortune (*E.E.* VIII 1249a 22–1249b 3)."[31] Aristotle also observes that the doctor is not helped in his art by knowing about the ideal (Platonic) "Good," but he insists that he is of use as himself a paradigm or standard for the application of measured human action in the world. The doctor's art represents a paradigmatic example of human beings channelling their minds in a particularly human way. One human being on this basis concentrates mentally and spiritually on another physical human being in order to restore the good of health to him.[32]

Aristotle's science must be similarly understood as a human possibility expressed as a doctor practices medicine. The world assessed through science does, however,

[31] Aristotle adds that: "Earlier we said, "in accordance with reasoning (*hōs ho logos*)"—but this, like saying in matters of diet "in accordance with medicine and medical reasoning (*hōs hē iatrikē kai ho logos tautēs*)," is true but uninformative (*E.E.* VIII 1249a21–1249b6)" which point he explains as that: "if someone should know that light meats are easily digestible and healthful, but is ignorant of what sorts are light, he will not produce health; rather, he who does know that poultry is light and healthful will to a greater degree produce health (*N.E.* VI 1141b 19–21)."

[32] The argument that we must recognise "good" and "health" in nature is remarkably controversial. Elliott Sober hence criticises Aristotle on the basis that: "It is no more part of human nature to be healthy than to be diseased ("Evolution, Population Thinking, and Essentialism" Philosophy of Science (1980) p 379)." Susan Sauvé Meyer explains Aristotle's position on "health" through "the example of a doctor curing a patient ([*GC* I] 324a29-b3) [on which] [Aristotle] claims that the unmoved mover of curing is the doctor, while the moved movers of the curing are the food and wine which the doctor administers to the patient ("Unmoved Movers, Form, and Matter" *Philosophical Topics* (1987) p 178)."

according to Aristotle, sit within a philosophical world which regards the whole without the practical focus. It is through philosophy that we understand "nature" itself and its "good" rather than merely its products and their goods.

8.2.6 "One Question Is, What Is the Smallest Number of Points at Which Animals Move; The Next Is, Why Red-Blooded Animals Move at Four Points, While Bloodless Animals Move at more than Four"

Aristotle's science breaks down into a range of powerful research projects. They are often hard to recognise since they differ radically from our own. Let us consider, however, if they are equally valuable and important to our own even if they treat different questions. The questions they treat are philosophical. The *Parts of Animals* and the *Generation of Animals* are, for instance, studies into two of the "four causes" as well as being studies in anatomy and embryology. All of Aristotle's studies consider the philosophical meaning of their subject areas as well as their scientific permutations.

One little appreciated work is the *Progression of Animals*. Its introductory passage informs us that it is a study into the "number of points (*sēmeiois*) at which animals move (*IA* 704a 10–11)" and it then reviews snakes, monopods, and centipedes as well as quadrupeds and human beings in order to draw its conclusions about "progression." It provides a development of and new departure for the principle of movement and motivation as set out in the *Movement of Animals*. Its avenue of approach is as follows: "One question is, what is the smallest number of points at which animals move; the next is, why red-blooded animals move at four points, while bloodless animals move at more than four; and, in general, why some animals are without feet, others biped, others quadraped, and others polypods, and why all that have feet at all have an even number of feet; and, in general, why the points at which movement is made are even in number (*IA* 704a 10–16)." With typical precision, Aristotle then runs through the key distinctions in respect to the flexion of limbs (and bodies) giving rise to the different movements of animal species. He provides, in the course of this project, a detailed study of the development of human movement from infancy to adulthood later in the treatise. Charles Darwin was drawn to attempt a similar physiological exercise in his *The Expression of the Emotions in Man and Animals* (1872).

Aristotle comments after stating the main questions being addressed in the work that: "At the beginning of the inquiry we must postulate the principles we are accustomed constantly to use for our scientific investigation of nature, that is we must take for granted principles of this universal character which appear in all nature's work. Of these one is that nature creates nothing without a purpose, but always the best possible in each kind of living creature by reference to its essential constitution. Accordingly if one way is better than another that is the way of nature. Next we must take for granted the different species of dimensions which inhere in various things; of these there are three pairs of two each, superior and inferior, before and behind, to the right and to the left. Further we must assume that the sources of movements in place are thrusts and pulls (*IA* 704b 12–22)." Aristotle's consideration moves from the general principle that "nature does nothing in vain" to the local principle that beings have a symmetrical referencing and directionality in space and then to the principle that local environmental forces and factors will also determine this movement through "thrusts and pulls." The modern approach has been to assert that meaning emerges from the "thrusts and pulls" of the world, as the result of "*bia*," but Aristotle argues that nature is itself meaningful and seeking to express itself at a distance through the buffering of local environments.

Aristotle then comes to the problem that: "it is evident that sanguineous animals like snakes, whose length is out of proportion to the rest of their dimensions, cannot possibly have limbs; for they cannot have more than four (or they would be bloodless), and if they had two or four they would be practically stationary; so slow and unprofitable would their movement necessarily be (*IA* 708a 14–20)." He puts forward a general principle regarding optimisation in nature and then, by comparing the snake and centipede, seeks to work out comparative principles behind these variations. We again observe in outline that the nature of organisms is constrained by their evolutionary pathways so that two organisms achieve the same end result in different ways.

Aristotle pays great attention to the centipede. He observes that centipedes move by undulating up and down whereas snakes undulate side by side. Their legs are used for stability and traction rather than for walking as octopus' suckers are also used. He observes that: "Polypods however, like the centipede, can indeed make progress on an odd number of limbs; for then the mutilation of one row of limbs is corrected by the number of limbs which remain on either side. Such mutilated creatures, however, drag the wounded limb after them with the remainder, and do not properly speaking walk (*IA* 708b 5–9)." We see here a reference to explicit experimentation, a feature of Aristotle's research that is often denied.[33] Aristotle also tests through experimentation the following hypothesis that: "no sanguineous animal if it be divided into more parts can live for any appreciable length of time, nor can it enjoy the power of locomotion which it possessed while it was a continuous and undivided whole. But some bloodless animals and polypods can live a long time, if divided, in each of the severed parts, and can move in the same way as before they were dismembered. Examples are what is termed the centipede and other insects that are long in shape, for even the hinder portion of all these goes on progressing in the same direction as the fore-part (*IA* 707a 24–707b 2)." Aristotle gives us the experimental program of cutting up various animals to see how long they will live in their "parts" before expiring.

[33] See Christopher E. Cosans "Aristotle's Anatomical Philosophy of Nature" *Biology and Philosophy* (1998).

He explains further in his *On Respiration* that: "In all cases destruction occurs owing to a failure of heat, but in the perfect animals the failure lies in that part in which is the source of their being. This, as has been said before, lies at the point at which the upper and lower parts meet: in plants, between the shoot and the root; in sanguineous animals, in the heart; and in bloodless animals, in the corresponding organ. Some of these creatures have potentially (but not actually) many vital sources. This is why some of the insects continue to live when divided, and among sanguineous animals those which have not a very lively nature live for a continuous time after the heart is removed; e.g. tortoises move upon their feet so long as their shells are on, because their nature is of a low order of construction, on much the same level as that of insects (*Resp.* 478b 33–479a 7)." Whereas in his *Progression of Animals* he is interested in the movement of animals, in his *On Respiration* he is concerned with the metabolism of animals. The focus of his account of metabolism is the heart which is the "centre" (*archē*) of a being with the energy involved in the *internal* movement of being needing, according to Aristotle, to be regulated by respiration. His ultimate philosophical concern here is with the nature and quality of a fully-integrated organic being. He rightly regards an increasing integrality in nature so that chemicals and plants can be easily cut up without losing their life and form whereas "higher" organisms like human beings do not regenerate or easily divide.

Aristotle makes the additional point above that life is an expression of "heat" or energy with his main interest here being expressed by reference to cold-blooded animals. He also explains regarding the chameleon that: "In its movements it is very sluggish, like the tortoise…It has blood only round about the heart, the eyes, the region above the heart, and in all the veins extending from these parts; and in all these there is but little blood after all…After being cut open along its entire length it continues to breathe for a considerable time; a very slight motion goes on in the region of the heart, and, while contraction is especially manifested in the neighbourhood of the ribs, a similar motion is more or less discernible over the whole body (*HA* 503b 9–27)." Aristotle considers the nature and role of (1) the heart, (2) blood, and (3) energy as life. Aristotle's research project here is to study the heart as an "*archē*," blood as matter, and life or energy as the soul or whole "being" of an organism.

Aristotle's concern throughout is clearly with the relation between the parts and the whole, this being one of his wider research programmes. We find another avenue of Aristotle's approach by considering the crab regarding which "As to right and left, crabs, too, show the distinction poorly, still they do show it. You can see it in the claw; the right claw is larger and stronger, as though the right and left sides were trying to get distinguished (*IA* 714b 17–19)" and the lobster regarding which "In the lobsters alone it is a matter of chance which claw is the larger, and this in either sex. Claws they must have, because they belong to a genus in which this is a constant character; but they have them in this indeterminate way, owing to imperfect formation and to them not using them for their natural purpose, but for locomotion (*PA* 684a 32–684b 2)." Aristotle looks at a natural phenomenon and tries to explain what it means. Here he argues that the crab's claws are imbalanced to give it a sense of disequilibrium which provides a sense of self whereas for the lobster the imbalance makes up for their "imperfect character" by helping their locomotion. As with the epiglottis, we find that nature finds a correction when necessary. The problem needing correction in respect to the lobster, following the crab and the tortoise, is that it has "a low order of construction, on much the same level as that of insects."

Aristotle provides us with a range of research projects with associated experimentation. He takes the crab, tortoise, chameleon, centipede etc. and experiments on them in order to establish how their parts cooperate with their wholes. We ultimately find that the overall principle that Aristotle wants to demonstrate is that as living beings become more advanced they also become more organic and integrated with a more sophisticated central control structure in respect to both body and mind.

8.2.7 "The Congenital Growths of Hair Never Fall Out, for a Eunuch Never Goes Bald"

Aristotle understands beings as "beginnings" (*archai*) moving through "means" (*mesa*) to their ends (*telē*). They manifest their own *logos* which traverses a path for their peculiar being. Their *logoi* are constituted from structured subordinate functions (*erga*) and principles (*archai*) on a psychic basis which runs parallel to how they physically possess organs (*organa*).

Aristotle explains these principles in one way, as above, by considering joints as points of movement. He explores these principles in another way by considering the principle of sex, as follows: "Some animals change their form and character, not only at certain ages and at certain seasons, but in consequence of being castrated; and all animals possessed of testicles may be submitted to this operation…Birds are castrated at the rump at the part where the two sexes unite in copulation. If you burn this twice or thrice with hot irons, then, if the bird be full-grown, his crest grows sallow, he ceases to crow, and foregoes sexual activity; but if you cauterise the bird young, none of these male attributes or propensities will come to him as he grows up. The case is the same with men: if you mutilate them in boyhood, the later-growing hair never comes, and the voice never changes but remains high-pitched; if they be mutilated in early manhood, the late growths of hair quite them except the growth on the groin, and that diminishes but does not entirely depart. The congenital growths of hair never fall out, for a eunuch never goes bald (*HA* 631b 20–632a 5)." Aristotle considers the effects of castration as further part of a research project covering young and mature boars, bulls, stags, and camels.

We return to Aristotle's wider research project looking at the morphology of hair growth with the overall hypothesis here being that the "principle" of the overall form of the body will determine the characteristics, i.e., the hairiness, of the various parts and shift them from one operating model to another. In this case, we find that the sexual status of the whole being will determine the sexual characteristics of its parts, with this principle here being observed in a number of mammals. The whole is demonstrated to exist *before* the parts. Aristotle observes further in respect to this idea that a being has an operating model or "form"—what we could call a *Bauplan*—that lizards and snakes regrow their tails and he adds, seemingly incorrectly, that snakes can regenerate their eyes; see *HA* II 508b 7–8. He also notes that sponges can regenerate and replicate their whole forms; see *HA* V 548b 17–18. I add both that this regenerative property is a basic feature of the health of *all* creatures and that modern research shows us that some animals can even regenerate hearts and brains.[34] We see on this basis that there is "form" in nature which represents the natural and ideal form of a being—its healthy development—which the creature *seeks* to attain naturally and biologically. This idea of formal being having goal directed activity is known as Aristotle's teleology and is a self-evident reality of nature.

Aristotle's position is not merely static and formal but also dynamic and environmental. He argues that sex is situational in the sense that: "Just as with all animals a change of action follows a change of circumstance, so also a change of character follows a change of action, and often some portions of the physical frame undergo a change, as occurs in the case of birds. Hens, for instance, when they have beaten the cock in a fight, will crow like the cock, and endeavour to tread him; the crest rises up on their head and the tail-feathers on the rump, so that it becomes difficult to recognise that they are hens; in some cases there is a growth of small spurs. On the death of a hen a cock has been seen to undertake the maternal duties, leading the chickens about and providing them with food, and so intent upon these duties as to cease crowing and indulging in his sexual propensities. Some cock-birds are congenitally so feminine that they will submit to other males who attempt to tread them (*HA* 631b 5–18)." Whether Aristotle is correct in these observations or not, we see that his intention is to consider the relationship between the spiritual character of a being—in this case its femininity or masculinity—and its physical manifestation. I make note of the blueheaded wrasse which becomes male or female depending upon the sexual distribution around it.[35]

We ultimately find a world of "beings" (*ousiai*) which have "parts" with their own "principles" (*archai*) with this being Aristotle's response to the aporiae of "the part and the whole" and "the one and the many." Aristotle explains the situation as he faced it as that: "the sense in which a whole can be regarded as being the sum of the parts into which it can be divided…brought our philosophers to a stand; and they had to admit multiplicity in unity, unwillingly, as though it were a paradox (*Phys.* I 185b 35–186a2)." His own response was that such things are quite easily understood metaphysically and ousiologically but not abstractly and mathematically. He

[34] See Celine J. Vivien et al. "Evolution, comparative biology and ontogeny of vertebrate heart regeneration" *npj Regenerative Medicine* (2016).

[35] Scott Gilbert observes that: "The sex of the blueheaded wrasse, *Thassoma bifasciatum*, is determined by the social structure into which the larva enters: if it enters a reef where there are no males, it becomes male; if there are other males present, a female. Usually, there is one male for each dozen females. When that male dies, the largest female develops testes within 24 h and, within 2 weeks, is making functional sperm and mating with the remaining females ("The Reactive Genome" in Gerd B. Muller & Stuart A. Newman (*eds.*) *Origination of Organismal Form; Beyond the Gene in Developmental and Evolutionary Biology* (Cambridge Mass., 2003) p 93)."

explains his own metaphysical representation of the world as that: "in every composite thing, where a plurality of parts, whether continuous or discrete, is combined to make a single common whole, there is always found a ruling and a subject factor (*to archon kai to archomenon*), and this characteristic of living things is present in them as an outcome of the whole in nature, since even in things that do not partake in life there is a ruling principle, as in the case of a musical scale (*Pol.* I 1254a28–33)." Aristotle utilises Pythagoras' emphasis upon musical scales to show that nature possesses various analogous sequences and structures in *all* its dimensions, and we will see in the sister volume to this work how the biosemiotician Jakob von Uexküll also envisages life by reference to "melodies" and "symphonies."

Aristotle philosophically seeks to understand what a principle or origin (*archē*) *is*. As regards his conclusions these seem to broadly align with Leibniz's monadology but premised on the biological basis that: "Genesis from seeds always starts from the middle (*ek tou mesou*). All seeds are bivalvular, and the place of juncture is situated at the point of attachment (to the plant), an intermediate part belonging to both halves. It is from this part that both root and stem of growing things emerge; the starting-point is a central position between them (*hē d' archē to meson autōn estin*). In the case of grafts and cuttings this is particularly true of the buds (*Juv.* 468b18–24)." Aristotle considers both the idea of a principle or source (*archē*) and also the fact of symmetry—both of radial and bilateral symmetry—in nature throughout his physical and biological work.[36] The idea that the heart is the centre of the body and that the human and animal body has bilateral symmetry is at the heart of his thought.[37] Aristotle applies on this basis a "branching out" model to our

[36] Aristotle argues that "symmetry" must be the symmetry of something on the basis that: ""disposition," (*diathesis*) "state" (*hexis*) and "[symmetry]" (*symmetria*)... these terms cannot possibly exist anywhere else except in the things in relation to which they are employed (*Top.* IV 125a35–37)." In respect to his study of "symmetry" he argues in his biological work that: "In fact, all of them [i.e., the organs] are double (*diphya*). And the reason for this is that the structure of the body is double (*diphyēs*), though its halves are combined under one source (*pros mian archēn*) (*PA* III 669b18–20)." This duality is assumed to inform all aspects of organic being and hence: "In all animals the brain is double. Beyond this, at the far end, is the cerebellum as it is called; its form is different from that of the brain, as can be both felt and seen (*HA* I 494b 31–34)." Aristotle adds that animals have a clear symmetry in respect to kidneys and lungs and he also tries, more disputably, to add the heart since its cavities make it bipartite (*PA* III 669b 22–26) and more obviously incorrectly the spleen as the "double" of the liver (*PA* III 669b 15–18 & 669b 36–670a 3). I note that Aristotle expresses doubts on this last matter just as he also doubts that the sense of touch can be easily explained in the terms of symmetry—though human beings do, of course, have two hands—as the other senses evidently can.

[37] Aristotle argues that the heart is the source and centre of life on the basis that: "The region (*topos*) in question is intermediate (*mesos*) between that where the food enters and that where excrement is discharged. In bloodless animals it has no name, but in the sanguineous class this organ is called the heart (*Resp.* 474b 1–3)." He explains the functional principle further as that: "An animal is defined by the fact that it possesses sensation: and the part of the body to have sensation first is the part that has blood [and hence nutrition and the principle of material growth] in it first—in other words, the heart, which is the source (*archē*) of the blood and the first part to have it (*PA* III 666a 34–666b 1)." We see that there is a centring of a being based upon the region occupied by the heart in "blooded animals" which we could also call the "*thymos*."

own animal limbs as follows: "they use their joints like a centre (*hōsper gar kentrō*), and the whole section containing the joint becomes both one and two (*kai hen kai duo*), both straight and bent, changing potentially and actually (*dynamei kai energeia*) by reason of the joint (*MA* 698a18–21)." It could perhaps be said that the idea that the heart is a "centre" or "measure" of life is simplistic and yet such a simplifying hypothesis is surely necessary for any philosophical inquiry.

Aristotle understands a foetus and a being by reference to a plant emerging from a seed and "branching out." As regards how Aristotle understands "evolution" we find that he treats animals and human beings as increasingly animated plants. He argues as regards our relationship with the plants that: "Just as in the case of plants which when divided are observed to continue to live though separated from each other (thus showing that in *their* case the soul of each individual plant was actually one, potentially many), so we notice a similar result in other varieties of soul, i.e., in insects which have been cut in two; each of the segments possesses both sensation and local movement; and if sensation, necessarily also imagination and appetition; for, where there is sensation, there is also pleasure and pain, and, where these, necessarily also desire. We have no evidence as yet about thought or the power of reflexion (*peri de tou nou kai tēs theōrētikēs dynameōs*); it seems to be a different kind of soul, differing as what is eternal from what is perishable; it alone is capable of being separated (*De An.* II 413b 17–27)." How, he asks, do we separate something with "soul," "life," and "mind"? He seeks to contextualise the human reflection of "mind" by reference to divided insects.

8.2.8 *"In Touch He [i.e., a Human Being] Is Much More Discriminating than the Other Animals. That Is Why He Is of All Living Creatures the most Intelligent"*

How does "mind" encompass and represent the whole integral being? Aristotle compares living being with chemical being as part of this study of divisibility and argues that: "each [body] is determined relatively to that part which is next to it by contact (*haphē*), for which reason each of them is in a sense many bodies [i.e., a composition or "*symphysis*"]. But the whole of which they are parts must necessarily be complete (*teleion*), and thus, in accordance with the meaning of the word, have being, not in some respect only, but in every respect (*De Cael.* I 268b 6–10)." Chemical being is achieved "by contact" but still has "being" as a unified mass on this basis. Living matter is in some sense a further articulation of chemical mass. Animal and human being is a further animation of this matter. In all cases, though, the being is by means of its mass "complete."

Chemical compounds are in a sense something complete and consistent in having their own formal structure, e.g. as carbon monoxide or carbon dioxide. Living beings morph from plants (and insects) which are divisible into "higher" organisms which possess an integrity, definiteness, and continuity which increasingly precludes division. We ultimately work through the world from passive chemical matter to active human being. Aristotle argues on this movement that the most critical transition here is the "power of reflexion" which "alone is capable of being separated." It is only this mental power of humanity which opens up a zone of being separate from the soul and body which are ordinarily combined as a joint biological function of animal being. Remarkably, though, Aristotle explains that mind is itself a transcendent extension of touch and of sense rather than being something other than it on the basis that: "in touch he [i.e., a human being] is much more discriminating than the other animals. That is why he is of all living creatures the most intelligent. Proof of this lies in the fact that among the human race men are well or poorly endowed with intelligence in proportion to their sense of touch, and no other sense; for men of hard skin and flesh are poorly, and men of soft flesh well-endowed with intelligence (*De An.* II 421a 22–27)." Aristotle assumes that the quality of the body will inform the quality of the soul and of the mind or intelligence.

Aristotle adds elsewhere that the quality of the food and of the blood will affect intelligence (see *PA* 650b 19–651a 15) and that the posture of the animal towards the world will determine an animal's thought (*dianoia*) and common sense (*koinē aisthēsis*) so that its uprightness will conduce towards a certain awareness allowing a corresponding intelligence whereas its being bent-over will make its intelligence sluggish; see *PA* 686a IV 21–32.[38] His concept of this "dwarf-like" contraction of a being is one of the most difficult and contentious elements of Aristotle's thought (and see also *Mem.* 453 a-b). It can be visualised as the unfolding of an animal from being a foetus or grub to being the fully formed, extended, and mature adult. He seems to envisage a constant evolutionary process resembling something like Haeckel's biogenetic laws through which every organism passes through an evolutionary pathway in its individual development. In Aristotle's case, he conceptualises an energy or "heat" which drives the development of an organism and makes it rise up to become a mature being, on both a speciate and an individual basis. What is perhaps of most interest to us here is that Aristotle assumes and insists that we must *physically* account for the "becoming" of an organism and that we do so by reference to the *material* nature of the organism; its flesh, blood, and energy.

Aristotle always operates *in the real world*. His human being is not an abstract mind but one animal among others and one physical being among others. His understanding is always situational and he hence expands upon the relationship between sense and mind further as that: "actual sensation corresponds to the exercise of knowledge; with this difference, that the objects of sight and hearing (and similarly those of the other senses), which produce the actuality of sensation, are external. This is because actual sensation is of particulars, whereas knowledge is of universals; these in a sense exist in the soul itself. So it lies in man's power to use his mind whenever he chooses, but it is not in his power to experience sensation; for the presence of the sensible object is essential. The same is true of our knowledge of sensible objects, and for the same reason, *viz.* that sensible objects are particular and external (*De An.* II 417b 19–27)." His *On the Soul* is powerfully concerned with

[38] See Claudia Zatta *Aristotle and the Animals: The Logos of Life Itself* (Abingdon, 2022) p 17.

how mind arises from out of sense and then out of being. It is concerned with the sense in which human beings transcend nature by moving to become different, distinguished or detached from it.

Aristotle ultimately argues that human beings depart from the "*automaton*" state of nature on the basis that: "In inanimate things the initiating source is single; but in living things it is multiplex, for desire and reason are not always in harmony (*symphōnei*). So in the other animals force is a simple matter, as it is in inanimate things, because they do not have a conflict between reason and desire, but live simply by desire. But in humans both elements are present, once they have reached the age at which we begin to assess their conduct. (We do not speak of the conduct of an infant, any more than of a wild beast, but only of someone who is old enough to act on reason.) (*E.E.* II 1224a23–30)." Aristotle argues that the *simple* forces of nature in a sense include animal beings along with material objects with only humanity being more complex because it *has* "*logos*" (*zōon echon logos*). His *logos* becomes in some sense detached and distinct and able to operate independently, as itself a "self-mover." This is the "*nous*" and "*theōrētikē dynamis*" referred to above.

He concludes as regards the relationship between "nature" and "mind" that: "Clearly the soul is also the cause in the final sense (*phaneron d'hōs kai hou heneka hē psychē aitia*). For just as mind acts with some purpose in view, so too does nature, and this purpose is its end (*hōsper gar ho nous heneka tou, ton auton tropon kai hē physis, kai tout' estin autē telos*). In living creatures the soul supplies such a purpose, and this in accordance with nature (*kata physin*), for all natural bodies are instruments of the soul (*panta gar ta physika sōmata tēs psychēs organa*); and just as is the case with the bodies of animals, so with those of plants. This shows that they exist for the sake of the soul (*De An.* II 415b 15–22)." Aristotle argues that "nature" is in a sense "mind" and "soul" expressed through "body" and "being." To us, these are taken to be two distinct things and yet if we look again we find that they are both constructive energies which can be understood together. Aristotle argues that whereas chemical being is an expression of the potentiality of matter through body, a living being is the expression of actuality through soul.[39] I outline this structuring as explained in Aristotle's *Ethics* in Fig. 8.4 below.

Aristotle's *Ethics* as represented above shows us a sophisticated, thorough, and nuanced philosophical and, in a real sense, scientific framing for human behaviour. His interpretation is observably a long distance from behaviourism. It addresses reality as we all understand it in actuality and represents a reality in which the human being is a structuring of power(s) actualised in various ways and which makes up the character of a person. One of the subtleties of Aristotle's account is, indeed, how various characteristics must go together as a set for a person to be of a certain type of character.[40] We cannot on this basis simply break human beings down into component events or activities. Aristotle's perspective regards the nature

[39] Aristotle argues that: "the actuality of each thing is naturally inherent in its potentiality, that is in its own proper matter…matter is potentiality, and form actuality; and since the compound is an animate thing, the body cannot be the actuality of the soul, but the soul is the actuality of some body (*De An.* II 414a 26–28)."

[40] See Paula Gottlieb "Aristotle on Dividing the Soul and Uniting the Virtues" *Phronesis* (1994), Fiasse Gaëlle "Aristotle's φρονησις: A True Grasp of Ends as Well as Means?" *The Review of*

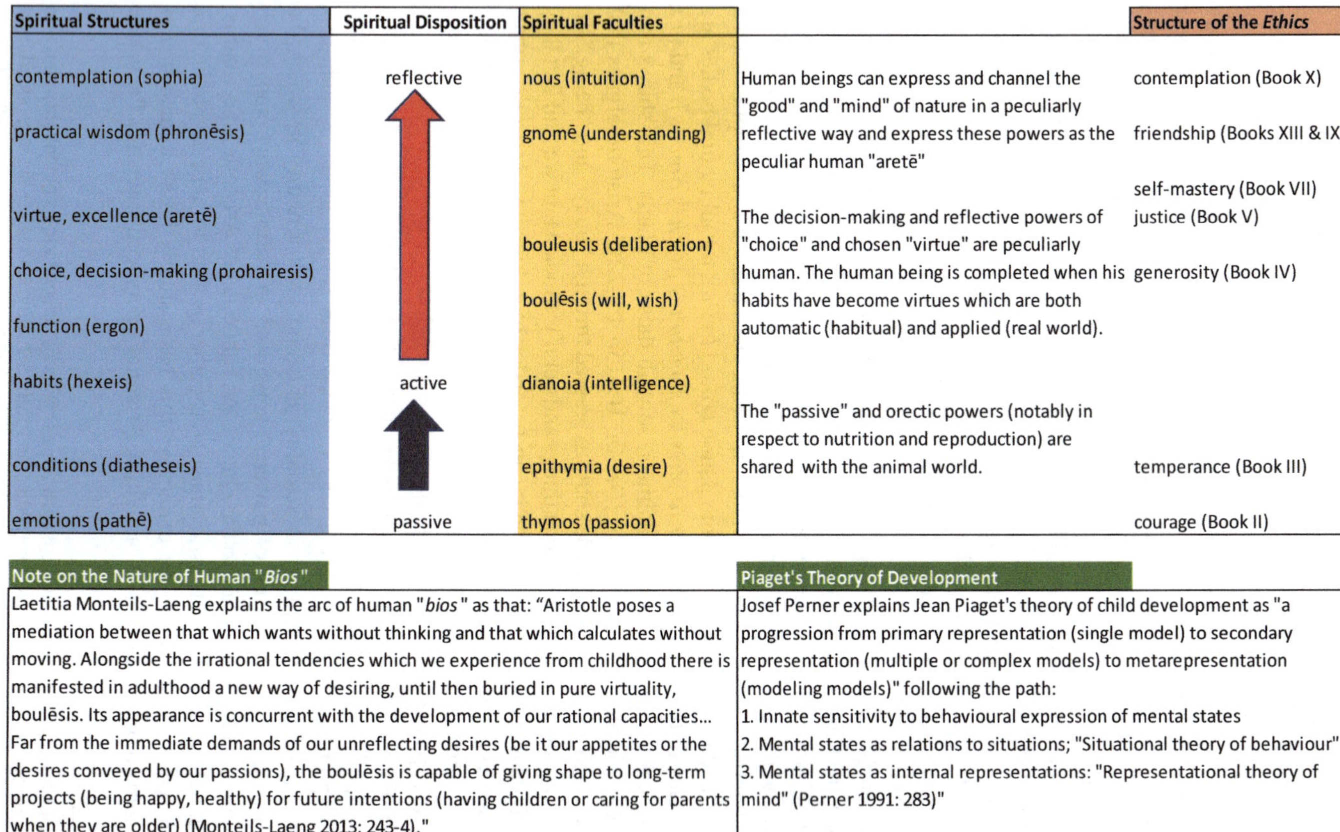

Fig. 8.4 Aristotle's mental/spiritual structuring of human being

of beings with powers rather than how beings can be trained or constrained to do x or y. We will explore this difference between Aristotle's thought and that of behaviourism in the sister volume to this work.

8.3 Is Aristotle's Science Valid Today?

How should we consider Bertrand Russell's decision to include Aristotle in his "hilarious catalogue of organized and individual stupidity"? I hope to have shown above that this charge is singularly inappropriate when applied to Aristotle. If so, why did a man as obviously capable as Russell make this false claim? Russell's lecture is remembered not because he criticises superstitious astrologers and magicians but because his take down of Aristotle classified him with them. Russell's real claim was that even the best thought of the intellectual and cultural tradition is really "intellectual rubbish." Karl Popper expanded upon this viewpoint 2 years later with his two volume *The Open Society and Its Enemies* which sought to systematically dismiss the Western philosophical canon from Plato (and Aristotle) to Hegel and Marx. Why, though, would Russell want to smear Aristotle? I suggest that Russell simply seeks to defend the anti-philosophy that I have sought to outline in Chapter Two. This is epitomised by the very real attempt to conflate philosophy with "bad grammar." Aristotle, the founder of metaphysics, is an obvious enemy in this context.

This movement of anti-philosophy could be said to begin with Francis Bacon who famously claimed that Aristotle's position that: "Both eyebrows and eyelashes exist to afford protection to the eyes (*PA* II 658b 14–15)" is somehow "impertinent."[41] Bacon supposed that Aristotle (and his followers) *was being impertinent to "God"* through his assertion that the "function" of the eyebrows is to prevent water, sweat, and debris falling into the eye and to provide an additional sensitivity to objects around the eye, with the eyebrows additionally forming a part of the face and enabling human expression.[42] We find, however, that Bacon's less theological objection to the concept of "function" is that: "to resolve nature into abstractions is less our purpose than to dissect her."[43] Bacon took Aristotle's own method of dissection employed, as we have seen, to understand nature by carving it "at the

---

Metaphysics (2001), and Marguerite Deslauriers "How to Distinguish Aristotle's Virtues" *Phronesis* (2002).

[41] Bacon, Francis "On the Advancement of Learning" in *Bacon's Essays and Advancement of Learning* (London, [*orig*. 1605] 1925) p 265.

[42] Aristotle also describes the process of blinking in terms of protection as follows: "The movement of the eyelids known as blinking is a natural and instinctive one, not dependent on the will (*ouk ek prohaireseōs*), and its object is to prevent things from getting into the eyes (*PA* II 657a 36–657b 2)."

[43] Bacon, Francis *Novum Organum* (1620) I, 51 (and see Etienne Gilson's chapter "The Mechanist Objection" in *From Aristotle to Darwin and Back Again: A Journey in Final Causality, Species, and Evolution* (Notre Dame, [*orig*. 1971] 1984)).

joints" and instead put her "on the rack." The strange gaming of modern philosophy we see here is clearly a strange and sad derivative of the apologetics of the Middle Ages and of the ideological (and religious) wars of the Early Modern Age. It is also, sadly, a revival of the clever sophistry of Aristotle's age, a gaming of thought that Aristotle (as also Plato) sought to address and counter through their own thought.

This gaming of thought went hand in hand with reductionism, with simplification, and with the dismissal of science (and religion) by practical men of science. It is hence that Bacon argued regarding the concept of "form" that: "forms are figments of the human mind, unless you call those laws of action forms"[44] and, in more detail, that: "When man contemplates nature working freely, he meets with different species of things, of animals, of plants, of minerals; whence he readily passes into the opinions that there are in nature certain primary forms which nature endeavours to educe…[But in] nature nothing really exists besides individual bodies [true particles], performing pure individual acts according to law."[45] We come to a sort of logical atomism, as we always do with this (anti-)philosophy.

As we might expect, anti-philosophy is not *good* philosophy. Its explicit aim is to obviate the need for philosophy, normally by reducing the world to what Russell calls "cold science." We can again consider W.V. Quine to draw out the situation. He explains "beings" as that: "To be assumed as an entity is, purely and simply, to be reckoned as the value of a variable"[46] and "being" as that: "Our acceptance of an ontology is, I think, similar in principle to our acceptance of a scientific theory, say, a system of physics: we adopt, at least insofar as we are reasonable, the simplest conceptual scheme into which the disorganised fragments of raw experience can be fitted and arranged. Our ontology is determined once we have fixed upon the over-all conceptual scheme which is to accommodate science in the broadest sense."[47] In all cases, philosophy is subjective and is just what is convenient to us and hence the use of such terms as "assumed, reckoned, acceptance, accommodate" etc.

In respect to content, Quine rejects the very existence of a "being" (*ousia*) and asserts that philosophy can and should be reduced to whatever can "accommodate science." Karl Popper encapsulates Quine "in a nutshell" by quoting him as follows: "The bodily states exist anyway; why add the others?"[48]

---

[44] Bacon, Francis *The New Organon* (1620) I. 51.

[45] Bacon, Francis *Ibid.* I.66, II 2, II 8.

[46] Quine, W.V. "On What There Is" in *From a Logical Point of View* (Cambridge, Mass., 1953) p 13.

[47] Quine, W.V. *Ibid.* p 18.

[48] Popper, Karl "Knowledge: Objective and Subjective" in *Knowledge and the Body-Mind Problem: In defence of interaction* (London, [*orig.* 1969] 1994) p 8.

8.3.1 An Aristotelian Critique of Our Modern World

Aristotle stands at the basis of our Western worldview. We can dismiss and ridicule him along with Russell, but he still needs to be addressed and his response is by no means risible. Indeed, Aristotle is both close to us and distant from us. We have traversed many avenues since he trod the agora in fourth century Athens. Our weird and wonderful spiritual peregrinations were, ironically, the result of the explosion occasioned by his pupil, Alexander, likely against his schoolmaster's wishes. Aristotle does in any case offer us a point of comparison and critique for how we see the world and for how we have transformed it from the position left to us by the Classical Greeks. If we are, as we proclaim, so keen on critique and on self-critique, this ready-made alternative reality provides a perfect counter-position to measure ourselves against.

I will argue in the sister volume to this work that our worldview based upon neo-Darwinian assumptions has been refuted by modern science itself. We will see there that neo-Darwinism is defended as a dogma today not on solid scientific but on dubious philosophical grounds. We could indeed suspect that the attribution of being to chance was always a commitment brought to science rather than derived from it. The ideological commitment behind the insistence upon existence as a result of "chance" is most obviously to pragmatism and technocracy. It is a commitment to the sort of nihilism that results from the misplaced reduction of philosophy to a scientific method that is quite properly unable to formulate and issue general value statements. By reducing philosophy to science, we merely negate philosophy into a sort of anti-philosophy and likely corrupt science itself.

What is the effect of using science as a sort of blocking device against values? We can consider this through W.V. Quine's thought that: "One part of the problem of induction, the part that asks why there should be regularities in nature at all, can, I think, be dismissed. *That* there are or have been regularities, for whatever reason, it is an established fact of science; and we cannot ask better than that. *Why* there have been regularities is an obscure question, for which it is hard to see what would count as an answer. What does make sense is this other part of the problem of induction: why does our innate spacing of qualities accord so well with the functionally relevant groupings in nature as to make our inductions tend to come out right? Why should our subjective spacing of qualities have a special purchase on nature and a lien on the future? There is some encouragement in Darwin. If people's innate spacing of qualities is a gene-linked trait, then the spacing that has made for the most successful induction will have tended to predominate through natural selection. Creatures inveterately wrong in their induction have a pathetic but praiseworthy tendency to die before reproducing their kind."[49] We simply find a mythology based upon science which barely pretends to be serious.

[49] Quine, W.V. "Natural Kinds" in *Ontological Relativity and Other Essays* (New Quinton, 1969) p 126.

As regards the philosophical robustness of the concept of "natural selection" itself, we can critique Mark Perlman's dedicated essay on the subject. Perlman takes the evolution of flight *as a paradigm example* of Darwinian evolution. He argues as regards how we can understand this "function" through "natural selection" that: "Flying comes in many varieties, and is actually a matter of degree…Perhaps we should say that anything short of falling at terminal velocity is, to some degree, a kind of flying. The baby birds can't take off into flight as they will in adulthood, but they don't drop like stones either…aerodynamic action is enough to give even those proto-wings [of ancient feathered reptiles] the function of some degree of "flight." Eons of evolution greatly increased the degree of aerodynamic lift, so now normally functioning adult birds can take off and really fly…So perhaps wings, even their early proto-wing stubs, have functioned all along to let the organism fly (to some degree)."[50] Perlman hypothesises that *all* wings have come into being through proto-bats, proto-bees, proto-birds, and *some* proto-ants happening to happen to fall habitually and adapt and breed themselves accordingly. Michael Ruse also seeks, equally inadequately, to explain bird migration on the basis of "natural selection."[51]

Aristotle would, of course, disagree. His alternative position observes as regards the whole of reality (rather than just a modelled trait) that: "No animal is merely able to fly (winged), as the fish is able to swim. Even the dermatous-winged animals can walk about, and bats have feet, and the seal has stunted feet. Some birds have feet which are not much use, and this accounts for their being called *apodes* (footless). The small bird concerned [probably the swift] is a very good flier (*HA* I 487b 22–26)." Aristotle notes that biological functioning is dependent upon use but that even when an evolutionary niche eliminates the need for a certain function and its corresponding organ, the residual organ may remain. Organs can wither with disuse but they do not, according to Aristotle, merely arise through use. Aristotle rather raises the problem of innateness that animals perform many tasks and functions (including metabolic) immediately and innately as the fish swims, and as the spider weaves its webs from birth (*HA* VIII 623a 30–31). He adds that all animals are *complex* in the sense that, for example, "Many animals combine ability to walk and ability to swim (*HA* I 487b 31–32)" which reiterates that there is (in the real world rather than in computer simulations) no simple "just-so" *explanations* for the *developed and integrated form and character of animals*.

When we look at the whole reality, it is more coherent to say, with Aristotle, that there is an optimising epigenetic principle at work in nature than that the world is

[50] Perlman, Mark "Traits have evolved to function the way they do because of a past advantage" in *Contemporary Debates in the Philosophy of Biology* (Blackwell, 2010) p 65.

[51] Ruse incoherently argues that: "Stephen Emlen (1970) was able to show that what the buntings learn is not some fixed map but the axis of rotation of the heavens. Exposed to heavens rotating around fictional axes, the birds "migrate" according to them. This plasticity in the birds' learning has an obvious evolutionary value, for, to take an instance, in 13,000 years the North Star will no longer be Polaris in the Little Dipper but Vega in the Lyra, 47° away. In the evolutionary time-scale 13,000 years is but little time, and, were the buntings to have a fixed orientation to the heavens, before long they would be migrating to all sorts of unsuitable places. As with the rats, natural selection has been at work here (*Sociobiology: Sense or Nonsense?* (2nd ed) (Dordrecht, 1985) p 184)."

created by accumulated fortuitous chance. Aristotle basically takes a structuralist view of nature which simply assumes the existence of structures in biology, as everywhere else. Happily, we find that there is increasing incredulity and resentment in respect to these "just-so" stories as it has become increasingly obvious that our increasingly dysfunctional modern worldview derives from such an odd and cryptic creation myth.[52]

Jerry Fodor and Massimo Piattelli-Palmarini provide us with a direct scientific response to Perlmn's neo-Darwinian hypothesis as regards the evolution of flight. They explain the hard science and reality of flight as that: "The angles of effective wing-stroke are extremely narrow…and one wants to question the process through which this narrow wedge of angles became fixated even before there was any real flight."[53] Their doubt is, of course, rhetorical. They argue regarding the observable optimisation of nature *which is observed everywhere* that: "a team of American and French biologists and physicists has recently determined by means of mathematical equations and artificial stimulations that the spontaneous biological "design" of leaves is perfect…the same optimisation criterion can be used to describe the placement of veins in leaves."[54] It seems unavoidably obvious on this basis that Aristotle's hypotheses regarding the natural optimising expression of nature based upon historical speciate pathways is actually the best hypothesis we have and not the idea that optimal organic functions are acquired by jumping off cliffs.

8.3.2 The Aristotelian Alternative

Should Aristotle be taken seriously today? Ironically, Myles Burnyeat, a notable classicist, thought not. He argued that when Aristotle argued that a being perceives and "takes on the form without the matter" he meant that: "the organ literally and physiologically undergoes an alteration: the eye-jelly turns red, the nose becomes smelly."[55] Is this *really* what Aristotle meant? There is little textual room for the assertion that the organ is not fitted to the world but somehow physiologically *becomes* what it engages with. The nose is not simply fitted to engage with the

---

[52] Richard Lewontin argues that: "Despite its claims to be above society, science, like the Church before it, is a supremely social institution, reflecting and reinforcing the dominant values and views of society at each historical epoch…The most famous case is Darwin's theory of evolution by natural selection (*Biology as Ideology* (New York, 1991) p 9)." See also Peter T. Saunders "Bricks Without Straw: Darwinism in the Social Sciences" *Theoria* (2003), Mary Midgely *The Solitary Self: Darwin and the Selfish Gene* (Durham, 2010) & *Evolution as a Religion: Strange hopes and stranger fears* (London, 1985), and Stephen Toulmin "Contemporary Scientific Mythology" in MacIntyre, Alasdair & Gregor Smith, Ronald (*eds.*) *Metaphysical Beliefs* (London, 1957).

[53] Fodor, Jerry & Piattelli-Palmarini, Massimo *What Darwin Got Wrong* (London, 2011) p 89.

[54] Fodor, Jerry & Piattelli-Palmarini, Massimo *Ibid.* p 83–4.

[55] Burnyeat Myles "Is an Aristotelian philosophy of mind still credible? (A Draft)" in Nussbaum, Martha & Rorty, Amelie O. (*eds.*) *Essays on Aristotle's* De Anima (Oxford, 1992) p 19.

relevant smells of its environment, as an eagle is fitted to the air and a fish to water, but it literally and physiologically "becomes smelly."

The subtext here is as follows. Burnyeat reveals his hand when he complains in the same essay that: "For Aristotle it is the existence of life which explains why animals have the physical constitutions they do, not the other way round. The unity of science is achieved from the top down, not the bottom up, which is the way we have seen it from the seventeenth century. Aristotle simply does not have our task of starting from the existence of matter as physics and chemistry describe it and working up to the explanation of the secondary qualities on one side and animal perceptual capacities on the other."[56] It is pretty evident that Burnyeat manufactures a straw man Aristotle so as to defend "the way we have seen it from the seventeenth century." Richard Rorty similarly supposedly defended "the seventeenth century," as we have witnessed earlier. We ultimately see that Burnyeat, the classicist, sacrifices Aristotle to defend out-of-date science and the bad philosophy based upon it. Remarkably, however, his critique did happen to produce a useful volume of essays in response showing the validity of Aristotle's consideration of "function," though unhelpfully considering Aristotle as a proto-functionalist when his philosophy is truly far beyond this and any other such—ism du jour.[57]

We can consider Patrick Suppes to consider what Aristotle likely actually meant in his consideration of the engagement between nose and world. Suppes engages with Aristotle and explains that: "I came to a much deeper understanding of Aristotle's theory of perception [and] clarified once and for all that the simple slogan "same form, different matter" was a succinct but pregnant way of describing the fundamental distinction between form and matter in Aristotle's philosophy. Many aspects of it are formulated by Plato, and to a lesser extent, by other Greek philosophers. But what was and is important is that I realised how fundamental this idea is in trying to face up to the nature of the accuracy and speed of human perception, and also of many animals as well. I reach into my pocket and hold up a key, I ask you, "What is this?" In a matter of milliseconds really, your brain has computed the answer. How could such rapid and accurate answers be computed when the number of objects I could exhibit of this kind is so large? From a formal standpoint of systematic theory, especially of the kind of mathematical psychology that has interested me all my academic career, the depth and clarity of this discussion of how perceptual recognition of objects and processes is computed, that is, by recognising the form of the object, in the way that Aristotle so clearly explains, has been a model of qualitative psychology and philosophy. A beautiful concept of isomorphism is behind this fundamental explanation."[58]

---

[56] Burnyeat, Myles *Ibid.* p 22.

[57] Martha Nussbaum and Hilary Putnam, for instance, defend Aristotle in their "Changing Aristotle's Mind" in *Essays on Aristotle's* De Anima (Oxford, 1992) on the basis that: "These are not outmoded questions; nor is Aristotle's way of pressing them at all remote from the methods and concerns of modern inquiries into substance and identity (*Ibid.* p 30)." I would rather contend that Aristotle's approach is different and deeper than similar modern approaches.

[58] Suppes, Patrick "[Commentary on] Dagfinn Føllesdal" in Crangle, Colleen E. *et al* (*eds.*) *Foundations and Methods from Mathematics to Neuroscience* (Stanford, 2014) p 266.

Suppes correctly emphasises the clarity and pregnancy of Aristotle's position—his "qualitative psychology and philosophy"—which, if we allow it, elegantly guides us into the depths of reality. Substantively, Suppes brings out Aristotle's insight regarding "sight" and more subtly he explores the significance of the event actualised by the instant or "*hama*." We glean from this a logical insight into what is known as the "*Augenblick*" that basic quotidian "reality" is not some sort of loose fluke or tight mechanism but is a pregnant possibility in which information and meaning can and normally does transfer between beings or objects through the medium of mind on an immediate or semi-automated basis.[59] It is not twisting Aristotle to regard his position generously, not assuming that it is "intellectual rubbish," but merely imputing the good sense to Aristotle that his work on the whole so obviously conveys. This does not, however, mean that we should ignore what seem to be errors such as Aristotle's use of the argument that when a woman looks into a mirror during her menstrual period, the mirror's surface takes on a cloudy red hue with this stain being more difficult to remove from a new mirror than an old one, for which see *Insomn.* 459b 24–460a 23. It has, however, even been questioned whether Aristotle is making use of a popular belief to make his point here or whether this is his opinion.[60]

Suppes observes further as regards this situation that: "Aristotle's doctrine of form as central to both perceiving and thinking provides already the key idea undoubtedly behind even the earliest developments of geometric representations, and is just as present in thinking about modern quantum mechanics or astrophysics. The key notion, put in modern terms, is that of structural isomorphism between, in the one case a perception and an object or process in the world, and in another, between a mental image and an "abstract" structure. Such isomorphisms are indeed as important, or perhaps even more important, to modern pure mathematics as they are to theoretical physics. How such isomorphisms are physically implemented in perceiving and thinking was beyond Aristotle and they have also been beyond most of modern psychology."[61] We see that Suppes takes Aristotle seriously and tries to make sense of his position in modern language. We will consider in the sister volume to this work how Aristotle translates meaningfully into and provides a deep

[59] On "*hama*" see Jacques Derrida "*Ousia* and *Grammē*: A Note on a Note from *Being and Time*" in *Margins of Philosophy* (Chicago, [*orig.* 1968] 1982). Maurice Merleau-Ponty explains the meaning of the "*Augenblick*" as that: "It is true that the instant is not a philosopher's fiction. It is the point at which one project is brought to fruition and another begun—the point at which my gaze is transferred from one end to another, it is the *Augen-Blick* [blink-of-an-eye]. But this break in time cannot occur unless each of the two spans is of a piece. Consciousness, it is said, is, though not atomized into instants, at least haunted by the spectre of the instant which it is obliged continually to exorcise by a free act (*Phenomenology of Perception* (London, [*orig.* 1945] 1962 p 508)."

[60] See Filip Radovic "The Case of Red-Stained Mirrors: Perception, Strange Phenomena, and the Role of Exemplification in Aristotle" in Bennett, D., Toivanen, J. (*eds.*) *Philosophical Problems in Sense Perception: Testing the Limits of Aristotelianism* (Cham, 2020).

[61] Suppes, Patrick "Why the Effectiveness of Mathematics in the Natural Sciences is not Surprising" in *Interdisciplinary Science Reviews* (1978) p 252–3.

philosophical basis for our best modern thinking on mathematics, psychology, physics, biology etc.

Aristotle's position is clearly reasonable, accurate yet expansive, and, fundamentally, true. Arguing for it is simply expanding our vision of the meaningful reality in which we evidently exist whereas to argue against it is to argue for ideology and (mathematical) abstraction. It is obvious which side of the argument one should support, and it is precisely this fact that causes the defensive manipulations of the dogmatic anti-Aristotelian side.

8.4 Conclusions

Aristotle's philosophy is an evidently true representation of "being" and "becoming" in the real world. It is as detailed and thorough as any philosophy and has a good claim to be the truest philosophy ever formulated. It is realism and eschews the ideology and idealism which imposes ideas rather than deriving truth from close study of the world. Its truth critically recognises the many gaps in respect to our understanding of the world. One of these admitted gaps recognises that we cannot conflate philosophy with science. We cannot simply reduce the world to mathematics or to any science since the world is not so reducible. Our world, according to Aristotle, is a rich topography of being and beings which is qualitative in nature and not purely quantitative. At the very outset of philosophical thought, Aristotle recognised the problem of reductionism and put measures in place to identify and deny it.

Perhaps even more remarkably, we find that Aristotle's science is not just defensible but rather provides us with a sophisticated holistic model that is eminently pertinent to us today. I will, indeed, show in the sister volume to this work that much cutting-edge modern science has either taken inspiration from or reached parallel positions to Aristotle, most notably biosemiotics. We will see that it has been a struggle to overcome reductionist science in order to return to a more subtle and theoretical science. The problem we face is that we have mythologised and fetishised science and transformed it into something monstrous by conflating it with philosophy. We have been left with a stunted philosophy and a monstrous science. A good example of this state of affairs is provided by the behaviourist B.F. Skinner who supposes that: "it should be possible to design such a world in which behaviour likely to be punished seldom or never occurs…What we need is a technology of behaviour. We could solve our problems quickly enough if we could just adjust the growth of the world's population as precisely as we adjust the course of a spaceship."[62] The pushing through humanity to some imagined technological end should be immediately and immensely troubling.

[62] Skinner, B.F. *Beyond Freedom and Dignity: A stunning new plan to alter human behaviour* (New York, 1971) p 62 & 3.

Skinner states that his philosophy should go "beyond freedom and dignity" on the basis that: "As a science of behaviour adopts the strategy of physics and biology, the autonomous agent to which behaviour has traditionally been attributed is replaced by the environment—the environment in which the species evolved and in which the behaviour of the individual is shaped and maintained."[63] This is car-crash philosophy based upon the application of philosophical reductionism—i.e., reducing the world to physics—and centred around a crude neo-Darwinism which denies the individual and reduces being to environmental forces. It is a technological vision with no sense of the philosophical tradition. It severs us from our intellectual roots, knowledge from meaning, and finds nothing of value in the world aside from itself. It is hence that Skinner explains that: "A culture is very much like the experimental space used in the analysis of behaviour. Both are sets of contingencies of reinforcement. A child is born into a culture as an organism is placed in an experimental place. Designing a culture is like designing an experiment; contingencies are arranged and effects noted."[64] Ironically, we find a worldview which argues entirely from design, though the great designer has now become the scientist himself.

[63] Skinner, B.F. *Ibid.* p 175.

[64] Skinner, B.F. *Ibid.* p 145.

Chapter 9
Can We Learn from Aristotle's Philosophy Today?

Abstract After several millennia of ink spilt on Aristotle's philosophy, is there still something we can learn from him? We find that much of what is positive in modern philosophy—from Leibniz to Heidegger—is influenced by or derived from Aristotle. Aristotle's broad philosophy of "being" is continued in the philosophy of such idealists as Leibniz and Hegel and by phenomenologists such as Heidegger and Husserl. Martin Heidegger's consideration of the "thingliness of the thing" is an essentially Aristotelian insight. The centrality of Aristotle's defence of "being" (*ousia*) has been set out in Chap. 6. The parallel tradition of anti-philosophy, as set out in Chap. 2, does by contrast seek to reduce meaning and hence denies "being," "form," and all "structure" and, ultimately, "meaning" itself. I argue here and throughout that the modern denial of "being" is not philosophically sound or required but is a product of what Aldous Huxley describes as a "philosophy of meaninglessness."

Keywords A defence of Aristotle's philosophy · The thingliness of the thing · The Aristotelian tradition in modern thought

I have suggested above that we can even learn from Aristotle's science and biology. In respect to philosophy, it seems evident that we can and should learn from Aristotle's philosophy. Aristotle was, after all, surely not named *the* Philosopher for nothing. The key issue, I would suppose, is how far modern philosophy is an improvement on Aristotle's. If the gentle reader has confident expectations on this score, he or she may sadly soon find themselves disappointed. The best modern philosophy is in the Aristotelian tradition, and not the watered-down neo-Aristotelianism of the scholastics but a reflection of the rich original Aristotelianism of Aristotle himself. The less said about the other modern philosophical traditions the better, though we do unfortunately need to address them here for our immediate purposes. Let us hence seek to diagnose the problem with modern philosophy, whether analytic or existentialist, and consider how "Aristotelian" philosophy exposes it.

We can begin with the excellent thought of Aldous Huxley. He explains our current intellectual posture as a "philosophy of meaninglessness" existing on the basis

P. N. Jackson, *Aristotle on Meaning in the Living World*, Biosemiotics 30,
https://doi.org/10.1007/978-3-032-00602-8_9

that: "For myself as, no doubt, for most of my contemporaries, the philosophy of meaninglessness was essentially an instrument of liberation…I had motives for not wanting the world to have a meaning; consequently assumed that it had none, and was able without any difficulty to find satisfying reasons for this assumption. Most ignorance is vincible ignorance. We don't know because we don't want to know. It is our will that decides how and upon what subjects we shall use our intelligence. Those who detect no meaning in the world generally do so because, for one reason or another, it suits their books that the world should be meaningless… Specialised meaninglessness has come to be regarded, in certain circles, as a kind of hall mark of true science…The scientific picture of the world is what it is because men of science combine this incompetence with certain special competences. They have no right to claim that this product of incompetence and specialisation is a complete picture of reality. As a matter of historical fact, however, this claim has constantly been made."[1] The ends and source of this meaninglessness is that it "in part…lends itself very effectively to furthering the ends of erotic or political passion; in part to a genuine intellectual error—the error of identifying the world of science, a world from which all meaning and value has been deliberately excluded, with ultimate value."[2]

Huxley explained the consequences of our pursuit of sexual liberation and scientific control further in a remarkable letter to George Orwell dated 21 October 1949, a few months before the unorthodox Etonian's untimely death. He wrote: "The first hints of a philosophy of the ultimate revolution—the revolution which lies beyond politics and economics, and which aims at total subversion of the individual's psychology and physiology—are to be found in the Marquis de Sade, who regarded himself as the continuator, the consummator, of Robespierre and Babeuf. The philosophy of the ruling minority in Nineteen Eighty-Four is a sadism which has been carried to its logical conclusion by going beyond sex and denying it. Whether in actual fact the policy of the boot-on-the-face can go on indefinitely seems doubtful. My own belief is that the ruling oligarchy will find less arduous and wasteful ways of governing and of satisfying its lust for power, and these ways will resemble those which I described in Brave New World. I have had occasion recently to look into the history of animal magnetism and hypnotism, and have been greatly struck by the way in which, for a 150 years, the world has refused to take serious cognizance of the discoveries of Mesmer, Braid, Esdaile, and the rest."[3] The ultimate risk was that our civilisation would conclude in "infant conditioning and narco-hypnosis…as instruments of government."[4] We see here a critique of the social conditioning or social engineering often associated with behaviourism and other scientific approaches to psychology.

[1] Huxley, Aldous "Beliefs" in *Collected Essays* (New York, [*orig*. 1937] 1971) p 363–8.

[2] Huxley, Aldous *Ibid.* p 362.

[3] Huxley, Aldous "1984 v. Brave New World" Letters of Note ([*orig*. 1949] online). It is noted that Huxley had previously taught Orwell French at Eton in 1917.

[4] Huxley, Aldous *Ibid.*

How are the problems highlighted by Huxley and Orwell relevant to Aristotle? They show us that we are in a mess. We are trapped in a philosophical, political, and spiritual quagmire. Even the very fact of meaning itself has been denied. Even science and technology, our greatest modern achievements, have been turned into a problem. We could hypothesise that what we are now missing is Aristotle's philosophy of care which saves the phenomenon and does not allow us to dispense with value and meaning for the sake of power and pragmatism. This overall framing transforms our world, even in respect to how we approach science. Whereas the science of Newton, Darwin, or Freud is itself anything but a problem, the attempt to reduce how we see the world to these scientific views upon the world most certainly is. The political and spiritual consequences of this philosophical problem is explained by Huxley and Orwell above. They are also of central concern to all the philosophers of our day who are worthy of the title.

9.1 Leibniz and the Yoke of Aristotle

Modernity is itself a problem or at least heralds an age of troubles and extremes. Its constant talk of a war between science and religion is merely the unwise endgame of Plato's Battle of the Gods and Giants from the fourth century B.C. This war would, of course, be wholly unnecessary if the various dimensions of human activity were kept to their proper spheres and in proper proportion. This is, of course, easier said than done though Aristotle does set out to achieve this very balance in his *Politics* and elsewhere. The principle here is not just that we need to avoid the extremes but that the extremes come to much the same thing *with the genuine alternative being the realistic and entangled alternative in the middle.*

Gottfried Leibniz shows us how it is the extremity of religion which reduces the world to the innate ideas of "God" that gives rise to the extremity of science and reduces the world to mathematical laws. Leibniz argues that to avoid, using Bacon's term, being "impertinent" to God, we *must* reduce the world to the hard laws of science. He hence argues that: "I don't assent to vulgar notions, that the images of things are conveyed by the organs (of sense) to the soul. For, it is not conceivable by what passage, or by what means of conveyance, these images can be carried from the organ to the soul. This vulgar notion in philosophy is not intelligible, as the new Cartesians have sufficiently shown. It cannot be explained how immaterial substance is affected by matter: and to maintain an unintelligible notion thereupon, is having recourse to the scholastic chimerical notion of I know not what inexplicable *species intentionales*, passing from the organs to the soul."[5] In order to avoid what

[5] Leibniz, Gottfried *The Leibniz-Clarke Correspondence* (Manchester, [*orig*. 1716] 1956) § 82–3. He continues that: "Those Cartesians saw the difficulty; but they could not explain it. They had recourse to a [very particular] concourse of God which would really be miraculous. But, I think, I have given the true solution to that enigma (*Ibid*. p 83)."

he calls "occult qualities,"[6] Leibniz denies the simple assumption common to Aristotle and modern science that we think by reference to the impressions we receive from our senses. Leibniz similarly denies the principle of gravity on the basis that it relies upon "action at a distance."

Leibniz's monodology is, of course, a profound and bold study of being and beings. It is, as we will see below, fundamentally Aristotelian in nature. Its fundamental problem is, however, in respect to the place of "God" within it. When push comes to shove, Leibniz argues that: "No primitive entelechy whatsoever can ever arise or be destroyed naturally, and no entelechy ever lacks an organic body. As far as my consideration of these matters goes, these things could not be otherwise; they are not derived from our ignorance of the formation of foetuses, but from higher principles."[7] Nature does not work through beings becoming but in accordance with "higher principles." These principles are mathematical and physical laws which in some sense represent the mind of "God." It is indeed upon this basis that the "positivism" of Auguste Comte becomes tenable that: "the fundamental character of positive philosophy is to regard all phenomena as subject to invariable natural laws the precise discovery of which and their reduction to the smallest possible number are the aim of all our efforts, with the search for what are called the causes, either primary or final, being absolutely inaccessible and meaningless for us."[8] The stress upon final causes being "inaccessible" reminds us, of course, of Bacon, Descartes, and Newton.

Leibniz was, however, the great iconoclast of his age because he found that he could not avoid the need to address Aristotle's "final causes." He argued on this point that: "There is an infinity of past and present shapes and motions that enter into the efficient cause of my present writing, and there is an infinity of small inclinations and dispositions of my soul, past and present, that enter into its final cause."[9] We can represent this causal situation through Fig. 9.1 below.

Leibniz pictures himself writing at his desk to explain the dynamics of the world. We see that the scholar has his own working space and tools, like the chef and shoemaker discussed above. He drives his own pathway through the world of knowledge, the noosphere, through his exertion and shapes himself as a scholar in the process through his power and object of concentration. This natural development or evolution of beings, powers, and activities is the reality that both Aristotle and

[6] Leibniz criticised the Aristotelian position which he argued: "preserved appearances by expressly fabricating suitable occult qualities or faculties, which were supposed to be like little demons or spirits able to do what was required of them out of hand ("New Essays on the Human Understanding" in *Philosophical Writings* (London, [*orig*. c. 1704] 1973) p 171)."

[7] Leibniz, Gottfried "Letter to de Volder 20 June 1703" *Philosophical Essays* (Indianapolis, 1989) p 176.

[8] Comte, Auguste *Cours de philosophie positive* (Paris, [*orig*. 1842] 1975) p 25–6.

[9] Leibniz, Gottfried "Monadology" in *Philosophical Essays* (Indianapolis, [*orig*. 1714] 1989) § 36 p 217.

Fig. 9.1 Images of scholars in their "world"

Leibniz seek to explain. Leibniz uses *his own mundane and extraordinary existence* to explain both his own monadology and also Aristotle's principles of causation and finality.

For Leibniz, then, as for Aristotle, the natural world is meaningful, full of meaning, and not meaningless. The world is a full, living world and not an empty, inert space. We cannot merely reduce the world to laws but must appreciate its lived reality. Leibniz explains this situation through his "conatus" which is a living principle or germ of being regarding which: "Dead forces perhaps always arise from living forces, as it appears from the fact that the conatus for receding from a centre, which ought to be counted among dead forces, arises from the living force of rotation."[10] Despite his caveats about Aristotle's supposed "barbarism,"[11] Leibniz reaches a similar result to the Philosopher since whatever his claims his "*conatus*" is effectively Aristotle's "*archē*."[12] His picturing of the world through the capturing and release of energy as centrifugal and centripetal forces also seems to look back to the idea that movements in the world can be split between circular and rectilinear movements, the circle and the straight line that we considered previously. Ivan Il'in explained on this point that: "Aristotle already pointed out that the problem of being is the problem of substantiality, and that the essence of substance lies in its distinctive centripetal self-activity. This conception of substance as something that is "per se" or "in se" was repeatedly developed by the scholastics and was then adopted by one of Hegel's most influential inspirers, Spinoza. "By substance," he [i.e., Spinoza] writes, "I understand that which is in itself and is conceived through itself.""[13]

---

[10] Leibniz, Gottfried "Letter to Johann Bernoulli, 18 November 1698" in *Philosophical Essays* (Indianapolis, 1989) p 169.

[11] Leibniz dismisses as "barbaric" a "return to chimeras, to *archae*, to certain plastic intelligences that attend to the formation of the foetus ("Against Barbaric Physics" in *Philosophical Essays* (Indianapolis, [*orig*. 1686] 1989) p 31)."

[12] On the "conatus" see Jasmin Özel & Nikos Psarros "Conatus, the Will to Persist: An Introduction" *Conatus* (2021).

[13] Il'in, Ivan *The Philosophy of Hegel as a Doctrine of the Concreteness of God and Humanity Vol. I* (Evanston, [*orig*. 1918] 2010) p 75.

Leibniz is the great precursor of Hegel and Heidegger in trying to rehabilitate and recover Aristotle. He explains himself on this point that: "At first, when I had freed myself from the yoke of Aristotle, I had believed in the void and atoms, for it is this which best satisfies the imagination. But returning to this view after much meditation, I perceived that it is impossible to find *the principles of a true unity* in matter alone, or in what is merely passive, since everything in it is but a collection or accumulation of parts *ad infinitum*. Now a multiplicity can be real only if it is made up of *true unities* which come from elsewhere and are altogether different from mathematical points, which are nothing but extremities of the extended and modifications out of which it is certain that nothing *continuous* could be compounded. Therefore, to find these *real unities*, I was constrained to have recourse to what might be called a *real and animated point* or to an atom of substance which must embrace some element of form or of activity in order to make a complete being…[regarding which] I found then that their nature consists of force and that from this there follows something analogous to feeling and to appetite; and that therefore it was necessary to form a conception of them resembling our ordinary notion of *souls*."[14] Leibniz the great mathematician rejects the idea that we can reduce the world to "mathematical points." The world rather needs to be "animated." I have discussed above Patrick Suppes' similar revelation in being able to stand back from the noise and appreciate what Aristotle is seeking to achieve through the character of his representation of nature.

I add that not only does Leibniz rehabilitate Aristotle's key concept of "entelechy" but that he also argues for two Aristotelian principles. He argues that: "Our reasonings are based on two great principles, [1] that of *contradiction*, in virtue of which we judge that which involves a contradiction to be false, and that which is opposed or contradictory to the false to be true"[15] with: "the principle of contradiction, [being] (what comes to the same) the principle of identity, as Aristotle has correctly pointed out."[16] Also, that: "And [2] that of sufficient reason, by virtue of which we consider that we can find no true or existent fact, no true assertion, without there being a sufficient reason why it is thus and not otherwise, although most of the time these reasons cannot be known to us."[17] This second argument of nature doing what is for the best, and doing nothing in vain, is recognisably "Aristotelian" in nature. Martin Heidegger expands upon this principle "nothing is without reason" in his *The Principle of Reason* (1956); from an Aristotelian perspective "being" must come both with its own "*logos*" and its own "good(s)." The point at stake here is whether finding meaning in the world, and not assuming that meaning is only a human mirage, is Panglossian or merely an acceptance of lived reality.

---

[14] Leibniz, G.W.F. "New System, and Explanation of the New System" in *ed.* G.H.R. Parkinson *Philosophical Writings* (London, [*orig*. 1695–6] 1973) p116–7.

[15] Leibniz, Gottfried "Monadology" (1714) § 31 in *Philosophical Essays* (Indianapolis, 1989) p 217.

[16] Leibniz, Gottfried "Critical Remarks Concerning the General Part of Descartes' Principles" *Monadology and Other Philosophical Essays* (Indianapolis, [*orig*. 1692] 1965) p 25.

[17] Leibniz, Gottfried "Monadology" (1714) § 32 in *Philosophical Essays* (Indianapolis, 1989) p 217.

9.2 Hegel's Premonition of "The Thingliness of the Thing"

Leibniz had the strength of mind to buck the fashion of his day, and to some degree its prevailing ideology, and to question whether we can indeed reduce the world to our own ideas or those of "God." He underpins the counter-tradition to Cartesian dualism which runs throughout modern European thought. I will show below that this counter-tradition is anything but subordinat. I will also contend that the philosophy that has come to be taken as "normal"—what A.N. Whitehead refers to as "the Hume-Newton situation [which is]...the primary presupposition for all modern philosophic thought. Any endeavour to go behind it is, in philosophic discussion, almost angrily rejected as unintelligible"[18]—is highly abnormal and utterly unsatisfactory.

Let us on this basis revisit Aristotle's conceptualisation of "*ousia*" from a Hegelian perspective. Starting with a "stone" Hegel argues that: "..."cause" and "effect" have a meaning simply and solely in terms of their relationship. The cause does not extend beyond the fact that it has an effect. The stone that falls has the effect of producing an impression on the object upon which it falls. Apart from this effect, which it has as a *heavy* body, it is physically particular and distinct from other similarly heavy bodies. Or, since it is a cause while it continues to produce this impression, we take its effect to be transitory, for example; then when it strikes against another body it ceases to this extent to be a cause, and outside of this relation it is just a stone, which it was before. This haunts representational thinking chiefly insofar as it characterises the cause as the "original thing" that continues to exist outside of its effect. Apart from its effect, the stone is undoubtedly a stone, but it is not a cause. It is a cause only in its effect, or, to introduce the category of time, *during* its effect."[19] Hegel stresses that the "stone" is more than its effect(s) or properties; it also has its own "being" (*ousia*) and its own "nature" (*physis*).[20]

Aristotle himself uses the stone as an example of an "*ousia*" or "*pragma*" of which we possess a "form" or image in our soul whilst we do not, of course, possess the "stone" itself there. Hegel explains this "objectification" of being as that: "man is not content with a bare acquaintance, or with the fact as it appears to the senses; he would like to get behind the surface, to know what it is, and to comprehend it. This leads him to reflect: he seeks to find out the cause as something distinct from the mere phenomenon: he tries to know the inside in its distinction from the outside. Hence the phenomenon becomes double, it splits into inside and outside, into force

[18] Whitehead, A.N. "Nature Lifeless" in *Modes of Thought* (New York, 1938) p 135.

[19] Hegel, G.W.F. *Lectures on the Proofs of the Existence of God,* (1829) §313.

[20] Michel Bastit argues that: "The stone that breaks the glass has a hardness and energy that the window does not have. This asymmetry is not only physical asymmetry, as in the case of energy transfer, but also ontological. It can be described as an active information transfer (*Le principe du monde: Le Dieu du philosophe* (Paris, 2016) p 183)."

and its manifestation, into cause and effect."[21] Hegel shows how our desire to understand and to occupy a worldly phenomenon leads us to deconstruct that phenomenon. This is, of course, a valuable deconstruction if we are willing to marry analysis with synthesis, deconstruction with reconstruction, whilst always recognising that the original being is itself something different from our mental construction of it.

Hegel correctly observes that the "being" (*ousia*) is ignored by modern analytical thought and he attributes the problem to Kant, arguing that: "The divorce between thought and thing is mainly the work of the Critical Philosophy, and runs counter to the conviction of all previous ages, that their agreement was a matter of course. The antithesis between them is the hinge on which modern philosophy turns. Meanwhile the natural belief of men gives the lie to it. In common life we reflect, without particularly reminding ourselves that this is the process of arriving at the truth, and *we think without hesitation, and in the firm belief that thought coincides with thing* [my italics]."[22] Hegel argues that we possess "the firm belief that thought coincides with the thing" and Aristotle also argues that we must indeed have the "form" of a thing "in our soul." What, on this basis, is the ontology of a stone? It is important to recognise that the "*ousiai*" of Aristotle's ousiology are not just organisms but also "beings." Indeed, the "sorites problem" is as much about the nature of the heap of sand as it is about the (other examples raised in respect to the) nature of man or the nature of the *Iliad*. Aristotle also uses the "stone" as a paradigmatic example of an elemental being from which "beings" (*ousiai*) or "things" (*pragmata*), e.g. houses, can be built.

What, then, *is* the stone in the world? The stone is, first, an example of an elemental "solid" which moves as a solid mass. The stone is, second, something in possession of the potentialities of a "solid" that enables it to be used as building materials, as a "boundary stone," a "foundation stone," and a "threshold," and also to be a material from which a statue can be shaped etc.[23] Aristotle hence ultimately argues that: "what we really mean by motion or change may be shown thus: build-

---

[21] Hegel, G.W.F. Hegel's *Logic* (Oxford, [*orig.* 1830] 1975) § 24. Kant also argues that an organism is "both cause and effect of itself (Critique of Judgement (1790) 64: 249)."

[22] Hegel, G.W.F. *Ibid.* § 24. On Hegel's *Logic* see Wendell Kisner "The Category of Life, Mechanistic reduction, and the uniqueness of Biology" *Cosmos and History* (2008).

[23] I note that Aristotle repurposes the meanings of everyday life such as "*ousia*" which meant "possessions" before "being" and he redeploys "*horos*" or boundary stone to mean "boundary"; see *E.E.* VIII 1249a 21–1249b 6. Stone was used as the material of meaning in Greece from public inscriptions to funeral monuments and statues; see *Phys.* I 188b13-21 & *Phys.* I 190b 5-11. Stone as a building material also frames our human lives so that Aristotle draws out the principle that: "that which produces the part of health, is the limiting-point (*eschaton*),—and so too with a house (the stones are the limiting-point here) and in all other cases (*Met.* Z 1032b 28-29)." He argues that the "being" of the house which we construct artificially is not a heap but a building with a stone structuring. The house is a fusion of matter and form; see *Met.* H 1043a15-21 & *PA* II 646a25-29. A house imitates the beings of nature by being physically and functionally structured; see *N.E.* X 1174a 19-27. The stone also acts when it is in the right place as a threshold for a house; see *Met.* H 1042b16-25.

ing material is actualising the potentialities in virtue of which we call it "building material" when it is in the act of being built into a structure, and this act is the process or "movement" of "building"; and so too with other processes—learning, healing, rolling, jumping, maturing, ageing (*Phys.* III 201a16-19)." Building and its materials are used as a paradigm that can be applied to a wide range of other processes. Building materials possess a quality of buildable-ness which in some way suggests the building-art and the building. There is an ontological and real-world relationship between the self of the human being and the objects and materials of his physical environment which enables him to live within the ecosystem ("*eco-*" being from "*oikos*" or home) of the world.

We most obviously perceive here Martin Heidegger's philosophical emphasis upon "the thingliness of the thing."[24] Heidegger explains this "thingliness" most famously by reference to a "jug" and to Vincent van Gogh's "peasant shoes." He argues in introductory outline, however, that: "The stone in the road is a thing, as is the clod in the field. A jug is a thing, as is the well beside the road. But what about the milk in the jug and the water in the well? These too are things if the cloud in the sky and the thistle in the field, the leaf in the autumn breeze and the hawk over the wood, are rightly called by the name of thing."[25] Heidegger gives us a stone/road, clod/field, well/road, milk/jug, cloud/sky, thistle/field, leaf/breeze, hawk/wood; he regards not only the "thing" but also its natural place or, as Aristotle would say, its natural "medium." I note that Jakob von Uexküll also argues in the field of biosemiotics that the "stone" is meaningful through its qualities as having a "path-quality" (*Weg-Ton*) and a "throw-quality" (*Wurf-Ton*).[26]

9.3 Hegel and the *Scala Naturae*

How do we meaningfully and logically get from the stone to the human being? Hegel argues that the earth is an objective, external, and dead "crystal of life," that such things as mould and mushrooms represent a transitional "inorganic-organic" state, that the plant represents an organism in which the "parts" possess only a loose and inorganic relation to the whole, and finally that a "true" organism is one in which the parts relate to the whole as "members."[27] This *scala naturae* is, of course, thoroughly Aristotelian with Aristotle being concerned in his work precisely with

[24] See Martin Heidegger "The Thing" in *Poetry, Language, Thought* (New York, [*orig.* 1950] 1971) & "The Origin of the Work of Art" in *Off the Beaten Track* (Cambridge: CUP [*orig.* 1935–6] 2002) (also see Paul Feyerabend "Art as a Product of Nature as a Work of Art" in *Conquest of Abundance: A Tale of Abstraction versus the Richness of Being* (Chicago, 1999)).

[25] Heidegger, Martin "The Origin of the Work of Art" in *Off the Beaten Track* (Cambridge: CUP [*orig.* 1935–6] 2002) p 20.

[26] Uexküll, Jakob von "The Theory of Meaning" *Semiotica* ([*orig.* 1940] 1982) p 27.

[27] Il'in, Ivan *The Philosophy of Hegel as a Doctrine of the Concreteness of God and Humanity Vol. I* (Evanston, [*orig.* 1918] 2010) p 238.

this *concretion* of "being" from sponges, plants, insects, to human beings. Aristotle observes that as beings become more sophisticated, they obtain an "organic unity" (*symphysis*) and a "growing together" (*sympephykenai*) (*Met.* Δ 1014b 23-25) which makes them more concentrated, organised, and having integrity. To this end Aristotle observes not only that plants are of a decentralised nature which can be easily cut up and grafted but also that tortoises (apparently) are of a nature to allow their dissection without immediate death; see *Resp.* 478b 33–479a 7. He implicitly argues that as "things" move to become "beings" (*ousiai*) from a heap of sand to a human being, they gain not only in greater mental projection but also in greater physical cohesiveness.

We see the remarkable perspicacity of Hegel's thought and the quality of his scientific speculation from his reference to the "crystal of life" which predicts the increasingly evident modern insight that living beings are in some sense "aperiodic crystals" or "liquid crystals."[28] Hegel also raises the potential importance of neoteny as a juvenile plasticity which provides humanity with the weakness to overcome its bodily limitations.[29] Aristotle also famously observes on this point that human beings are the weakest animals in nature but that they make up for their weakness by the plasticity of their natures, including their use of tools including language and mind. They also have no fixed mating season and are omnivorous. They are animals that can swing between the un-natural extremes of "beast" and "god."[30] It is in this sense that Aristotle argues that human beings are from birth a "blank slate" needing to be filled, with it being this very gap that allows human beings in some sense to tap into some deeper well of nature and have some access to its "*nous*."

I represent the human power of reflection and how it transforms an animal into a human world through Fig. 9.2 below.

Hegel explains this situation in detail as that: "Animal, qua animal, does not exist: it is merely the universal nature of the individual animals, while each existing animal is a more concretely defined and particularized thing. But to be an animal—the law of kind which is the universal in this case—is the property of the particular animal, and constitutes its definite essence. Take away from the dog its animality, and it becomes impossible to say what it is. All things have a permanent inward nature, as well as an outward existence. They live and die, arise and pass away. [But M]an is a thinker, and is universal; but he is a thinker only because he feels his own

---

[28] Erwin Schrödinger famously posited in his "*What is Life?*" that organisms could be "aperiodic crystals" and see also Mae-Wan Ho "Organism and Psyche in a Participatory Universe" in D. Loye (*ed.*) *The Evolutionary Outsider: The Impact of the Human Agent on Evolution, Essays in Honour of Ervin Laszlo* (Praeger, 1998) and P.W. Anderson "More is Different: Broken symmetry and the nature of the hierarchical structure of science" *Science* (1972).

[29] See Emmanuel Chaput "Hegel and Bichat on Symmetry and Life: Between Physiology, Aesthetics, and Freedom" *Cosmos and History* (2021).

[30] Aristotle explains that: "he who is unable to live in society, or who has no need because he is sufficient for himself, must be either a beast or a god: he is no part of a state. A social instinct is implanted in all men by nature, and yet he who first founded the state was the greatest of benefactors. For man, when perfected, is the best of animals, but, when separated from law and justice, he is the worst of all (*Pol.* I 1253a 28-33)."

Fig. 9.2 The animal and human sense of "self"

universality. The animal too is by implication universal, but the universal is not consciously felt by it to be universal: it feels only the individual. The animal sees a singular object, for instance, its food, or a man. For the animal all this never goes beyond an individual thing. Similarly, sensation has to do with nothing but singulars, such as this pain or this sweet taste. Nature does not bring its νους [*nous*] into consciousness: it is man who first makes himself double so as to be a universal for a universal. This first happens when man knows that he is 'I'. By the term 'I' I mean myself, a single and altogether determinate person."[31] Hegel argues that it is through "ego" and its awareness that human beings transform all their activities through the power of their reflective thought to enter in to a dialogue with nature and transform it in their image.

This is how life rises to humanity but how also, let us ask, does nature rise up to humanity according to Hegel? Hegel explains that: "The world comprises within itself human beings as much as natural things. When the world is taken as this aggregate, even only of natural things, it is not represented as *nature*, by which we understand something that is inwardly a systematic whole, a system of arrangements and levels, and especially of laws. The term "world" merely expresses the aggregate in a way that, resting as it does simply on the existing mass of things, it has no advantage, at least no qualitative advantage, over worldly things."[32] The critical distinction here according to Hegel is that we must choose between seeing the world before us *either* (i) as *an "aggregate" of things in the "world"* and (ii) as a *"system" of "nature."* He explains this organic nature further as that: "the necessity of something requires two sorts of opposed qualities: its independence, in which, however, it is isolated and whether it is or not a matter of indifference; and its being based on and contained in a complete relation to everything else by which it is surrounded and through the nexus of which it is sustained, so that it is not independent."[33]

[31] Hegel, G.W.F. Hegel's *Logic* (Oxford, [*orig*. 1830] 1975) § 24.

[32] Hegel, G.W.F. *Lectures on the Proofs of the Existence of God* (1829) §274.

[33] Hegel, G.W.F. *Ibid.* §284.

Any "*ousia*" must, on this basis, be something with a viable independent being and yet also fit into and express the world around it.

As regards where Hegel goes *beyond* Aristotle, we find that he observes that: "Just as the totality of life is contained to an equal degree in the nature of the polyp and the nature of the nightingale and the lion, so too the world spirit revealed in each shape (*Gestalt*) its absolute self feeling (*Selbstgefühl*)—be it more obscure or more developed—and enjoyed its own essence and its own self in each nation, in each ensemble of mores and laws."[34] Whereas Aristotle happily and reasonably travels from physics and biology to human politics, Hegel's objective is to move from the culture of biology to the culture of religion. His ultimate conclusion on the "world spirit" is that: "Until true religion has entered the world and is made supreme in the state, the true principle of the state cannot be realised in actuality."[35] Hegel clearly seeks to move through a single progression or evolution of "spirit" which takes us from a "culture" of mould growth to the highest human "culture" which is "true religion." Everything in between is a phenomenology of Spirit. Aristotle would agree with the idea that everything has meaning but would surely object to the desire to locate meaning in another world than our own, a speculation that he clearly regarded as unsupported and distracting.

9.4 Hegel Against the Hume-Newton(-Darwin) Situation

Hegel agrees with Aristotle on the essential point regarding the significance of nutrition that the organism is not simply produced *by* matter but is a "being" (*ousia*) that extracts matter *from* the world to feed (and become) *its* constitution according to its *needs*. A being is not produced by matter but is an expression of nature by being enmattered according to its nature within nature. We could picture this scientific, philosophical, and theological situation as per Fig. 9.3 below:

We consider in the above mentioned diagram the comparison between Hegel's conception of a "circular" system of "beings" as *evolved* forms of "nature" and Hume's idea that the "world" arises through "chance" and material cause-and-effect. We can represent Hume here through his argument that: "upon the whole, there appears not, throughout all nature, any one instance of connexion which is conceivable by us. All events seem entirely loose and separate. One event follows another; but we never can observe any tie between them. They seem *conjoined* but never *connected*…the necessary conclusion *seems* to be that we have no idea of connexion or power at all, and that these words are absolutely without any meaning, when employed either in philosophical reasonings or common life."[36] The

---

[34] Il'in, Ivan *The Philosophy of Hegel as a Doctrine of the Concreteness of God and Humanity Vol. I* (Evanston, [*orig*. 1918] 2010) p 206–7.

[35] Il'in, Ivan *The Philosophy of Hegel as a Doctrine of the Concreteness of God and Humanity Vol. II* (Evanston, [*orig*. 1918] 2010) p 179.

[36] Hume, David *An Enquiry Concerning Human Understanding* (1757) §58.

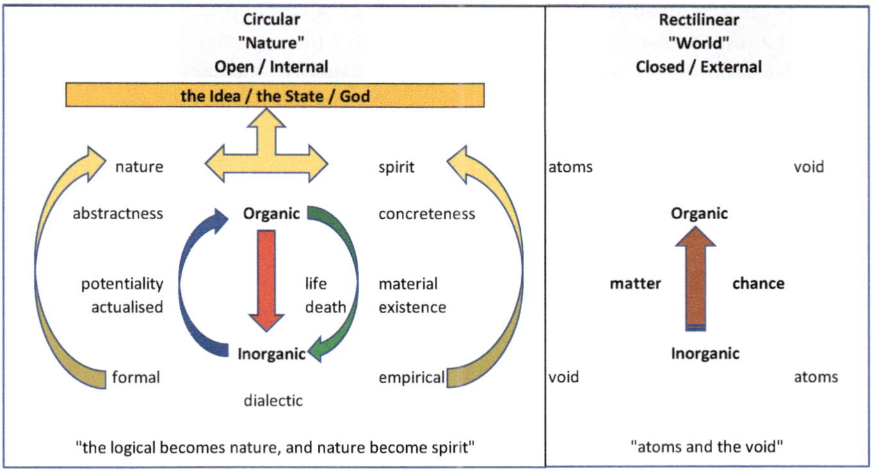

Fig. 9.3 Hegel's nature vs. Hume's world

difference between an empirical and a metaphysical approach should be evident from this, and their relative merits.

I suggest that the philosophical conflict exposed above is between one vision of the world based upon the chance history of "atoms and the void" and another based upon the expressed meaning of a plenum populated by entangled quanta. Hegel effectively invents the concept of "open" and "closed" systems to explain this distinction. He explains that: "Everything organic is in an active relation to inorganic nature, which is out there as something independent. In one respect the organism is infinite, since it forms a circuit of pure return into itself, but there is at the same time a tension between it and the external inorganic nature—it has *needs*. The means for meeting these needs come from outside; human beings need air, light, and water; and they also consume animals and other living things, which they thereby make into inorganic matter, as a means."[37] It is notable that this "biological" explanation occurs in his *Lectures on Religion*; how we philosophically imagine our biological state informs how we theologically approach the world.

Hegel clearly understood the conflict of worldviews we still face today a generation before the publication of Darwin's *Origin of Species* in 1859. He hence observed in 1831 that there was (as there still is today) a clash between two *Weltanschauungen* premised respectively (i) upon the hypothesis that the world can be understood *as a product of a historical chain of events* in the sense that: "what we call inorganic nature as such is represented on its own account as self-subsistent, while the organic is represented as an external accretion, so that it is a matter of chance whether it finds in what stands over against it the necessary conditions for its existence"[38] and (ii) upon the hypothesis that *the things of the world need to be considered as parts*

[37] Hegel, G.W.F. *Lectures on the Philosophy of Religion* (1831) §600.

[38] Hegel, G.W.F. *Ibid.* §602.

of nature in the sense that: "humanity is not an accidental accretion upon what comes first; on the contrary, what comes first is, for the organic, itself; and the inorganic only has a semblance of being in regard to it."[39] In the second variant we are returned to Leibniz's idea of life as an impulse, the conatus, and the material world as its residue.

We again see the conflict between philosophy and anti-philosophy as outlined in Chap. 2 of this work. The biologist Theodosius Dobzhansky expresses this situation particularly well in his address "Two Contrasting World Views" (1974).[40]

9.5 Heidegger and the Nature of "World"

Let us move from Hegel to Heidegger. Martin Heidegger's lecture series *The Fundamental Concepts of Metaphysics: World, Finitude, Solitude* from 1928–9 is utterly remarkable as a consideration of "metaphysics" from the perspective of world ("*Welt*"), finitude (separation, limits), and solitude (self, ego). Heidegger's work is so profoundly attuned to Aristotle's that his consideration of "metaphysics" does not merely conclude with Aristotle's *Metaphysics* but rather draws out the metaphysics inherent in Aristotle's semiotics (using the *On Interpretation* as the source text) alongside a consideration of the philosophical basis of biology. Heidegger essentially defines Aristotle the metaphysician as a biosemiotician.

In these lectures Heidegger expounds a philosophical consideration of "being" and "life" by reference to the increase in, or evolution of, world ("*Welt*") in beings (*ousiai*) as represented in Fig. 9.4 below:

The above mentioned diagram shows us Heidegger's three chosen examples of "being"—the stone, the animal, and the human being—through which he posits that: "the stone is worldless, the animal poor in world, and man world-forming."[41]

Even regarding this framing of the problem I observe that the *analogical* method (from "*ana*" + "*logos*") which works by comparing paradigms is quintessentially Aristotelian, e.g., his sorites problem compares a heap of sand, a human being, and the *Iliad*.[42] The *scala naturae* is also quintessentially Aristotelian with Heidegger

[39] Hegel, G.W.F. *Ibid.*, §603.

[40] Dobzhansky, Theodosius "Two Contrasting World Views" in Lewis, John (*ed.*) *Beyond Chance and Necessity* (London, 1974).

[41] Heidegger, Martin *The Fundamental Concepts of Metaphysics* (Bloomington, [*orig*. 1929–30] 1995) p 185.

[42] Aristotle uses both analogy ("*ana*" + "*logia*") and metaphor ("*meta*" + "*phora*") in a highly technical and also in a literal or original sense. Pierre Aubenque explains on analogy that: "The notion of analogy intervenes in the same context as the doctrine of the multiple meanings of being. But analogy is nowhere invoked to unify these meanings. On the contrary, its radical nature presupposes rather than attenuating or correcting the positive plurality of the meanings of being since only the irreducible character of this plurality requires the recourse to analogy to show a basic unity—albeit "general"—in our discourse on principles and causes ("Les Origines de la Doctrine de l'analogie de l'être: Sur l'histoire d'un contresens" *Les Études philosophiques* (1978) p 6)." On

Fig. 9.4 Heidegger's categorisation of "ousia"

explicitly referring to the scale of life—inorganic object, plant, animal, man—put forward by Aristotle most philosophically in his *On the Soul.* I have additionally included Jakob von Uexküll's "tick" in the above picture on the grounds that Heidegger cites the influence of this thinker on his own thought in this work. I have also used Leonardo da Vinci's drawing of "Vitruvian Man" since it symbolises the peculiar proportionality and nature of the human being. This consideration of the structure and position of human being could be said to go back to Aristotle who concentrated on the idea of the human being as an upright framing of being facing the world, seeing it, and using his hands and head to engage with it. I suggest that we should consider these positionalities of being (*ousia*) in the world by comparison to the "moments" of human approach to the world set out as Foucault's "moments" in Fig. 1.1 of this work.

I suggest that we see above a *real* questioning of the meaning and nature of evolution from stone to man. This is what Heidegger puts forward as "metaphysics." In respect to the stone, Heidegger comments that: "A stone either has a property or does not have it. We, on the contrary, can have something and at the same time not have it, that is, not know of it. We speak, after all, of the unconscious."[43] In respect to the animal, in this case a "tick," he comments that: "The animal does have access to beings…[but] the character of the world remains underdetermined here [and] captivation is the condition of the possibility of poverty in world….as the essence of animality."[44] He adds, following Uexküll, that: "Every animal and every species of animal actively encircles itself in its own way with this encircling ring, with which it circumscribes and adapts itself to a certain domain. The ring that encircles the sea-urchin is quite different from that of a bee, and that of a bee quite different again

"metaphor" as a transference of meaning see *Rhet.* II 1405b 5-6 & *Poet.* 1457b ff. and Sara Newman "Aristotle's Notion of "Bringing-Before-The-Eyes": Its Contributions to Aristotelian and Contemporary Conceptualizations of Metaphor, Style, and Audience" *Rhetorica: A Journal of the History of Rhetoric* (2002).

[43] Heidegger, Martin *The Fundamental Concepts of Metaphysics* (Bloomington, [*orig.* 1929–30] 1995) p 61.

[44] Heidegger, Martin *Ibid.* p 269–271.

from that of the great tit, and this different from the squirrel and so on."[45] This is the basis of Uexküll's famous reflection on the space or place of a being, its environment, "encircling ring" or "functional circle," or "*Umwelt*."

The stone lacks reflection. The animal has access to the world but is *passively* captured by it in an "encircling ring." As regards the *actively* "world-forming" nature of the human being Heidegger explains that his world does not externally encircle him but arises in some way internally *from* him on the basis that: "man exists in a peculiar way *in the midst* of beings. In the midst of beings means: living nature holds us ourselves captive as human beings in a quite specific way, not on the basis of any particular influence or impression that nature exerts or makes on us, but rather from out of our essence, whether we experience that essence in an originary relationship or not."[46] He explains further in his difficult but profound way that: "The events which man experiences in himself: procreation, birth, childhood, maturing, aging, death, are not events in the narrow, present-day sense of a specifically biological process of nature. Rather, they belong to the prevailing of beings, which comprehends within itself human fate and its history."[47] Heidegger echoes Hegel when he concludes that: "Man is history, or better, history is man."[48]

The human being is peculiar by being "world forming" in the sense that he truly forms his own world, self, and history. This seems on its surface to be a mundane and obvious point but with these supposedly mundane facts actually being set forth as "fundamental concepts of metaphysics." We find another such obvious statement of reality at the beginning of Aristotle's *Politics* as that: "man, when perfected, is the best of animals, but when separated from law and justice, he is the worst of all; since armed injustice is the more dangerous, and he is equipped at birth with arms, meant to be used by intelligence and excellence, which he may use for the worst ends. That is why, if he has not excellence, he is the most unholy and the most savage of animals, and the most full of lust and gluttony. But justice is the bond of men in states; for the administration of justice, which is the determination of what is just, is the principle of order in political society (*Pol.* I 1253a 31-39)." Aristotle stresses the unformedness and extremity of human nature ("*physis*") which must be corrected and shaped through artificial human civilisation ("*nomos*"). His *Ethics* and *Poetics* also explain quite how a human being *forms* his own world.

Heidegger followed Aristotle's recognition that man is by nature, through his full(er) adaptation to a life based on thought, not just one animal of many but also a wholly different genus of being.[49] I suggest that the lazy charges of speciesism and

[45] Heidegger, Martin *Ibid.* p 277.

[46] Heidegger, Martin *Ibid.* p 278.

[47] Heidegger, Martin *Ibid.* p 26.

[48] Heidegger, Martin *Ibid.* p 365.

[49] Aristotle does, of course, argue that man is an animal in an immediate sense whereas he is a plant or a mineral in an indirect or derivative sense. He argues in respect to this scale of being that the ape "dualises" between man and quadraped in part in respect to their standing positionality and physical placement but also in part because it is related to both. The ape dualises in form between the two on the basis that: "The ape is, in form (*morphē*), intermediate (*epamphoterizein*) between

anthropocentrism often pushed around as regards this distinction are spurious since we do indeed see clear and obvious evidence that man is fitted to live by the mind in a generically peculiar way with this position being supported by modern science.[50]

9.6 Heidegger and the Symbolism of Logos

Heidegger argues (following Aristotle) that man is world-forming not only on the bodily basis of his upright stance but also on the mental basis of his power of "*logos*" which Heidegger represents as follows:

Heidegger's own title for the diagram reproduced in Fig. 9.5 is "The pointing out (*apophansis*) pertaining to assertion as letting beings be seen as what and how they are." What may seem obtuse and hard to understand is on closer inspection profound and informative. We find here a precise interpretation of how we do as human beings engage with the processes of reality.

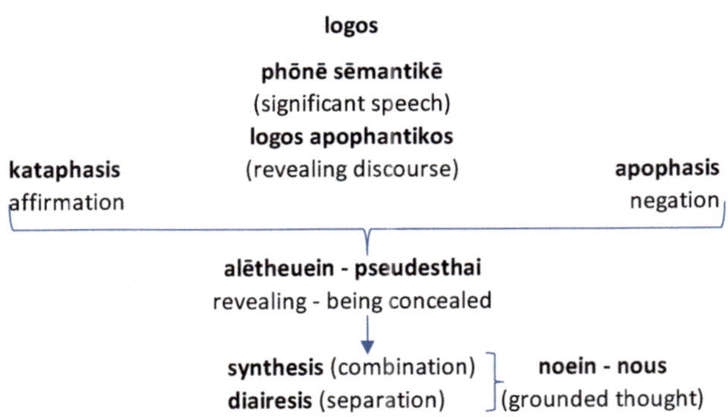

Fig. 9.5 Heidegger's pathway from "logos" to "nous"

---

the two, man and quadruped, and belongs to neither (*mēdeterōn*), or to both (*amphoterōn*), and consequently he has no tail, qua biped, and no buttocks, qua quadruped (*PA* IV 689b33–36)." Although he recognises the complexity of this situation in which man is an animal with a character shared with other animals, he also argues that human beings possess a peculiarity which makes them a separate genus of being. Aristotle argues that human beings are both a separate genus and hence: "there is always a genos of men, of animals, of plants (*GA* II 732a 1)" and in other places he classifies man as another type of animal; see for example *HA* I 490b 31–34.

[50] Washburn & Howell comment on the massive difference between man and ape that: "the uniqueness of modern man is…the result of a technical-social life which tripled the size of the brain, reduced the face, and modified many of the structures of the body" and that: "the large size of the brain of certain hominids was a relatively late development due to new selection pressures after bipedalism and consequent upon the use of tools" ("Human Evolution and Culture" in S. Tax (*ed.*) *The Evolution of Man* (Chicago, 1960) p 49 ff.). I note that human motion may superficially seem the same as that of other primates but that it is significantly different. For a philosophical consideration of the "upright" man see Erwin W. Straus "The Upright Posture" *The Psychiatric Quarterly* (1952).

Heidegger's above framing is drawn from a close and productive reading of Aristotle's *On Interpretation* in which Aristotle explains that: "As at times there are thoughts in our minds unaccompanied by truth and by falsity, while there are others at times that have necessarily one or the other, so also it is in our speech, for combination and division (*synthesin kai diairesin*) are essential before you can have truth and falsity (*to pseudos kai to alēthes*) (*De Int.* 16a 10-13)" He explains further that: "A sentence (*logos*) is significant speech (*phōnē sēmantikē*), of which this or that part may have meaning—as something, that is, that is uttered but not as expressing a judgement of a positive or negative character (*hōs kataphasis ē apophasis*) (*De Int.* 16b 27-28)." We see not just empty speculation or sophistry about the nature of man but a close scientific analysis of that which sets him apart from other natural objects. There is a twofold difference; the detached logic which allows the recognition of true and false statements and the further reflective "significant" thought and speech which finds qualitative depths of meaning beyond truth and falsehood.

Aristotle's most famous comment on the human being as a peculiarly logical being—as holding "*logos*" as something between thought, speech, and direction—is found at the beginning of his *Politics*. Heidegger observes, however, that it is his *On Interpretation* that Aristotle explicitly explores how human beings represent the world through language. He picks up here upon Aristotle's emphasis upon the power of negation, of being "capable of understanding what is *not*,"[51] as the remarkable basis through which the human being can reveal and understand what *is*. This duality of being—of being and not-being—is Aristotle's development of the principle of diairesis into a distinction that can provide a mechanism for us to engage with the world. There is, as Hegel noted, something destructive and delusive about this ability to analyse the world, though it is precisely this access to objective detachment that allows us to affirm or reject facts, ascertain truth or falsehood, and to separate out (analysis) or synthesise (synthesis) reality in our own minds. It is this power that allows us as human beings to imitate nature in ourselves in a way not open to other animals. I note that Aristotle's "*apophansis*" (from "*phaos*"; light) is "categorical" and hence bears reference to Aristotle's "categories" which are (originally legal) assertions made which shed light on and reveal the truth.

Heidegger ultimately argues, following Aristotle, that human language represents a full reflective engagement with the world which is more than merely living within a bubble in the world. Human beings have the detachment to have a worldview rather than merely living within the "world." This is ultimately *reflective* reason or "*nous*" which is famously and brilliantly explained by Aristotle as "thought thinking itself."[52] *Logos* is, in full, the meaningful direction or movement inherent

[51] Heidegger, Martin *The Fundamental Concepts of Metaphysics* (Bloomington, [*orig*. 1929–30] 1995) p 299.

[52] Aristotle gives us the sophisticated and detailed philosophical account of thought as that: "mind thinks itself (*auton noei ho nous*) by sharing in the object of thought (*kata metalepsin tou noētou*), for it becomes an object of thought by coming into contact with and thinking its objects (*noētos gar gignetai thinganōn kai noōn*), so that mind and the object of thought are the same (*tauton nous kai noēton*). For the mind is that which is capable of receiving the object of thought, i.e., the

in things as their moving force which can also be channelled by human beings as "mind" (*nous*) and "language" (*logos*) so that it can be abstracted or detached from nature and reinvented as a parallel human world. Aristotle hence uses "*dianoia*" as the "mind" which is common with animals as how organisms "think" and move themselves *through* (*dia-*) the world as opposed to "*nous*" which is a higher power of dispassionate reason accessible only to human beings.

Heidegger explains regarding the peculiarity of human being that: "There is language only in the case of a being that by its essence *transcends*."[53] Heidegger here follows Aristotle's emphasis upon the fact that human beings are meaning-making because they are "capable of learning grammar" (*Top*. I 102a 20-22). Language and the structured and active thought that it projects is not a fluke of nature but is rather a feature and expression of human being as the gliding flight qualifies the eagle and the web making qualifies the spider.[54] As regards the nature of human "*logos*" Heidegger explains that: "Aristotle states: The *logos* is not *physei*, is not some product of a physical event or process; it is not anything like digestion or the circulation of the blood, but has its genesis in something quite different: not *physei*, but *kata syntheken*"[55] and that: "What does *semantikos* mean? Aristotle says: This event is not a natural occurrence, but something that happens on the basis of an *agreement*."[56] What is this "agreement"? Heidegger argues (following Aristotle) that: "the word, indeed every word, is *hotan genetai symbolon*, whenever a *symbol* occurs."[57] Both thinkers agree that the world is (for those who can see it) symbolic in nature—a coming together of two sides on the basis that: "Whatever is held together, fits together so that the two parts prove to belong together, is *symbolon*."[58]

The human world is obviously symbolic in nature even in its most mundane aspects with the critical question being how far we can read back from the human world to the natural world itself. We have already seen on this point that the symbol, the "throwing together" of two sides, is at the very core of Aristotle's thinking as the

---

essence (*to gar dektikon tou noētou kai tēs ousias nous*); and it is active when it possesses the object (*energei de echōn*) (*Met*. Λ 1072b19-24)." I comment that this account covers all thought, e.g., a lion seeking its dinner, as "*dianoia*," but that we see a different dimension emerge in respect to the nature of human thought as "*nous*."

[53] Heidegger, Martin *The Fundamental Concepts of Metaphysics* (Bloomington, [*orig*. 1929–30] 1995) p 308.

[54] Aristotle uses "sight" as a paradigm for potentiality and actuality with his view being that sight is not just something that happens but, rather, that it is a living expression of the engagement of the seeing organism with the world. It is through human "sight" and then through the human "soul" that we can actually engage with "forms" in the world and then reflect upon them. Aristotle develops Plato's question: "which answer is more correct, that our eyes are that by which we see or that through which we see (*Theaetetus* 184C)" and argues that the *physical* structure of being is clearly derived from the *spiritual* needs of the being and not the reverse; see *PA* I 640a 20–640b 4.

[55] Heidegger, Martin *The Fundamental Concepts of Metaphysics* (Bloomington, [*orig*. 1929–30] 1995) p 309.

[56] Heidegger, Martin *Ibid*. p 311.

[57] Heidegger, Martin *Ibid*. p 307.

[58] Heidegger, Martin *Ibid*. p 307.

expression of potentiality becoming actuality. The rejection of this symbolism and this relationship is actually an attack on "life" and "meaning" by philosophers who wish to extirpate these things with Martin Heidegger commenting on this situation that: "we still have the feeling that violence has long been done to the thingly element of things and that thought has played a part in this violence, for which reason people disavow thought instead of taking pains to make it more thoughtful."[59] People throw the baby—in this case thought, meaning, and being—out with the bathwater. We find, however, something more than simple neglect. We find a turning of mind against its human owner which is, as Aristotle noted, most obvious in respect to money ("*nomisma*" from "*nomos*"; see *N.E.* V 1133a 31) which can transform into an end from being a means if we do not carefully restrain its power.

Aristotle argues that our goal in life should be to first come to a position of intellectual detachment, to then understand the limitations as well as the powers of this detachment, and to then restore, reintegrate, and reinvest this power of mind in our own humanity. This is also perhaps the subtext of Hegel's "dialectic."[60] This conflict between abstraction (technology) and life (humanity) is the great philosophical battleground of our day fought over by technologists and humanists.

9.7 The War on "Being"

Martin Heidegger concludes his thoughts on the fundamental concepts of metaphysics by defending the basic principle of metaphysics (and ontology) that something "is" in the context of the insistent claim of modern philosophy to be able to "get by without being!"[61] Heidegger makes the point that "being" itself—which he calls "*Dasein*"—is denied and under attack and that we need to understand what this attack on existence itself *means* and what its consequences are.

Who on earth would want to get by without being? Our opponents in this new battle are not primarily the philosophers who seek to "deconstruct" the world but those who claim only to "analyse" it. To understand this situation let us consider Bertrand Russell's position: "Let us suppose that we have a group of rods, all equally long. It is easy to suppose that there is a certain quality, called their length, which they all share. But all propositions in which this supposed quality occurs will retain their truth-value unchanged, if instead of "length of the rod x" we take "membership of the group of all those rods which are as long as x." In various special cases—e.g.,

[59] Heidegger, Martin "The Origin of the Work of Art" in *Off the Beaten Track* (Cambridge: CUP [*orig*. 1935–6] 2002) p 24.

[60] See Ivan Il'in "Dialectic" in *The Philosophy of Hegel as a Doctrine of the Concreteness of God and Humanity Vol. I* (Evanston, [*orig*. 1918] 2010).

[61] Heidegger, Martin "The Origin of the Work of Art" in *Off the Beaten Track* (Cambridge: CUP [*orig*. 1935–6] 2002) p 355.

the definition of real numbers—a simpler construction is possible."[62] The *meaning* of this approach is that "beings" disappear beneath a dull (and incoherent) techno-speak on the basis that: "An important consequence of the theory of descriptions is that it is meaningless to say "A exists" unless "A" is (or stands for) a phrase of the form "the so-and-so." If the so-and-so exists, and x is the so-and-so, to say "x exists" is nonsense. Existence, in the sense in which it is ascribed to single entities, is thus removed altogether from the list of fundamentals. The ontological argument and most of its refutations are found to depend on bad grammar."[63]

What does it mean to reduce a real being to a "phrase of the form "the so-and-so"? Russell explains this denial of speciate being, reduced to a placeholder, as that: "The explanation of *denotation* is now as follows. Every proposition in which "the author of *Waverley*" occurs being explained as above, the proposition "Scott was the author of *Waverley* (i.e., "Scott was identical with the author of *Waverley*") becomes "One and only one entity wrote *Waverley* and Scott was identical with that one.""[64] The subtext of this argument is Russell's attack on Aristotle's "*ousia*" that: "The conception of "substance" like that of "essence" is a transference to metaphysics of what is only a linguistic convenience. We find it convenient, in describing the world, to describe a certain number of occurrences as events in the life of "Socrates," and a certain number of others in the life of "Mr Smith." This leads us to think of "Socrates" or "Mr Smith" as denoting something that persists through a certain number of years, and as in some way more "solid" and "real" than the events that happen to him…"Substance," in fact, is merely a convenient way of collecting events into bundles."[65] Russell's position stands on the shocking and distasteful basis that he is denying the very existence of "Socrates" (to say nothing of "Mr Smith"). Russell's attack is not just against "essence" but against (human) "being" (*ousia*) itself.

---

[62] Russell, Bertrand "Logical Atomism" in Ayer, A.J. (*ed.*) *Logical Positivism* (New York, [*orig.* 1924] 1959) p 34–5.

[63] Russell, Bertrand *Ibid.* p 36.

[64] Russell, Bertrand "On Denoting" in *Logic and Knowledge* (London, [orig. 1905] 1956). Alex Oliver and Timothy Smiley give the additional example that: "When sets are used, "is one of" becomes membership. For example, "Whitehead is one of Whitehead and Russell" becomes "Whitehead is a member of {Whitehead, Russell}" (*Plural Logic* (Oxford, 2013) p 40)."

[65] Bertrand Russell explains his position on "substance" (*ousia*) as: "What can we know about Mr. Smith? When we look at him, we see a pattern of colours; when we listen to him talking, we hear a series of sounds. We believe that, like us, he has thoughts and feelings. But what is Mr. Smith apart from all these occurrences? A mere imaginary hook, from which the occurrences are supposed to hang. They have in fact no need for a hook, any more than the earth needs an elephant to rest upon. Anyone can see, in the analogous case of geographical region, that such a word as "France" (say) is only a linguistic convenience, and that there is not a *thing* called "France" over and above its various parts. The same holds for "Mr Smith"; it is a collective name for a number of occurrences. If we take it as anything more, it denotes something completely unknowable, and therefore not needed for the expression of what we know. "Substance" in a word, is a metaphysical mistake, due to the transference to the world-structure of the structure of sentences composed of a subject and a predicate. I conclude that the Aristotelian doctrines in this chapter are wholly false (*History of Western Philosophy* (New York, 1945) p 193)."

Fig. 9.6 A good "ontological argument"

Michael Dummett makes this position more explicit as that: "There is no such thing as a "plurality," which is the misbegotten invention of a faulty logic: it is only as referring to a concept that a plural phrase can be understood, because only a concept-word admits a plural…"The Kaiser's carriage is drawn by four horses" [must take] the form "There are four objects each of which is a horse that draws the Kaiser's carriage.""[66] What is the idea behind this denial of (individual) being? Russell himself asserts that his "fundamental doctrine upon which all rests" is "the doctrine that the subject of a proposition may be plural, and that such plural subjects are what is meant by classes which have more than one term"[67] with "classes" being explained (or not explained) further on the basis that: "Fortunately all the propositions in which classes appear to be mentioned can be interpreted without supposing that there are classes."[68] We see here a philosophical model that reduces the world to the "participation" in forms or, as Russell puts it, "membership" in classes or patterns of being. Russell's logical atomism reduces the world to a particularly obnoxious form of Platonism.

George Boolos explains the strange and obvious consequence of Russell's error as that: "It is haywire to think that when you have some Cheerios, you are eating a set—what you're doing is: eating THE CHEERIOS." Let us represent this situation in Fig. 9.6.

The problem that Boolos raises is in respect to whether the child, the Cheerios, the milk, bowl, table, kitchen, parents etc. *really exist and actually matter*. Does it matter that the child exists and eats her breakfast? Does it matter if the child ate the

[66] Dummett, Michael quoted in Alex Oliver & Timothy Smiley *Plural Logic* (Oxford, 2013) p 66.

[67] Russell, Bertrand *The Principles of Mathematics* (London, [*orig*. 1903] 1937) p 517.

[68] Russell, Bertrand "Logical Atomism" in *Logical Positivism* ed. Ayer, A. J. (New York, [*orig*. 1924] 1959) p 35.

Cheerios or if the photograph was merely taken for advertising without any such action? Does it matter if the child has now grown up and the packet of Cheerios is long gone? Does it matter if Socrates actually existed? Does anything matter? I suggest that we should see this line of thought as a *reductio ad absurdum* arising from but not intended by the Cartesian philosophy. Whatever its provenance, however, Oliver and Smiley explain the fallacy of this position well as that "we cannot play fast and loose with the sense of "eat." You cannot eat abstract objects. Sets are abstract objects, so you cannot eat one. Collective predicates also cause trouble. Tom and Dick weigh over 200 kg, but the set of the two has no weight."[69]

I take it as obvious that this positivistic position is, to put the matter mildly, erroneous by omission. The real question is not whether it is true but as regards why on earth it could be taken seriously. Jaakko Hintikka explains that the whole problem goes back to the philosophical compromise or evasion performed by Immanuel Kant. He comments that: "What Kant argues is that the concept of a hundred dollars—sorry, thalers—is not changed in the least if we add the notion of existence to it"[70] and he quotes Kant himself on "the small word "is"" that: "The small word "is" adds no new predicate, but only serves to posit the predicate in its relation to the subject….The content of both must be one and the same; nothing can have been added to the concept, which expresses merely what is possible, by my thinking its object (through the expression "it is") as given absolutely."[71] What is the problem with this philosophical tweak? Can we get rid of this existential indicator? The problem is that Kant is not interested in children or Cheerios but "God" and the proposition "God is almighty" and the "money" and "number" of "one hundred dollars" which are both explicitly not "natural" beings.[72] It is only in this way, by circumventing physical reality and "nature," that Kant can dismiss the disciplines of ontology and metaphysics which consider the basis upon which beings come-to-be and then be.

Aristotle himself remarkably identifies the problem inherent in the linguistic evasion (or what Hintikka calls a calculated "ambiguity") of the word "is." He commented that: "some, like Lycophron [a sophist], were led to omit "is," others to change the mode of expression and say "the man has been whitened" instead of "is white," and "walks" instead of "is walking," for fear that if they added the word "is"

[69] Oliver, Alex & Smiley, Timothy *Plural Logic* (Oxford, 2013) p 37.

[70] Hintikka, Jaakko "On Aristotle's Notion of Existence" in *Analyses of Aristotle* (New York, [*orig.* 1999] 2004) p 8.

[71] Kant, Immanuel *Critique of Pure Reason* (Cambridge, [*orig.* 1781] 1998) A599.

[72] Kant explains further that: "If, then, I take the subject (God) with all its predicates (including that of almightiness), and say, God is, or there is a God, I do not put a new predicate to the concept of God, but I only put the subject by itself, with all its predicates, in relation to my concept, as its object. Both must contain exactly the same kind of thing, and nothing can have been added to the concept, which expresses possibility only, by my thinking its object as simply given and saying, it is. And thus the real does not contain more than the possible. A hundred real dollars do not contain a penny more than a hundred possible dollars (*Ibid.* A599)." See Elliott Sober on "line-drawing problems" in his "Evolution, Population Thinking, and Essentialism" *Philosophy of Science* (1980).

they should be making the one to *be* many—as if "one" and "being" were always used in one and the same sense. What "is" may be many either in definition (for example "to be white" is one thing, "to be musical" another, yet the same thing may be both, so the one is many) or by division, as the whole and its parts (*Phys*. I 185b 28-34)." Aristotle also argued against the Megarian philosophers that we must understand that a builder exists who then builds, i.e., the builder actualises the potentiality that he has already acquired, rather than that the building simply happens. Heidegger usefully explores this conflict and see also *Met*. Γ 1008b 10-20.[73] The subtext of all this argumentation is the expression of anti-reductionism and the refusal to reduce the beings of the real world to the empiricism of what factually happens or the sophism concerned only with how we speak about it. We see from Aristotle's thought and work that a real conversation took place in his day over what mathematics and language actually are as elements of a "third world" of representative thought.[74] On the importance of Aristotle's "is" see also Kurt Riezler's *Physics and Reality* (1940).

Martin Heidegger observes (following Plato and Aristotle) as regards the nature of the sophism that Aristotle is addressing that: "sophistry is characterised by an unconcern with substantive content, an unconcern in a quite determinate sense, not one that is haphazard…We could determine their [i.e., the sophists'] unconcern in a better way by calling it emptiness, a lack of substantive content."[75] He observes that sophism is an attack on "substance" and "content." Paul Feyerabend also observes: "Needless to say the knowledge produced by the sophists, though "sophisticated," lacked depth. In a sense this was progress—deep knowledge creates deep respect and deadens our critical abilities. The sophists, we might say, led people out of a religious approach to knowledge into a more mundane approach. This was an advantage, but not entirely."[76] Feyerabend puts his finger (here as elsewhere) on the problem of technology and the conformism and philosophical reductionism that it engenders which was, of course, also a deep concern for Heidegger and Foucault.

9.8 Conclusions

It should be obvious that the Aristotelian position is meaningful and that its modern opponent is meaningless. Strangely enough, this last meaninglessness is in some part by design. If we reduce the world to science, we necessarily make our representation of the world operational, instrumental, pragmatic, and transactional. This is a

[73] See Martin Heidegger *Aristotle's* Metaphysics *Θ 1–3* (Bloomington, [*orig*. 1931] 1995).

[74] Aristotle himself rightly observes that both "word" and "number" are themselves forms of reductionism which disfigure as well as give shape and meaning to our thought; see *Cat*. 4b28-39.

[75] Heidegger, Martin *Plato's* Sophist (Indiana, [*orig*. 1925] 1997) p 159.

[76] Feyerabend, Paul *The Tyranny of Science* (Cambridge, [*orig*. 1992] 2011) p 102. See also his brilliant *Conquest of Abundance: A Tale of Abstraction versus the Richness of Being* (Chicago, 1999).

choice we make or take. We alter our vision of the world to fit with the value-neutral mechanicism of science.

These two alternatives are obvious and constantly appear in the literature. There is the meaningful, holistic position and then there is "atoms and the void." Aldous Huxley, with whom we began, explains this dichotomy as that: "In science and philosophy there are probably two main methods of explaining reality. One is the method of concentrating attention on the atomic elements of reality…the basic methodology of modern physics and chemistry starting with Galileo and Newton. We see it applied on the psychological level in behaviourism. The other method is the formal one of concentrating attention on the gestalten of nature, on the forms which are presented on a large scale. In the classical period this formal approach was represented, in different ways, by Plato and Aristotle, and in modern science we see it in taxonomy, in comparative anatomy, and in morphology. Incidentally, the word "morphology" was invented by Goethe, so it has a profoundly artistic and poetic overtone to it; in modern psychology we see it represented in the Gestalt school."[77] The same contrast is made over and over again, as I will document in the sister volume to this work. The question is why the obviously false and dangerous "analytic" position is so dominant in our society. It is surely because it is convenient to break the world down into manipulable masses. This returns us to the letter Huxley wrote to Orwell, quoted earlier, and its sentiments.

It seems simple and convenient to reduce the world to science, or to what people instruct in the name of science. Or religion. The Aristotelian alternative is the "realism" of a being in the world described by Jan Patočka as that: "the world is not a chaos dependent on chance. The world does in truth depend on something necessary, as Plato believed, but on something profoundly different from Platonic Ideas. Ideas are without any vigilance, dead beings. That on which the visible world depends is, however, the source of all movement—movement which, as we already know, is in some way the life of all things."[78] He explains Aristotle's position further as regards how "sense" engages with "form" as that: "of all the virtues that make man happy, the most excellent is that of intellectual sight. Its organ is philosophy. In it is concentrated all the essential content of human life which from the vision of the eyes and other sensory impressions passes to ἐμπειρία and μνήμη, with the property of retaining the past in the present, and finally rises to the level designated by the

---

[77] Huxley, Aldous "Art" in *The Human Situation* (St Albans, [*orig.* 1923]1980) p 184. Huxley also exhibits a close awareness of the fact that the artistic medium refers back to the biological world on the basis that: "The movement of the heavenly bodies, the cycle of growth, the rhythms of breathing, of heart activity, of peristalsis, and so on, and the more irregular rhythms such as hunger and satisfaction, all find their analogy in the various arts which contain an element of time. Man looks at the external world, sees the cosmic and physiological rhythms around him, makes an analogue of them in his temporal arts (*Ibid.* p 187)."

[78] Patočka, Jan *Aristote, Ses Devanciers, Ses Successeurs* (Paris, [*orig.* 1964] 2011) p 225. He adds that: "*Aisthēsis* is the ability to grasp shapes independently of material, though only singly. Humans in addition have logos, the ability to grasp an order, not only individual shapes but shape as a manifestation of the eternal logos, eternal and universal shapes (*Body, Community, Language, World* (Chicago, [*orig.* 1968–9] 1998) p 7)."

terms τέχνη and λόγος allowing a direct grasp of the universal, the necessary, the eternal. Φιλοσοφία then carries this grasp to the highest degree by seeking, among the universal, the necessary and the eternal, that which is supremely elevated. It is by this an inquiry into the being of the being, into that which makes the being be. Philosophy is therefore not only the realisation of what is most properly human—it is at the same time the culmination of the whole movement of the world towards that which, eternal and imperishable, is."[79]

It is hence that we get back to an idea of the "whole" which is so studiously and fundamentally negated by the "analytic" position. Once we accept this whole, this system, the world must make sense *within that system*. The existence of being and of meaning is the very basis of philosophy *and also of reality as we experience it*. The alternative is to argue that considering "the whole" is without merit since only concentrating upon operationalising the parts is pragmatically useful. This is where the profit is. We could, however, ask: "For what is a man profited, if he shall gain the whole world and lose his own soul? Or what shall a man give in exchange for his soul?"[80] If there is a value, it is surely more than analytic philosophy—and its latest fads of posthumanism and transhumanism—can or are willing to offer.

[79] Patočka, Jan *Aristote, Ses Devanciers, Ses Successeurs* (Paris, [*orig*. 1964] 2011) p 260.

[80] Matthew 16:26.

Chapter 10
The Battle of the Gods and Giants

Abstract The Battle of the Gods and Giants is Plato's famous picturing of the schism within philosophy between materialism (matter) and idealism (mind). This schism has shaped philosophy in various ways, culminating in Cartesian dualism. Plato argues that we must find a middle way and Aristotle identifies this "middle" with the reality that contains both mind and matter. This is how Aristotle develops and naturalises Plato. As regards how modern philosophy treats this impasse we consider both modern "Gods" such as Husserl, Peirce, Cassirer, Whitehead, Merleau-Ponty, Heidegger, Derrida, Baudrillard etc. and the terrible cottage-industry of reductionism churned out by the modern "Giants." Despite any advances made, we see that we are still trapped within the false paradigm of Cartesian dualism and within Plato's "interminable" battle of the Gods and Giants.

Keywords The battle of the gods and giants · Aristotle and modernism and post-modernism

I have shown above that Aristotle provides us with a lucid, systematic, and nuanced philosophical representation of being and becoming in the world. His particular emphasis upon "becoming" leads him to concentrate upon the world of "coming-to-be and passing-away," i.e., the *bio*-logical world. Aristotle does not just use the world to validate a philosophical idea or ideology. He rejects any such idealism and invents a realism which locates value in the beings of the world. Aristotle invented the very concept of science and established the various special sciences to pursue *this* exploration of reality. His philosophy maintains that it is by exploring and examining the various objects of the world that we can take meaning from the world and develop ourselves by reflection upon and engagement with it. It is hence that Aristotle's philosophical centrepieces are his *Physics*, his philosophical study on "nature" (*physis*), and his *Metaphysics*, his pursuant study to establish *the* being (*ousia*) as the foundational basis for being (*to on*) in this world.

P. N. Jackson, *Aristotle on Meaning in the Living World*, Biosemiotics 30,
https://doi.org/10.1007/978-3-032-00602-8_10

Aristotle is concerned with beings and being and the becoming of beings within the framing of being. He hence explains in his *Metaphysics* that: "the question is *why* the matter is some definite thing; e.g. why are these materials a house (*dēlon dē hoti tēn hylēn zētei dia ti ti estin: oion oikia tadi dia ti*)?...And why is this individual thing, or this body having a form, a man (*kai anthrōpos todi, ē to sōma touto todi echon*)? Therefore what we seek is the cause (*hōste to aition zēteitai*), i.e., the form, by reason of which the matter (*tēs hylēs touto d' esti to eidos*) is some definite thing (*hō ti estin*); and this is the substance of the thing (*touto d' hē ousia*) (*Met.* Z 1041b 4-9)." He adds that: "it is the *cause* (*aition*) [i.e., the form] which makes *this* thing flesh and that a syllable (*Met.* Z 1041b 26-27)." Cause is not general but something specific, *the cause of something.* The world is therefore made up from different forms or types of being or thing though *all exist in the world on a parallel basis*, e.g. the syllable, the flesh, and the house are all a particular something (*tode ti*), as "beings" which cannot be conflated with or reduced to a single (mathematical) law of being. Being, for Aristotle, is distinctly manifested in beings (*ousiai*) and their forms (*eidē*) though also with a need for a general natural law, a platform for "being" without which those beings would be "as mice walk(ing) upon loose soil."

It is upon this basis that Aristotle engages in biology, with an example of this approach being as follows. Aristotle observes that: "Heat and cooling (which is deprivation of heat (*sterēsis thermotētos estin*)) are both employed by nature. Each has the faculty, grounded in necessity (*echousi men dynamin ex anankēs*), of making one thing into this and another thing into that (*to men todi to de todi poiein*); but in the case of the forming of the embryo it is for a purpose that their power of heating and cooling is exerted and that each of the parts is formed, flesh being made soft—as heating and cooling make it such, partly owing to necessity (*ex anankēs*), partly for a purpose (*heneka tinos*),—sinew solid and elastic, bone solid and brittle (*GA* II 743a 37–743b 6)." Aristotle's four causes—the material, formal, moving, and final—are *together* the most complete way of representing the becoming of beings. They are a necessary nexus of causation which prove that the world is complex and meaningful since the form must find the matter and move to a given end, as we could imagine a task finding a program and a code to achieve itself. This reality may be inconvenient to various reductionist philosophers, but it is not only obvious in reality but incontrovertibly established *philosophically* as reality by Aristotle.

We must, and this is Aristotle's own emphasis, accept the fact that we live in a meaningful and complex world of beings. To think that we can find an idea or a mathematical law to explain this richness of being is to miss the very point of existence. We should not seek to reduce nature and try to form it in our image but should rather seek to express the full richness of nature and to express it in ourselves. This is Aristotle's "realism."

10.1 The Path of Philosophy

Today, our standard philosophy claims that the world is an "enormous lottery presided over by natural selection, blindly picking the rare winners from among numbers drawn at random"[1] and that we are merely "robot vehicles blindly programmed to preserve the selfish molecules known as genes."[2] This biology is at the core of our *philosophical* worldview even when, as I will show in the sister volume of this project, the *science* has long since disproved this perspective. Most tragically, this philosophy, or *anti*-philosophy, seems to have been internalised by the man on the street. The claim of the intellectual system in which we live (or are trapped!) is that this nihilism is unavoidable and that we must accept its hard truth, whereas we see, in reality, that it is firmly disproved not only in the field of science but also in the field of philosophy. Let us here finally concentrate on showing how and why neo-Darwinism is *not* true philosophically but has a deep power as a sort of ideological scholasticism pumped out to prop up a terrible new religion. This is my hypothesis. Let us consider what I take to be philosophy below and then what I take to be modern ideological scholasticism in the following subsection.

Edmund Husserl (1859–1938) was the founder of the school of phenomenology. The great man saw in his times not a great upward path of progress through science and technology but a crisis of Western civilisation coming to a head. He explained this "*Krisis*" in 1936 on the basis that: "Scientific culture in accord with ideas of infinity means…a revolutionising of all culture, a revolution that affects man's whole manner of being as a creator of culture"[3] and that: "the reason for the downfall of a rational culture does not lie in the essence of rationalism itself but only in its exteriorisation, its absorption in "naturalism" and "objectivism.""[4] He testified to a *problem* of science and a *problem* of rationalism and on this situation he was willing to appeal to: "Rationality, in that high and genuine sense of which alone we are speaking, the primordial Greek sense which in the classical period of Greek philosophy had become an ideal, [which] still requires, to be sure, much clarification through self-reflection; but it is called in its mature form to guide [our] development."[5]

We do not, of course, necessarily need to accept Husserl's solution, *viz.* to return to the "genuine" rationality of the Greeks. It is, however, useful for us to recognise how a modern thinker can break from the treadmill of supposed progress and *look backwards*. On this point, however, Husserl clearly looks to Plato rather than

[1] Monod, Jacques *Chance and Necessity* (London, 1971) (subtitled as "a philosophy for a universe without causality") p 138.

[2] Dawkins, Richard *The Selfish Gene* (Oxford, [*orig.* 1973], 1989) p xxi.

[3] Husserl, Edmund "Philosophy and the Crisis of European Humanity" in *The Crisis of European Sciences and Transcendental Philosophy: An Introduction to Phenomenological Philosophy* (Evanston, [*orig.* 1936] 1970) p 279.

[4] Husserl, Edmund *Ibid.* p 299.

[5] Husserl, Edmund *Ibid.* p 290.

Aristotle. He diverges from the Aristotle presented above but his philosophy (as also Plato's) still contains enough "realism" to run parallel with Aristotle's as regards the world we need to recognise and account for. Aristotle's sublunary world is Husserl's "Lifeworld" which Husserl himself explains as that: "People and animals *have* material Bodies, and to that degree they have spatiality and materiality. According to what is specifically human and animal, that is, according to what is psychic, they *are*, however, not material, and, consequently, taken *also as concrete totalities*, they are *not material* realities in the proper sense. Material things are open to fragmentation, something which accompanies the extension that belongs to their essence. But men and animals cannot be fragmented. Men and animals are *spatially localised*; and even what is psychic about them, at least in virtue of its essential foundedness in the Bodily, partakes of the spatial order. We will even say that much of what is included under the broad—and, at first, unclarified—heading of the psychic has something like spread (although not extension in space). In principle, however, *nothing* on this side is *extended in the proper sense*, in the specific sense of the extension we described."[6]

Husserl provides us with a brilliant defence of life as the separate existence of an organic whole. Jenny Slatman explains his position on this matter further as that: "Humans and animals…cannot be fragmented: they are indivisible unities. It is this unity that also characterises the spatiality of psychic reality. Psychic reality is not extended (*ausgedehnt*), nor is it non-extended. Its spatiality is formed by a spreading out (*Ausbreitung*)…Whenever I touch someone or am touched by something, I have localised sensations somewhere in my body, for instance in my hand. These localised sensations (*Empfindnisse*)—however, are not limited to my hand. They spread out in such a way that I have not only the experience of *my* hand, but also the indivisible unity that forms *my* body. We could therefore say that the phenomenon of spreading out constitutes the body as a unity or the body according to a corporeal scheme."[7] Husserl unpacks the philosophical consequences of the human experience, the human situation as a worldly reality. He explains what it means to be an animal and what it means to reflect upon this situation. As regards the idea of touching as sympathy, Aristotle himself notes that we can be emotionally touched by somebody by having empathy for them.[8]

Husserl lucidly explains as regards this holistic being of the "animal" that: "Animal spirituality, that of the human and animal "souls," to which all other spirituality is referred, is in each individual instance causally based on corporeality"[9]

---

[6] Husserl, Edmund *Ideas II* (New Delhi, [*orig*. 1928] 2013) p 36.

[7] Slatman, Jenny "The Sense of Life: Husserl and Merleau-Ponty on Touching and Being Touched" in *Life and Individuation* (Memphis, 2005) p 311.

[8] Aristotle argues that: "if anything causes motion without being itself moved, it might touch that which is moved, though not itself touched by anything; for we say sometimes that a man who grieves us "touches" us, though we ourselves do not "touch" him. So much for our definition of contact in the realm of Nature (*GC* I 323a 32–35)."

[9] Husserl, Edmund "Philosophy and the Crisis of European Humanity" in *The Crisis of European Sciences and Transcendental Philosophy: An Introduction to Phenomenological Philosophy* (Evanston, [*orig*. 1936] 1970) p 271.

with this body being causally essential as an "organ of the spirit."[10] He does not understand mind, spirit, or ego as being something intangible in our head but as something that explains real action in the world on the basis that: "in a certain general sense, the Ego directs itself in every case to the Object, but in a more particular sense at times an Ego-ray, launched from the pure Ego, goes out toward the Object, and, as it were, counter-rays issue from the Object and come back to the Ego. This is how it is that in desire I find myself attracted by the desired Object; I am directed to it, but in such a way that I reach out toward it, without, however, attaining it in mere desire."[11] Husserl recognises that it is insufficient to say that we simply see or act since the reality of this seemingly mundane situation is the irruption of a complex, technical, mystical, and physical pathway taken through reality by a being.

Husserl explains the external world itself on the basis that an object can "assume psychic significance" since: "ultimately *everything* Bodily from no matter what point of view can assume psychic significance… even where at the outset it is not phenomenally the bearer of a soul."[12] He provides the example to demonstrate this point of a stone becoming something that can be recognised as such with the implication being that *the stone becomes something by becoming our artefact*. He concludes as regards the nature of spirit in space as that: "the human being is a human being in nature and is in nature only because, first and foremost, the Body is a material thing in spatial nature. Psychic reality is constituted as reality only through psychophysical dependencies."[13] We are, however, critically limited by the localisation of our living body from understanding the world in an ideal sense on the basis that: "it is difficult to send rays of reflection into the realm of obscurities and gain certitude in this sphere of givens."[14] This subtlety reminds us of Aristotle, for whom the whole need to biologically reproduce in the sublunary world is due to the localised detachment in matter of a being from its "eternal" source. We humans do in some profound and difficult ontologically sense strive "to transmit [our] rays of reflection into the realm of obscurities."

John C. McCarthy also usefully explains Husserl on this situation (in an explicitly Aristotelian context) as that: "living forms move through time in an essentially different way than do the nonliving. Husserl concludes that animate beings alone have a "history." Their past is taken up into themselves and is thereby able to affect their present. Not despite but because living things form themselves, they are formed by their times. It follows that changes in them are irreversible. Tides ebb and flow again, but no animate being is ever what it was ([*Logical Investigations*] §33)."[15] He

[10] Husserl, Edmund *Ideas II* (Isha, [*orig*. 1928] 2013) p 102.

[11] Husserl, Edmund *Ibid.* p 104.

[12] Husserl, Edmund *Ibid.* p 102.

[13] Husserl, Edmund *Ibid.* p 146.

[14] Husserl, Edmund *Ibid.* p 114.

[15] McCarthy, John C. "Parts, Wholes, and the Forms of Life: Husserl and the New Biology" in Hardy, L. & Embree, L. (*eds.*) *Phenomenology of Natural Science* (Springer 1992) p 145.

adds that: "To know an animal, one must know its habitat (*Ideas* II, §15c). Proof of longstanding acquaintance with an acquaintance with an animate form is, however, what we might call categorial "synecdoche," the ability to intuit the whole in a part. A seasoned hunter can "see" the deer in a broken branch or a tuft of fur. Indeed, the wholeness of a whole is never given through an exhaustive apprehension of its parts: in a sense, categorial intuition always "jumps to conclusions." Otherwise stated, there is always the possibility of a richer understanding of a being's being one."[16] We see in this consideration of the signs of the interactions between beings and the world the sign of a semiotic perspective on the world.[17] Husserl's philosophy is also teleological in nature and seeks to "provide the scientific foundations for a reconciliation between the naturalistic world-view that dominated the epoch just expired and the teleological world-view. But the teleological world-view is the definitively true one."[18] Husserl provides us, in his difficult lengthy way, with a holistic, teleological picturing of the world which is compatible with Aristotle's but still trapped to some degree within a Cartesian-Kantian paradigm that renders it more concerned with self and sense than with world and being.

Charles Sanders Peirce (1839–1914) is both the founder of the science of semiotics and of the philosophy of pragmatism. Following Peirce, I will suggest below that his pragmatism both opened up new philosophical vistas and produced yet another roadblock. Peirce distanced himself from the developments of his own philosophy, as many other philosophers would likely distance themselves from the later developments of their thought. Aristotle would, for instance, likely not have been fully enthused by the later "neo-Aristotelian" scholastic tradition. It would also be interesting to know what Plato made of the philosophy of his most famous student. The themes of his later work certainly suggest a dialogue with Aristotle or with the critique that he came to represent, though he did not choose to name Aristotle as his successor at the Academy! We must also address the parallel problem that even the most original thinkers cannot fully shake themselves free of old paradigms. I will contend below in the case of Peirce that he—as also Husserl—could certainly critique the scientific reduction but not fully break free from it.

Let us begin here by considering the problem of pragmatism and then move on to consider the problem of scientific reductionism that will ultimately conclude with the "pragmatism" of Richard Rorty, previously considered. For this, we can begin with the "pragmatic maxim" stated in Peirce's influential paper "How to Make our Ideas Clear" (1878) as "Consider what effects, which might conceivably have

---

[16] McCarthy, John C. *Ibid.* p 151.

[17] Husserl explains his symbolism as that: "Perception is without exception a *unitary accomplishment* which arises essentially out of the playing together of two *correlatively related functions* (*Ideas II* (New Delhi, [*orig.* 1928] 2013) p 63)." On the "synecdoche" see Yuri Lotman "Excerpts from Universe of the Mind: A Semiotic Theory of Culture (1990)" in Favareau, Donald (*ed.*) *Essential Readings in Biosemiotics: Anthology and Commentary* (Dordrecht, 2010).

[18] Husserl, Edmund "Letter to Albrecht; August 2 1917" (and see his "Universal Teleology" *Telos* (1969)).

practical bearings, we conceive the object of our conception to have."[19] Peirce defines his own philosophy as a "laboratory philosophy"[20] in accordance with which "there is absolutely no difference between a hard and a soft thing so long as they are not brought to the test"[21] which the Stanford Encyclopaedia explains as that: "what we mean by calling something hard is that "it will not be scratched by many other substances." In this way, then, I can use the concept hard in certain contexts when I am wondering what to do, and absent such contexts, the concept is empty. The principle has something of a verificationist character: "our idea of anything is our idea of its sensible effects" (EP1: 132)."[22] A diamond is not hard unless we have tested it as such.

John Dewey quotes Peirce as follows: "The rational meaning of every proposition lies in the future. How so? The meaning of a proposition is itself a proposition. Indeed, it is no other than the very proposition of which it is the meaning: it is a translation of it. But of the myriads of forms into which a proposition may be translated, which is that one which is to be called its very meaning? It is, according to the pragmaticist, that form in which the proposition becomes applicable to human conduct, not in these or those special circumstances nor when one entertains this or that special design, but that form which is most applicable to self-control under every situation and to every purpose."[23] The idea that "The meaning of a proposition is itself a proposition" is another philosophical slippery slope down which modern thought has gleefully slipped and one clearly enjoyed by Peirce himself. We need to understand the problems that remain in Peirce's philosophy which inform his tradition and which potentially influence biosemiotics.

It is useful to consider Peirce's influences in order to understand the problems we see with his philosophy *from an Aristotelian perspective*. We first see the residual influence of Descartes in his seeking after "clear and distinct ideas."[24] We also find that Peirce himself commented that he formulated the term pragmatism (and then pragmaticism) by reference to the thought of Kant.[25] As regards "truth" we find that Peirce clarified against his fellow pragmatist William James' assertion that: "The true is the name of whatever proves itself to be good in the way of belief, and good,

[19] Peirce, Charles *The Essential Peirce* (Bloomington, 1997) [EP] 1: 132.

[20] See "Pragmatism" in *Stanford Encyclopedia of Philosophy* (Winter 2024 Edition).

[21] Peirce, Charles EP 2: 131.

[22] "Pragmatism" in *Stanford Encyclopedia of Philosophy* (Winter 2024 Edition).

[23] Peirce, Charles "What Pragmatism Is" *The Monist* (1905) 173–4. Dewey describes Peirce's thought as "an empirical rendering of the Kantian generality of moral action ("The Pragmatism of Peirce" *The Journal of Philosophy, Psychology and Scientific Methods* (1916) p 711)."

[24] "Peirce's description of his maxim as a logical principle is reflected in passages where he presents it as a development of a distinction that had been a staple of traditional logic texts, and familiar to readers of Descartes, between ideas that are clear and distinct (EP1: 126f) ("Pragmatism" in *Stanford Encyclopedia of Philosophy*, Winter 2024 Edition).

[25] See John Dewey "The Pragmatism of Peirce" *The Journal of Philosophy, Psychology and Scientific Methods* (1916).

too, for definite assignable reasons"[26] that: "The opinion which is fated to be ultimately agreed to by all who investigate, is what we mean by the truth, and the object represented in this opinion is the real. That is the way I would explain reality."[27] Whereas "true" is for James the "name" for "belief," it is for Peirce what is "fated" to be the "representation" of collective "opinion" that is "real." Despite the great value of the thought of Peirce and James in restoring the matter of "meaning" to philosophy, we see only a chronic dearth of depth and meaning in their thought from an Aristotelian perspective. Their philosophy is clearly both utilitarian and trapped within the linguistic paradigm. It is also Kantian in its universalism as seeking universal rules supposedly applicable to human conduct.

Peirce certainly represented an intellectual paradigm shift in centring both the sign and praxis and he gave birth to biosemiotics from this application of meaning to the living world. This is the huge benefit of Peirce's massive labour. The problem we encounter, however, is how to avoid the breakdown or ossification of a philosophy, especially when, as in Peirce's case, the source material is fragmentary and to some degree open to (re-) interpretation. His philosophy was, for better or worse, subject to interpretation (without evident misrepresentation) by John Dewey. Dewey quotes Peirce as that: "What a thing means is simply what habits it involves"[28] which he explains by reference to Peirce's "pragmatic maxim" as that: "Summing up, we may say that Peirce's pragmaticism is a doctrine concerning the meaning, conception, or rational purport of objects, namely, that these consist in the "effects, which might conceivably have practical bearings, we conceive the object of our conception to have."[29] Ernst Cassirer would also argue that: "An object consists solely of that, as which it gives itself to us: a sum of actual and possible effects."[30] The problem here is "solely." W.V. Quine would make the same argument a generation later that: "To be assumed as an entity is, purely and simply, to be reckoned as the value of a variable"[31] Whatever Peirce intended, his philosophy ended in anti-ontological relativism.

I suggest that Charles Peirce offers us a salutary warning in respect to the problem of the slippery slope. The dead hand of orthodoxy constantly pulls us back to what Whitehead calls the "Hume-Newton situation…[which is]…the primary presupposition for all modern philosophic thought. Any endeavour to go behind it is, in philosophic discussion, almost angrily rejected as unintelligible."[32] It is in this context that we find that Aristotle's "*ousia*" is lost or dismissed in the pragmatic tradi-

---

[26] James, William *Pragmatism: A New Name for some Old Ways of Thinking* (Cambridge, MA: [orig. 1907] 1975) p 42.

[27] Peirce, Charles EP1: 139.

[28] Dewey, John "The Pragmatism of Peirce" *The Journal of Philosophy, Psychology and Scientific Methods* (1916) p 714.

[29] Dewey, John *Ibid.* p 714.

[30] Quoted in John Michael Krois "More than a Linguistic Turn in Philosophy: the Semiotic Programs of Peirce and Cassirer" *Sats—Nordic Journal of Philosophy* (2004) p 18.

[31] Quine, W.V. "On What There Is" in *From a Logical Point of View* (Cambridge, Mass., 1953) p 13.

[32] Whitehead, A.N. "Nature Lifeless" in *Modes of Thought* (New York, 1938) p 135.

tion as it is, in fairness, lost or ignored by most other later philosophical traditions. It could, of course, be said that Peirce is far better than most in interpreting and using Aristotle and in creating a meaningful modern philosophy that reintroduces meaning to philosophy and fights the good fight against scientific reductionism whilst still giving appropriate space and value to the *modern* scientific tradition. All this is, of course, true. Our problem, however, is whether we need to push beyond Peirce and through his own preconceptions shaped by Descartes and Kant *in order to fully give birth to a new paradigm.* This new paradigm is that of a meaningful world which is mapped through but not reducible to science. In order to break through to this new world in the fields both of biology and philosophy, I contend that we need such additional thought as provided by such figures as Aristotle and Heidegger. We need to break through modernity to a productive form of post-modernity and we will try to achieve this task below.

Ernst Cassirer (1874–1945) was a leading philosopher of his time in the neo-Kantian tradition. His philosophy of symbolic forms can best be understood as a phenomenology of culture which posits that all the forms of representation used by human beings—language, myth, art, religion, history, science—are symbolic. The world for Cassirer is symbolic rather than, as for Peirce, semiotic. As regards Aristotle, Cassirer begins his interesting work *An Essay on Man*, from 1944, by considering the Western tradition within which he describes Aristotle on the basis that: "he speaks as a biologist when he denies this Platonic severance between the ideal and the empirical world. He attempts to describe the ideal world, the world of knowledge, in terms of life."[33] Cassirer also astutely comments on Aristotle's theory of evolution that: "The characteristic and fundamental distinction between the Aristotelian and the modern version of evolution consisted in the fact that Aristotle gave a formal interpretation whereas the moderns attempted a material interpretation."[34] Cassirer also, however, spent his career attacking what he called Aristotle's "abstractionism" and offering his own philosophy of "symbolic forms" in its stead.

By Aristotle's supposed "abstractionism" Cassirer meant the construction of universals through empirical particulars, an avenue he rightly associates with Aristotle's logic. He instead asserted that we can and should approach the world through the constructivist mathematics which Robert Rosen critiques as that: "[the] essence [of constructivist mathematics] is to deny *existence* to anything that cannot be generated from a finite set of symbols by a finite sequence of rote operations (i.e., an algorithm). According to this view, most of the familiar properties and objects with which mathematics deals do not even exist; they are to be excluded from the mathematical universe. From this point of view, it is thus generic not to exist. Indeed, taking such a constructivist view, every infinite set requires a separate axiom to

[33] Cassirer Ernst *An Essay on Man* (New York, [*orig*. 1944] 1970) p 17.

[34] Cassirer Ernst *Ibid.* p 36.

allow it to exist, its ticket of admission into the constructivist universe."[35] We are yet again very much returned to a Cartesian situation. Reducing the world to symbols is best achieved by defining it as ego, mathematics, and plastic matter.

Cassirer hence presents us with two variants of "abstraction." His own *mathematical* abstraction versus Aristotle's *empirical* abstraction. His argument of *An Essay of Man* presents history as that the Greeks began with the study of physics and moved through Aristotle's metaphysics to an understanding that philosophy should be concerned with the human subject. His symbolic forms then lead us to the world of religion in an echo of Pythagoras. We will see on this basis how symbolic becomes subjective.

As regards Cassirer's conclusion on the human subject, he argues that: "There is, therefore, only one approach to the secret of human nature: that of religion. Religion shows us that there is a double man—the man before and after the fall. Man was destined for the highest goal, but he forfeited his position. By the fall he lost his power, and his reason and will were perverted. The classical maxim, "Know thyself," when understood in its philosophic sense, in the sense of Socrates, Epictetus, or Marcus Aurelius, is therefore not only ineffectual, it is misleading and erroneous."[36] Whereas Aristotle rejected Plato's appeal to mythology, Cassirer is enabled by it. He is caught up in a symbolic story about man rather than a naturalistic assessment as regards how we can best know ourselves. Most famously Karl Jaspers but also Martin Heidegger similarly demanded that we reinstate myth to a defining place in human thought. William James' *The Varieties of Religious Experience* is an excellent study of religion, but it also reflects his desire to push ideas about beliefs into philosophy. Cassirer explains our world as that it is through the symbology of mythical thought that art develops, and through the symbols of language that science develops. All meaning is therefore mythical and mental.

For Cassirer, the true abstraction was to be found in the fact that the peculiarly human world is thoroughly "symbolic." He argued as regards these symbols that: "A symbol has no actual existence as a part of the physical world; it has a "meaning." In primitive thought it is still very difficult to differentiate between the two spheres of being and meaning. They are constantly being confused: a symbol is looked upon as if it were endowed with magical or physical powers. But in the further progress of human culture the difference between things and symbols becomes clearly felt, which means that the distinction between actuality and possibility also becomes more and more pronounced."[37] As with Plato, Cassirer presents us with a supposed layer of meaning that is not attached to or derived from the natural world. This world is invented as opposed to being a "magic" epiphenomenon of the world itself. Stanley Rosen explains this Kantian hypostatisation of "meaning" well as that: "Meanings are the replacements for Platonic Ideas and Aristotelian species-forms

[35] Rosen, Robert "Biological Challenges to Contemporary Paradigms of Physics and Mimetics" in *Essays on Life Itself* (New York, 2000) p 38.

[36] Cassirer Ernst *An Essay on Man* (New York, [*orig.* 1944] 1970) p 28.

[37] Cassirer Ernst *Ibid.* p 80.

provided by those who deny that we are in direct contact with things, and who interpose a domain of representations between the intellect and the objects of cognition, which thus becomes objects of subjects, and in particular, of discursive subjects."[38] Symbolism easily becomes subjectivism.

Cassirer ultimately represents an *anti*-Aristotle. His approach is mathematical and mythological. It is most obviously Platonic but is also Kantian and Cartesian in its breakdown of the world into the mathematical and the ego or into "the starry heavens above me and the moral law within me."[39] What is perhaps most interesting for our purposes is that Cassirer's philosophy does not take the side of a worldview based upon "natural selection" in any meaningful sense. We find, as G.A. Esparza Urzúa explains, that Cassirer's "symbolic system" is rather a reimagining of Jakob von Uexküll's "functional circle" with Cassirer also employing Kurt Goldstein's Gestalt psychology for the philosophical purposes of "present[ing] phenomenological findings to describe the distance that exists between the immediate world of the sick person and the world of possibilities offered by the symbol."[40] Cassirer considers the body only in the sense that it is a potential physical restraint upon the symbolic and formal operation of the human mind. We see here how a semiotic interpretation of biology is typically philosophically Kantian, as centred upon the mind acting on world, but I will show in the sister volume of this work why Aristotle's fuller empirical approach is more satisfactory.

Alfred North Whitehead (1861–1947) provides us with a philosophical and terminological structure for the world in his *Modes of Thought* lectures from 1937–8. I take his thought to represent an important attempt to recover philosophical meaning. He explains, for instance, our mundane concept of "importance" as that: "the notion of existence involves the notion of an environment of existences and of types of existences. Any one instance of existence involves the notion of other existences, connected with it and yet beyond it. This notion of the environment introduces the notion of "more and less," and of multiplicity. The notion of "importance" also refers to grades of importance and types of importance. Here again we reach the notion of "more and less." Also something has to be "important." There is no importance in a vacuum."[41] Whitehead flips the script. With his "notion of existence" he assumes that beings exist. He assumes that "existence" assumes a system or "environment" in which it is a part or member. Such being emerges through "grades of importance" and the "more and less"[42] with "importance" being both a real-world fact and a technical philosophical term. Seeing the world through the lens of "atoms and the void" does not recognise this importance.

---

[38] Rosen, Stanley "Austin and Ordinary Language" in *The elusiveness of the ordinary: Studies in the possibility of philosophy* (New Haven, 2002) p 194.

[39] Kant, Immanuel *Critique of Practical Reason* (1788) 5:161.

[40] Esparza Urzúa, G.A. "Body, Illness and Symbol in Ernst Cassirer" *Filozofia* (2024) p 375.

[41] Whitehead, A.N. "Importance" in *Modes of Thought* (New York, 1938) p 6–7.

[42] Aristotle treats animals by reference to their relation to the environment based upon the "more and less"; for example, see *PA* I 644b 8-16.

As regards these "grades of importance" Whitehead explains that: "in nature we find four types of aggregations of actualities: the lowest is [1] the non-living aggregation, in which mutual influence is predominantly of a formal character expressible in formal sciences, such as mathematics. The inorganic is dominated by the average. It lacks individual expression in its parts. Their flashes of selection (if any) are sporadic and ineffective. Its parts merely transmit average expressions; and thus the structure survives. For the average is always there, stifling individuality. [2] The vegetable grade exhibits a democracy of purpose issuing from its parts…the nature of this average is dominated by the intricacies of its own bodily formation, [3] The animal grade includes at least one central actuality, supported by the intricacy of bodily functioning, [4] The human grade of animal life immensely extends this concept."[43] Whitehead provides us with a sophisticated and elegant version of "evolution." We begin with a mathematically analysable world and move through a biological world to a human "concrescence" which is the full evolution or actualisation of living being. Whitehead notably disputed with Frederick Gowland Hopkins, the discoverer of vitamins, as regards whether we must accept the whole as being more than the sum of its parts.[44]

Whitehead recognises that the "world" simply cannot be understood on an accidental or materialistic basis and then that there is an evolution of consciousness through the various levels of actuality—inorganic, vegetable, animal, and man—which is also an increased complexity, expression, and individuality which moves us from mathematical to biological to human models of being. We can easily relate this to Aristotle's *scala naturae* as per Fig. 10.1 below.

Whitehead further defines "expression" as that: "Expression is founded on the finite occasion. It is the activity of finitude impressing itself on its environment…The laws of nature are large average effects which reign impersonally. Whereas, there is nothing average about expression. It is essentially individual. In so far as an average dominates, expression fades. Expression is the diffusion, in the environment, of something initially entertained in the experience of the expressor. No conscious determination is necessarily involved; only the impulse to diffuse. This urge is one of the simplest characteristics of animal nature. It is the most fundamental evidence of our presupposition of the world without."[45] Expression is not merely human expression in the world but the expression of an organic being to act through an "activity of finitude impressing itself upon its environment." As with Husserl, Whitehead provides us with a quite physical description of what expression in the world actually entails. For Whitehead, life is defined as an "impulse to diffuse"—in Aristotelian terms the "*hormē*" or "*archē*"—which emerges through "something initially entertained in the experience of the expressor."

Whitehead understands that the critical question is not "What is Life?" but "Why is Life?" He poses this question on the basis that: "In fact life itself is comparatively deficient in survival value. The art of persistence is to be dead. Only inorganic things persist

[43] Whitehead, A.N. "Expression" in *Modes of Thought* (New York, 1938) p 27.

[44] See Hans-Joachim Niemann *Karl Popper and the Two New Secrets of Life* (Tübingen, 2014) p 63.

[45] Whitehead, A.N. "Expression" in *Modes of Thought* (New York, 1938) p 21.

for great lengths of time. A rock survives for eight hundred million years; whereas the limit for a tree is about a thousand years, for a man or an elephant about fifty or one hundred years, for a dog about twelve years, for an insect about one year. The problem set by the doctrine of evolution is to explain how complex organisms with such deficient survival power ever evolved. They certainly did not appear because they were better at that game than the rocks around them."[46] His answer to the question is that we find in nature a "growth of emphasis" characterised by "selection" and "concentration" as follows: "The growth of consciousness is the uprise of abstractions. It is the growth of emphasis. The totality is characterized by a selection from its details. That selection claims attention, enjoyment, action, and purpose, all relative to itself. This concentration evokes an energy of self-realization. It is a step towards unification with that drive towards realization which discloses the unity of aim in the historic process."[47] It is not the what of life that matters but its why, not its quantity but its quality that counts.

The above analysis of our worldly reality and being states the obvious in a necessarily complex way and hence philosophically explains reality. Reality is complex as human reality and consciousness and all the terms we take from granted—importance, expression, selection—are both obvious and complex. As regards "final causes," Whitehead argues that: "The analogy of the animal body suggests that the extreme rejection of final causation from our categories of explanation has been fallacious. A satisfactory cosmology must explain the interweaving of efficient and of final causation. Such a cosmology will obviously remain an explanatory arbitrariness if our doctrine of the two modes of causation takes the form of a mere limitation of the scope of one mode by the intervention of the other mode. What we seek is such an explanation of the metaphysical nature of things that everything determinable by efficient causation is thereby determined, and that everything determinable by final causation is thereby determined. The two spheres of operation should be interwoven and required, each by the other. But neither sphere should arbitrarily limit the scope of the alternative mode."[48] The understanding of the complexity of the "movement" (as "becoming") of the "animal body" led Whitehead to appreciate the importance of "finality."

Maurice Merleau-Ponty (1908–1961) explores and expands upon the nature of (human) "being" as that: "The flesh (of the world or my own) is not contingency, chaos, but a texture that returns to itself and conforms to itself"[49] and that: "We have to reject the age-old assumptions that put the body in the world and the seer in the body, or, conversely, the world and the body in the seer as in a box."[50] His emphases upon the importance of the flesh or matter of the body and also upon seeing and the seer as the perceiving being recapitulates Aristotle's vision of the enveloping by a "*periechon*" of a "*periechomenon*" which is both something with its flesh or substance—its "proper matter"—and also with formal limits and boundary condi-

[46] Whitehead, A.N. *The Function of Reason* (Princeton, 1929) p 2.

[47] Whitehead, A.N. "Civilised Universe" in *Modes of Thought* (New York, 1938) p 123.

[48] Whitehead, A.N. *The Function of Reason* (Princeton, 1929) p 23.

[49] Merleau-Ponty, Maurice *The Visible and the Invisible* (Evanston, 1968) p 146.

[50] Merleau-Ponty, Maurice *Ibid.* p 138.

Elements (stoicheia)	Plant (phyton, growing thing)
 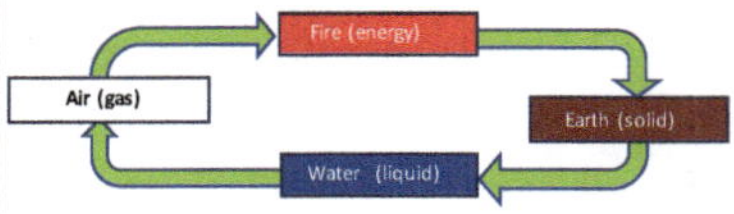	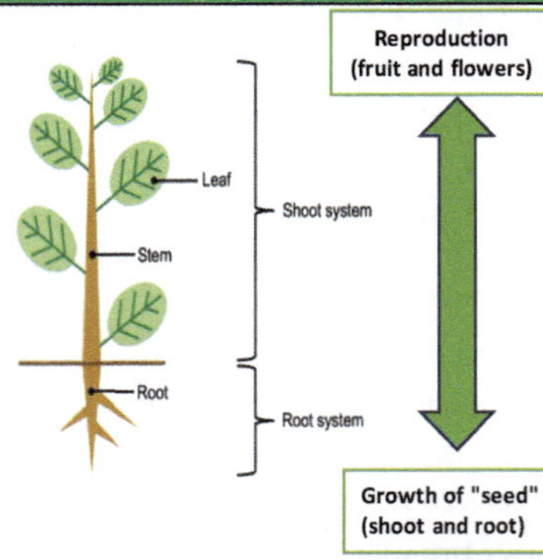
Key Texts	
Meteorology ***On Generation and Corruption***	***On Longness and Shortness of Life*** ***On the Soul***
Comments:	
Aristotle's "four elements" are an essential basis for his "*ousia*." The "*ousia*" is a higher principle which must constrain its chemical basis. When it loses its ability to keep itself together, it dies. This situation provides us with the essential distinction between entropy and negentropy. I note that Aristotle's "elements" should be understood as phase states - liquid, solid, energy, gas - rather than, literally, as water, earth, fire, and air. Through this account Aristotle founds our sublunary world on the concepts of cycles, plurality, elements, etc. Aristotle takes elements to be a paradigm for "circular" movement and of natural transformation; see GC II 338a 4-19.	Aristotle describes "movement" as that: "...all things that go through the process of becoming acquire locomotion last. It is this that accounts for the fact that some living things, e.g. plants and many kinds of animals, owing to their lack of the requisite organ, are entirely without motion, whereas others acquire it in the course of their being perfected ...the degree in which things possess locomotion corresponds to the degree in which they have realised their natural development (*Phys*. VIII 261a 13-18)." He argues that the soul is "arranged in a series" of the "nutritive," "sensitive," and "intellective": see De An. II 414b33 - 415a 8. The plant only has the "nutritive" whereas the human being has "evolved" all faculties.
Whitehead Modes of Thought (1938) 27	
"in nature we find four types of aggregations of actualities: the lowest is [i] the non-living aggregation, in which mutual influence is predominantly of a formal character expressible in formal sciences, such as mathematics. The inorganic is dominated by the average. It lacks individual expression in its parts. Their flashes of selection (if any) are sporadic and ineffective. Its parts merely transmit average expressions; and thus the structure survives. For the average is always there, stifling individuality."	"[ii] The vegetable grade exhibits a democracy of purpose issuing from its parts...the nature of this average is dominated by the intricacies of its own bodily formation"

Fig. 10.1 The "evolution" of the consciousness of being in the world

Animal (zōon, living thing)	Man (anthrōpos)
Reproduction Nutrition Self-movement Intelligence (dianoia) Desire (epithymia) 	 Individuation Thought: Nous (intuition), Dianoia (intelligence), Gnomē (understanding) Desire: Boulēsis (will, wish), Epithymia (desire), Thymos (passion) Self-movement Nutrition Reproduction
Movement of Animals *Progression of Animals* *Generation of Animals*	*Nicomachean Ethics* *Politics*
Aristotle's biological work is overwhelmingly concerned with the increasingly complex, individuated, and mobile beings of nature. His scale naturae begins with chemicals and plants and concludes with human beings on this basis. The true natural revolution was, for Aristotle, the transformation of a rooted plant into an animal which he explains as that: "those that move about from place to place , have a stomach, or bag - as it were an earth inside them...they must have some instrument corresponding to the roots of a plant (PA II 650a 21-27)."	Aristotle argues that: "All substances [or "beings"] are by nature sources of a certain sort, so that each is able to produce many things of its kind, as a human begets other humans, an animal other animals, and a plant other plants. In addition to this a human being, alone among animals, is a source of conduct, for we would not ascribe conduct to any of the others (E.E. II 1222b16-20)." His theory of evolution moves from the natural cycles of his Meteorology to the development of his Nicomachean Ethics. It is essential to recognise that all of these things are natural objects or phenomena and should hence be treated on a common basis, i.e. human conduct is based upon our biological and chemical basis. Aside from conduct we find that human beings transform and transfigure all aspects of animal life (whilst also building upon our animality) whilst animals represent a quantum leap over the life of plants etc. On the scale of nature, see Top. 137b 24-26, Rhet. I 1362a 23-24, and GA I 731b 1-8.
"[iii] The animal grade includes at least one central actuality, supported by the intricacy of bodily functioning"	"[iv] The human grade of animal life immensely extends this concept."

Fig. 10.1 (continued)

tions.[51] Merleau-Ponty argues further that: "the body itself is not grasped as a material and inert mass or as an external instrument but as the living envelope of our actions, the principle of these actions has no need of being a quasiphysical force. Our intentions find their natural clothing or their embodiment in movements and are expressed in them as the thing is expressed in its perspectival aspects. Thus, thinking can be "in the throat," as the children questioned by Piaget say it is."[52]

Merleau-Ponty adds on the fascinating point that we can "think" with our body that: "it is the body which "understands" in the acquisition of habit"[53] and he provides such examples of this as dancing, driving a car, operating a typewriter, and walking with an exotic hat. The person conforms himself and shapes himself according to his accoutrements, tools, and their functions, requirements, and styles.[54] It is hence that humanity is an essential and transformative element of the world on the bodily basis that: "Our own body is in the world as the heart is in the organism: it keeps the visible spectacle constantly alive, it breathes life into it and sustains it inwardly, and with it forms a system. When I walk round my flat, the various aspects in which it presents itself to me could not possibly appear as views of one and the same thing if I did not know that each of them represents the flat seen from one spot or another, and if I were unaware of my own movements, and of my body as retaining its identity through the stages of those movements."[55] We do not exist as an accident or as a "point" or "atom" in the world but as a "whole" (and a "visible spectacle") which interacts with wholes but with these wholes only ever being "seen" incompletely.

Our being transforms the world around us, as presumably with Maurice's flat, and *shapes it in our image through our needs, culture, style, and personality*. This is the true understanding of the idea of "house" having a "final cause" as a place to live constructed and shaped by human beings. Merleau-Ponty adds on the very difficult point regarding the precise quality of something being a being-in-itself which can be

---

[51] Aristotle explains the "envelope" as that: "if what we mean by the "place" of a body is its immediate envelope (*to prōton periechon*), then "place" is a limiting determinant, which suggests that it is the specifying or moulding "form" by which the concrete quantum, together with its component matter, is "determined" (*Phys.* IV 209b2-4)" which Michel Bastit explains as that: "The place is separable; the boundary, on the contrary, belongs to the body; it traces the outline and is in some way the radiance of the form and is the focus and centre of the thing. It is therefore necessary to distinguish between the envelope and the enveloped. The place remains relative and the "ποῦ" is a predicate of the substance ("Aristote et la separation" *Revue Philosophique de Louvain* (1992) p 301)."

[52] Merleau-Ponty, Maurice *The Structure of Behaviour* (Boston, [*orig*. 1942] 1967) p 188. Merleau-Ponty's position on "seeing" is that: "I direct my gaze upon a sector of the landscape, which comes to life and is disclosed, while the other objects recede into the periphery and become dormant, while, however, not ceasing to be there (*Phenomenology of Perception* (London, [*orig*. 1945] 1962 p 78)."

[53] Merleau-Ponty, Maurice *Ibid.* p 165.

[54] Martin Heidegger makes this point in respect to "peasant shoes" in his "The Origin of the Work of Art" in *Poetry, Language, Thought* (New York, [*orig*. 1935–6] 1971) p 28 *ff*. See also Ernst Kapp *Elements of a Philosophy of Technology: On the Evolutionary History of Culture* (Minneapolis, [*orig*. 1877] 2018).

[55] Merleau-Ponty, Maurice *Phenomenology of Perception* (London, [*orig*. 1945] 1962 p 235.

perceived from various aspects that: "Perspective does not appear to me to be a subjective deformation of things but, on the contrary, to be one of their properties, perhaps their essential property. It is precisely because of it that the perceived possesses in itself a hidden and inexhaustible richness, that it is a "thing.""[56] Far from being an inadequate or partial rendering of a being or a merely subjective or imaginative experience of it, we see by rendering a "thing" as more than an "object" that we both express its meaning and also that of ourselves in respect to our relationship with it.

Merleau-Ponty draws out the ontological and ousiological sense of being. The world is not reducible to the logical or symbolic structures of mind but manifests in rich, material beings. A.N. Whitehead also identifies and explains this point as that: "Nothing is more astonishing in the history of philosophic thought than the naive way in which our association with our human bodies is assumed"[57] and Aldous Huxley makes the complementary point that: "It is worthy of note that the most "spiritual" religions have been the ones to pay the closest, most scientific attention to the body…There can be no spirituality except on a basis of well-informed materialism."[58] It is a point well appreciated by these philosophers that profundity and reality are both based upon a "*well-informed* materialism." We can see how and why Aristotle can be understood as a materialist *on this basis*.

Martin Heidegger (1889–1976) is the greatest philosophical exponent of Aristotle in the modern world. His divergences from both Husserl and Cassirer (and Peirce) can be attributed to his move from Kant, Descartes, and Plato back to Aristotle. It is a shift from philosophy based on subjectivity to one based on objectivity and from epistemology to ontology. This movement is epitomised by the famous 1929 disputation between Cassirer and Heidegger. Cassirer rested his neo-Kantian position on mathematics and morality, both of which were irrelevant to Heidegger's ontological approach. He also maintained that we have no access to the "thing-in-itself." We cannot even have access to the human person but we, as Simon Truwant notes, "only know what it *produces*. This means that his [i.e., Cassirer's] entire philosophy of culture is grounded in a transcendental subjective principle or mechanism that itself can be neither directly nor fundamentally understood."[59] This "transcendental subjective principle or mechanism" is ultimately Cassirer's "symbolic system" which, as we have noted, is derived from von Uexküll's biosemiotic "functional circle." The root dispute between Cassirer and Heidegger was over whether we approach the world through (Aristotelian) finitude or (Newtonian) infinitude and is ultimately a dispute over whether we work *from* "being(s)" (*ousia*(*i*)) or *from* mathematical laws.

Heidegger himself expands upon this situation at length in his "What Is a Thing?" lectures from 1935–6. He begins his programme here with a thorough consideration of why we should need to ask such an obvious philosophical question. He argues

---

[56] Merleau-Ponty, Maurice *The Structure of Behaviour* (Boston, [*orig*. 1942] 1967) p 188.

[57] Whitehead, A.N. "Forms of Process" in *Modes of Thought* (New York, 1938) p 114.

[58] Huxley, Aldous "The Oddest Science" in *Collected Essays* (New York, [*orig*. 1957] 1971) p 323–324.

[59] Quoted in Paul Gottfried "Showdown at Davos" *European Conservative* (2024).

that philosophy is a seeing of the world that is not a cut-and-dried objective mechanism as is science and therefore requires subtle contemplation and an unpacking of the meaning and value of concepts and ways of approaching the beings of the world. As regards what he takes to be "things," he concludes that: "As we ask "What is a thing?" we now mean the things around us. We take in view what is most immediate, most capable of being grasped by the hand... A thing in the sense of being present-at-hand: a rock, a piece of wood, a pair of pliers, a watch, an apple, and a piece of bread. All inanimate and all animate things such as a rose, shrub, beech tree, spruce, lizard, and wasp."[60] It is clear that Heidegger fundamentally shares Aristotle's emphasis upon the fundamentality of the "*ousia*" (being)—a word originally used to mean possessions—and also the "*pragma*" (thing; also meaning act or fact). The significance of these things being "present-at-hand" is beautifully explored by Maurice Merleau-Ponty in his Parisian flat, as outlined previously.

Heidegger provides us with some interesting, if perhaps eccentric, thoughts on the nature of "mathematics." He first argues that mathematics is not necessarily symbolic or quantitative but that it is rather an established framework of cognisance that we can immediately grasp and use rather than needing to contemplate and learn. He explains this mathematical framing as that: "The mathematical is, as *mente concipere*, a project (*Entwurf*) of thingness (*Dingheit*) which, as it were, skips over the things."[61] He adds as regards the dualism inherent in this mathematical approach that: "It is natural to arrange the whole state of modern metaphysics in the following way. For this form of metaphysics two concepts are essential: (1) the Christian conception of entities as *ens creatum* and (2) the basic mathematical character. The first instance concerns the content of metaphysics, the second its form."[62] Heidegger regards modern thought as expressing a collapse into the Christian and the mathematical. He attributes this situation to Descartes' "ego" as follows: "Until Descartes every thing present-at-hand for itself was a "subject"; but now the "I" becomes the special subject, that with regard to which all the remaining things first determine themselves as such. Because—mathematically—they first receive their thingness only through the founding relation to the highest principle and its "subject" (I), they are essentially such as stand as something else in relation to the "subject," which lie over against it as *objectum*. The things themselves become "objects.""[63] Heidegger concurs with Husserl as regards the perils of "objectivism" and "exteriorisation" and perhaps "naturalism" in the reduction of the world to the facticity of a mundane extension.

Hegel had similarly understood on this situation (as far back as 1829) that the problem of modernity is subsequent to the Christian posture. He observed that: "There can be no doubt that the essential Christian definition of freedom and of individuality, which as free is infinite within itself and is personality, has misled the

[60] Heidegger, Martin *What Is A Thing?* (Chicago, [*orig*. 1935–6] 1967) p 6–7.

[61] Heidegger, Martin *Ibid.* p 92.

[62] Heidegger, Martin *Ibid.* p 110.

[63] Heidegger, Martin *Ibid.* p 105.

understanding into conceiving the individualisation of finitude in terms of the category of a *subsisting* unchangeable atom, and of overlooking the element of the negative that resides in *power* and its general system."[64] He complained that the modern world has returned to "atoms and the void." He excellently observed that: "it is immediately evident that they [i.e., faith and reason] have reached formal extremes in which all content is abstracted"[65] with its consequences for "faith" being that it suffers an "intensive withdrawal of self-consciousness into itself, the development of the *content* into a mere *determinacy* of feeling"[66] and for "thought" that: "Thought supposedly is capable of comprehending one thing only, its incapacity to grasp the truth and see into it, and of proving to itself its own nothingness, with the result that suicide is its highest calling."[67] We see in this Cartesian position a collapse into a dualism between overdetermined objectivism and underdetermined subjectivism. It is an important point of note that Hegel observes that the collapse of "being" we witness in modern thought has an ostensibly *religious* origin.

Heidegger followed Hegel in emphasising the spiritual crisis that has come about in modernity through Christianity and the collapse into ego. He hence critiques Descartes and Kant (and, by association, Cassirer) on the basis that: "Metaphysical thought moves in the variously defined domain of subjectivity. Later Kant therefore says: All questions of metaphysics, i.e., those of the designated disciplines, can be traced back to the question: What is man? In the priority of this question there is concealed the priority of method coined in Descartes' *Regulae*."[68] The very mastery of method leads us to become technical specialists of our subject areas as egos or as atoms or monads within a mathematical world that we control. The "ego" increasingly controls and regulates the extended world around it. The alternative is to reject this constructed technological world and to face the entangled, situational world of reality. Heidegger explains that this suspended reality is a creative fusion between the (human) being and its world and is eminently emergent. We could say that we see a bookshelf at a distance from one perspective and then see it in a different light and in a different way when we approach it and see its stock. This is not, according to Heidegger, relativism or nihilism but a statement of the emergent and rich emergent meaningfulness of our reality.[69]

It is noteworthy that this emergent meaning-making is of an altogether deeper ontological nature than Cassirer's "symbolic system" which is a more sophisticated form of mechanicism. Ungrounded in beings, we find that: "mathematical logic and symbolic logic necessarily place themselves outside of every sphere of logic…as a

[64] Hegel, G.W.F. *Lectures on the Proofs of the Existence of God*, (1829) §311.

[65] Hegel, G.W.F. *Ibid.* §233.

[66] Hegel, G.W.F. *Ibid.*, §245.

[67] Hegel, G.W.F. *Ibid.*, §232.

[68] Heidegger, Martin *What Is A Thing?* (Chicago, [*orig.* 1935–6] 1967) p 110.

[69] See Martin Heidegger *Ibid.* p 29.

mere combination of representations."[70] For Heidegger, symbolism is no substitute for the meaning of the world and is deceptive when detached from it.

Jacques Derrida (1930–2004) as the great deconstructor is in a sense antithetical to the great constructor Aristotle. He does, however, write at some length on the Philosopher and even frames his post-structuralism, his denial of structuralism, in Aristotelian terms. He explains this point in his essay "Force and Signification" as that: "If there are structures, they are possible only on the basis of the fundamental structure which permits totality to open and overflow itself such that it *takes on meaning* by anticipating a telos which here must be understood in its most indeterminate form. This opening is certainly that which liberates time and genesis (even coincides with them), but it is also that which risks enclosing progression toward the future—becoming—by giving it form. That which risks stifling force under form."[71] Derrida gives us a useful but typically opaque explanation of how a being takes on meaning in accordance with its *telos*, its "whence and whither."

Derrida defines structuralism as that: "To be a structuralist is first to concentrate on the organization of meaning, on the autonomy and idiosyncratic balance, the completion of each moment, each form; and it is to refuse to relegate everything that is not comprehensible as an ideal type to the status of aberrational accident. The pathological itself is not the simple absence of structure. It is organized. It cannot be understood as the deficiency, defect, or decomposition of a beautiful, ideal totality. It is not the simple undoing of telos."[72] Structuralism suggests that meaning permeates all of nature, even its pathologies. Derrida adds, however, that the problem of structuralism is that it renders "meaning" as a subtraction, an extraction from nature that in a sense narrows and distorts it. He hence argues that: "light is menaced from within by that which also metaphysically menaces every structuralism: the possibility of concealing meaning through the very act of uncovering it. *To comprehend* the structure of a becoming, the form of a force, is to lose meaning by finding it."[73] Derrida explains the limitation of structuralism as being its reduction of the world to structures since meaning is itself destroyed by its construction and deconstruction.

If the above makes some constructive sense, how do we explain Derrida's own deconstructionism? We can begin here by observing that Derrida recognises the problem posed by number and word that was perhaps most fully elucidated by Aristotle. The Bible perhaps also picks up on this point with its statement that "the letter kills, but the spirit gives life."[74] Christianity solves the problem of formalism

[70] Heidegger, Martin *Ibid.* p 156.

[71] Derrida, Jacques "Force and Signification" in *Writing and Difference* (Chicago, 1978) p 26.

[72] Derrida, Jacques *Ibid.* p 26.

[73] Derrida, Jacques *Ibid.* p 26. He adds that: "The meaning of becoming and of force, by virtue of their pure, intrinsic characteristics, is the repose of the beginning and the end, the peacefulness of a spectacle, horizon or face. Within this peace and repose the character of becoming and of force is disturbed by meaning itself. The meaning of meaning is Apollonian by virtue of everything within it that can be seen (*Ibid.* p 26)."

[74] 2 Corinthians 3:6.

and legalism through the Christian message whereas Aristotle builds up an aporetic picture of reality which both develops and valorises analogical structures and leaves critical space for gaps. Derrida's response is less constructive than either of these approaches. John Sturrock hence describes Derrida's philosophy as a denial of what he calls a "metaphysics of presence" on the basis that: "The traditional priority accorded to speech comes about, Derrida claims, because of the deep "phonocentrism" of Western thought. We are sentimental towards the human voice and deluded by it into forgetting that even when we speak, or speak inwardly with ourselves, i.e. think, we still have recourse to signs. We cannot be fully "present" even to ourselves in so far as we must of necessity commune with ourselves in a system of signs that is not ours alone but a social institution. "Hearing ourselves speak" is the illusory model of intimacy and immediacy which Derrida suggests has enabled us conveniently to ignore the true nature of signs."[75]

What is the "true nature" of signs? Because we cannot apparently be "fully present even to ourselves" and must depend upon "signs," we must, according to Derrida, dismantle our thought and deconstruct the world. Sturrock hence observes that: "Derrida's ambitions…are to dismantle the whole system of Western thought since the time of Plato, because he believes it has been led astray by its reprehensible failure to grasp the nature of language and of meaning. In the work of Derrida the principles laid down by Saussure serve the sweeping and sceptical purpose of undermining a prevailing and generally unconscious "idealism," which asserts that language does not create meanings but reveals them, thereby implying that meanings pre-exist their expression. This for Derrida is a nonsense. For him there can be no meaning which is not formulated: we cannot reach outside language."[76] We see a sort of sophisticated derangement. To be generous, Derrida seems, as previously Diogenes, to be "a Socrates gone mad." He recognises the limitations of structuralism, of the reduction of the world to its mechanisms, but then offers a sort of apologia for a subversive nihilism in its place.

Is Derrida's philosophy a strange pathology of modern thinking? Strangely enough, Derrida is a flagship for modern thought. His ultimate achievement is to deny the existence of the "author," i.e., the human being. All we have instead is a "text," i.e., the meaning of something is just what is there, not what is intended. This deconstruction, this denial of "being" (*ousia*), is implicit in *all* leading schools of modern thought. They achieve this denial is various opaque, cryptic ways. The world is reduced to "effects," "masses", and "data." Derrida could have been vastly clearer, but he could not have been more on message.

Jean Baudrillard (1929–2007) is the final brilliant philosopher on our rostrum. He seems to embody the most excellent, bold, and lucid thought of our age. He creates a whole new modality to present modernity and postmodernity as a situationist "society of the spectacle" though he does so not to extol but to expose the tragic bankruptcy of this brave new world. Situationism does not free us to create our own

[75] Sturrock, John *Structuralism* second ed. (Oxford, [*orig*. 1986] 2003) p 127.

[76] Sturrock, John *Ibid.* p 123.

world but unmoors us from the world and places us into a hyperreal simulacrum of reality.[77] Baudrillard transforms Cassirer's symbolic world through the dark lens of such works as *For a Critique of the Political Economy of the Sign*, *Symbolic Exchange and Death*, and *Simulacra and Simulation*. He recognises that we have moved into a world *beyond reality* but clearly wishes to *re*-ground our abstractions in reality even if he perhaps entertains no hope for any such return to normality.

For Baudrillard, we have lost a sense of our own body and of our own being. As regards the human body in the modern world, he argues that: "What makes the obese obscene is not that there is too much body, but that the body is superfluous, *de trop*."[78] The body loses its form and being or identity in the same way as all other beings in the post-modern world. He explains this situation as that: "There are no obese animals, just as there is no obscene animal. Could it be that the animal is never confronted with the scene, nor with its own image? Not being subjected to this scenic obligation, it could not be obscene. For man, on the other hand, this obligation is absolute, but for the obese there is a kind of release from this obligation, from all pride of representation, from any seductive impulse—the loss of the body as face. The pathology of the obese is not endocrine, but a pathology of the scene and the obscene."[79] As Aristotle argued that only man can be a beast, Baudrillard argues that only man has the self-awareness and calculation to be obscene.

Baudrillard recognises that the loss of beings and the loss of limits ultimately represents a destruction of meaning itself. Rather than a healthy rhythm of human growth and maturation, we encounter a transformed "logic of excresence and inertia" operating on the basis that: "There is an incredible overpotentiality of strategic weapons, equalled only by the demographic excresence of the world. As paradoxical as it seems, both are of the same kind and correspond to the same logic of excresence and inertia. This is a triumphant anomaly: no principle of law or measure can temper either one or the other—they reciprocally imply each other. And the worst thing is that there is no Promethean challenge involved here, no excess of passion or pride. It simply seems that the species has crossed some specific, mysterious point, from which it is impossible to retreat, decelerate, or slow down."[80] This new logic is a case study for how the logic forming our worldview can "overflow its own reality principle and…be refracted in another logic."[81] Heidegger had already

[77] Guy Debord, founder of the revolutionary group "Situationist International," explains that: "Situationist theory resolutely supports a noncontinuous conception of life. The notion of unity must cease to be seen as applying to the whole of one's life (where it serves as a reactionary mystification based on the belief in an immortal soul and, in the final analysis, on the division of labour); instead, it should apply to the construction of each particular moment of life through the unitary use of situationist methods. In a classless society there will no longer be "painters," but only situationists who, among other things, sometimes paint (*Report on the Construction of Situations and on the International Situationist Tendency's Conditions of Organisation and Action* (online, 1957))."

[78] Baudrillard, Jean *Fatal Strategies* (South Pasadena, [*orig*. 1983] 2008) p 53.

[79] Baudrillard, Jean *Ibid.* p 51.

[80] Baudrillard, Jean *Ibid.* p 32.

[81] Baudrillard, Jean *Ibid.* p 211.

commented on this false logic of unreality. We find as regards the new logic and worldview of postmodernism that: "Things rush into it at the rhythm of a technology, including soft and psychedelic technologies, which drag us ever further away from any reality, any history, any destiny."[82] As regards how we now engage with ourselves, we find that: "As with television, the entire surrounding world, and our own body, becomes a control screen."[83]

Far from being elevated, liberated, and enlightened, we face the strange thought that our modernity is worse than a sham, is a fraud, a deceit, a malicious game. Baudrillard explains that our movement has now become a sick frenzy of activity rather than progression to a goal such that: "Movement does not disappear as much into immobility as into speed and acceleration—into the more mobile than movement, so to speak, which pushes it to the limit while stripping it of sense."[84] The "*telos*," now desubstantiated, becomes a "hypertely" such that: "Plunged by chance into an abnormal uncertainty, we have responded with an excess of causality and finality. Hypertely is no accident in the evolution of a few animal species, but a challenge of finality to a growing indeterminacy. In a system where things are increasingly governed by chance, finality turns into delirium, and elements develop that know only too well how to exceed their end – until they wind up invading the whole system."[85] We ultimately see how the marriage of "chance" and "necessity," so dear to the worldview arising out of neo-Darwinism, necessarily results in "an attempt to exhaust phenomena back to their infinitesimal causes."[86]

We unavoidably arrive at Nietzsche at his "last man" (""We invented happiness"—say the last human beings, blinking."). Are we the last men? Have we invented or inverted happiness? This is the question that Baudrillard's brutal anti-anti-philosophy poses to the anti-philosophy of modernity. Innocent, without the need for Baudrillard's irony, Aristotle tackled and analysed out the inherent problems of power and illusion as exemplified by his consideration of tyranny and sophism. Aristotle was perhaps the "first man" to find the need to ground humanity in reality on a human scale, to determine and impose limits, and, if we are to return to reality, the philosophers, from Husserl and Heidegger to Ellul and Camus, agree that we must bite our lips and return to the Greeks.[87]

---

[82] Baudrillard, Jean *Ibid.* p 46.

[83] Baudrillard, Jean *Ibid.* p 91.

[84] Baudrillard, Jean *Ibid.* p 29.

[85] Baudrillard, Jean *Ibid.* p 30.

[86] Baudrillard, Jean *Ibid.* p 31.

[87] Camus draws a contrast between "German dreams and Mediterranean traditions, between the violence of eternal adolescence and virile strength…between history and nature (*The Rebel* (London, 1953) p 262)" with the Mediterranean tradition being "the long tradition that can be called solitary thought and in which, since the time of the Greeks, nature has always been weighed against evolution (*Ibid.* p 262)." He adds that: "German ideology…has come into an inheritance. It consummates two centuries of abortive struggle against Nature, first in the name of an historic god and then of a deified history…When Nature ceases to be an object of contemplation and admiration it can then be nothing more than material for an action which aims at transferring it (*Ibid.* p 262–3)." See also Jacques Ellul *The Technological Society* (New York, 1964).

10.2 The False Path of Scholasticism

Having reviewed what I designate as philosophy, let us now consider what I suggest should be designated as modern ideological scholasticism. I take this to be the academic philosophy put forward by the universities to support the ruling ideology or religion of our day. One example of this in the modern world has been the cottage industry of philosophy to abolish philosophy and make it merely a defence of scientific method taken in lieu of philosophy. W.V. Quine stated this objective as that: "Our ontology is determined once we have fixed upon the over-all conceptual scheme which is to accommodate science in the broadest sense."[88] Much of our modern scholasticism competes to achieve this dubious goal.

A further downstream scholastic competition concerns proving the ideological claims of neo-Darwinism, *viz.* that the world is the result of (random) chance and (material) necessity. This dispensing with beings and values in favour of (random) atoms and mathematics is perhaps a prerequisite for the elevation of science over philosophy. Even Karl Popper could not help but recognise on this situation that: "The fear of using teleological terms reminds me of the Victorian fear of speaking about sex, because teleological terms are constantly used in biology and are utterly unavoidable…How can we avoid speaking of the roots of a tree *seeking* water, *seeking* better conditions? If we avoid it, then we deceive ourselves and speak in an artificial language when there is no need to introduce an artificial language?"[89] A.N. Whitehead also pithily and correctly commented on this venture in professional obfuscation that: "Scientists animated by the purpose of proving that they are purposeless constitute an interesting subject for study."[90] Our hungry scholars have not, however, typically exhibited any willingness to listen to any such criticism that might put their goals under question.

Scott Sehon provides us with a good first example of these academic apologetics in his book on teleology in which he concludes that: "If we could say that the entire world is composed of physical particles, we would be back in business; the physical constitution of the world as a whole would explain the global supervenience of mental properties on physical properties…because we are composed of physical particles, supervenience is no longer mysterious. In fact, it seems obvious to me that the supervenience of the mental on the physical can be explained on the fact that we are composed of physical particles."[91] To explain this position he uses the example of a "chessplus" computer program, as follows: "Just as the chess positions were constituted by the pixels, so too are we constituted by elemental physical particles, and hence it should be unsurprising that our mental properties supervene on physi-

---

[88] Quine, W.V. "On What There Is" in *From a Logical Point of View* (Cambridge Mass., 1953) p 18.

[89] Popper, Karl "A New Interpretation of Darwinism" in Niemann, Hans-Joachim *Karl Popper and the Two New Secrets of Life* (Tübingen, [*orig.* 1986] 2014) p 124–5.

[90] Whitehead, A.N. *The Function of Reason* (Princeton, 1929) p 12.

[91] Sehon, Scott *Teleological Realism: Mind, Agency, and Explanation* (Cambridge Mass., 2005) p 133–4.

cal properties. However, the analogy between us and the chessplus screens is not perfect…anything that is capable of having effects is composed of physical particles. So, given also that everything inside the skin is composed of physical particles, it follows that everything capable of determining the content of mental states is composed of physical particles…the only available alternative theories might make the agent even more irrational or might create large mysteries in our theory of the rest of the world."[92]

To supposedly avoid "large mysteries" and the "irrational," Sehon does not even get to the starting block for explaining the "mind-body problem." Instead, as with theological apologetics, we find that the logic is wholly circular. Beneath the pseudo-scientific jargon, we find no substance but only ideology. The ideology we find is, here as typically, that the world is made up of little bits of matter that we can clump together and take apart to reach "scientific" conclusions. Bertrand Russell explained this dream, in 1914, as that: "If the mechanical account of matter were complete, the whole physical history of the universe, past and future, could be inferred from a sufficient number of data concerning an assigned finite time, however short."[93] It is in this context that Sehon claims that bodies can be defined as "material particles" which are "embedded in the body of an agent" on the reductionistic basis that: "We are constituted by material particles, and these material particles don't suddenly cease to follow the laws of nature just because they are embedded in the body of an agent."[94] Sehon does not provide any meaningful sense in respect to how or why these particles are "embedded" or what an "agent" actually is or signifies. Typically, there is no attempt at philosophical subtlety and no attempt to engage with the philosophical tradition on these matters.

We have above just one example from an industry, or perhaps a desert, of scholastic apologetics in a similar vein. Let us consider three further examples of these scholastic apologetics for dispensing with teleology in biology.

1. Graham Macdonald argues regarding *biological function* that: "Sperm surely have as their function the fertilisation of eggs, but since very few of them perform this function, on a probabilistic basis it would be foolhardy to attribute to sperm the function of fertilising eggs."[95] Note the scholastic tone and evasiveness of this proposition.
2. Scott Sehon argues regarding *human action* that: "Explaining the notion of agency in terms of conscious choice may not seem to have gotten us very far; worse yet, it is highly misleading, since many fully intentional actions are not performed with anything like a conscious viewing of reasons. For example, when I signal before making a turn in my car, I may not be consciously aware of

[92] Sehon, Scott *Ibid.* p 133.

[93] Russell, Bertrand *Our Knowledge of the External World* (London, [*orig.* 1914] 1926) p 224.

[94] Sehon, Scott *Teleological Realism: Mind, Agency, and Explanation* (Cambridge Mass., 2005) p 231.

[95] Macdonald, Graham "Reduction and Evolutionary Biology" in David Charles & Kathleen Lennon (*eds.*) *Reduction, Explanation, and Realism* (Oxford, 1992) p 80.

having done it at all, much less am I consciously contemplating some light-giving reason for signalling."[96]

3. Andrew Woodfield argues regarding *biological ecosystems* that: "he [i.e., Aristotle] thinks that every natural thing will, if the opportunity arises, perform *some* movement which is end-directed, namely, that which is natural to it. He thinks this about both inorganic and organic things."[97] Woodfield then accuses Aristotle of a "spurious invocation of a natural tendency, whether it is a tendency to be, to become, or to move. Suppose that an explanation is required for the fact that ripe apples fall off trees. A possible place to look is the apples themselves. If we could discover something about apples which causes them to fall when certain circumstances occur, this would count as a perfectly good explanation, on a par with explaining why sugar dissolves in water in terms of its molecular structure. Consider, however, the claim that apples fall because they have a natural tendency to fall."[98]

Why are the above positions obviously wrong according to Aristotle and the current author? On the first two points we find:

1. As regards Macdonald's "sperm," that he falls into the Megarian error of assuming that any something only has a "power" when it is being actively exercised. We see that sperms do unquestionably (i) possess the active and highly specialised positive power of inseminating ova and (ii) exercise their function *en masse* rather than as a one-off chance individual. The contention that the sperm exists in the world only as a chance, functionless occurrence (since not *every* sperm will go on to fertilise an ovum) is evidently ridiculous since no other random thing could fertilise a specific ovum.
2. As regards Sehon's driver, driving, and car, we see that the idea of driving is actually a perfect example of a "goal directed" activity. The idea that an action being a non-conscious (and habituated) goal-directed action somehow refutes rather than supports and validates the existence of teleology in nature seems to represent a fundamental confusion. It is a basic premise of Aristotle's *Ethics* that "habit" (*hexis* or *ethos*) is something *formed* which *becomes* part of us. This is an observable truth about behaviour that Aristotle seeks to explain. He compares habit to a stone that we can choose to throw but which once released will follow its own path. He maximises the influence of formed habit whilst never losing sight of the fact that the action is in an important sense, even as the action of animal x, *in* the individual and expressed by it. Although he is less concerned in his work with the individual and the spiritual, i.e., that which gives us our creative individuality, than with generalisable, scientific rules of human action, he leaves space for creative choice and action within his framing of behaviour.

[96] Sehon, Scott *Teleological Realism: Mind, Agency, and Explanation* (Cambridge Mass., 2005) p 176.

[97] Woodfield, Andrew *Teleology* (Cambridge, 1976) p 6.

[98] Woodfield, Andrew *Ibid.* p 6–7.

Aristotle deemphasises the creative and individual dimension of man on pedagogical grounds whereas Sehon and his colleagues seem to wish to deny it on ideological grounds.

I note that Sehon seems to take us down the path towards B.F. Skinner's behaviourism which operates on the basis that: "If you will be reinforced by reaching Boston, you will be reinforced if you follow Route 1"...[which is equivalent to] "You should (you ought to) read *David Copperfield*," which may be translated, "You will be reinforced if you read *David Copperfield*."[99] Aristotle himself insists that human action is *always* meaningful (again, by reference to Megara) on the basis that: "if he [i.e., a man] makes no judgement but "thinks" and "does not think" indifferently, what difference will there be between him and a vegetable?...why does a man walk to Megara and not stay at home, when he thinks he ought to be walking there?... Evidently, he judges one thing to be better and another worse. And if this is so, he must also judge one thing to be a man and another to be not-a-man, one thing to be sweet and another to be not-sweet (*Met.* Γ 1008b 10-20)." Even the most quotidian human action refutes behaviourism, let alone reading *David Copperfield*.

On the third point we find:

3. Regarding Woodfield's "apple" we should note that his full argument is that it is spurious to say that *any* organism or organic product has a "tendency to be, become, or to move." This actually seems to dismiss the very possibility of any organic being and to run completely against observed reality. On the cosmic situation in respect to the "apple" we find that Aristotle did believe that such a thing as an apple would "fall" but not *qua* apple but on the basis that it is solid matter. He would also likely argue that that the seed within the apple is a potentially living being seeking out life and sustenance so that it is properly "placed" within the earth. We could say that the seed (or the tree) seeks out the earth through the apple as a mechanism on this basis. The seed and fruit is an organ of the tree which has a place in the world and an innate directionality on this basis. Dandelion seeds are dispersed on the basis of a very different directionality.[100]

I note that the "apple" is not the paradigm that Aristotle himself uses in his work although he does give many examples which Woodfield *could have* used and critiqued. Aristotle rather explains *his* own position as regards *the functions, placement, and directionality of organs* as that: "The parts of plants are organs too, though very simple ones: e.g., the leaf protects the pericarp, and the pericarp protects the seed; the roots are analogous to the mouth, for both these absorb food (*De An.* II 412b 1-4)." He also puts forward *as a paradigm of a mechanistic explanation* the example that leaves fall only from deciduous trees because of "coagulation of

[99] Skinner, B.F. *Beyond Freedom and Dignity: A stunning new plan to alter human behaviour* (New York, 1971) p 106.

[100] See for instance Caroline Brogan "Engineers uncover secret 'thinking' behind dandelions' seed dispersal" *Imperial* (2022) and Lauren Sullivan "Seed Dispersal: Deciding when to move" *eLife* (2023).

sap at the junction of the leaf-stalk (*Post. An.* II 99a29)," i.e., he gives an example of organic matter falling for purely mechanistic reasons. Today, we can, however, ourselves provide a "reason why" (*dia ti*) for this mechanism of nature, *viz.* that deciduous trees shed their "broad-leaved" leaves because they are thin and unable to function in cold weather.[101] Aristotle also offers the example of a stone falling from a roof (as opposed to being thrown) as a paradigm of a *chance* "falling" event at *Phys.* II 197b 30-33. We see that Aristotle does not have strange animistic commitments but does recognise the functionality operating both in organic beings and in ecosystems which is the real basis of both his science and ours.

Why, then, does Woodfield provide us with a dubious example of "Aristotelian" teleology when Aristotle himself gives us other authentic and contradicting examples? I suggest that Woodfield likely chooses the "apple" because he wants his account to be identified as scientific and modern through the reflected glory of (the myth identified with) Newton. He wants to draw upon the mythology associated with animism. It is also no coincidence that Woodfield ultimately seems to propound the "behaviourist" agenda of B.F. Skinner who himself puts forward the calumny in his *Beyond Freedom and Dignity* from 1971 that: "Aristotle argued that a falling body accelerated because it grew more jubilant as it found itself nearer home."[102] We find regarding this "behaviourist" position that the promotional material for Skinner's book is that it is a "million copy bestseller" which argues that: "We can no longer afford freedom, says B.F. Skinner, the most influential and controversial psychologist of our time; we must design our culture to shape the behaviour needed for survival." Is this race to the bottom what the denial of teleology and final causes entails? Is this the price we must pay for Woodfield's apple?

We find regarding the overall picture of this approach that one of the most interesting points of commonality between Macdonald and Woodfield is their shared emphasis upon the spuriousness of the "natural tendency, whether it is a tendency to be, to become, or to move." For Macdonald, the birds and the bees are redundant; reproduction simply happens as some sort of chance happening. Woodfield dismisses the very idea of cybernetic systems and feedback loops etc. on the basis that they suggest processes higher than the atomic individual. His argument on this point is that: "Only the organisms which make up the species do things. What actually happens is that individuals of one generation reproduce differentially, and the consequences of this, when aggregated, determine the constitution of the next generation. The way in which the distribution of characteristics throughout a population alters as a result of natural selection depends on haphazard environmental changes."[103] He defends this position but only against such ridiculous teleological

[101] On "broad leafed" see *Post. An.* II 98b 1. On the mechanics of leaf fall see Melissa Petruzzello "Why Do Leaves Fall in Autumn?" *Britannica* (online).

[102] Skinner, B.F. *Beyond Freedom and Dignity* (New York, 1971) p 6.

[103] Woodfield, Andrew *Teleology* (Cambridge, 1976) p 199. *Cf.* Ludwig van Bertalanffy "General Theory of Systems: Application to Psychology" in *A Systems View of Man* (Boulder, [*orig.* 1967] 1981) and Denis Noble *The Music of Life: Biology Beyond Genes* (Oxford, 2006).

arguments as that: "Noses exist in order to support spectacles"[104] though no-one would refer to this as a "function" leading the olfactory organ *to exist* even if the sense organs are (in actual fact) clearly "arranged" in a pattern according to which it has the functional sense of having the nose and its sense below the eyes and above the mouth.

10.3 Keeping Reductionism Safe from Reality

Jason Scott Robert provides us with a useful, and sadly rare, critique of this strange industry of anti-meaning. He counters Alexander Rosenberg's claim that "the developmental geneticist [should expect to be able] to "compute" the eye from nucleic acids and proteins alone" with the riposte that: "Rosenberg implausibly contends that a full explanation of development will have "no room" for any reference to cell physiology, or anything else above the level of the "molecular processes that subserve development.""[105] Robert also replies to Rosenberg's claim that: "The future cannot cause the past" with his own that: "Of course, no one is claiming that causation works against the arrow of time; but even if molecular processes do indeed occur before (and concurrently with, and after) cellular processes, it is an open question whether cellular processes and structures are in fact explicable (or even predictable) from a description of molecular processes and structures."[106] Robert correctly defends the fact of structure, whole, and hierarchy although he perhaps unreasonably cedes ground on the claim that: "the future cannot cause the past." The whole idea of a genetic program, of development, and perhaps even of evolution suggests that many biological events do occur in the present by reference to future events, including the apple tree having its seeds distributed. We simply do not know how this natural "program" exists and operates but this does not mean that we should ignore or dismiss it.

What can we infer from these strange attempts to deny (contrary to all appearances) the existence of "life"? Here we can consider Gunnar Babcock and Daniel McShea's article "An Externalist Teleology." The authors argue that they can "naturalise" teleology on the basis that: "For Aristotle, artifacts are guided by something external to themselves, human intentions, while natural entities are guided by an internal nature. We aim to show that there is, in fact, only one type of legitimate teleological explanation, what Aristotle would have considered a variant of an artefact model, where entities are guided by external fields."[107] How do they prove this point? They appeal, remarkably enough, to Aristotle's hypothesis on

104 Woodfield, Andrew *Teleology* (Cambridge, 1976) p 32.

105 Robert, Jason Scott *Embryology, Epigenesis, and Evolution: Taking Development Seriously* (Cambridge, 2004) p 15.

106 Robert, Jason Scott *Ibid.* p 14.

107 Babcock, Gunnar & McShea, Daniel W. "An Externalist Teleology" *Synthese* (2021) p 8755.

"spontaneous generation": "Recall that Aristotle believes that spontaneous generation occurred when the right concoction of earth, water and air came together and were heated in the right way to generate certain animals. If such a process were to take place, field theory would merely note that the combination of environmental conditions that directed the critical materials is responsible for spontaneous generation. Without the NAD [Natural/Artificial Distinction], without a commitment to internal guiding factors for natural entities, the solution is straightforward."[108] The goal of this whole approach is to devise a form of materialism or physicalism that can obviate the need for "a commitment to internal guiding factors for natural entities."[109]

What is wrong with this position? Or, what is right about it? Aristotle's hypothesis of "spontaneous generation" is one of the few of his hypotheses that has been broadly disproved. However simple, organisms *do not directly arise from out of matter*. The reality of simple or complex organisms is that they possess a structure, organisation, and a genetic "program." Matter and materialism, even as qualified by Aristotle as a pregnant possibility, is a *sine qua non* but not a sufficient explanation for the meaning of life. This is also not how artifacts come about. Artifacts are created and constructed and are not the result of random material interaction. What, then, is the *point* of this argument? Babcock and McShea's concluding sentence is that: "An externalist teleology may be not just safe for biology, but essential."[110] Their aim is supposedly to make biology "safe." They are, in other words, engaged in the same dismal task as the other examples given above of trying to construct a plausible narrative which can be claimed to be the "simplest conceptual scheme…to accommodate science in the broadest sense."[111]

I note that another disappointing example of sophistry to make biology supposedly "safe" is provided by Tom Vinci and Jason Scott Robert in their "Aristotle and Modern Genetics" (2005) which attacks Ernst Mayr's defence of "teleonomy" and "genetic programs" in nature on the basis (i) that such ideas (of Mayr) are "vitalistic claptrap" and (ii) that they were not concepts held by Aristotle. It seems strange to challenge either the idea that Aristotle was a "vitalist" or that Mayr made advances in rehabilitating his vitalistic position.[112]

---

[108] Babcock, Gunnar & McShea, Daniel W. *Ibid.* p 8766.

[109] See Tim Crane & D.H. Mellor "There is No Question of Physicalism" *Mind* (1990).

[110] Babcock, Gunnar & McShea, Daniel W. *Ibid.* p 8777.

[111] Quine, W.V. "On What There Is" in *From a Logical Point of View* (Cambridge Mass., 1953) p 18.

[112] On Aristotle's vitalism see Denis M. Walsh "Objectcy and Agency: Towards a Methodological Vitalism" in Nicholson, D. & Dupré, J. (*eds.*) *Everything Flows: Towards a Processual Philosophy of Biology* (Oxford, 2018) and Armando Aranda Anzaldo "Aristotle and the search of a rational framework for biology" *Organisms. Journal of Biological Sciences* (2019) & "Back to the Future: Aristotle and Molecular Biology" *Ludus Vitalis* (2007).

10.4 Conclusions

What do we conclude about *our* Battle of the Gods and Giants? Its existence is obvious. We find the "Gods" providing us with subtle and elegant profundities often in language so convoluted and arcane as to almost parody medieval scholasticism. Perhaps the archetypal culprit of this convolutedness is Jacques Derrida whose strange crypsis merely hides a nihilism, an awful attempt to negate meaning and being itself. Jean Baudrillard presents us with the only truly serious philosophy we are capable of, exposing and critiquing the corruption of the "last man." The "Gods" can not only not "save the phenomenon," they cannot even save themselves. The "Giants" represent the mirror opposite of this helplessness leading to a sort of managed collapse. Here the *intention* is to attack and break down being in the name of "analysis" and "science" generally. This is a more toxic deconstruction of reality even than Derrida because it seems less obviously subversive owing to its exaggerated tediousness. It *appears* useful, boring, and inevitable. In reality, it is simply more toxic ideology. Perhaps Bertrand Russell could be made the patron saint of this school of "intellectual rubbish."

What is the aim of the "Giants"? The worldview put forward by neo-Darwinism is designed to support the externalist emphasis of Hume and Newton. What is being kept "safe" is more a philosophical dogma than a scientific conclusion. The subtext here is to keep alive some imagined "seventeenth century" and, per Richard Rorty, "the conviction that seventeenth-century science discovered that everything could be explained by atoms and the void, and that philosophy has a moral duty to preserve this insight…[and] if there was supposed to be answers to such general questions as "How is abstraction (recognition, constancy) possible?" [we] can say, there is *no* answer, except the pointless remark that nature has evolved suitable hardware to get the job done."[113] This approach is embodied in Babcock and McShea's emphasis upon forming behaviour through external forces which represents a defence of behaviourism and hence of technocratic control over society, social engineering etc. Their position is therefore political as well as philosophical.

We find the most evident source of the problems of modernity in the traditional dogmas of positivism. These are simply stated by Karl Popper (early in his career) as that: "scientific or nominalist definitions do not contain any knowledge whatever, not even any "opinion"; they do nothing but introduce new arbitrary shorthand labels; they cut a long story short"[114] and that: "science does not use definitions in order to determine the meaning of its terms, but only in order to introduce handy shorthand labels."[115] It is notable both that even Popper radically rejected this position later in his career and that his philosophy was first motivated by the political rejection of collectivism and then by the further contrary rejection of neoliberalism. This intrusion of politics into philosophy and science is itself a problem of some

[113] Rorty, Richard *Philosophy and the Mirror of Nature* (Princeton, 1979) p 217 & 235.

[114] Popper, Karl R. *The Open Society and Its Enemies vol. 2* (London, 1945) p 14.

[115] Popper, Karl R. *Ibid.* p 18.

note. As regards the state of the science itself, Jason Scott Robert notes that "even a cursory look at developmental biology today"[116] disproves Alexander Rosenberg's reductionism. We will explore these refutations of reductionism from within science—from quantum physics to biosemiotics—and the philosophy of science in the sister volume for this work. Here I simply note the fundamental problem with all these apologetics.

[116] Robert, Jason Scott *Embryology, Epigenesis, and Evolution: Taking Development Seriously* (Cambridge, 2004) p 15.

Chapter 11
Conclusions

Abstract The problem we face today, as yesterday, is that philosophy has been ruptured between an existentialistic spirituality and a materialistic rationality. Science has become the authoritative source of all worldly meaning, without the means or mandate to be such, and has replaced philosophy which has been reduced to a fag-end of subjective human feelings. Hegel rightly describes this situation as a kind of "suicide." By contrast, Aristotle philosophically locates human beings in nature as the prime movers of any scientific or philosophical envisioning of the world. Science is as much under the eye of philosophy and its metaphysics as anything else in the world. Through Aristotle, we join up the world again and give it meaning rather than just the technique of technology.

Keywords The problem of scientific rationalism · The problem of conflating science with philosophy · The value of Aristotle's philosophy

The obvious (Hobson's) choice we seem to face is between Plato's "Gods" and "Giants"? I have suggested above that the various phenomenological and existentialist schools of philosophy, from Hegel to Heidegger and Baudrillard, clearly add to the Western philosophical tradition in a meaningful sense. We must, however, consider within the context of this exercise, quite *what* they add.

The problem we face is that these modern thinkers are "Gods" and contribute as such. They explore and explain the psychological and spiritual situation in which the human being finds himself. They represent the *spiritual* tradition of the "care of the soul" so eloquently analysed by Michel Foucault in the latter years of his life. In his final lecture series, Foucault sought to unpack the "parresiastic discourse and standpoint in philosophy"[1] of the ancient world as a "parresiastic modality [which]

[1] Foucault, Michel *The Courage of Truth: The Government of Self and Others II* (Basingstoke, [*orig*. 1983–4] 2008) p 68.

P. N. Jackson, *Aristotle on Meaning in the Living World*, Biosemiotics 30,
https://doi.org/10.1007/978-3-032-00602-8_11

has, I believe, precisely disappeared as such."[2] Foucault explained further that: "For a Greek, human freedom has to be invested not so much, or not only in the city-state, the law, and religion, as in this *teknē* (the art of oneself) which is practiced by oneself. It is, then, within this general form of the *teknē tou biou* that the principle or precept "take care of yourself" is formulated."[3] Foucault ultimately concentrates on the question: "What is Man?" in the Stoic and Christian traditions and then in the Cartesian and Kantian traditions.

We find that even Michel Foucault is a "God." By recognising this fact, we can understand his seeming bemusement with or lack of sympathy for Aristotle. As noted earlier, Foucault described Aristotle only as: "the one who is called "the" philosopher, because he was no doubt the only philosopher in Antiquity for whom the question of spirituality was least important; the philosopher whom we have recognised as the founder of philosophy in the modern sense of the term: Aristotle. But as everyone knows, Aristotle was not the pinnacle of Antiquity but its exception."[4] Why do I think that Foucault (and, according to him, "everyone") is in error on this point? He is in error on the point that Aristotle is "modern" because he is uninterested in "spirituality." I rather suggest that Aristotle does not pursue the "modern" obsession with inner feelings or proffer advice about how we should feel about the world but takes an objective approach that is *more* spiritual than any such subjective speculation. He defines and defends a framework of reality in which our feelings are represented not merely as the capricious operations of our egos but as a real engagement of a human being with the meaningful world around him.

His *Ethics* do not tell us how to behave or how to be a valuable member of society, but they rather exhibit the space in which meaning is made and people and societies come to be. His theology similarly creates a space for "God" based upon the recognition of a need for "God." This is not a failing but a truly spiritual understanding of value and good as something that can only be recognised as something ineffable, special, unquantifiable, emergent, and transcendent.

11.1 The Value of Aristotle

Aristotle is, in fact, a deeply spiritual thinker who can look beyond images to the meaning and significance of those images. It is hence that he explicitly dispenses with Plato's mythology on the basis of its crudity and literalness as confusing divinity with "eternal men (*Met.* B 997b 11)" and he puts philosophy and theology in its place. The "Unmoved Mover" is on this basis a space for the ineffable meaning of the world.

[2] Foucault, Michel *Ibid.* p 30.

[3] Foucault, Michel *The Hermeneutics of the Subject* (New York, [*orig.* 1981–2] 2005) p 447.

[4] Foucault, Michel *Ibid.* p 17.

Aristotle generally observes that Plato's forcing of fixed content onto an emergent aspect of reality degrades that reality, something ideologues and idealists have systematically failed to understand ever since. It is to valorise the everyday meaning of the real world that Aristotle replaces idealism with realism. It is hence that Aristotle breaks out the "higher" meaning of the world between "*Nous*," the meaningfulness of being, and the "Unmoved Mover," the stability of being. He also adds that the sun has a special place in this framework and additionally posits the existence of 47 or 55 cosmic spheres shaping our world.[5] None of this is silly or inconsistent; it simply represents the fragments of our knowledge of the "eternal" and the gaps between them. As regards "*Nous*" we have already noted Aristotle's argument that: "thinking is actuality (*hē noesis energeia*) (*Met.* Θ 1051a 31)" and that: "the actuality of *nous* is life (*hē gar nou energeia zōē*) (*Met.* Λ 1072b 27)." Aristotle reenvisages Plato's emphasis upon the power of ideas as that the actuality of life is manifested in thought.

We also find that Aristotle's politics are completely consistent with his cosmology and not in a forced or unnatural way but as a product of a holistic, properly calibrated and subtle approach to the world. It is through the confident and powerful lens of his *Politics* that Aristotle looks at the dangerous errors of misapplied rationalism hiding beneath the wisdom and power of Plato's *Republic*. He explains here that: "there are two motives that most cause men to care for things and be fond of them, the sense of ownership and the sense of preciousness (*to te idion kai to agapēton*); and neither motive can be present with the citizens of a state so constituted [as in Plato's *Republic*] (*Pol.* II 1262b 23-25)." He explains further that: "just as putting a little sugar into a quantity of water makes the mixture imperceptible, so it also must come about that the mutual relationship (*oikeiotēta*) based on these names [i.e., "my father" and "my son"] must become imperceptible, since in the republic described by Plato there will be the least possible necessity for people to care for one another (*Pol.* II 1262b 18-21)."[6] Aristotle understands perfectly well about the meaning of "care" and morality but he does not prescribe it for fear that to do so would destroy it. He hence observes that humanity and its values are natural and emergent and not artificial and rational and cannot be captured and formularised on this basis.

11.2 Should We Be Gods or Giants?

It is obvious that Aristotle is not an idealist or a "God" but nor is he a "Giant." Aristotle is an "empiricist" but it would be ridiculous to associate him with the dreadful monsters "trying to drag everything down to earth out of heaven and the

[5] On the importance of the sun see *Met.* Λ 1071a14–18 and on the heavenly spheres see *Met.* Λ 1073b 1–1074a 13.

[6] On Aristotle's politics vis-à-vis Plato's see especially Darrell Dobbs' "Aristotle's Anticommunism" *American Journal of Political Science* (1985).

unseen, literally grasping rocks and trees in their hands; for they lay hold upon every stock and stone and strenuously affirm that real existence belongs only to that which can be handled and offers resistance to the touch. They define reality as the same thing as body, and as soon as one of the opposite party asserts that anything without a body is real, they are utterly contemptuous and will not listen to another word (*Soph.* 246A-B)." The philosophy of these "Giants," such as it is, is clearly crude and awful. It is, as we have seen above, the reduction of human beings to "robot vehicles blindly programmed to preserve the selfish molecules known as genes"[7] and Aristotle clearly and rightly rejects *this*.

Can we therefore simply dismiss these "Giants"? The problem we face here is, ironically, a problem arising from the dereliction of duty by the "Gods" who were "very wary in defending their position somewhere in the heights of the unseen, maintaining with all their force that true reality consists in certain intelligible and bodiless Forms (*Soph.* 246B)." The "Giants" *may* have an awful philosophy, *but they have been left in control of the world.* The "Gods" *may* have better arguments, but these mean nothing if they are disconnected from reality, and their arguments will be all the weaker for not connecting with reality. One side grabs the world and the other ignores it.

Plato himself had been prescient in describing the Battle of the Gods and Giants as an "interminable battle [which] is always going on between the two camps…on this issue…over being [substance] (*ousia*) (*Soph.* 246A)." As we have hopefully seen above, Aristotle correctly identifies our fundamental problem as being how we see and approach a "being" (*ousia*) and "beings" (*ousiai*). Hegel himself explained Aristotle's empiricism on the basis that he occupied himself with one being and then with another, as follows: "He [i.e., Aristotle] gets the sensuous phenomenon before him in its entire completeness, and omits nothing, be it ever so common. All sides of knowing enter his mind, all interest him; all are handled by him with depth and exhaustiveness…[and] Aristotle…abandons a determination only when he has traced it to another sphere wherein it retains no longer its former shape…[and] sometimes Aristotle does not aim to reduce all to unity, or at least to a unity of antithetic elements; but, on the contrary, to hold fast each one in its determinateness, and thus to preserve it."[8] He observed further that: "Aristotle's conception of nature is…nobler than that of today, for with him the principle point is the determination of the end as the inward determination of natural things. Thus he comprehended nature as life."[9] Heidegger also rightly complained about the dissolving and distorting effect of technology, but it is only Aristotle who in his work rightly simply refuses to cede "science" itself to the mechanicists and technologists. The success of modern science has progressively caused the impotence of modern philosophy.

[7] Dawkins, Richard *The Selfish Gene* (Oxford, [*orig.* 1973], 1989) p xxi.

[8] Hegel, G.W.F. "The Philosophy of Aristotle" *The Journal of Speculative Philosophy* ([*orig.* 1825–6] 1871) p 73–75.

[9] Hegel, G.W.F. *Lectures on the History of Philosophy* (New York, [orig. 1840] 1974) II p 157.

Martin Heidegger also explained Aristotle's philosophical *intent* well as: "did we not assert, during the first enumeration of the four meanings of being in the Aristotelian sense, that the unity of these four meanings remains obscure in Aristotle? We did. However, this does not rule out but, for a philosopher of Aristotle's stature, precisely entails that this unity be troubling in view of its multiplicity. We need only observe how Aristotle explains the πολλαχῶς [i.e., the manifold]. Thus he says on one occasion (*Met.* K 1060b32f): τὸ δ' ὂν πολλαχῶς καὶ οὐ καθ' ἕνα λέγεται τρόπον. "Beings are manifold and so not articulated according to one way." But he also sees immediately and clearly the result that this view, when taken out of context, could generate, namely the dispersion of ὂν into many τρόποι, a dissolution of the ἕν. In contrast, Aristotle states: παντὸς τοῦ ὄντος πρὸς ἕν τι καὶ κοινὸν ἡ ἀναγωγὴ γίγνεται (1061a10f). "For each being, for all beings in whatever sense, there is a leading up and back to a certain one and common"; and at 1060b35: κατά τι κοινόν: "to some sort of common." We are always encountering this cautious and (as to what the encompassing one may be) open-ended τι (of some sort). Aristotle speaks of the final and highest unity of being in this fashion; see 1003a27 in *Met.* Γ 1003a27 (and many other passages): τὸ ὂν ᾗ ὄν, τὸ εἶναι as φύσις τις—a sort of governing from out of and in itself."[10]

Heidegger brilliantly explains Aristotle. Aristotle's four causes break down the world into its fundamental constituents in respect to "becoming" but they need to be synthesised to explain any given being. On this principle, Hegel lays stress upon how Aristotle manages to review the world by "hold[ing] fast to each one [subject] in its determinateness, and thus to preserve it" and Heidegger similarly lays stress upon Aristotle's "caution and open-ended τι [subject]."

Where we do find a significant divergence between Aristotle and his illustrious commentators is not in respect to "beings" (*ousiai*) but in respect to the nature of the underlying "being" (*to on*). On this matter, Hegel criticises Aristotle on the basis that: "Aristotle always moves in the speculative…he seems always to be philosophising only on the individual, the special, and not to arrive at what is absolute, universal or God…the one Absolute, the Idea of God, appears in Aristotle's Philosophy, but as a particular somewhat, side by side with the others."[11] Heidegger makes the similar complaint that: "in spite of his tendency to radicality he [i.e., Aristotle] did not press on into the ultimate originality of the Being of the world."[12] In both cases, our philosophers commend Aristotle for managing to "save" the beings of the world by keeping them "open-ended" whilst also complaining that he did not progress to *their* unity of "Being" (*Dasein*) or "the one Absolute, the Idea of God."

Pierre Aubenque explains, however, quite how both these great thinkers miss Aristotle's point as that: "The notion of analogy intervenes in the same context as

---

[10] Heidegger, Martin *Aristotle's* Metaphysics Θ *1–3: On the Essence and Actuality of Force* (Indiana, [*orig*. 1931] 1995) p 23.

[11] Hegel, G.W.F. "The Philosophy of Aristotle" *The Journal of Speculative Philosophy* ([*orig*. 1825–6] 1871) p 77–8.

[12] Heidegger, Martin *Plato's Sophist* (Indiana, [*orig*. 1924–5] 1992) p 59.

the doctrine of the multiple meanings of being. But analogy is nowhere invoked to unify these meanings. On the contrary, its radical nature presupposes rather than attenuating or correcting the positive plurality of the meanings of being since only the irreducible character of this plurality requires the recourse to analogy to show a basic unity—albeit "general"—in our discourse on principles and causes."[13] Aubenque argues that we should recognise that Aristotle goes some way towards recognising the principles which inhere in "beings" and that he *deliberately* leaves "gaps" which do not allow him to pass on towards a unified and fully "filled in" idea of "being." I posit on this basis that Aristotle's philosophy should not be understood as being incomplete, cautious, or speculative but rather as complete, bold, and considered. It is an explicit consideration of the problem of "the one and the many" which identifies the problem of the freedom of the particular and the commonality of the collective.

11.3 Aristotle and the Good Man

We have noted previously that Aristotle seems less concerned with exploring the inner feelings of men than with defining how we philosophically understand and frame these feelings as worldly phenomena. The human being must always be understood as a human *being*.

Pierre Aubenque embodies this emphasis (following Aristotle himself) through the case study of the *phronimos* as a "personification of the norm."[14] He explains Aristotle's thought that: "Virtue then is a settled disposition of the mind as regards the choice of actions and feelings, consisting essentially in the observance of the mean relative to us, this being determined by principle, that is, as the prudent man would determine it (*N.E.* II 1106b 36-1107a 2)" as that: "Virtue consists in acting according to the happy medium, and the criterion of the happy medium is the right rule. But what is the right rule? Aristotle gives us no means of recognizing it, except the appeal to the prudent man's judgement…Here is a man who, notwithstanding all his attenuations, is not only the interpreter of the right rule, but who *is* the right-hand rule, the living bearer of the norm."[15] The human being is a necessary basis for any picture of human action which cannot be abstracted out. Aubenque explains as regards the type of man described as the *phronimos* that: "Not being endowed with any special familiarity with the transcendent, he moves to the level of the particular

[13] Aubenque, Pierre "Les Origines de la Doctrine de l'analogie de l'être: Sur l'histoire d'un contresens" *Les Études philosophiques* (1978) p 6. Suzanne Mansion also argues that Plato conceptualises "*ousia*" as "Being" whereas Aristotle's does so as "beings" at "La première doctrine de la substance: la substance selon Aristote" *Revue Philosophique de Louvain* (1946) p 351.

[14] Aubenque, Pierre *La Prudence Chez Aristote* (Paris, 1963) p 41.

[15] Aubenque, Pierre *Ibid.* p 40–41.

and fixes to each the right middle which answers to its particularity."[16] The human being is always the basis for the manifestation of any particular human being.

The "*phronimos*" is a *channel* for the "particular" (and is grounded in the lived and practical world) whereas the "*sophos*" is a *channel* for the "universal" (and can rise above the practical world and see the wider principles animating it). Aristotle's position here is not anthropocentric or anthropomorphic but rather uses the human situation to explain the situation of being in general. Pierre Aubenque hence explains Aristotle's ontology as an "ontology of contingency" on the basis that: "prudence moves in the domain of the contingent, that is to say, in that which can be other than it is, the τὸ ἐνδεχόμενον ἄλλως ἔχειν. It is precisely on this basis that prudence is distinguished from wisdom which is concerned with necessity in so far as it is a science and which insofar as it is the highest of sciences deals with the most immutable realities and ignores the world of becoming. The theory of prudence is therefore connected with a cosmology and, more fundamentally, with an ontology of contingency."[17] Aubenque's further position on Aristotle's "contingency" is that: "There is a tragedy of moral life [in Aristotle's thinking], which is due to the fact that the union of happiness and virtue is not, so to speak, analytical, as the Socratics believed, but always synthetic because it depends, in an irreducible proportion, on chance. If this trait does not strike the commentators so much, it is because this tragic fact has a residual aspect in Aristotle, so that we can see in it the survival of a popular prudence…[and t]his Aristotelian contingency is not fundamental, constitutive, as will be the contingency of the world for *the Christians*. It is, one might say once again, *residual*."[18] For Aristotle, we are placed in a precarious position in the world through which we can and must try to steer an equally precarious course.

What, then, *is* our choice? It is clearly between idealism and realism. It is between reductionism and subtlety. It is also in a sense between the mechanisms of science and religion. Plato agrees with Aristotle when he argues that we should avoid "the champions either of the one or of the many forms of the doctrine that all reality is changeless [i.e., the "Gods"], and he must turn a deaf ear to the other party who represent reality as everywhere changing [i.e., the "Giants"] (*Sophist* 249C-D)." Plato was right to argue that we must find a pathway through the Gods and the Giants. Aristotle corrected him only in respect to the execution of this pathway. Alexander's less than great peregrinations suspended both their agreements and disagreements by clearing away their *poleis*, their republics, and subsuming them into an Empire. Pierre Aubenque explains the spiritual and political revolution that resulted as that: "It is the moment when the liberty of the free man, which up to that point was combined with the exercise of civil rights, transforms itself, for want of something better, into internal freedom… the internal ascetic generating ataraxy or pure speculation as a refuge and as a substitute for an impeded action. This atmosphere of retreat or, as has been said, of "abstraction" is characteristic of Hellenistic

[16] Aubenque, Pierre *Ibid.* p 41.

[17] Aubenque, Pierre *Ibid.* p 65.

[18] Aubenque, Pierre *Ibid.* p 82–5.

philosophies."[19] Even in Aristotle's own lifetime, the "bottom up" was systematically replaced by the "top down."

We are the grandchildren of Plato and Aristotle. We have the honour of being born from them even if we may have lost sight of them. We may be their successors, but we are by no means their superiors. We have travelled down paths of religion and then technology that have allowed us in some sense to surpass them. To acquire these undoubtedly remarkable spiritual and physical techniques, we have tortured and distorted ourselves. We have ended up building systems and methods *around us* to control us, whether religious, political, or technological or all three together. Aristotle deeply understood this problem, paradigmatically represented in his caution against the role of money as an end rather than as a means. The question has rightly been asked: "For what shall it profit a man, if he shall gain the whole world, and lose his own soul?"[20] We should perhaps also ask: "If we sacrifice the whole world to some alien machine, are we not truncating our own humanity?" Perhaps setting our limits and simply accepting that we are meaningful beings in a meaningful world is needed to negate the artificial "noble lie" of all such manipulation dressed up as meaning. We simply need to accept and embrace the "reality" as it appears to us. But to do this we need to put aside all the ideological castles in the air we have spent centuries and millennia constantly constructing to contain and constrain us.

If we can overcome our own past, our great civilisational founder, and the culture which bore him, offers us a meaningful science, a thoroughgoing rationality, a humanistic sensibility, a pluralistic politics, and all else valuable on a human scale. He eminently meets our own claims for modernity and is hence, as Foucault suggested, a true "modern" and he is indeed one who can guide us and teach us how to be most meaningfully modern today.

[19] Aubenque, Pierre *Ibid.* p 15–16.

[20] Matthew 16:26.

Postscript

I have sought to show above through Aristotle that we need to induce a full paradigm shift in philosophy. We don't just need to spin out a new -ism, -ology, or method but rather need to completely alter our approach to the world. We need to reinstall philosophy *as our framing of the world supported by science* as opposed to Quine's principle that: "Our ontology is determined once we have fixed upon the over-all conceptual scheme which is to accommodate science in the broadest sense."[1] The idea that we are aiming towards a unity of sciences is not just overly optimistic but is a positive distortion of our thought and of reality. It moves us in the direction of idealism rather than realism. It is the opposite of the truth and what we need since it supposes that we can and should reduce reality to laws or algorithms whereas the reality is that these mental tools are only our access points to the fuller meaning of nature.

The problem we face is not only how we look at philosophy but also how we look at science. We need to achieve a paradigm shift on both. We need, more particularly, to reconsider the place we are supposed to give *to scientific method* in our thought. As we would all accept, having a methodology in science is valuable and perhaps essential. It can, however, itself become a restriction, though hopefully a *useful* one, and is only ever a groundwork or frame of approach for any substantive content. Leibniz hence commented on Descartes' method that it is "like the precepts of some chemist; take what you need and do what you should, and you will get what you want."[2] Despite their vast philosophical differences, we could say much the same about Bacon's method. Both treat the world before us as a passive, plastic mass, a *res extensa*, that only has meaning to us and not in itself. This engineering approach has been immensely fruitful and yet it is not *philosophically* complete. Paul

[1] Quine, W.V. "On What There Is" in *From a Logical Point of View* (Cambridge Mass., 1953) p 18.

[2] Leibniz, Gottfried "Untitled Essay" in *Die philosophischen Schriften* (*ed.*) C.I. Gerhardt IV (Berlin, 1875–90) 329. On the problem of the idea of a scientific method per se see Paul Feyerabend *For and Against Method* (Chicago, 1999).

P. N. Jackson, *Aristotle on Meaning in the Living World*, Biosemiotics 30,
https://doi.org/10.1007/978-3-032-00602-8

Feyerabend expands upon this situation when he explains that Aristotle produced "the only reasonable empiricism that ever existed [whereas] Bacon...was an anti-empiricist plus an anti-rationalist ("rationalism" in the sense of Descartes) and in this respect the father of science."[3] Even our philosophical position as regards science, something absolutely central to our modern worldview, may be mistaken.

It is, of course, eminently natural for human beings to protect what they perceive to be the "golden goose" responsible for their existence and wellbeing. It is difficult to make a man understand something if his paycheck depends upon his not understanding it. Turkeys don't and, unless they are of a peculiarly objective and philosophical persuasion, *shouldn't* contemplate voting for Christmas. The problem here, however, is quite whether the scientific method from the seventeenth century ever was and still remains a golden goose. Joseph de Maistre early on explained the straw man argument involved in ruling out the arguments of "the theologian, the moralist, the metaphysician etc., [on the basis that they] could never put one of their demonstrations in a jar, put it through a filter, or under a hammer."[4] We can see on this basis that our worldview is based upon a vulgar scientism and materialism. We also see here the problems of ideology or mythology and of the misuse of the power of authority that is often *only* ascribed to religion. The problem of mythology, again part of Aristotle's critique of Plato, is flagged up by Feyerabend when he questions whether the dispute between Galileo and the Church was quite as one sided as normally claimed. We find that the losers in history—whether scholastic or Lamarckian—are often made into cartoon characters. The victors, by contrast, are lauded and given authority.

The modern *philosophical* elevation of science is still based on the authority of Newton, but we now know that Newton's science and the philosophy based upon it is fundamentally incomplete and erroneous. We will see in the sister volume to this work that, as A.N. Whitehead explained, it has been the case up to the current day that the "Hume-Newton situation...[is]...the primary presupposition for all modern philosophic thought. Any endeavour to go behind it is, in philosophic discussion, almost angrily rejected as unintelligible."[5] This is the dead hand of dogma *currently* hanging over our thought that we need to examine and clear up if we are as intellectually honest as we claim to be. We must recognise and challenge our own founding mythology.

We can consider the problem inherent in a scientific philosophy by reviewing the behaviourist B.F. Skinner's account of his dinner with the philosopher Alfred North Whitehead. Skinner proudly recounts the story in his *Verbal Behaviour* as proof of the purported value and validity of his work that: "He [i.e., A.N. Whitehead] brought the discussion to a close with a friendly challenge: "Let me see you," he said, "account for my behaviour as I sit here saying, "No black scorpion is falling upon

[3] Feyerabend, Paul *For and Against Method* (Chicago, 1999) p 197.

[4] Maistre, Joseph de *An Examination of the Philosophy of Bacon* (Montreal, [*orig.* 1836] 1998) p 18 (and see Tim Crane & D.H. Mellor's "There is No Question of Physicalism" *Mind* (1990)).

[5] Whitehead, A.N. "Nature Lifeless" in *Modes of Thought* (New York, 1938) p 135.

this table." The next morning I drew up the outline of the present study."[6] Skinner's supposed solution to Whitehead's challenge is as follows: "Perhaps it is time to consider Professor Whitehead's challenge. Can we indeed account for the fact that he said, "No black scorpion is falling upon this table"? As a particular instance of verbal behaviour, emitted under a set of circumstances now long since largely forgotten, we cannot. It is as unfair to ask a science of behaviour to do this as to ask the science of physics to account for the changes in temperature which were taking place in the room at the same time. Suppose a thermographic record had been made from which we could now reconstruct those changes, at least as accurately as I have reconstructed Professor Whitehead's verbal behaviour. What could now be done with it? …Unfortunately we have been led to expect something else in verbal behaviour."[7] Far from being an inevitable conclusion of science, this is childishly bad philosophy. I suggest that much modern philosophy is just a more respectable version of this common-sense utilitarianism including much "folk" philosophy.

Let us unpack this story. We first see that there is something vague and fishy about the very idea of "verbal behaviour." It is clearly philosophically dubious, and Noam Chomsky explains its basic error well in his famous review of this work.[8] We find, second, that Skinner's account of "verbal behaviour" assumes and concludes that *it doesn't matter what Whitehead says or what he intends by saying*. Skinner contends that a science should not care about content, rather as Bertrand Russell argued that "Socrates" was only a bundle of moments and not a person. How, then, can his account possibly explain any form of "verbal behaviour"? The philosophical implication of Skinner's intervention is simply to strip meaning from any content and even on this point his position is confused and contradictory. We strangely find that Skinner *did* understand the content and meaning of what Whitehead had verbally put forward and intended, *viz.* "The black scorpion was behaviourism"[9] and in this very moment of awareness and insight he disqualified his whole science and philosophy without even recognising the fact. As regards the wider implications of this situation, we see quite how philosophically naïve and simplistic his position is. We could accept Skinner's science, as also Pavlov's, but we should question whether we can extrapolate these scientific observations or even laws formulated through the study of dogs and pigeons to produce a full philosophical worldview. I suggest that this has been done and that—as with other such extrapolations in the past—it has led to catastrophic results.

How do we address this problem of science? We can see with Aristotle that science and philosophy are to an extent complementary whilst also being distinct disciplines with different objectives. Both should be manifested to their fullest and richest degree without being conflated and reduced. That is the basic conclusion of

---

[6] Skinner, B.F. "The Validity of the Author's Verbal Behaviour" in *Verbal Behaviour* (New York, 1957) p 456.

[7] Skinner, B.F. *Ibid.* p 456.

[8] See Noam Chomsky "A Review of B.F. Skinner's Verbal Behaviour" *Language* (1959).

[9] Skinner, B.F. "The Validity of the Author's Verbal Behaviour" in *Verbal Behaviour* (New York, 1957) p 458.

this work. We can see that Skinner's research into habits and "conditioning" retains some seemingly valuable applications today in educational and occupational psychology. Behind it, however, is a considerable amount of bad philosophy on the need for social engineering, the denial of the existence of free will etc. We could take the science and ignore the philosophy, but this raises an issue of *caveat emptor*. We see here the fundamental problem of technology. Can we separate science and philosophy? Jacques Ellul suggests on this subject that: "In Greece a conscious effort was made to economise on means and to reduce the sphere of influence of technique. No one sought to apply scientific thought technically, because scientific thought corresponded to a conception of life, to wisdom. The great preoccupation of the Greeks was balance, harmony and moderation; hence, they fiercely resisted the unrestrained force inherent in technique, and rejected it because of its potentialities. For these same reasons, magic had relatively little importance in Greece."[10]

It would certainly be quite something if the Greeks had identified and avoided the problems of modernity. Whether they were indeed so Aristotelian is also perhaps an interesting point for further consideration. The problem we now have, however, is whether we would want to sacrifice our "golden goose" to such ideals. We value our science, and we could also imagine that Aristotle and Plato would have also valued our science if they have had access to it. I suggest we can redefine this situation as that it is more accurate and certifiable to say that the Greeks recognised and sought to some degree to avoid the problems of scientific reductionism along with other forms of sophism. Remarkably, we also find that the great scientific figures of modernity also did not themselves subscribe to the various dogmas and mythologies of scientific reductionism. We will explore this fact in the sister volume to this work. The very strange fact we must recognise there is that scientific reductionism is refuted by modern science itself. *Reductionism is the forcing of a strange and unnecessary ideology of chance and necessity upon science.* It is a monstrous intellectual fraud, likely not the first or last of its kind. We must also question, however, whether, even if we intellectually and spiritually recognise the philosophical limits of technology, we do, when we subject ourselves to it, reduce ourselves to its mechanisms despite our best intentions. Does technology have a life of its own? We will unravel this relationship between life, mind, and meaning in this later work.

[10] Ellul, Jacques *The Technological Society* (New York, 1964) p 29.

Appendix: A Sample of Aristotle's Biological Terminology

The claim that Aristotle is philosophically correct directionally seems evidently reasonable. We can, however, say much more than this. It is not only clear that Aristotle had some reasonable ideas, but we also find behind his general concepts a full technical framework of structure and representation that explains reality on a parallel and likely deeper basis than any of the other approaches mentioned above. Let us consider a few key examples of Aristotle's terminology to prove this point taken from Fig. 4.6 replicated below in Fig. A.1:

Let us consider the substance of Aristotle's position to show how his philosophy, his philosophical position, informs not only his concepts but also the very nature and quality of his concepts vis-à-vis our own. Ladislav Kvasz explains this situation in principle as that: "the ancient notions of *apeiron*, *tyché*, *kenón*, and *kinesis* were much broader than our modern notions of infinity, randomness, space, and motion,

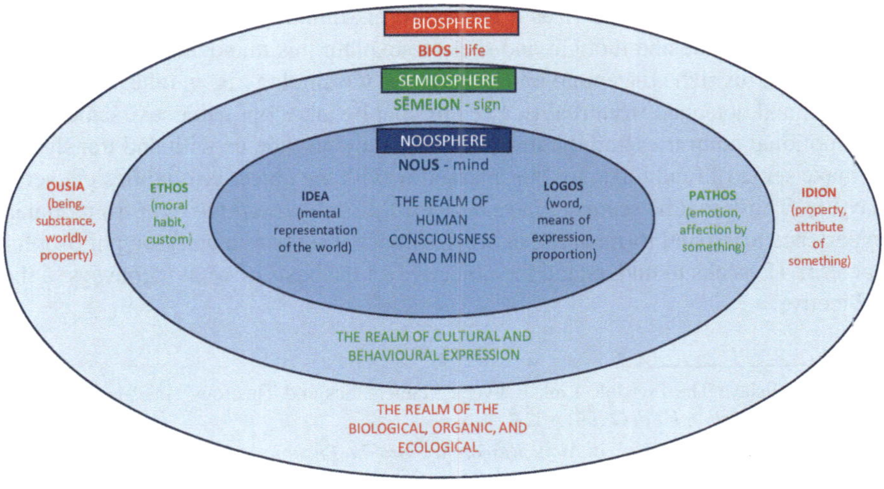

Fig. A.1 An Aristotelian framing for Biosemiotics

P. N. Jackson, *Aristotle on Meaning in the Living World*, Biosemiotics 30,
https://doi.org/10.1007/978-3-032-00602-8

which became the bases of the new mathematical disciplines. Today we strictly discriminate between the infinite and the indeterminate, between randomness and fate, between emptiness and space, between motion and change. Thus from the ancient notions, which were broad and ambiguous, narrow and specific parts were separated, and it was only these narrower notions that were mathematicised."[1] He observes that the mathematisation of terminology goes hand in hand with dualism and reductionism. Conversely, Aristotle's broad thought is holistic, full, and un-mathematisable.

Our first critical but little appreciated term is "*pathos*." This is Aristotle's basic element or unit of possibility which is best though not comfortably translated as "affection" following the Latin word "*affectio*" (from ad "to" + facere (past participle factus) "to make, do"). The original and common meaning of the word "*pathos*" in Aristotle's day seems to have been its use in the theatre to explain human emotion and Aristotle notes this provenance as that: "Misfortunes and painful experiences when on a large scale are called affections (*pathē*) (*Met.* Δ 1022b21)." Aristotle gives "pity, fear, anger (*Poet.* 1456b 1-2)" as examples of "*pathē*" in his *Poetics*. As "affection" the term *pathos* denotes the disposition towards something produced in a person. As regards why emotions should be understood as passive rather than as active Carl Jung explains that: "Emotion, incidentally, is not an activity of the individual but something that happens to him."[2] *Pathē* are the "bottom up" qualities of animal life the force of which are, as we will see, developed to support the rich tapestry of life.

Aristotle generalises this "bottom up" aspect of the term "*pathos*" as that it is: "a quality in respect of which a thing can be altered (*alloiousthai*), e.g. white and black, sweet and bitter (*Met.* Δ 1022b15-16)." He removes the local context, i.e., human emotion, from the term to make it into a fundamental qualitative *unit* or qualia of the sensory affection produced by *any* worldly being, e.g. sugar. We see three things here. First, that it is a feature of his thinking that Aristotle *listens to* popular experience and thought and seeks to explain this mundane world and integrate it into his rich conceptual philosophy. His terminology is, in other words, not abstract and detached from reality. Second, that he takes our subjective experience of emotional contraries and the mimetic basis they provide us with and transfers it to make sense of (and relate to) our interaction with the objective qualities of everyday life. Third, that he seems to have *deliberately redeployed the term* from human experience to expand its meaning in order to make it serve as a unifying philosophical term. He seeks to understand the objective on the basis of what it conveys to the subjective.[3]

---

[1] Kvasz, Ladislav "The Invisible Link Between Mathematics and Theology" in *Perspectives on Science and Christian Faith* (2004) p 114.

[2] Jung, Carl G. "The Shadow" in *Aion: Researches into the Phenomenology of the Self* (London, 1959) p 8–9.

[3] Aristotle argues that "*pathē*" can *both* be visualised (i) as "emotions" that exist by reference to "pleasure and pain" and hence: "by [affections] (*pathē*) I mean…the feelings that are accompanied by pleasure (*hēdonē*) and pain (*lypē*) (*N.E.* II 1105b23)" and *also* (ii) as the "passive" side of all

Aristotle explains the "*pathos*" in respect to intellectual engagement as that: "every disposition and every affection (*pasa gar diathesis kai pan pathos*) naturally comes into being in that of which it is a disposition or affection, for example, knowledge in the soul, since it is a disposition of soul (*diathesis ousa psychēs*) (*Top.* VI 145a35-37)." The *pathos* is thus a qualitative attribute of the soul which relates to the world as, for instance, the knowledge of something. He adds as regards the soul that it is conditioned by the world on the basis that: ""knowledge" is said to be of the "knowable," but is a "state" (*hexis*) or "disposition" (*diathesis*) not of the "knowable" but of the "soul" (*Top.* IV 124b34-35)." The apparatus of *pathos*, *diathesis*, and *hexis* is "of the soul" in relation to our knowledge of the object and not of the object itself or of knowledge of the object. This structuring of self and of the "soul" is described *in full* along with all the other ethical and mental structures in Aristotle's *Ethics*.

There is also a strictly biological application of the "*pathos*" which Aristotle explains as that: "*Peri de tōn pathēmatōn* …I mean such conditions of the parts as the following: blue and dark colour of the eyes, high and deep pitch of the voice, and differences of colour and of hair or feathers (*GA* V 778a16-21)." He intends "*pathē*" here to describe non-essential qualitative or semiotic characteristics of animal being which David Depew describes as a "trait-vocabulary." Aristotle adds regarding the associated term "*diathesis*" (disposition) that: "one country differs from another…and one water from another, on account of the same causes, for the quality of the nourishment especially and of the bodily condition of a person (*tou sōmatos hē diathesis*) depends upon the blend of the surrounding air (*dia te tēn krasin tou periestōtos*) and of the foods that the body takes up, and especially upon the nourishment supplied by the water, since this is what we take the most of (*GA* IV 767a29-33)." The "*diathesis*" is in this example the bodily condition of a person derived from his "blend" (*krasis*) which can be expected to correspond with the "*krasis*" of his environment.[4] Aristotle's hypothesis here is that the "matter" determines the specific character and constitution of an organism with this factor

being as follows: "passive qualities (*pathētika poiotētes*) and affections (*pathē*)… [are] "Just as honey itself contains sweetness and, therefore is said to be sweet"…we call honey sweet, as we said; but we do not imply that the honey itself is in some way affected. And so with all similar cases. Again, if we take heat and cold, though we call all such things passive, we do not imply that the things which admit or possess them are passive. We mean that the qualities mentioned can, one and all, *cause* a sensation. The [affection] (*pathos*), for example, of taste is affected by sweetness and sourness, by coldness and warmth that of touch (*haphē*) (*Cat.* 9a28–9b8)." The "*pathos*" therefore represents the affection of a being, its sense and feeling, which is affected by (the beings of) its environment. This account naturalises Plato's "third man" as the breakdown between the honey itself, our sensing of the honey and its attributes, and the affected "soul" which is the platform for this sensing. The soul takes the place of the mind as the place for the "forms."

[4]Aristotle also argues that: "each condition (*pasan diathesin*) is destroyed by a particular application of the very same things that produce it, as health is by food and exertion and climate. These matters are clear from induction…[although] we must observe that in every divisible continuum (*en pananti synechei kai diairetō*) there exists excess, deficiency, and a mean, which may be either relative to each other or relative to us (*kai tauta ē pros allēla ē pros hēmas*)—in gymnastics, for instance, in medicine, in architecture, in navigation, and in any practical pursuit …[and] our nature

explaining the various qualitative differences found in the same species in different places and different individuals.

Aristotle gives the further example of sounds as "*pathē*" on the basis that: "Now spoken sounds are symbols of affections in the soul (*tōn en tē psychē pathematōn symbola*), and written marks symbols of spoken sounds. And just as written marks are not the same for all men, neither are spoken sounds. But what these are in the first place signs of—affections of the soul—are the same for all, and what these affections are likenesses of—actual things—are also the same (*kai hōn tauta homoiōmata, pragmata ēdē tauta*) (*De Int.* 16a 4-8)." He argues, further, that "truth" is a "*pathos*" on the basis of the profound analysis that: "As to that which "is" in the sense of being true (*hōs alēthes*) or of being by accident (*kata symbebēkos*), the *former* depends on a combination in thought and is an affection of thought (*en symplokē dianoias kai pathos*) (which is the reason why it is the principles, not of that which "is" in this sense, but of that which is outside and can exist apart, that are sought (*peri de to exō on kai chōriston*)); and the *latter* is not necessary but indeterminate (*to d' ouk anankaion all' aoriston*) (*Met.* K 1065a21-25)." Aristotle argues that the "signs" and "symbols" of language relate to the world on the presumption that: "affections are likenesses (*homoiōmata*) of—actual things (*pragmata*)." He also finds an "indeterminate" world of "accident" alongside a structured world of the transfer of meaning between "separate" inside and outside.

We ultimately determine that "*pathos*" is employed everywhere in various senses. Does this mean that it is a vacuous term? We rather find that Aristotle's terminology is broad and analogical and also qualitatively precise and focussed. It is both *broad and focussed*.

Let us also consider the less well-known Aristotelian term "*idion*," meaning property or something proper to something. This is most obviously manifested in the term "idiot" which is, charmingly, used for a person concerned only with *his or her own* affairs. Aristotle himself defines "*idion*" generally as that: "A property (*Idion*) is something which does not show the essence of a thing (*to ti ēn einai…tou pragmatos*) but belongs to it alone and is predicted convertibly of it. For example, it is a property of man to be capable of learning grammar; for if a certain being is a man, he is capable of learning grammar, and if he is capable of learning grammar, he is a man (*Top.* I 102a 18-22)." It is notable that Aristotle refers to "things" (*pragmata*) here rather than "beings" (*ousiai*). I suggest the following situation: whereas the "property" is of the "thing" itself, an "affection" is the expression of that property in the "being." Aristotle is interested here, as throughout, with the "*symbolē*," the bringing together of the objective outside with the subjective inside.

Aristotle brings the above picture together as that the "affection" relates to the "sign" (*sēmeion*) by reference to what is proper to things, as follows: "if a peculiar affection applies to any individual class, e.g. courage to lions, there must be some corresponding sign of it; for it has been assumed that body and soul are affected

---

from the outset is not in every case equidistant from the mean; we are not sufficiently keen on exercise, and are excessively keen on self-indulgence (*E.E.* II 1220a26–*E.E.* II 1222a39)."

together…a sign is peculiar in the sense that the affection is peculiar to the class as a whole (*to gar sēmeion outōs idion estin, hoti holou genous idion esti to pathos*), and not to it alone, as we are accustomed to use the term. Thus the same affection will be found in another class also, and man or some other animal will be brave. Therefore he will have the sign; for *ex hypothesi* there is one sign of one affection (*hen gar enos ēn*). If, then, this is so, and we can collate signs of this kind in the case of animals which have only one particular affection, and if each affection has a sign, since it necessarily has only one sign, we shall be able to judge their character by their appearance (*Pr. An.* II 70b 14-26)." He seeks to understand how meaning is expressed in the world through signs and assumes, likely not wholly incorrectly but simplistically, that each "*pathos*" will be expressed across species through one common "*sēmeion*." See also Aristotle's treatise *Physiognomics* and Charles Darwin's *The Expression of the Emotions in Man and Animals* (1872) on this matter.

Finally, let us consider the term "*logos*" which famously means not only "word" or "sentence" but also "reason" and in a wider sense the "ratio" and "proportion" that we find in a meaningful world. "*Logos*" brings these significant structures under an umbrella of "that by which the inward thought is expressed" with "analogy" being "*ana-logia*" signifying movement along or upon the "*logos*" between genera.

We can begin our consideration of this single term by considering what language *means*, with *how it has meaning*. Aristotle argues that: "A sentence is significant speech, of which this or that part may have meaning (*Logos de esti phōnē sēmantikē hēs tōn merōn ti sēmantikon esti kechōrismenon*) (*De Int.* 16b 26)." The "*logos*" emerges in every area in which we can have "significant speech." Hence Aristotle points in his reflection on biochemistry to "the *logos* in direct consequence of which one thing is flesh and another bone (*GA* II 734b 34)." In his ethics, he argues that: "In all the states of character we have mentioned, as in all other matters, there is a mark to which the man who has the *logos* looks (*esti tis skopos pros hon anoblepōn ho ton logon echōn*), and heightens or relaxes his activity accordingly (*N.E.* VI 1138b19-25)." The "*logos*" appears in "all other matters." Aristotle argues further that "*logos*" is the meaning which motivates something in the sense that: "in one respect blood is essentially hot, and in another respect it is not. Heat will be included in the *logos* of blood, just as fairness is included in the *logos* of a fair man, and in this way blood is essentially hot; but in so far as it is hot owing to external influence, blood is not essentially hot (*PA* II 649b 24-29)." Blood is *not essentially* hot in its material self, but it must be *made hot* to serve its purpose as blood in mammals. For Aristotle, "*logos*" is almost the same as "*ousia*" and its "essence" (*to ti ēn einai*) since it is the inner logic which motivates or carries the directed meaning or "*ergon*" of the being (see *GA* I 715a 4-5). The "work" or "function" of a being—in part its "essence" (*to ti ēn einai*) or that which it is to be—supports its "ends" (*telē*).

We hopefully see that Aristotle's biology is theoretical and bio-logical in a way that the neo-Darwinian approach is not. Whereas the neo-Darwinian approach works backwards from fitness through the idea that chance adaptations lead to reproductive success, Aristotle assumes that there are natural structures which manifest in the world according to their principles. Aristotle's biology is on this basis far closer to modern physics than to what passes as mainstream biology. Aristotle does, for instance, explain sexual reproduction as a basic principle through the *logos* that: "They [i.e., male and female] differ in their *logos* (*<diapherei> kata ton logon*), because the male is that which has the power to generate in another (*eis heteron*)…while the female is that which can generate in itself (*eis auto*) (*GA* I 716a 20-22)." He expands upon this *logos* further as that: "it will surely be for the sake of generation (*heneka tēs geneseōs*) that "the male" and "the female" are present in the individuals which are male and female. And as the proximate motive cause (*tēs aitias tēs kinousēs prōtēs*), to which belong the *logos* and the form (*hē ho logos hyparchei kai to eidos*), is *better* and more divine in its nature (*beltionos de kai theioteras tēn physin*) than the matter (*tēs hylēs*), it is *better* also that the superior one should be separate from the inferior one (*GA* II 732a 3-7)." We see an emphasis upon "separation" / "external" and "continuity" / "internal"—as a sort of qualitative physics.

I represent the logic of the "separation" involved in sexual dimorphism in Fig. A.2 below.

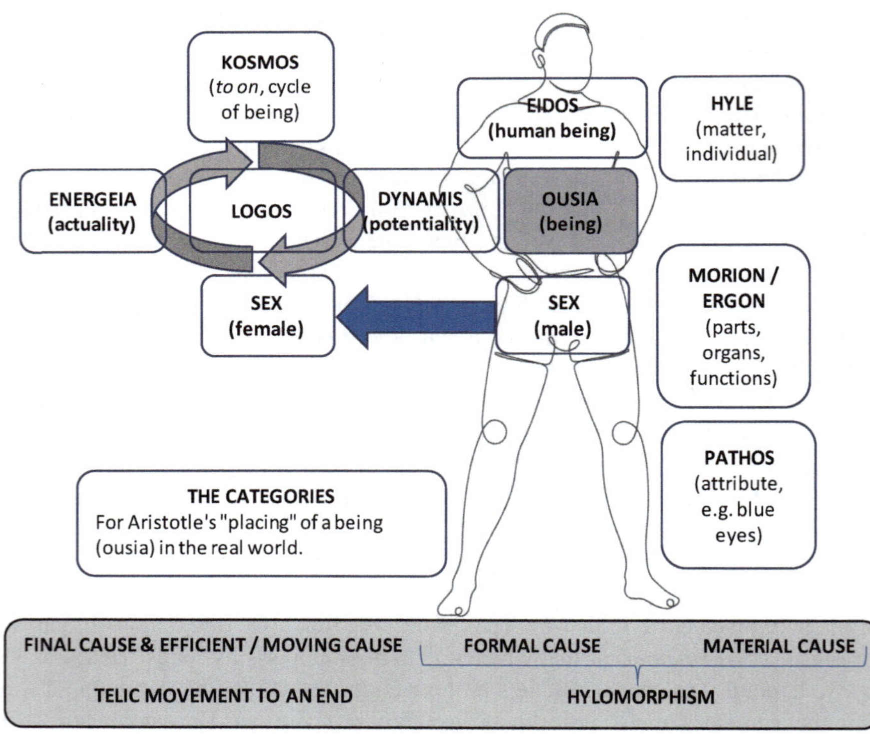

Fig. A.2 Aristotle's philosophical representation of sexual reproduction

What is the meaning of Aristotle's approach here? We see that he looks at sexual reproduction and looks for its meaning, its "*dia ti*"? He assumes that the world is cosmic or universal, that the world and its beings are deeply structured, and that there is a "good" that suffuses this structure as a pregnant emergent potentiality. This "good" as meaning and as increasing complexity and consciousness can be understood as Aristotle's "theory of evolution." The emergence of sexual reproduction—along with such things as the narrowing and heightening of life to the highly conscious and high investment human being—can be understood in this context. A reasonable argument can be made that it can *only* be explained on this basis. The alternative explains this situation away with the hypothesis that the thorough structuring of nature can be understood as the product of "chance."

I argue that we find here a sophisticated statement of biosemiotics, i.e., as an attempt to adequately assess meaning in life, within what I would contend is one of the most brilliant philosophical systems in history. I note by reference to Fig. A.2 above that Aristotle's philosophy is not just a statement of neo-Aristotelian hylomorphism which is also a form of reductionism; the burden of Aristotle's system is the introduction of the efficient and final causes to explain the various movements and evolution of natural beings. Aristotle's original philosophy is far more nuanced and sophisticated than most of the modern neo-Aristotelian attempts to explain and utilise it. To showcase this sophistication I have taken a sample of this original Greek philosophy—just *three* terms from our diagram "*pathos*," "*idion*," and "*logos*"—to show its clarity, structure, and depth. We could also take the other terms referred to such as "*ousia*," "*idea*," "*ethos*," "*nous*," "*sēmeion*," and "*bios*" and show their depth and range on a similar basis. For the many other terms in Aristotle's technical vocabulary, see Aristotle's *Metaphysics* Book Δ and the Glossary to this work.

Select Glossary of Aristotelian Terms[1]

adiairetos (ἀδιαίρετος) indivisible, undivided
aei (ἀεί) always, eternally; root of **aiōn (αἰών)** always being, era, lifetime and **aithēr (αἰθήρ)** always running; see *De Cael.* I 270b 23 & *De Cael.* I 279a 23
aiōn (αἰών) period of existence, lifetime; era, epoch of time; see also **periodos (περίοδος)**—a going round, a course of life and **bios (βίος)—**life, a course of life, manner of living.
aisthēsis (αἴσθησις) perception by the senses; common name of the senses—touch, hearing, sight, taste, smell; also **anaisthēsia (ἀναισθησία)**—insensibility to pleasure or pain, **aisthēsis koinē (αἴσθησις κοινή)**—common sense, perception through the five individual senses
aition (αἴτιον) & aitia (αἰτία) cause; [*orig.*] a charge, accusation; see also **synaitios (συναίτιος)**—joint cause, co-responsibility; see also Plato *Phaedo* 99b-c
alloiōsis (ἀλλοίωσις) a change, alteration, modification; an "othering"
akmē (ἀκμή) the highest point of anything; the bloom, flower, prime of a man's age
alētheia (ἀλήθεια) truth, reality, as opposed to appearance, "revealing" of being; MH "revealing the order at the start"
allos (ἄλλος) another, some other

[1] Detail is given below of some core Greek terms employed by Aristotle. Only a limited selection of the many variants and forms of terms has been given. A fascinating aspect of Aristotle's thinking is in respect to how he repurposes the rich Greek language for philosophical purposes; hence we see below that Aristotle's technical language is drawn from terms expressing common reality. Throughout we see an emphasis on the seeing, showing, and handling of being derived from the retention of worldly meaning in the technical terms. I note that Aristotle's own glossary of *Metaphysics* Book Δ is a remarkable piece of work which represents his own philosophy in the form of an ontological lexicon. Many definitions in the text below have been extracted from Liddell & Scott, some comments have been added from Martin Heidegger as "MH" and some detail has been taken from Hippocrates Apostle's glossary to his translation of the *Metaphysics*. Philosophical terms typically used by other Greek thinkers have not been included.

P. N. Jackson, *Aristotle on Meaning in the Living World*, Biosemiotics 30,
https://doi.org/10.1007/978-3-032-00602-8

amphoteros (ἀμφότερος) both of two; also **ep' amphotera (ἐπ' ἀμφότερα)**—from both ways, see *N.E.* VI 1143a 35–1143b 3

analogia (ἀναλογία) proportion, analogy; see also **logos (λόγος)**—ratio

analysis (ἀνάλῦσις) a loosening up, a breaking down something along the lines of its meaning; often contrasted with **synthesis (σύνθεσις)**—"seeing together"

anankē (ἀνάγκη) necessity, by necessity

anchinoia (ἀγχίνοια) quick wit, readiness of mind; derived from **nous (νοῦς)**—intuition; see *Post. An.* I 89b10-20 and Kosman (1992)

anhomoiomerē (ἀνομοιομερή) "non-uniform" parts of the body *viz.* face, hand, foot; contrasted with **homoiomerē** (**ὁμοιομερή**)—"uniform parts" such as flesh, bone, blood *etc.*; see *GA* I 715a 9-12 & Thom (1990a)

aoristos (ἀόριστος) indeterminate, limitless, without boundaries; often contrasted with **hōrismenon (ὡρισμένον)**—boundedness; something which has a boundary

apeiros (ἄπειρος) boundless, infinite, countless; see **peras (πέρᾰς)**—limit

apeiros (ἄπειρος) without trial or experience of a thing; see **peira (πεῖρα)**—trial, attempt

aphairesis (ἀφαίρεσις) abstraction, to take away the matter leaving the form; also **di' aphaireseōs (δι' ἀφαιρέσεως)**—by abstraction; see *N.E.* VI. 1142a 16-19

apodeixis (ἀπόδειξις) proof, demonstration; setting forth, showing forth, exposition; from **apodeiknumi** (**ἀποδείκνυμι**) point out, exhibit, appoint or assign

aporia (ἀπορία) difficulty of passing, perplexity; also **aporos (ἄπορος)**—unable to pass

archē (**ἀρχή**) a beginning, origin, first cause, the cause of change and the principle of something being something such and behaving so; also power, sovereignty, dominion; also **archōn (ἄρχων)**—ruler or commander, **hyparchein** (**ὑπάρχειν**)—to arise, spring up, and **to archon kai to archomenon** (**τὸ ἄρχον καὶ τὸ ἀρχόμενον**)—the ruling and the ruled part or principle in a composite being; see *Pol.* I 1254a28-33

architektōn (ἀρχιτέκτων) master-craftsman; guiding artist who understands his materials (as a scientist understands his field of investigation); also **architektonikē** (**ἀρχιτεκτονική**)—master artistry

aretē (ἀρετή) excellence (of all *ousiai*), virtue (of man)

arithmos (ἀριθμός) number; amount, sum, extent; the science of numbers

atelēs (ἀτελής) incomplete, imperfect; see also **telos (τέλος)**

atopos (ἄτοπος) without place, an absurdity; see also **topos (τόπος)**

automatos (αὐτόματος) self-moving, self-acting, spontaneous, instinctual; without apparent cause, accidental; acting of one's own will; see *Phys.* II 197b 7-8

autos (αὐτός) oneself, one's true self, the soul not the body; also **kath' auto (καθ' αὐτό)**—in itself, by virtue of itself, essentially; see *Top.* I 103a6 *ff*

autotelēs (αὐτοτελής) having an end in itself; see also **telos (τέλος)**

axia (ἀξία) the worth or value of a thing; price, value

bia (βία) force, violence, power for hindering (stopping from continuing) or preventing something from proceeding according to its natural tendency

bios (βίος) life, a course of life, manner of living; a living, livelihood; the fact of life rather than its structure is called **zōē** (**ζωή**)

boulēsis (βούλησις) willing, will, wish; a form of **orexis (ὄρεξις)—**appetite
bouleusis (βούλευσις) deliberation about worldly affairs; often contrasted with the pursuit of **epistēmē (ἐπιστήμη)—**scientific knowledge; for detail, see **nous (νοῦς)**—intuition, mind
chaos (χάος) chaos
chōra (χώρα) the space in which a thing is; a land, country; one's place in life; also Plato's term for "receptacle," the form by which a being is contained; for Aristotle's alternative see **topos (τόπος)**—place, **dektikon (τὸ δεκτικόν)—** that which is receptive, with the contrast between **periechon (περιέχον)**— the environment; that which surrounds something and **periechomenon (περιεχόμενον)**—the being circumscribed
chōristos (χωριστός) separate or separable in being or thought; from **chōra (χώρα);** see also **diakrisis (διάκρισις)**—separation, dissolution and **synkrisis (σύγκρισις)**—intermingling, compounding
dektikon (τὸ δεκτικόν) what is receptive, the matter which receives the form; a term that Aristotle uses rather than Plato's **chōra (χώρα)**—receptacle; see Keyt (1961)
diagōgē (διαγωγή) "passing across," way of living, life *chosen* for its own sake
diakosmēsis (διακόσμησις) the orderly arrangement of the universe, a coming into order; see *GA* II 740a6-24; see also **kosmos (κόσμος)—**the ordering of the universe
diakrisis (διάκρισις) separation, dissolution; used by Presocratics and often contrasted with **synkrisis (σύγκρισις)**—intermingling, compounding; see also **krisis (κρίσις)**—judgement, selection
dialektikē (διαλεκτική) dialectics, a discipline dealing with any problem whether defending or attacking a thesis, starting from commonly accepted beliefs
dianoia (διάνοια) thought, intellect, intelligence; from dia (through) + nous (intuition); see also **gnomē (γνώμη)**—understanding; see *N.E.* Book VI
diaphora (διαφορά) difference, differentia (in logic), distinction; variance, disagreement
diathesis (διάθεσις) a disposition, arrangement, bodily disposition; placed between **hexis (ἕξις)**—ethical state and **pathos (πάθος)**—feeling, emotion; see Aristotle's *Ethics*
dia ti (διὰ τί) the cause of something
dynamis (δύναμις) power, potentiality, faculty; often compared with **energeia** (ἐνέργεια)—activity
eidōlon (εἴδωλον) an image in the mind, a phantom; see also **εἶδος**—that which is seen, form
eidos (εἶδος) that which is seen, form, shape, figure; a class, species; often used with **genos (γένος)**—kind, genus
einai (εἶναι) to be, to exist; also **to on (τὸ ὄν)**—being; and **to ti ēn einai (τὸ τί ἦν εἶναι)**—essence, that which enables a thing come to be what it becomes
ekeininon (ἐκείνινον) "thaten", a term coined by Aristotle to denote the material medium of a thing's being; see Beere (2006)

empeiria (ἐμπειρία) experience, experience in or acquaintance with something; from **peira (πεῖρα)**—trial; basis for the term "empiricism"
enantion (ἐναντίον) the contrary state of a possibility, e.g. see *Long*. 465b17-19; compared with **enantiōsis (ἐναντίωσις)**—complete difference, opposite
endechesthai (ἐνδέχεσθαι) admitting of being or of not being, having possibility of being other than it is
endoxos (ἔνδοξος) common opinion, accepted view
energeia (ἐνέργεια) actuality, activity; work; "en" (in) + "ergon" (function); often compared with **dynamis (δύναμις)—**power, potentiality
enkrateia (ἐγκράτεια) strength of character, self-control or mastery; from "en" (in) "kratos" (power); see **kratos (κράτος)—**governing power
en parergō (ἐν παρέργῳ) "on the side"; something arising or emerging due to some other process or function; see *Met*. Λ 1074b 35-36
entelecheia (ἐντελέχεια) entelechy, goal pointing principle (of specific type of *ousia*); from telos (τέλος) + en (in) and echein (having)—end; often used alongside but distinguished from **energeia (ἐνέργεια)**—activity; see *Met*. Θ 1047a30-1047b2
enyparchein (ἐνυπάρχειν) "en" + "hyparchein," to be innate, to inhere, be inherent
epagōgē (ἐπαγωγή) induction, knowing the world through the process of reasoning from particulars to other particulars and also to universals (i.e., to what is **katholou**); compare with **diagōgē (διαγωγή)**
epamphoterizein (ἐπαμφοτερίζειν) to admit a double sense, to "dualise" between species, a technical term used by Aristotle in his biology, e.g. for the mule and the ape
ephexēs (ἐφεξῆς) one after the other; successive, in succession; often contrasted with **synechēs (συνεχής)**—continuous
epistēmē (ἐπιστήμη) knowledge, scientific knowledge; knowledge of the "why" as well as the "what"; understanding of the principles of some subject matter, a coming to a resolution; contrasted in its various forms with the various forms situational practical wisdom involving **bouleusis (βούλευσις)—**human deliberation
epiteleō (ἐπιτελέω) complete, bring to perfection
epithymia (ἐπιθυμία) desire, longing; from "epi" (upon) and "thymos" (spiritedness); a form of **ὄρεξις (appetite)**
ergon (ἔργον) work, function of something; activity, MH "the manner of presencing"; also **en parergō (ἐν παρέργῳ)**—"on the side," a function arising as a consequence of something else; see *Met*. Λ 1074b 35-36
eschaton (ἔσχατον) most extreme example, point, instance of a phenomenon; see *Mete*. IV 390a 4-12, *Met*. Z 1032b 28-29, & *N.E.* VI 1143a 35
ethnos (ἔθνος) a number of people accustomed to live together; flocks; a nation, people, tribe
ethos (ἔθος) an accustomed place, a custom, mores; related to **hexis (ἕξις)—**habit

ēthos (ἦθος) moral habit; also **ēthikos (ἠθικός)**—of or for morals, and **ta ēthika (τὰ ἠθικά)**—a treatise on morals see *N.E.* II 1103a 17-18
eu (εὖ) good, well, well being, right
eudaimonia (εὐδαιμονία) happiness, good fortune, having a beneficial **daimōn (δαίμων)**—spirit
euergetēma (εὐεργέτημα) a kindness, service done; see *Rhet.* I 1367a 1
eutychēs (εὐτυχής) having good fortune; contrasted with **dystychēs (δυστυχής)** unfortunate; see *Rhet.* I 1361b 39
exōthen (ἔξωθεν) from outside; see also **thurathen (θύραθεν)**—"from outside the door"
genesis (γένεσις) an origin, source, productive cause; manner of birth; creation, production; race, descent; from **gignomai (γίγνομαι)**—to be born, come to be
genos (γένος) race, stock, family; a class, sort, kind; an abstract grouping which covers but does not directly refer to species, e.g. animal to pig; from **gignomai (γίγνομαι)**—to be born, come to be; often contrasted with **eidos (εἶδος)**—form, species
gnomē (γνώμη) understanding, a means of knowing, mind, thought, judgement, intelligence; see also **dianoia (διάνοια)**—intelligence and **nous (νοῦς)**—intuition
grammē (γραμμή) line, a continuous quantity; often contrasted in different ways with **stigmē (στιγμή)**—point and **kyklos (κύκλος)**—circle
hama (ἅμα) instantaneous, simultaneous, "in the blink of an eye"; see *Rhet.* I 1363b 31 ff.
hamartia (ἁμαρτία) a failure, fault, sin; from **hamartanō (ἁμαρτάνω)**—to miss, miss the mark
haphē (ἁφή) a touch, contact, touching, the physical contact between two different objects, the animal sense or sensation of physical contact
harmonia (ἁρμονία) harmony, as a concord of sounds; a covenant, agreement; a settled government; a fastening to keep ship-planks together; see also **symphōnia (συμφωνία)**
hēgoumenon (ἡγούμενον) ruling part, governing principle
hekastos (ἕκαστος) particular; every, every one, each, each one, also **kath' hekaston (καθ' ἕκαστον)** individual; often compared with **heteros (ἕτερος)**—the other one
hen (ἓν) one, a unity; of the state, condition, position, in which one is; often contrasted as "the one and the many" with **pollachōs (πολλαχῶς)**—manifold, also with **on (ὄν)**—being
heneka (ἕνεκα) for the sake of, on account of; see **hou heneka (οὗ ἕνεκα)**—for the sake of something; the "final cause"
hexis (ἕξις) a having, possession; an ethical state, *esp.* a good habit; see also **ethos (ἔθος)**—custom; often contrasted with **diathesis (διάθεσις)**—disposition and **pathos (πάθος)**—feeling, emotion
holos (ὅλος) whole, a whole; often used with **morion (μόριον)** and **meros (μέρος)**—part

homoiomerē (ὁμοιομερή) "uniform parts" of the body such as flesh, bone, blood *etc.*; contrasted with **anhomoiomerē (ἀνομοιομερή)**—"non-uniform" parts of the body such as face, hand, foot *etc.*; see *GA* I 715a 9-12

homoios (ὅμοιος) like, resembling

homoiotēta (ὁμοιότητα) likeness, resemblance; see *Top.* I 108a 7

hōrismenon (ὡρισμένον) boundedness; something which has a boundary, is determinate; a bounded and determined something; often contrasted with **aoristos (ἀόριστος)**—indefinite

horismos (ὁρισμός) marking out by boundaries, limitation; the definition of a word

hormē (ὁρμή) tendency, impulse, violent movement onward, the first stir or start in a thing

horos (ὅρος) definition; boundary, landmark, limit, standard; see *E.E.* VIII 1249a 21-1249b 6; see also **peras (πέρᾰς)**—limit and **eschaton (ἔσχατον)**—final instance

hōs epi to polu (ὡς ἐπὶ τὸ πολύ) for the most part, things which happen for the most part but admit exceptions

hothen (ὅθεν) source, whenceforth something; **hothen hē kinēsis (ὅθεν ἡ κίνησις)**—moving cause, source of motion

hou heneka (οὗ ἕνεκα) that for the sake of which; the final cause; see also **telos (τέλος)**—end and **archē (ἀρχή)**—origin

hylē (ὕλη) a wood, forest, woodland; firewood, fuel; matter (or "energy"); a subject matter; also **oikeia hylē**—proper matter

hyparchein (ὑπάρχειν) (from **archē**) to begin, make a beginning, to begin to be, come into being, arise, spring up—MH "presencing which rules from what already; **hyparchontōn (ὑπαρχόντων)** existing thing with attributes, see *Met.* E 1026a 32; also **enyparchon (ἐνυπάρχον)**—to inhere, be a beginning or source for, see *Met.* Δ 1014b 15

hypokeimenon (ὑποκείμενον) substratum; that which lies present as enformed or enformable matter

hypothesis (ὑπόθεσις) A premise posited as being true, not yet proven, i.e., an assumption

hysteros (ὕστερος) posterior; following, coming after; often contrasted with **proteros (πρότερος)**—prior

idea (ἰδέα) objects of knowledge posited as existing by Plato over and above objects of being, outward appearance; used interchangeably with **eidos (εἶδος)**—form

idios (ἴδιος) ones' own, pertaining to oneself, peculiar, property

isotēs (ἰσότης) oneness in quantity; opposed to **symmetria (συμμετρία)**—symmetry see *Met.* Γ 1004b 11-13

kairos (καιρός) due measure, proportion, fitness; the right point in time, the proper time or season of action, the exact or critical time; see also **kosmos (κόσμος)** and **aiōn (αἰών)**

katēgoria (κατηγορία) category, predicate; [*orig.*] accusation, charge [in a lawcourt]

katholou (καθόλου) on the whole; "universal" in the sense of being what the human mind can grasp about the forms of the world through particulars by means of **epagōgē (ἐπαγωγή)**—induction

kenon (κενόν) the void, often compared with **plēres (πλῆρες)**—plenum

kinoun (κινοῦν) the mover, the acting being in motion

kolobos (κολοβός) mutilated, truncated

kosmos (κόσμος) the world universe; the perfect order of being; good behaviour, decency, arrangement, (also fashion or dress); compared with **taxis (τάξις)**—particulate order

krasis (κρᾶσις) biological blending, compounding, combination; contrasted with **mixis (μῖξις)**—chemical mixing, mingling; also **synkrasis (σύγκρασις)**—intermingling, compounding

kratos (κράτος) governing power over something; often compared with **akrasia (ἀκρασία)**—incontinence, lack of guiding principle

krisis (κρίσις) judgement, selection

kyklos (κύκλος) ring, circle, round; any circular motion, an orbit of the heavenly bodies, revolution of the seasons, cycle of events; also **kyklophoria (κυκλοφορία)** from kyklos + phora

kyrios (κύριος) authoritative, dominant, decisive power, principle, or person

logos (λόγος) the word, sentence, or that by which the inward thought is expressed; ratio, proportion; see also **analogia (ἀναλογία)**—analogy, ratio

mathēmatika (μαθηματικά) for Plato (and, in a sense, Aristotle) these are "Intermediate Objects" between the ideas and sensible objects; see *Met.* M 1077b 18 ff.

meros (μέρος) part; often compared with **holos (ὅλος)**—whole

mesos (μέσος) the middle, moderate, mediating; also **mesotēs (μεσότης)** mean or middle and **to meson (τὸ μέσον)**; see *Post. An.* I 84b 35-6

metabolē (μεταβολή) a change, changing, transition, e.g. conversion of food through the metabolic process; a "going across"; see *Phys.* V 225a2-3

metalēpsis (μετάληψις) participation, a key Platonic term, compared with Pythagoras' **mimēsis (μίμησις)**—imitation

metaphora (μεταφορά) metaphor, carrying across of meaning; see *Poet.* 1457b 6

metaxy (μεταξύ) intermediate, between, between-whiles, a medium for

methodos (μέθοδος) method, inquiry as a path to nature

metrios (μέτριος) moderate, holding to the mean; temperate; proportionate, fitting

metron (μέτρον) a measure or rule, that by which anything is measured; measure, length, size

mixis (μιξις) mixing, mingling; intercourse with others; chemical as opposed to the biological **krasis (κρασις)**—blending

moira (μοῖρα) human fate, destiny

monas (μονάς) a unit of a number, that which is indivisible in every way with respect to quantity and has no position; see *Met* Δ 1016b 17-31

morion (μόριον) part of the body, a part of a **holon (ὅλον)**—whole

noēma (νόημα) that which is perceived, perception, thought; also **noētos (νοητός)**—intelligible, **noēsis (νόησις)**—intelligence, thought, thinking; and **ennoēma (ἐννόημα)**—notion, conception

nomos (νόμος) anything assigned, a usage, custom, law, ordinance; anything sanctioned by custom, usage, institution; often contrasted with **physis (φύσις)—nature**; also **nomisma (νόμισμα)—money**

nous (νοῦς) intuition, mind; mental perception; natural reason; derivatives are **agnoia (ἄγνοια)**—ignorance, **anchinoia (ἀγχίνοια)**—quickness of wit, and **dianoia (διάνοια)**—intelligence; in Book VI of the *Ethics* broken down as rational mind and contrasted with **orexis (ὄρεξις).** In this book, Aristotle considers the aspects of the rational generally defined as **nous (νοῦς)**—intuition as being **dianoia (διάνοια)**—intelligence, **gnomē (γνώμη)**—understanding, **synesis (σύνεσις)**—judgement, and **bouleusis (βούλευσις)**—deliberation about worldly affairs. He considers in the *Ethics* both how we achieve **epistēmē (ἐπιστήμη)**—scientific knowledge and **sophia (σοφία)**—intellectual wisdom and also **phronēsis (φρόνησις)**—practical wisdom

nun (νῦν) now

oikeios (οἰκεῖος) originally of the house; proper, fitting, suitable for a thing and private to it; conformable to the nature of a thing; see also **idios (ἴδιος)**—one's own

on (ὄν) being; compared with **mē on (μὴ ὄν)**—not being

orexis (ὄρεξις) appetite, likely a term coined by Aristotle to group all the desiring human faculties, *viz.* **thymos (θυμός)**—spiritedness, **epithymia (ἐπιθυμία)**—desire, yearning, and **boulēsis (βούλησις)**—will; contrasted with the rationalising mind in Book VI of the *Ethics*; for which contrast see **nous (νοῦς)**

ousia (οὐσία) that which is one's own, one's substance or property or belongings; the being, essence, nature of a thing; MH "presencing, unconcealedness"—see also **alētheia (ἀλήθεια)**—truth, revealedness

paradeigma (παράδειγμα) pattern, model, a sculptor's or painter's model

parrēsia (παρρησία) freespokenness, openness, frankness, truth-telling

pathēma (πάθημα) a suffering, a calamity; a passive emotion or condition; also **pathos (πάθος)**

pathos (πάθος) a passive characteristic of something; an emotion; a passive state or condition; an affection; the incidents or changes to which a thing is liable; often contrasted with **hexis (ἕξις)**—habit

peira (πεῖρα) trial, attempt, essay, experiment; see also **empeiria (ἐμπειρία)**—experience

pepsis (πέψις) concoction, mastering of food or foetal matter for the maintenance of an organism or its reproduction; see *GA* IV 765b 15-16

peras (πέρᾶς) an end, limit, boundary; see also **horos (ὅρος)**—definition, limit

periechon (περιέχον) the environment; that which surrounds something (the "form," for e.g., surrounds the "matter"); used with **periechomenon (περιεχόμενον)** the being which is circumscribed or enformed

periodos (περίοδος) a going round; circuit, circumference; cycle or period of time; a prescribed course of life; also **aiōn (αἰών)**

periousia (περιουσία) abundance, that which is over and above; see *Top.* III 118a8

phainomenon (φαινόμενον) observed facts, things seen or brought to light by being seen; also "setting down the phenomena" for which see *N.E.* VII 1145b 4 and Owen (1975). Used with "legomena" ("λεγόμενα") or things said as the data and (first) interpretation ready for further analysis and further philosophical treatment. Aristotle does not use the phrase "saving the phenemona" but does take this approach, hence his routine references to **sōtēria (σωτηρία)**—safety, saving

phantasia (φαντασία) imagination, the power by which an object is presented; also **phantasma (φάντασμα)**—imaginative visions, impressions

phora (φορά) a carrying, a motion or movement, locomotion, a crop brought forth, used by Aristotle for "local movement" in space; also **periphora (περιφορά)**—revolution, from "carrying round" of food, movement of planets and **kyklophoria (κυκλοφορία)**—circular movement of the planets. Other derivatives **metaphora (μεταφορά)**—metaphor and **diaphora (διαφορά)**—difference

phasis (φάσις) an assertion; from **phēmi (φημί)** to declare, make known, affirm; also **kataphasis (κατάφασις)** affirmation, **apophasis (ἀπόφασις)** statement that something does or does not belong to something, **antiphasis (ἀντίφασις)** contradiction

phronēsis (φρόνησις) practical wisdom, prudence; the ability to manage one's affairs and to act effectively and well in the world; often compared with **sophia (σοφία)**—intellectual wisdom; see the *Ethics*

physis (φύσις) the nature, natural qualities, powers, constitution, condition, of a person or thing; from **phyō (φύω)**—to bring forth, put forth leaves, etc. see *Met.* Δ 1014b 16; often contrasted with **nomos (νόμος)**—human law, convention; things are referred to as **kata physin (κατὰ φύσιν)**—in accordance with nature and **para physin (παρὰ φύσιν)**—against nature

plēres (πλῆρες) full, the opposite of **kenon (κενόν)**—empty or void, plenum

pneuma (πνεῦμα) breeze, influence, breath, that which is breathed forth, spirit

polis (πόλις) city, republic, body of citizens; see also **politeia (πολιτεία)**—citizenship

politeia (πολιτεία) the condition or constitution of a state; the body of citizens; civic life, the rights of a citizen; citizenship, rights of a citizen (civitas)

pollachōs (πολλαχῶς) the manifold, the manifoldness of being

pragma (πρᾶγμα) an action which has been done, an actualised thing, an alleged fact

praxis (πρᾶξις) an action, act, doing; often contrasted with **poiēsis (ποίησις)**—the act of making

prohairesis (προαίρεσις) a choosing one thing before another, a deliberate choice, a resolution; a principle of action, a commitment to something

proteros (πρότερος) prior; before, former, sooner; often contrasted with **hysteros (ὕστερος)**—posterior

psychē (ψῦχή) soul; informing essence of something

schēma (σχῆμα) form, shape

scholē (σχολή) free time to spend as one wishes, normally in a desired activity

sēmeion (σημεῖον) sign, mark, token; a sign from the gods, an omen; a sign or signal; a sign or proof; also **tekmērion (τεκμήριον)**—necessary sign see *Rhet.* I 1357a35-1357b11
skopos (σκοπός) the object upon which one fixes one's eye, a look-out watchman stationed on a **skopia (σκοπιά)**
sōma (σῶμα) the living body
sophia (σοφία) intellectually guided wisdom; intuitive wisdom combined with knowledge which enables true thinking and **philosophia (φιλοσοφία)**—the systematic treatment of a subject and proper investigation of truth and nature; often compared with **phronēsis** (**φρόνησις**)—practical wisdom
sōros (σωρός) heap; often contrasted with **holos (ὅλος)**—whole
sōzō (**σῴζω**) preserve; to keep, observe, maintain laws; see for e.g. *Juv.* 469b 18-20
sperma (σπέρμα) seed, origin
spoudaios (σπουδαῖος) term used to specifically reference a good man as opposed to general terms such as **kalos (καλός)**—fine and **agathos (ἀγαθός)**—good
stasis (στάσις) rest; standing, the posture of standing; a locked factionalism; often compared with **kinēsis (κίνησις)**—movement, change
sterēsis (στέρησις) privation, deprivation
stigmē (στιγμή) point, that which is indivisible in respect to quantity and has position; often contrasted with **grammē** (**γραμμή**)—straight line and **kyklos (κύκλος)**—circle
stoicheion (στοιχεῖον) element, both chemical and metaphysical; letter
symbebēkos (συμβεβηκός) a chance event, accident, contingency; a happening
symbolē (συμβολή) a coming together, a contract
symmetria (συμμετρία) commensurability; symmetry, due proportion
symphōnia (συμφωνία) concord or unison of sounds, harmony; from **phōnē (φωνή)**—voice
symphuesthai (συμφύεσθαι) to grow together as in the parts of a hand or a foetus or abnormally like Siamese twins; also **symphysis (σύμφυσις)** "organic unity"; see *Met.* Δ 1014b 22-26 & *Met.* Z 1040b 13-16
symphytos (σύμφυτος) congenital, innate, natural, born with; from **physis (φύσις)**—nature; also
symphysis (σύμφυσις) a "growing together"
synaitios (συναίτιος) being the cause of a thing jointly with another
syndesmos (σύνδεσμος) that which is bound together, bundled
synechēs (συνεχής) continuous, continuous with or contiguous to, in a line with (in space), continuous, unintermitting (in time); constant, persevering (in persons), holding together; contrasted with **ephexēs (ἐφεξῆς)**—successive
synesis (σύνεσις) judgement, a coming-together, union; quick comprehension, mother-wit, intelligence
syngenēs (συγγενής) innate; kinfolk
synistēmi (συνίστημι) to put together, unite, to be made out of material parts excluding the form, to bring together friends or form an order of battle
synholon (σύνολον) a whole which is a composite, a concretum

synthesis (σύνθεσις) a seeing together, combination; often contrasted with **analysis (ἀνάλυσις)**—"a breaking down"; see *Top.* VI 151a 20 *ff.*
syntheton (σύνθετον) composite, complex thing; see *Top.* VI 151a 20 *ff.*
systasis (σύστασις) composition, bringing together; guardianship
systoikha (σύστοιχα) coordinates, a preservative relationship between just man and justice; see *Top.* II 114a 27 *ff.*
tautos (ταὐτός) one in number, or kind, or substance, or formula; see *Top.* I 103a 6 *ff.*
taxis (τάξις) order; an arranging, order of battle, the order or disposition of an army; compared with **kosmos (κόσμος)**—cosmic order
technē (τέχνη) art, skill; an art, i.e., a system or method of making or doing; often contrasted with **epistēmē (ἐπιστήμη)**—scientific understanding
teleios (τέλειος) complete, having reached its end, perfect
teleutē (τελευτή) finishing, completion
telos (τέλος) the fulfilment or completion of anything; also **teleutē (τελευτή)**—finishing, completion, **atelēs (ἀτελής)**—incomplete, **teleios (τέλειος)** complete, **autotelēs (αὐτοτελής)**—having an end in itself, and **epiteleō (ἐπιτελέω)**—bring to completion
theōria (θεωρία) seeing, view, investigation, speculation
thermotēs (θερμότης) heat
thesis (θέσις) placing, setting, arrangement; position, situation; relative place, that which is posited; money paid in advance
thingano (θιγγάνω) to touch upon or apprehend by the mind, see *Met.* Λ 1072b 21
thymos (θυμός) passion, spirit, courage, seat of self; a form of **orexis (ὄρεξις)**—appetite
thurathen (θύραθεν) from outside, from outside the door; see *GA* II 736b 27-29
tithēmi (τίθημι) to posit; to place or put down; see "tithenai ta phainomena" and Owen (1975)
tode ti (τόδε τι) the this itself, a defined individual "this"
toionde (τοιόνδε) suchness, in contrast to an individual a universal or kind
tokos (τόκος) money interest, childbirth; see *Pol.* I 1258b 7-8
topos (τόπος) place; a place, occasion, opportunity; subject, topic; see also **chōra (χώρα)**—space; on the use as "topic" see *Rhet.* I 1358a 31-33
trophē (τροφή) nourishment, food; a mode of life, bringing up; also **syntrophos (σύντροφος)**—brought up with, reared with, fed in common
tychē (τύχη) chance, contingency; see also **automatos (αὐτόματος)—**a mechanical happening and **symbebēkos (συμβεβηκός)**—contingent accident, non-essential attribute
zētēsis (ζήτησις) a searching, seeking for the truth
zōē (ζωή) life, animal life; compared with **bios (βίος)**—system of life

Bibliography of Classical Resources[1]

Achtenberg, D. (2002). *Cognition of value in Aristotle's ethics: Promise of enrichment, threat of destruction*. SUNY.

Achtenberg, D. (1995). Human being, beast and god: The place of human happiness according to Aristotle and some twentieth-century philosophers. In M. Sim (Ed.), *The crossroads of norm and nature: Essays on Aristotle's ethics and metaphysics* (pp. 29–50). Rowman & Littlefield. [*orig*. 1988] (*).

Alloa, E. (2009). Metaxu. Figures de la médialité chez Aristote. *Revue de Métaphysique et de Morale, 62*, 247–262. (*).

Ambuel, D. (2007). *Image and paradigm in Plato's sophist*. Las Vegas.

Annas, J. (1989). Self-love in Aristotle. *The Southern Journal of Philosophy, 27*, 1–18. (*).

Annas, J. (1982). Aristotle on inefficient Causes. *The Philosophical Quarterly, 32*, 311–326.

Annas, J. (1980). Aristotle on pleasure and goodness. In A. O. Rorty (Ed.), *Essays on Aristotle's ethics* (pp. 285–300). University of California.

Annas, J. (1975). Aristotle, number, and time. *The Philosophical Quarterly, 25*, 97–113.

Anton, J. P. (1957). *Aristotle's theory of contrariety*. Routledge.

Anzaldo, A. A. (2019). Aristotle and the search of a rational framework for biology. *Organisms. Journal of Biological Sciences, 3*(2), 54–64. (*).

Anzaldo, A. A. (2011). Assuming in biology: The reality of real virtuality (a come back for entelechy?). *Ludus Vitalis, XIX*, 333–342. (*).

Anzaldo, A. A. (2007). Back to the future: Aristotle and molecular biology. *Ludus Vitalis, XV*, 195–198. (*).

Armstrong, J. M. (1998). Aristotle on the philosophical nature of poetry. *The Classical Quarterly, 48*, 447–455.

Aubenque, P. (2011). *Kairos* et Prudence de la Détermination Temporelle de L'Acte Morale Selon Aristote. In *Problèmes Aristotéliciens: Philosophie pratique* (pp. 29–36). Vrin.

Aubenque, P. (1978). Les Origines de la Doctrine de l'analogie de l'être: Sur l'histoire d'un contresens. *Les Études Philosophiques*, 3–12.

Aubenque, P. (1963). *La prudence chez Aristote*. PUF.

Aubenque, P. (1962). *Le problème de l'être chez Aristote*. PUF.

[1] Some additional useful resources are provided over those directly referred to in the above text to provide a fuller resource for the exploration of Aristotle for those so minded.

(*) = available online as open access on authors' homepages, internet journals, academic site, Google Scholar etc.

P. N. Jackson, *Aristotle on Meaning in the Living World*, Biosemiotics 30,
https://doi.org/10.1007/978-3-032-00602-8

Baghdassarian, F. (2016). *La question du divin chez Aristote: Discours sur les Dieux et science du principe*. Peeters.

Baghdassarian, F. (2014). (*)). Aristote, Métaphysique H 5: La Génération des Contraires et la Matière. *Elenchos, 35*, 61–88.

Baghdassarian, F. (2013). La lecture aristotélicienne du mythe: nature, fonctions et usages du mythe en philosophie. *Revue de philosophie ancienne, 31*, 119–147. (*).

Baghdassarian, F. (2011a). Aristote, De Caelo, I, 9: l'identité des "êtres de là-bas". *Philosophie Antique, 11*, 175–203. (*).

Baghdassarian, F. (2011b). *La question du divin dans la philosophie aristotélicienne*. HAL Thesis. (*).

Baghdassarian, F., & Trego, K. (Eds.). (2023). *L'être et ses degrés: Histoire de l'ontologie scalaire*. Hermann.

Balaban, O. (1995). The modern misunderstanding of Aristotle's theory of motion. *Journal for General Philosophy of Science, 26*, 1–10.

Balme, D. M. (1990). Matter in the definition: A reply to G.E.R. Lloyd. In D. Devereux & P. Pellegrin (Eds.), *Biologie, Logique et Métaphysique chez Aristote* (pp. 49–55). CNRS.

Baracchi, C. (2008). *Aristotle's ethics as first philosophy*. CUP.

Bastit, M. (2016a). *Le Principe du monde: Le Dieu du philosophe*. I'IPC.

Bastit, M. (2002). *Les quatre causes de l'être dans la philosophie première d'Aristote*. Peeters.

Bastit, M. (2016b). La Science Aristotélicienne des Animaux et la Biologie Contemporaine: D'Empédocle et Démocrite à Aristotle. *Noema, XV*, 195–213. (*).

Bastit, M. (2014). Causality: From contemporary scientific conceptions of causality to the classical four Causes. *Angelicum, 91*, 37–58.

Bastit, M. (1994). La Marche des Animaux chez Aristote et Jakob von Uexküll. In *Actes du 25e Congrès de l'Association des Sociétés de philosophie de langue française* (*).

Bastit, M. (1992). Aristote et la séparation. *Revue Philosophique de Louvain, 90*, 297–316.

Beere, J. (2009). *Doing and being: An interpretation of Aristotle's metaphysics theta*. OUP.

Beere, J. (2010). *Thinking thinking thinking: On god's self-thinking in Aristotle's metaphysics Λ.9* (Online) (*).

Beere, J. (2008). The priority in being of *Energeia*. In M. Crubellier et al. (Eds.), *Dunamis: Autour de la Puissance Chez Aristote* (pp. 429–456). Peeters. (*).

Beere, J. (2006). Potentiality and the matter of composite substance. *Phronesis, 51*, 303–329. (*).

Bell, J. (1993). Aristotle as a source for Leonardo's theory of colour perspective after 1500. *Journal of the Warburg and Courtauld Institutes, 56*, 100–118.

Bell, J. L. (2005). Aristotle. In *The continuous and the infinitesimal in mathematics and physics* (pp. 20–30). Polimetrica. (*).

Bénatouïl, T. (2004). L'usage des analogies dans le *De Motu Animalium*. In A. Laks & M. Rashed (Eds.), *Aristote et le mouvement des animaux* (pp. 81–114). Septentrion.

Bénatouïl, T. (2005). Échelle de la nature et division des mouvements chez Aristote et les stoïciens. *Revue de Métaphysique et de Morale, 48*, 537–556. (*).

Berryman, S. (2003). Ancient automata and mechanical explanation. *Phronesis, 48*, 344–369.

Berti, E. (2011). The contemporary relevance of Aristotle's thought. *Iris, 3*(6), 23–35. (*).

Berti, E. (2001). Multiplicity and unity of being in Aristotle. *Proceedings of the Aristotelian Society, 101*, 185–207.

Bigwood, J. M. (1993). Aristotle and the elephant again. *The American Journal of Philology, 114*, 537–555.

Blair, G. A. (1993). Aristotle on entelexeia: A reply to Daniel Graham. *The American Journal of Philology, 114*, 91–97.

Bogaard, P. A. (2006). After substance: How Aristotle's question still bears on the philosophy of chemistry. *Philosophy of Science, 73*, 853–863.

Bogaard, P. A. (1979). Heaps or wholes: Aristotle's explanation of compound bodies. *Isis, 70*, 11–29.

Bos, A. P. (2003). *The soul and its instrumental Body: A reinterpretation of Aristotle's philosophy of living nature*. Brill.

Bos, A. P. (2009). Aristotle on soul and soul-"parts" in semen (GA 2.1, 735a4-22). *Mnemosyne, 62*, 378–400.

Bos, A. P. (2000a). Why the soul needs an instrumental body according to Aristotle (Anim. I,3, 407B13-26). *Hermes, 128*, 20–31.

Bos, A. P. (2010a). Aristotle on god as principle of *genesis*. *British Journal for the History of Philosophy, 18*, 363–377.

Bos, A. P. (2010b). Aristotle on the differences between plants, animals, and human beings and on the elements as instruments of the soul (De Anima 2.4.415b18). *The Review of Metaphysics, 63*, 821–841.

Bos, A. P. (2000b). Basilides as an Aristotelianising gnostic. *Vigiliae Christianae, 54*, 44–60.

Bos, A. P., & Divina, P. (1976). *The theme of divine Pronoia in Plato and Aristotle*. Van Gorcum.

Boutot, A. (1998). Les théories de la morphogenèse et le dilemme de la substance. *Revue de Métaphysique et de Morale, 2*, 189–204.

Boutot, A. (1993a). *L'inventio\n des formes: Chaos—Catastrophes—Fractales—Structues dissipatives—Attracteurs étranges*. Odiles Jacob.

Bowin, J. (2009). Aristotle on the order and direction of time. *Apeiron, 42*, 33–62.

Brague, R. (2015). Necessity of the good: Why western culture needs to return to Plato. *First Things* (Online) (*).

Brague, R. (1990). Aristotle's definition of motion and its ontological implications. *Graduate Faculty Philosophy Journal, 13*, 1–22. (*).

Brague, R. (2013). On Aristotle's formula ὄ ποτε ὄν: Physics IV.11, 14. In C. Baracchi (Ed.), *The Bloomsbury companion to Aristotle* (pp. 75–88). Bloomsbury.

Brague, R. (1988). *Aristote et la question du monde*. PUF.

Brague, R. (2008). *Having heart: Sport and θυμός* (Online) (*).

Brague, R. (1997). *Prudence, foresight, providence* (Online) (*).

Brague, R. (1976). Note sur le Concept D' ἡδονὴ chez Aristote. *Les Études philosophiques, 1*, 49–55.

Broadie, S. (2006). Aristotle and contemporary ethics. In R. Kraut (Ed.), *The Blackwell guide to Aristotle's ethics* (pp. 342–361). Blackwell.

Brewer, T. (2006). Three dogmas of desire. In T. Chappell (Ed.), *Values and virtues: Aristotelianism in contemporary ethics* (pp. 257–284). OUP.

Brogan, W. A. (2005). *Heidegger and Aristotle: The twofoldness of being*. SUNY.

Brook, R. (2004). Aristotle on sperm competition in birds. *Classical Quarterly, 54*, 277–278.

Brown, L. (1997). What is "the mean relative to us" in Aristotle's *ethics*? *Phronesis, XLII/1*, 77–93.

Brunschwig, J. (1996). Rule and exception: On the Aristotelian theory of equity. In M. Frede & G. Striker (Eds.), *Rationality in Greek thought*. OUP.

Buddensiek, F. (2012). Does good fortune matter? *Eudemian ethics* VIII.2 on *Eutuchia*. In F. Leigh (Ed.), *The Eudemian ethics on the voluntary, friendship, and luck* (pp. 43–76). Brill.

Burgin, M. (2018). Platonic triangles and fundamental triads as the basic elements of the world. *Athens Journal of Humanities & Arts, 5*(1), 29–44. (*).

Burke, R. J. (1967). Aristotle on the limits of argument. *Philosophy and Phenomenological Research, 27*, 386–400.

Burkert, W. (1997). Impact and limits of the idea of progress in antiquity. In A. Burgen et al. (Eds.), *The idea of progress* (pp. 19–46). de Gruyter.

Burkert, W. (1977). Air-imprints or eidola: Democritus' Aetiology of vision. *Illiniois Classical Studies, 2*, 97–109. (*).

Burnyeat, M. F. (1992). Is an Aristotelian philosophy of mind still credible? A draft. In M. C. Nussbaum & A. O. Rorty (Eds.), *Essays on Aristotle's De Anima* (pp. 15–26). OUP.

Byl, S. (1968). Note sur la place du coeur et la valorisation de la ΜΕΣΟΤΗΣ dans la Biologie d'Aristote. *L'antique classique, 37*, 467–476. (*).

Byrne, P. H. (1997). *Analysis and science in Aristotle*. SUNY.

Campbell, B. J. (1944). Aristotle's natural theology. *Franciscan Studies, 4*, 134–150.

Cartwright, N., & Pemberton, J. (2013). Aristotelian powers: Without them, what would modern science do? In J. Greco & R. Gross (Eds.), *Powers and capacities in philosophy: The new Aristotelianism* (pp. 93–112). Routledge.

Caston, V. (2002). Aristotle on consciousness. *Mind, 111*, 751–815.

Caston, V. (1999). Aristotle's two intellects: A modest proposal. *Phronesis, XLIV/3*, 199–227.

Caston, V. (1996). Why Aristotle needs imagination. *Phronesis, XLI/1*, 20–55.

Cavalcante-Schuback, M. S. (2013). The hermeneutic slumber: Aristotle's reflection on sleep. In C. Baracchi (Ed.), *The Bloomsbury companion to Aristotle* (pp. 128–143). Bloomsbury.

de Cesaris, A. (2011). Aristotle on non-contradiction. *Metalogicon, XXIV*, 1–57. (*).

Chamberlain, C. (1984a). Why Aristotle called ethics ethics: The definition of ηθος *Eudemian ethics* 2,2. *Hermes, 112*, 176–183.

Chamberlain, C. (1984b). The meaning of *Prohairesis* in Aristotle's *ethics*. *Transactions of the American Philological Association, 114*, 147–157.

Charles, D. (2015). Aristotle on practical and theoretical knowledge. In D. Henry & K. M. Nielsen (Eds.), *Bridging the gap between Aristotle's science and ethics* (pp. 71–93). CUP.

Charles, D. (2010). Actuality and potentiality in *metaphysics* Θ. In J. G. Lennox & R. Bolton (Eds.), *Being, nature, and life in Aristotle* (pp. 168–197). CUP.

Charles, D. (2011a). Akrasia: The rest of the story? In M. Pakaluk & G. Pearson (Eds.), *Moral psychology and human action in Aristotle* (pp. 187–210). OUP.

Charles, D. (2011b). Desire in action: Aristotle's move. In M. Pakaluk & G. Pearson (Eds.), *Moral psychology and human action in Aristotle* (pp. 75–94). OUP.

Charles, D. (2012). The *Eudemian ethics* on the "voluntary". In F. Leigh (Ed.), *The Eudemian ethics on the voluntary, friendship, and luck*. Brill.

Charles, D. (2003). Wittgenstein's Builders and Aristotle's Craftsmen. In R. W. Sharples (Ed.), *Perspectives on Greek philosophy: S.V. Keeling Lectuers in ancient philosophy 1992–2002* (pp. 103–126). Ashgate.

Charlton, W. (2006). The human soul as form: The relationship between Aristotle and Catholic teaching. *FAITH Magazine*. (*).

Charlton, W. (1987). Aristotelian powers. *Phronesis, XXXII/1*, 277–289.

Charlton, W. (1972). Aristotle and the principle of individuation. *Phronesis, 17*, 239–249.

Chase, M. (2011). Teleology and final causation in Aristotle and in contemporary science. *Dialogue, 50*, 511–536. (*).

Chroust, A. H. (1968). The psychology in Aristotle's lost dialogue Eudemus or on the soul. *Acta Classica, 9*, 49–62. (*).

Clark, S. R. L. (1975). *Aristotle's man: Speculations on Aristotelian anthropology*. OUP.

Cleary, J. J. (1998). "Powers that be": The concept of potency in Plato and Aristotle. *Méthexis, 11*, 19–64.

Cleary, J. J. (1988). *Aristotle on the many senses of priority*. Jnl. Hist. Phil.

Cleary, J. J. (2002). Abstracting Aristotle's philosophy of mathematics. In B. E. Babich (Ed.), *Philosophy of science, van Gogh's Eyes, and god: Hermeneutic essays in honour of Patrick a. Heelan, S.J.* Springer. (*).

Cleary, J. J. (1985). On the terminology of "abstraction" in Aristotle. *Phronesis, 30*, 13–45.

Code, A., & Moravcsik, J. (1992). Explaining various forms of living. In M. C. Nussbaum & A. O. Rorty (Eds.), *Essays on Aristotle's De Anima* (pp. 129–146). OUP.

Code, A. (2008). Aristotelian colours as causes. In D. Follesdall & J. Woods (Eds.), *Festschrift for Julius Moravcsik* (pp. 235–242). College. (*).

Code, A. (1987). Soul as efficient cause in Aristotle's embryology. *Philosophical Topics, XV*, 51–58.

Code, A. (1986). Aristotle's investigation of a basic logical principle: Which science investigates the principle of non-contradiction? *Canadian Journal of Philosophy, 16*, 341–357.

Code, A. (1976). The persistence of Aristotelian matter. *Philosophical Studies, 29*, 357–367.

Cohen, S. M. (1996). *Aristotle on nature and incomplete substance*. CUP.

Cohen, S. M. (1994). Aristotle on elemental matter. *Phronesis, 39*, 150–159.

Cohen, S. M. (1985). Aristotle and individuation. *Canadian Journal of Philosophy, X*, 41–65. (*).

Corcilius, K. (2014). Activity, passivity, and perceptual discrimination in Aristotle. In J. F. Silva & M. Yrjönsurri (Eds.), *Active perception ancient to modern*. Springer. (*).

Connell, R. J. (1988). *Substance and modern science*. St Thomas.

Corkum, P. (2008). Aristotle on ontological dependence. *Phronesis, 53*, 65–92.

Cosans, C. E. (1998). Aristotle's anatomical philosophy of nature. *Biology and Philosophy, 13*, 311–339.

Crittenden, P. (1991). Ethics and aesthetics in Aristotle's *poetics*. *Literature & Aesthetics, 1*, 15–27. (*).

Crombie, I. M. (1963). *An examination of Plato's doctrines: Plato on knowledge and reality*. Humanities.

Crubellier, M. (2004). Le "syllogisme pratique" ou Comment la pensée meut le corps. In A. Laks & M. Rashed (Eds.), *Aristote et le mouvement des animaux* (pp. 9–26). Septentrion.

Curd, P. (2007). *Anaxagoras of Clazomenae*. Univ. of Toronto Press.

Curzer, H. J. (2012). *Aristotle & the virtues*. OUP.

Dasen, V. (2007). "All children are dwarves." Medical discourse and iconography of children's bodies. *Oxford Journal of Archaeology, 27*, 49–62. (*).

Davies, H. (1899). Psychological experiences implicating the concept of substance. *The Philosophical Review, 8*, 604–621. (*).

Decaen, C. A. (2004). Aristotle's Aether and contemporary science. *The Thomist, 68*, 375–429.

Defillipo, J. G. (1994). Aristotle's identification of the prime mover as god. *The Classical Quarterly, 44*, 393–409.

Delcomminette, S. (2017). *Aristote et la nécessité*. Vrin.

Delbrück, M. (1971). Aristotle-totle-totle. In J. Monod & E. Borek (Eds.), *Of microbes and life* (pp. 50–55). Columbia University Press. (*).

Depew, D. J. (1995). Humans and other political animals in Aristotle's "history of animals". *Phronesis, XL/2*, 156–181.

Derrida, J. (1982). Ousia and Grammē: A note on a note from being and time. In *Margins of philosophy* (pp. 29–68). Harvester. [*orig*. 1968] (*).

Deslauriers, M. (2002). How to distinguish Aristotle's virtues. *Phronesis, XLVII/2*, 101–126.

Destreé, P. (2011). Aristotle on responsibility for one's character. In M. Pakaluk & G. Pearson (Eds.), *Moral psychology and human action in Aristotle* (pp. 285–318). OUP.

Devereux, D. T. (1986). Particular and universal in Aristotle's conception of practical knowledge. *The Review of Metaphysics, 39*, 483–504.

Di Muzio, G. (2000). Aristotle on improving one's character. *Phronesis, XLV/3*, 205–219.

Dobbs, D. (1996). Family matters: Aristotle's appreciation of women and the plural structure of society. *The American Political Science Review, 90*, 74–89. (*).

Dobbs, D. (1994). Natural right and the problem of Aristotle's defense of slavery. *The Journal of Politics, 56*, 69–94. (*).

Dobbs, D. (1985). Aristotle's anticommunism. *American Journal of Political Science, 29*, 29–46. (*).

Dobbs, D., & Grafstein, R. (1988). Rationalism or revelation? *The American Political Science Review, 82*, 579–587. (*).

Dokic, J. (2014). Common sense and metaperception: A practical model. *Res Philosophica, 91*, 241–259. (*).

Dow, J. (2011). Aristotle's theory of the emotions: Emotions as pleasures and pains. In M. Pakaluk & G. Pearson (Eds.), *Moral psychology and human action in Aristotle* (pp. 47–74). OUP.

Dryer, D. P. (1983). Aristotle's conception of "Orthos logos". *The Monist, 66*, 106–119.

Dye, J. W. (1974). Heraclitus and the future of process philosophy. *Tulane Studies in Philosophy, 23*, 13–31.

van der Eijk, P. (2000). Aristotle's psycho-physiological account of the soul-body relationship. In J. Wright & P. Potter (Eds.), *Psyche and Soma: Physicians and metaphysicians on the mind-Body problem from antiquity to enlightenment*. OUP.

van der Eijk, P. (1999). On sterility (HA X), A medical work by Aristotle? *The Classical Quarterly, 49*, 490–502.

Elden, S. (2005). Reading logos as speech: Heidegger, Aristotle and rhetorical politics. *Philosophy & Rhetoric, 38*, 281–301.

Ellrod, F. E. (1982). *Energeia* and process in Aristotle. *International Philosophy Quarterly, 22*, 175–181. (*).

El Murr, D. (2021). Platonic 'desmology' and the body of the world animal (*Tim.*30c-34a). In R. Salles (Ed.), *Cosmology and biology in ancient philosophy: From Thales to Avicenna* (pp. 46–71). CUP. (*).

Engelbretson, G. (1985). Quine on Aristotle on identity. *Crítica Revista Hispanoamericano de Filosofia, 7*.

Falcon, A. (2005). *Aristotle & the science of nature: Unity without uniformity*. CUP.

Feibleman, J. K. (1953). Aristotle as a finite ontologist. *Tulane Studies in Philosophy, 2*, 39–58.

Ferejohn, M. T., & Causes, F. (2013). *Definition, explanation, and primacy in Socratic and Aristotelian thought*. OUP.

Ferejohn, M. T. (1981). Aristotle on necessary truth and logical priority. *American Philosophical Quarterly, 18*, 285–283.

Ferrarin, A. (2005). Aristotle on ΦΑΝΤΑΣΙΑ. *Proceedings of the Boston Colloquium in Ancient Philosophy, 21*, 89–112.

Fiasse, G. (2001). Aristotle's φρονησις: A true grasp of ends as well as means? *The Review of Metaphysics, 55*, 323–337.

Fine, G. (1981). Aristotle on determinism: A review of Richard Sorabji's necessity, cause, and blame. *The Philosophical Review, 90*, 561–579.

Flakne, A. (2005). Embodied and embedded: Friendship and the sunaisthetic self. *Epoché, 10*, 37–63. (*).

Forbes, T. M. (1930). The silkworm of Aristotle. *Classical Philology, 25*, 22–26.

Fortier, T. L. (1971). Aristotle on social friendships. *Laval théologique et philosophique, 27*, 235–250. (*).

Franklin, J. (1989). Species in Aristotle. *Philosophy, 64*, 107–108.

Frappier, G. (1977). L'art dialectique dans la philosophie d'Aristote. *Laval théologique et philosophique, 33*, 115–134. (*).

Frede, D. (1992a). The cognitive role of *Phantasia*. In M. C. Nussbaum & A. O. Rorty (Eds.), *Essays on Aristotle's De Anima*. OUP.

Frede, M. (1992b). On Aristotle's conception of soul. In M. C. Nussbaum & A. O. Rorty (Eds.), *Essays on Aristotle's De Anima* (pp. 93–108). OUP.

Freeland, C. A. (1994). Aristotle on perception, appetition, and self-motion. In M. L. Gill & J. G. Lennox (Eds.), *Self-motion: From Aristotle to Newton* (pp. 35–64). Princeton.

von Fritz, K. (1981). Aristotle's anthropological ethics and its relevance to modern problems. *Journal of History of Ideas, 42*, 187–207.

von Fritz, K. (1945). Nous, Noein, and their derivatives in pre-Socratic philosophy (excluding Anaxagoras). *Classical Philology, 40*, 223–242.

Fuller, B. A. G. (1907). The theory of god in book Λ of Aristotle's *metaphysics*. *The Philosophical Review, 16*, 170–183.

Furth, M. (1990). Specific and individual form in Aristotle. In D. Devereux & P. Pellegrin (Eds.), *Biologie, Logique et Métaphysique chez Aristote* (pp. 85–111). CNRS.

Gabriel, M. (2009). God's transcendent activity: Ontotheology in "metaphysics" 12. *The Review of Metaphysics, 63*, 385–414.

Garcia-Marqués, A., & Ballester, M. (1994). Le Problème *Phusis-Nomos* chez Aristote. In *Actes du 25e Congrès de l'Association des Sociétés de philosophie de langue française (ASPLF)* (pp. 416–422) (*).

Gaukroger, S. (1982). The one and the many: Aristotle on the individuation of numbers. *The Classical Quarterly, 32*, 312–322.

Gelber, J. (2018). Females in Aristotle's embryology. In A. Falcon & D. Lefebvre (Eds.), *Aristotle's generation of animals: A critical guide* (pp. 171–187). CUP.
Georgiou, C. D. (2004). Analogies between Aristotle's ontology and biological ideologies of human nature. *Nature, Society, and Thought, 17*, 47–65.
Gilson, E. (1984). *From Aristotle to Darwin and back again: A journey in final causality, species, and evolution*. Notre Dame Press. [*orig*. 1971].
Gill, M. L., & Lennox, J. G. (Eds.). (1994). *Self-motion: From Aristotle to Newton*. Princeton.
Goldin, O. (2016). Aristotle, The Pythagoreans, and structural realism. *Philosophy Faculty Research and Publications, 558*. (*).
Golluber, M. (1999). Aristotle on how one becomes what one is. *The Review of Metaphysics, 53*(2), 363–382.
Gonzalez, F. J. (1991). Aristotle on pleasure and perfection. *Phronesis, XXXVI*, 141–159.
Gonzalez, F. J. (2007). Wax tablets, aviaries, or imaginary pregnancies? On the powers in Theaetetus' soul. *Études platoniciennes, 4*, 273–293.
Gooch, P. W. (1983). Aristotle and the happy dead. *Classical Philology, 78*, 112–116.
Goodman, L. E., & Talisse, R. B. (Eds.). (2008). *Aristotle's politics today*. SUNY.
Gotthelf, A. (2012). *Teleology, first principles, and scientific method in Aristotle's biology*. OUP.
Gotthelf, A. (1999). Darwin on Aristotle. *Journal of the History of Biology, 32*, 3–30.
Gotthelf, A. (1987). Aristotle's conception of final causation. In J. G. Lennox & A. Gotthelf (Eds.), *Philosophical issues in Aristotle's biology*. CUP.
Gottlieb, P. (2009). *The virtues of Aristotle's ethics*. CUP.
Gottlieb, P. (1994). Aristotle on dividing the soul and uniting the virtues. *Phronesis, XXXIX/3*, 275–290.
Graham, D. W. (1986). Some myths about Aristotle's biological motivation. *Journal of the History of Ideas, 47*, 529–545.
Granger, H. (1993). Aristotle on the analogy between action and nature. *The Classical Quarterly, 43*, 168–176.
Grene, M. (1972). Aristotle and modern biology. *Journal of the History of Ideas, 33*, 395–424.
Greene, W. C. (1946). Aristotle on metaphor. *The Classical Weekly, 39*, 94–95.
Gregorić, P., & Grgić, F. (2006). Aristotle's notion of experience. *Archiv für Geschichte der Philosophie, 88*, 1–32. (*).
Gregory, A. (2001). Aristotle, dynamics and proportionality. *Early Science and Medicine, 6*, 1–21.
de Groot, J. (2008). *Dunamis* and the science of mechanics: Aristotle on animal motion. *Journal of the History of Philosophy, 46*, 43–68.
Guthrie, W. K. C. (1933). The development of Aristotle's theology I. *The Classical Quarterly, 27*, 162–171.
Guthrie, W. K. C. (1934). The development of Aristotle's theology II. *The Classical Quarterly, 28*, 90–98.
Hahn, R. (1979). Aristotle as Ontologist or theologian? Or, Aristotelian form in the context of the conflicting doctrines of being in the "metaphysics". *The Southwestern Journal of Philosophy, 10*, 79–88.
Haldane, J. B. S. (1955). Aristotle's account of bees "dances". *Journal of Hellenic Studies, 75*, 24–25.
Hammond, W. A. (1902). The significance of the creative reason in Aristotle's philosophy. *The Philosophical Review, 11*, 238–248.
Hardie, W. F. R. (1964–5). Aristotle's doctrine that virtue is a "mean". *Proceedings of the Aristotelian Society, 65*, 183–204.
Hartmann, N. (1993). Le concept mégarique et aristotélicien de possibilité. Contribution à l'histoire du problème ontologique de la modalité. *Laval théologique et philosophique, 49*, 131–146. [*orig*. 1937] (*).
Hatab, L. J. (2015). A story of unrequited love; the tragic character of Aristotle's philosophy. *Epoché, 19*, 289–298.
Heath, M. (2009). Cognition in Aristotle's poetics. *Mnemosyne, 62*, 51–75.

Heinaman, R. (Ed.). (1995). *Aristotle and moral realism: The keeling colloquia I*. Westview.

Heinaman, R. (1993). Rationality, Eudaimonia and Kakadaimonia in Aristotle. *Phronesis, XXXVIII/1*, 31–56.

Heinaman, R. (1990). Aristotle and the mind-Body problem. *Phronesis, XXXV/1*, 83–102.

Heller, M. (2006). Aristotle on the mechanism of inheritance. *Journal of the History of Biology, 39*, 425–455.

Henn, M. J. (2001). God as both ideal and real being in the Aristotelian metaphysics. *Auslegung, 24*, 89–115. (*).

Henry, D. (2018). Aristotle on epigenesis: Two senses of epigenesis. In A. Falcon & D. Lefebvre (Eds.), *Aristotle's generation of animals: A critical guide* (pp. 89–107). CUP.

Henry, D. (2015). Holding for the most part: The demonstrability of moral facts. In D. Henry & K. M. Nielsen (Eds.), *Bridging the gap between Aristotle's science and ethics* (pp. 169–190). CUP.

Henry, D. (2013). Optimality and teleology in Aristotle's natural science. *Oxford Studies in Ancient Philosophy, XLV*, 225–263. (*).

Henry, D. (2011). Aristotle's pluralistic realism. *The Monist, 94*, 197–220.

Henry, D. (2007). How sexist is Aristotle's developmental biology? *Phronesis, 52*, 251–269.

Henry, D. (2005). Embryological models in ancient philosophy. *Phronesis, 50*, 1–42.

von Heyking, J. (2008). "Sunaisthetic" friendship and the foundation of political anthropology. *International Political Anthropology, 1*, 179–193. (*).

Hintikka, J. (2004a). On Aristotle's notion of existence. In *Analyses of Aristotle*. Springer. [*orig.* 1999] (also *The Review of Metaphysics, 52*, 779–805 (1999)).

Hintikka, J. (2004b). Introduction. In *Analyses of Aristotle*. Springer.

Hintikka, J. (1997). Three dogmas of quine's empiricism. *Revue Internationale de Philosophie, 51*, 457–477.

Hintikka, J. (1967). Some conceptual presuppositions of Greek political theory. *Scandinavian Political Studies, Bind 2*, 11–25. (*).

Hintikka, J. (1964). The once and Future Sea fight: Aristotle's discussion of future contingents in *De Interpretione* IX. *The Philosophical Review, 73*, 461–492.

Horowitz, M. C. (1976). Aristotle and woman. *Journal of the History of Biology, 9*, 183–213.

Houlihan, T. (1970). A point of rapport between Piaget and Aristotle: From vagueness and confusion to clarity and distinctness in the acquisition of human knowledge. *Laval théologique et philosophique, 26*(3), 233–261. (*).

Hutchinson, D. S. (1986). *The virtues of Aristotle*. Routledge.

Irwin, T. (2010). The sense and reference of *Kalon* in Aristotle. *Classical Philology, 105*, 381–396.

Irwin, T. (2001). Vice and reason. *The Journal of Ethics, 5*, 73–97.

Jackson, P. (2016). *Aristotle on the meaning of man: A philosophical response to idealism, positivism, and Gnosticism*. Peter Lang.

Jacobs, W. (1980). The existential presuppositions of Aristotle's logic. *Philosophical Studies: An International Journal for Philosophy in the Analytic Tradition, 37*, 419–428.

Jansen, L. (2009). Aristotle's theory of dispositions: From the princple of movement to the unmoved mover (Online précis of *Tun und Können. Ein systematischer Kommentar zu Aristoteles' Theorie derVermögen im neunten Buch der Metaphysik*. Frankfurt: Hänsel-Hohenhausen, 2002). (*).

Jiménez, E. R. (2013). Mind and body in Aristotle. In C. Baracchi (Ed.), *The Bloomsbury companion to Aristotle* (pp. 91–105). Bloomsbury.

Joachim, H. H. (1951). *The Nicomachean ethics: A commentary by the late H.H. Joachim*. OUP.

Johansen, T. K. (2012). *The powers of Aristotle's soul*. OUP.

Johansen, T. K. (1996). Aristotle on the sense of smell. *Phronesis, 41*, 1–19.

Johnson, M. R. (2005). *Aristotle on teleology*. OUP.

Johnson, M. R. (2015a). Aristotle's architectonic sciences. In D. Ebrey (Ed.), *Theory and practice in Aristotle's natural science* (pp. 163–187). CUP. (*).

Johnson, M. R. (2015b). Luck in Aristotle's *physics* and *ethics*. In D. Henry & K. M. Nielsen (Eds.), *Bridging the gap between Aristotle's science and ethics* (pp. 254–275). CUP. (*).

Johnson, M. R. (2011). The medical background of Aristotle's theory of nature and spontaneity. *Proceedings of the Boston Colloquium in Ancient Philosophy, XXVII*, 105–152. (*).

Karaca, Ç. (2021a). Life's organization between matter and form: Neo-Aristotelian approaches and biosemiotics. *History and Philosophy of the Life Sciences, 43*. (*).

Kahn, C. H. (1985). The place of the prime mover in Aristotle's teleology. In A. Gottheld (Ed.), *Aristotle on nature and living things* (pp. 183–205). Mathesis. (*).

Kavanaugh, L. J. (2007). *The architectonic ofphilosophy: Plato, Aristotle, Leibniz*. Amsterdam Univ.

Kaitz, E. (2010). Aristotle's warning. *American Thinker.* (Online) (*).

Keizer, H. M. (2010). 'Eternity' revisited: A study of the Greek word αἰών. *Philosophia Reformata, 65*, 53–71. (*).

Keyt, D. (1961). Aristotle on Plato's receptacle. *The American Journal of Philology, 82*, 291–300.

Klein, J. (1968). *Greek mathematical thought and the origin of algebra*. MIT.

de Koninck, T. (1994). Aristotle on god as thought thinking itself. *The Review of Metaphysics, 47*, 471–515.

Koslicki, K. (2015). In defense of substance. *Grazer Philosophische Studien, 91*, 59–80. (*).

Koslicki, K. (2014). The causal priority of form in Aristotle. *Studia Philosophica Estonica*, 113–141. (*).

Koslicki, K. (2006). Aristotle's mereology and the status of form. *The Journal of Philosophy, 103*, 715–736. (*).

Kosman, A. (2013). *The activity of being: An essay on Aristotle's ontology*. Harvard.

Kosman, A. (2014a). Understanding, explanation, and insight in Aristotle's *posterior analytics*. In *Virtues of thought: Essays on Plato and Aristotle* (pp. 7–26). Harvard. [*orig.* 1973].

Kosman, A. (2014b). Perceiving that we perceive: *De Anima 3.2*. In *Virtues of thought: Essays on Plato and Aristotle* (pp. 43–61). Harvard. [*orig.* 1975].

Kosman, A. (2014c). Necessity and explanation in Aristotle's *analytics*. In *Virtues of thought: Essays on Plato and Aristotle* (pp. 77–93). Harvard. [*orig.* 1987].

Kosman, A. (2014d). Acting: Drama as the *mimesis* of *praxis*. In *Virtues of thought: Essays on Plato and Aristotle* (pp. 94–118). Harvard. [*orig.* 1992].

Kosman, A. (2014e). Saving the phenomena: Realism and instrumentalism in Aristotle's theory of science. In *Virtues of thought: Essays on Plato and Aristotle* (pp. 138–156). Harvard. [*orig.* 1992].

Kosman, A. (2014f). Aristotle on the desirability of friends. In *Virtues of thought: Essays on Plato and Aristotle* (pp. 157–182). Harvard. [*orig.* 2004].

Kosman, A. (2014g). Justice and virtue: The *Republic*'s inquiry into proper difference. In *Virtues of thought: Essays on Plato and Aristotle* (pp. 183–203). Harvard. [*orig.* 2007].

Kosman, A. (2014h). Male and female in Aristotle's *generation of animals*. In *Virtues of thought: Essays on Plato and Aristotle* (pp. 204–226). Harvard. [*orig.* 2010].

Kosman, A. (2014i). Self-knowledge and self-control in Plato's *Charmides*. In *Virtues of thought: Essays on Plato and Aristotle* (pp. 227–245). Harvard.

Kosman, A. (2014j). Aristotle on the virtues of thought. In *Virtues of thought: Essays on Plato and Aristotle* (pp. 280–298). Harvard.

Kosman, A. (2014k). Platonic love. In *Virtues of thought: Essays on Plato and Aristotle* (pp. 27–42). Harvard. [*orig.* 1974].

Kosman, A. (1994a). Aristotle's prime mover. In M. L. Gill & J. G. Lennox (Eds.), *Self-motion: From Aristotle to Newton* (pp. 135–154). Princeton.

Kosman, A. (1992). What does the maker mind make? In M. C. Nussbaum & A. O. Rorty (Eds.), *Essays on Aristotle's De Anima* (pp. 343–358). OUP.

Kosman, A. (2014l). What does the maker mind make? In *Virtues of thought: Essays on Plato and Aristotle* (pp. 119–137). Harvard.

Kosman, A. (1980). Being properly affected: Virtues and feelings in Aristotle's ethics. In A. O. Rorty (Ed.), *Essays on Aristotle's ethics* (pp. 103–116). Univ. Cal..

Kosman, A. (2014m). Being properly affected: Virtues and feelings in Aristotle's ethics. In *Virtues of thought: Essays on Plato and Aristotle* (pp. 62–76). Harvard.

Kosman, A. (1994b). The activity of being in Aristotle's *metaphysics*. In T. Scaltsas, D. Charles, & M. L. Gill (Eds.), *Unity, identity, and explanation in Aristotle's metaphysics* (pp. 195–214). OUP.
Kosman, A. (1969). Aristotle's definition of motion. *Phronesis, 14*, 40–62.
Kouremenos, T. (2003). Aristotle on geometric perfection in the physical world. *Mnemosyne, 56*, 463–479.
Kullmann, W. (1990). Bipartite science in Aristotle's biology. In D. Devereux & P. Pellegrin (Eds.), *Biologie, Logique et Métaphysique chez Aristote* (pp. 335–347). CNRS.
Kuo, P. (1931, March). The humanism of Confucius and Aristotle. *The Bookman*, 29–34. (*).
Labarrière, J.-L. (1993). Aristote et la question du langage animal. *Mètis, 8*, 247–260. (*).
Labarrière, J.-L. (1990). De La Phronesis Animale. In D. Devereux & P. Pellegrin (Eds.), *Biologie, Logique et Métaphysique chez Aristote* (pp. 405–428). CNRS.
Laks, A. (2016). Le Moteur Immobile. In E. Berti & M. Crubellier (Eds.), *Lire Aristote* (pp. 145–158). PUF.
Lang, H. S. (1998). *The order of nature in Aristotle's physics: Place and the elements*. CUP.
Lang, H. S. (2013). Aristotle on sensible objects: Natural things and Body. In C. Baracchi (Ed.), *The Bloomsbury companion to Aristotle* (pp. 59–74). Bloomsbury.
Lawrence, G. (2011). Acquiring character: Becoming grown-up. In M. Pakaluk & G. Pearson (Eds.), *Moral psychology and human action in Aristotle* (pp. 233–284). OUP.
Lawrence, G. (2001). The function of the function argument. *Ancient Philosophy, 21*, 445–475.
Lawrence, J. P. (1988). The hidden Aporia in Aristotle's self-thinking thought. *Journal of Speculative Philosophy, 2*, 155–174.
Lear, J. (1982). Aristotle's philosophy of mathematics. *The Philosophical Review, 91*, 161–192.
Lee, W. L. (2007). Aristotle's ecological conception of living things and its significance for feminist theory. *Diametros, 14*, 68–84. (*).
Lefebvre, D. (2012). La Question de L'Unité D'Une Science des Substances: Interpretation de Métaphysique, Λ, 1, 1069A36-B2. In M. Bonelli (Ed.), *Physique et Métaphysique Chez Aristote* (pp. 133–174). Vrin. (*).
Lefebvre, D. (2008). L'usage de la Série des Contraires en *Métaphysique* Γ 2, 1004B27-1005A6. In A. Stevens (Ed.), *Aristote Métaphysique Gamma: Édition, Traduction, Études*. Leuven.
Lefebvre, D. (2005). Les non-êtres de le temps (Physique, IV, 12, 221b23-222a9). *Le Temps Philosophique, 11*, 173–196. (*).
Lefebvre, R. (1988). "Le Miroir de L'Âme" (Aristote, *De Insomnis*, 459 b 23–460 a 26). *Les Études Philosophiques, 2*, 195–206. (*).
Leighton, S. R. (1982). Aristotle and the emotions. *Phronesis, XXVII*, 144–174.
Lennox, J. G. (2015). Aristotle on the biological roots of virtue: The natural history of natural virtue. In D. Henry & K. M. Nielsen (Eds.), *Bridging the gap between Aristotle's science and ethics* (pp. 193–213). CUP.
Lesher, J. H. (2009). Aristotle's considered view of the path to knowledge. In M. Boeri (Ed.), *Festschrift for Alfonso Gomez-Lobo*. (Online) (*).
Lesher, J. H. (1973). The meaning of ΝΟΥΣ in the *Posterior Analytics*. *Phronesis, XVIII*, 44–68.
Leunissen, M. (2010a). *Explanation and teleology in Aristotle's science of nature*. CUP.
Leunissen, M. (2018). Order and method in Aristotle's *generation of animals* 2. In A. Falcon & D. Lefebvre (Eds.), *Aristotle's generation of animals: A critical guide* (pp. 56–74). CUP.
Leunissen, M. (2015). Aristotle on knowing natural science for the sake of living well. In D. Henry & K. M. Nielsen (Eds.), *Bridging the gap between Aristotle's science and ethics* (pp. 214–231). CUP.
Leunissen, M. (2010b). Nature as a good housekeeper: Secondary teleology and material necessity in Aristotle's biology. *Apeiron, 43*, 117–142.
Leunissen, M., & Gotthelf, A. (2010). "What's teleology got to do with it?" A reinterpretation of Aristotle's *Generation of Animals* V. *Phronesis, 55*, 325–356.
Leunissen, M. (2007). The structure of teleological explanations in Aristotle: Theory and practice. *Oxford Studies in Ancient Philosophy, XXXIII*, 145–178.

Lewis, F. A. (2008). What's the matter with prime matter. *Oxford Studies in Ancient Philosophy, 34*, 123–146.

de Ley, H. (1980). Pangenesis versus panspermia Democritean notes on Aristotle's *generation of animals. Hermes, 108*, 129–153.

von Leyden, W. (1964). Time, number, and Eternity in Plato and Aristotle. *The Philosophical Quarterly, 14*, 35–52.

Lindsay, J. (1921). Aristotle and the criterion of truth. *The Monist, 31*, 470–475.

Lindsay, T. (1992). Aristotle's defense of democracy through "political mixing". *The Journal of Politics, 54*, 101–119.

Long, C. P. (2015). On touch and life in the *De Anima*. In A. Cimino & P. Kontos (Eds.), *Phenomenology and the metaphysics of sight* (pp. 69–94). Brill. (*).

Long, R. T. (2006). Realism and abstraction in economics: Aristotle and Mises *versus* Friedman. *The Quarterly Journal of Austrian Economics, 9*, 3–23. (*).

Long, R. T. (2000). *Reason and value: Aristotle versus Rand*. Objectivist Centre.

Lorenz, H. (2006). *The brute within: Appetitive desire in Plato and Aristotle*. OUP.

Lorenz, H. (2007). The assimilation of sense to sense-object in Aristotle. *Oxford Studies in Ancient Philosophy, XXXIII*, 179–220.

Losin, P. (1987). Aristotle's doctrine of the mean. *History of Philosophy Quarterly, 4*, 329–341.

Lowe, M. F. (1983). Aristotle on kinds of thinking. *Phronesis, XXVIII*, 17–30.

Ludlam, I. (2000). Defining terms in Aristotle's "Topics": ΟΡΟΣ or ΟΡ<ΙΣΜ>ΟΣ? *Mnemesyne, 53*, 267–287.

Machamer, P. K. (1978). Aristotle on natural place and natural motion. *Isis, 69*, 377–387.

Maher, D. P. (2011). Aristotle on mathematical and eidetic number. *Hermathena, 190*, 29–51.

Manicas, P. T. (1965). Aristotle, dispositions and occult powers. *The Review of Metaphysics, 18*, 678–689.

Mansion, A. (1913). *Introduction à la physique aristotélicienne*. Inst. Phil.

Mansion, S. (1984). *Études Aristotéliciennes: Recueil d'Articles*. De Wulf. (note: all page numbers in brackets below refer to this work).

Mansion, S. (1981). La signification de l'universel d'après An. Post. I 1. In *Aristotle on Science: the "Posterior Analytics" (Studia Aristotelica, 9)* (pp. 329–342 [155–168]). Antenore.

Mansion, S. (1968). Notes sur la doctrine des categories dans les *Topiques*. In G. E. L. Owen (Ed.), *Aristotle on dialectic* (pp. 189–201 [169–181]). Clarendon.

Mansion, S. (1941). *La rôle de la connaissance de l'existence dans la science aristotélicienne* (pp. 183–203). Louvain.

Mansion, S. (1947). Quelques réflexions sur le réalisme d'Aristote. *Tijdschrift voor Wijsbegeerte en Psychologie, 39*, 129–135 [205–211].

Mansion, S. (1979a). "Plus connu en soi," "plus connu pour nous." Une distinction épistémologique importante chez Aristote. *Pensamiento, 35*, 161–170 [213–222].

Mansion, S. (1964). Aristotle's theory of knowledge and French phenomemology. *International Philosophical Quarterly, 4*, 183–199 [223–239].

Mansion, S. (1955). Les apories de la Métaphysique aristotélicienne. In *Autour d'Aristote* (pp. 141–179 [243–281]). Univ. Louvain.

Mansion, S. (1946). La première doctrine de la substance: la substance selon Aristote. *Revue Philosophique de Louvain, 44*, 349–369 [283–303]. (*).

Mansion, S. (1948). La doctrine aristotélicienne de la substance et le traité des Catégories. In *Proceedings of the tenth international congress of philosophy, Amsterdam* (pp. 337–340 [305–308]).

Mansion, S. (1971). Sur la composition ontologique des substances sensibles chez Aristote (Métaphysique Z 7-9). In R. B. Palmer & R. Hamerton-Kelly (Eds.), *Philomathes. Studies and essays in the humanities in memory of Philip Merlan* (pp. 75–87 [309–321]). Martinus Nijhoff.

Mansion, S. (1979b). La notion de matière en "Métaphysique" Z 10 et 11. In P. Aubenque (Ed.), *Études sur la "Métaphysique" d'Aristote, Actes du VIe Symposium Aristotelicum* (pp. 185–205 [323–343]). Vrin.

Mansion, S. (1969). Tò σιμόν et la définition physique. In I. Düring (Ed.), *Naturphilosophie bei Aristoteles und Theophrast* (pp. 124–132 [347–355]). Stiehm.
Mansion, S. (1973). Deux definitions différentes de la vie chez Aristote? *Revue Philosophique de Louvain, 71*, 425–449 [365–390].
Mansion, S. (1978). Soul and life in the De Anima. In G. E. R. Lloyd (Ed.), *Aristotle on the mind and the senses: Proceedings of the seventh symposium aristotelicum* (pp. 391–410). CUP.
Mansion, S. (1960). Contemplation and action in Aristotle's "Protrepticus". In I. Düring & G. E. L. Owen (Eds.), *Aristotle and Plato in the mid-fourth century: Papers of the symposium aristotelicum* (pp. 56–75 [413–432]). Almqvist and Wiksell.
Mansion, S. (1976). Le plaisir et la peine, matière de l'agir moral selon Aristote. In *Images of man in ancient and medieval thought* (pp. 37–51 [447–461]). Leuven Univ.
Marmodoro, A. (2013). Aristotle's hylomorphism without reconditioning. *Philosophical Inquiry, 36*, 5–22. (*).
Marmodoro, A. (2019). Structure versus whole versus one. In *Third Pisa colloquium in logic, language and epistemology* (pp. 239–249). Editzioni ETS. (*).
Martin, C. (2006). Experience of the new world and Aristotelian revisions of the earth's climates during the renaissance. *History of Meteorology, 3*, 1–15. (*).
Martineau, E. (1979). Aiōn chez Aristote "De Caelo," I, 9: Théologie cosmique ou cosmo-théologie. *Revue de Métaphysique et de Morale,* 84e Année. (*).
Massie, P. (2013). The irony of chance: On Aristotle's physics B, 4–6. *International Philosophy Quarterly, 43*, 15–28.
Matthen, M., & Hankinson, R. J. (1993). Aristotle's universe: Its form and matter. *Synthese, 96*, 417–435.
Matthews, G. B. (1986). Aristotelian explanation. *Illinois Classical Studies, 11*, 173–179. (*).
May, H. (2010). *Aristotle's ethics: Moral development and human nature*. Continuum.
Mayhew, R. (2007). Aristotle on prayer. *RHIZAI, IV.2*, 295–309. (*).
Mayhew, R. (2004). *The female in Aristotle's biology: Reason or rationalization*. Univ. Chic.
Mayhew, R. (1997). Part and whole in Aristotle's political philosophy. *The Journal of Ethics, 1*, 325–340.
Mayhew, R. (1995). The communism of property: A note on Aristotle, *politics* 1263a8-15. *The Classical Quarterly, 45*, 566–569.
McCabe, M. M. (2012). With mirrors or without? Self-perception in *Eudemian ethics* VII.12. In F. Leigh (Ed.), *The Eudemian ethics on the voluntary, friendship, and luck* (pp. 43–76). Brill.
McClymont, J. D. (2010). Reading between the lines: Aristotle's views on religion. *Acta Classica, LIII*, 33–48.
McFadyen, J. E. (1904). Hellenism and Hebraism. *The American Journal of Theology, 8*, 30–47.
McNeil, W. (1999). *The glance of the eye: Heidegger, Aristotle, and the ends of theory*. SUNY.
Mele, A. R. (1981). Choice and virtue in the *Nicomachean ethics*. *Journal of the History of Philosophy, 19*, 405–423.
Mendell, H. (1987). Topoi on Topos: The development of Aristotle's concept of place. *Phronesis, XXXII/2*, 206–231.
Menn, S. (2012). The intellectual setting. In *The Cambridge history of seventeenth-century philosophy* (pp. 22–86). CUP. (*).
Menn, S. (2002). Aristotle's definition of the soul and the Programme of the *De Anima*. *Oxford Studies in Ancient Philosophy, 22*, 83–139.
Menn, S. (1994). The origins of Aristotle's concept of 'Ενεργεια: Ενεργεια and Δυναμις. *Ancient Philosophy, 14*, 73–114.
Menn, S. (1992). Aristotle and Plato on god as *nous* and as the good. *The Review of Metaphysics, 45*, 543–573.
Merlan, P. (1960). *From Platonism to neo-Platonism* (2nd ed.). Martinus Nijhoff.
Merlan, P. (1946). Aristotle's unmoved movers. *Traditio, 4*, 1–30.
Meyer, S. S. (1994). Self-movement and external causation. In M. L. Gill & J. G. Lennox (Eds.), *Self-motion: From Aristotle to Newton* (pp. 65–80). Princeton.

Meyer, S. S. (1993). *Aristotle on moral responsibility: Character and cause*. OUP.
Meyer, S. S. (1992). Aristotle, teleology, and reduction. *The Philosophical Review, 101*, 791–825.
Meyer, S. S. (1987). Unmoved movers, form, and matter. *Philosophical Topics, 15*, 171–196.
Mesquita, A. P. (2015). Individual substances and individual accidents in the "categories" of Aristotle. *Revista Portuguesa de Filosofia, 71*, 388–421. (*).
Miller, A. E., & Miller, M. G. (2004). Aristotle's *metaphysics* as the ontology of being-alive and its relevance today. *Proceedings of the Boston Area Colloquium in Ancient Philosophy, 2004*, 1–96.
Miller, F. D. (1999). Aristotle's philosophy of soul. *The Review of Metaphysics, 53*, 309–337.
Mirus, C. V. (2012). Order and the determinate: The good as a metaphysical concept in Aristotle. *The Review of Metaphysics, 65*, 499–523.
Mix, L. J. (2016). Nested explanation in Aristotle and Mayr. *Synthese, 193*, 1817–1832.
Moline, J. (1981). *Plato's theory of understanding*. Univ. Wisconsin.
Monteils-Laeng, L. (2013). Aristote et l'invention du désir. *Archives de Philosophie, 76*, 441–457. (*).
Moravcsik, J. M. E. (1974). Aristotle on adequate explanations. *Synthese, 28*, 3–17.
Morison, B. (2004). Self-motion in *physics* VIII. In A. Laks & M. Rashed (Eds.), *Aristote et le mouvement des animaux* (pp. 67–79). Septentrion.
Morsink, J. (1979). Was Aristotle's biology sexist? *Journal of the History of Biology, 12*, 83–112.
Murray, G. (1935). *Five stages of Greek religion*. Watts. [*orig*. 1912].
Murray, O. (1990). Cities of reason. In O. Murray & S. Price (Eds.), *The Greek City from Homer to Alexander* (pp. 1–25). Clarendon.
Nagel, T. (1980). Aristotle on Eudaimonia. In A. O. Rorty (Ed.), *Essays on Aristotle's ethics*. Univ. of California.
Natali, C. (2015). The search for definitions of justice in *Nicomachean ethics* 5. In D. Henry & K. M. Nielsen (Eds.), *Bridging the gap between Aristotle's science and ethics* (pp. 169–190). CUP.
Needham, J. (1934). *"Aristotle" in a history of embryology* (pp. 19–41). CUP.
Newman, S. (2002). Aristotle's notion of "bringing-before-the-Eyes": Its contributions to Aristotelian and contemporary conceptualizations of metaphor, style, and audience. *Rhetorica: A Journal of the History of Rhetoric, 20*, 1–23.
Newsome, D. (1972). *Two classes of men: Platonism & English Romantic thought*. John Murray.
Nidditsch, P. H. (1983). The first stage of the idea of mathematics: Pythagoras, Plato, Aristotle. *Midwest Studies in Philosophy, VIII*, 3–34.
Nieuwenburg, P. (2004). Learning to deliberate: Aristotle on truthfulness and public deliberation. *Political Theory, 32*, 449–467.
Nussbaum, M. C. (1984). Plato on commensurability and desire. *Proceedings of the Aristotelian Society, LV111*, 55–80.
Nussbaum, M. C., & Putnam, H. (2003). Changing Aristotle's mind. In M. C. Nussbaum & A. O. Rorty (Eds.), *Essays on Aristotle's De Anima* (pp. 27–56). OUP.
Nussbaum, M. C., & Rorty, A. O. (Eds.). (1992). *Essays on Aristotle's De Anima*. OUP.
O'Gorman, N. (2005). Aristotle's "Phantasia" in the "Rhetoric": "Lexis," appearance, and the epideictic function of discourse. *Philosophy & Rhetoric, 38*, 16–40. (*).
O'Rourke, F. (2004). Aristotle and the metaphysics of evolution. *Review of Metaphysics, 58*, 3–59.
O'Rourke, F. (2005). Aristotle and the metaphysics of metaphor. *The Boston Area Colloquium in Ancient Philosophy, 21*, 155–177.
Oehler, K. (1974). Aristotle on self-knowledge. *Proceedings of the American Philosophical Society, 118*, 493–506.
Oele, M. (2012). Heidegger's reading of Aristotle's concept of *pathos*. *Philosophy, 16*, 389–406. (*).
Olshewsky, T. M. (1995). Self-movers and unmoved movers in Aristotle's *physics* VII. *The Classical Quarterly, 45*, 389–406.
Owen, G. E. L. (1975). Tithenai ta Phaineomena. In J. Barnes et al. (Eds.), *Articles on Aristotle* (Vol. 1, pp. 113–126). Duckworth.
Owens, J. (2007). *Aristotle's gradations of being in metaphysics E-Z*. St Augustine.

Owens, J. (1981a). Aristotelian ethics, medicine, and the changing nature of man. In *Aristotle: The collected papers of Joseph Owens* (pp. 169–180). SUNY. [*orig.* 1977].
Owens, J. (1981b). Aristotelian soul as cognition of sensibles, intelligibles, and self. In *Aristotle: The collected papers of Joseph Owens* (pp. 81–98). SUNY.
Owens, J. (1981c). Aristotle—Cognition a way of being. In *Aristotle: The collected papers of Joseph Owens* (pp. 74–80). SUNY.
Owens, J. (1981d). Aristotle's definition of soul. In *Aristotle: The collected papers of Joseph Owens* (pp. 109–121). SUNY. [*orig.* 1976].
Owens, J. (1981e). The grounds of ethical universality in Aristotle. In *Aristotle: The collected papers of Joseph Owens* (pp. 48–58). SUNY. [*orig.* 1966].
Owens, J. (1981f). The universality of the sensible in the Aristotelian noetic. In *Aristotle: The collected papers of Joseph Owens* (pp. 59–73). SUNY. [*orig.* 1971].
Owens, J. (1988). The self in Aristotle. *The Review of Metaphysics, 41*, 707–722.
Owens, J. (1960/1). Aristotle on categories. *The Review of Metaphysics, 14*, 73–90.
Owens, J. (1982). Aristotle on common sensibles and incidental perception. *Phoenix, 36*, 215–236.
Owens, J. (1951). *The doctrine of being in the Aristotelian metaphysics*. Pontifical Institute.
Page, C. (1985). Predicating forms of matter in Aristotle's "metaphysics". *The Review of Metaphysics, 39*, 57–82.
Papachristou, C. S. (2013). Three Kinds of *Phantasia* in Aristotle's *De Anima*. *Journal of Ancient Philosophy, 7*, 19–48. (*).
Parker, H. N. (2012). Aristotle's unanswered questions: Women and slaves in *politics* 1252a–1260b. *EuGeStA, 2*. (Online) (*).
Patočka, J. (2002). *Plato and Europe*. SUP. [*orig.* 1973].
Patočka, J. (2011). *Aristote, Ses Devanciers, Ses Successeurs ; Études d'Histoire de la Philosophie d'Aristote à Hegel*. Virin. [*orig.* 1964].
Patsioti-Tsacpounidi, I. (2005). Commentary on O'Rourke. In *The Boston area colloquium in ancient philosophy* (pp. 178–190).
Pearson, G. (2012). *Aristotle on desire*. CUP.
Pearson, G. (2011). Aristotle and Scanlon on desire and motivation. In M. Pakaluk & G. Pearson (Eds.), *Moral psychology and human action in Aristotle* (pp. 95–118). OUP.
Pellegrin, P. (2009). *La vocabulaire d'Aristote*. Ellipses.
Pellegrin, P. (1986). *Aristotle's classification of animals: Biology and the conceptual unity of the Aristotelian corpus*. Univ. of Cal.
Pérez Álvarez, M. (2009). The four causes of behaviour: Aristotle and skinner. *International Journal of Psychology and Psychological Therapy, 9*, 45–57. (*).
Peterson, S. (1988). Horos (limit) in Aristotle's "Nicomachean ethics". *Phronesis, 33*, 233–250.
Pettigrew, R. (2009). Aristotle on the subject matter of geometry. *Phronesis, 54*, 239–260.
Preus, A. (1990). Man and cosmos in Aristotle: *Metaphysics* Λ and the biological works. In D. Devereux & P. Pellegrin (Eds.), *Biologie, Logique et Métaphysique chez Aristote* (pp. 471–490). CNRS.
Price, A. (2011). Aristotle on the ends of deliberation. In M. Pakaluk & G. Pearson (Eds.), *Moral psychology and human action in Aristotle* (pp. 135–158). OUP.
Quigley, P. (2015). Aristotle and real possibility. In G. Tsianikas et al. (Eds.), *Reading, interpreting, experiencing: an inter-cultural journey into Greek letters* (pp. 51–61). Mod. Greeek Studies of New Zealand. (*).
Radovic, F. (2020). The case of red-stained mirrors: Perception, strange phenomena, and the role of exemplification in Aristotle. In D. Bennett & J. Toivanen (Eds.), *Philosophical problems in sense perception: Testing the limits of Aristotelianism* (pp. 77–89). Springer.
Reale, G. (1980). *The concept of first philosophy and the unity of the metaphysics of Aristotle*. SUNY. [*orig.* 1967].
Reeve, C. D. C. (2012). *Action, contemplation, and happiness*. Cambridge Mass.

Rehmann-Sutter, C. (1994). La Vie de L'Organisme est comme la Vision de L'Oeil: Vers une autre biologie d'Aristote: entelecheia en termes de poises et de praxis. In *Actes du 25e Congrès de l'Association des Sociétés de philosophie de langue française (ASPLF)* (pp. 428–433). (*).

Riezler, K. (1940). *Physics and reality: Lectures of Aristotle on modern physics at an International Congress of Science 679 Olymp. Cambridge 1940 A.D.* Yale.

Ritchie, A. D. (1946). A defence of Aristotle's logic. *Mind, 55*, 256–262.

Roessler, M. (1994). Natura Artifex: Le Sens de l'analogie nature-technique chez Aristote. In *Actes du 25e Congrès de l'Association des Sociétés de philosophie de langue française (ASPLF)* (pp. 411–415). (*).

Rogers, K. (1994). Aristotle on loving another for his own sake. *Phronesis, XXXIX/3*, 291–302.

Roochnik, D. (2009). What is *Theoria*? *Nicomachean ethics* book 10.7-8. *Classical Philology, 104*, 69–82.

Roochnik, D. (2016). Aristotle's topological politics: Michael Sandel's civic republicanism. In G. C. Kellow & N. Leddy (Eds.), *On civic republicanism*. Univ. Toronto.

Rorty, A. O. (Ed.). (1980a). *Essays on Aristotle's ethics*. Univ. Cal.

Rorty, A. O. (1984). Aristotle on the metaphysical status of *Pathe*. *The Review of Metaphysics, 38*, 521–546.

Rorty, A. O. (1980b). Akrasia and pleasure: Nicomachean ethics book 7. In A. O. Rorty (Ed.), *Essays on Aristotle's ethics* (pp. 267–284). Univ. Cal..

Ross, W. D. (1939–1940). Some thoughts on Aristotle's logic: The presidential address. *Proceedings of the Aristotelian Society, 40*, i–xviii.

Ross, W. D. (1925). Platonic philosophy and Aristotelian metaphysics. In *Philosophy and metaphysics; Proceedings of the Aristotelian Society* (Vol. V).

Rovelli, C. (2015). Aristotle's physics: A physicist's look. *Journal of the American Philosophical Association, 1*, 23–40. (*).

Runciman, W. G. (1990). Doomed to extinction: The *polis* as an evolutionary dead-end. In O. Murray & S. Price (Eds.), *The Greek City from Homer to Alexander* (pp. 347–367). Clarendon.

Russon, J. E. (1996). Self-consciousness and the tradition in Aristotle's psychology. *Foi et Raison, 52*, 777–803. (*).

Ryan, E. E. (1973). Pure form in Aristotle. *Phronesis, 18*, 209–224.

Sachs, J. (2005). Aristotle: Motion and its place in nature. *Internet Encyclopaedia of Philosophy*. (Online, Section 1) (*).

Salles, R. (Ed.). (2021). *Cosmology and biology in ancient philosophy: From Thales to Avicenna*. CUP.

Sanday, E. (2013). *Phantasia* in *De Anima*. In C. Baracchi (Ed.), *The Bloomsbury companion to Aristotle* (pp. 106–127). Bloomsbury.

Schall, J. V. (1989). Nature and finality in Aristotle. *Laval théologique et philosophique, 451*, 73–85. (*).

Scharle, M. (2009). A synchronic justification for Aristotle's commitment to prime matter. *Phronesis, 54*, 326–345.

Scheiter, K. M. (2012). Images, appearances, and *Phantasia* in Aristotle. *Phronesis, 57*, 251–278. (*).

Schiefsky, M. (2007). Galen's teleology and functional explanation. *Oxford Studies in Ancient Philosophy, XXXIII*, 369–400.

Schiller, F. C. S. (1914). Aristotle's refutation of "Aristotelian" logic. *Mind, 23*, 1–18.

Schindler, D. C. (2009). Why Socrates didn't charge: Plato and the metaphysics of money. *Communio, 36*(3), 394–426.

Schmitt-Pantel, P. (1990). Collective activities and the political in the Greek City. In O. Murray & S. Price (Eds.), *The Greek City from Homer to Alexander* (pp. 199–213). Clarendon.

Schoen-Nazzaro, M. B. (1978). Plato and Aristotle on the ends of music. *Laval théologique et philosophique, 343*, 261–273. (*).

Schofield, M. (2011). Phantasia in *De Motu Animalium*. In M. Pakaluk & G. Pearson (Eds.), *Moral psychology and human action in Aristotle* (pp. 119–134). OUP.

Schollmeier, P. (2003). Aristotle and women: Household and political roles. *Polis, 20*, 22–42. (*).

Schollmeier, P., & Selves, O. (1994). *Aristotle on personal and political friendship*. SUNY.

Schollmeier, P. (1991). Practical intuition and rhetorical example. *Philosophy and Rhetoric, 24*, 95–104. (*).

Schollmeier, P. (1989). Aristotle on practical wisdom. *Zeitschrift für Philosophische Forschung, 43*, 124–132. (*).

Scolnicov, S. (forthcoming). Plato against logical atomism. In J. Hartmann (Ed.), *Festschrift Jan Wolenski* (Krakow).

Scolnicov, S. (2007). Plato's two faces of platonic knowledge. *Journal of the International Plato Society*.

Scolnicov, S. (2003). *Plato's parmenides*. Univ. Cal.

Segvic, H. (2011). Deliberation and choice in Aristotle. In M. Pakaluk & G. Pearson (Eds.), *Moral psychology and human action in Aristotle* (pp. 159–186). OUP.

Sellars, W. (1957). Substance and form in Aristotle. *The Journal of Philosophy, 54*, 688–699.

Sellars, W. (n.d.). *Aristotle and emergent evolution*. (Online, Unpublished undated typescript) (*).

Sharvy, R. (1983). Aristotle on mixtures. *The Journal of Philosophy, 80*, 439–457.

Shaw, E. (2014). Aristotle on the fullness of social living. *The Imaginative Conservative*. (Online) (*).

Shaw, J. R. (1972). Models for cardiac structure and function in Aristotle. *Journal of the History of Biology, 5*, 355–388.

Sherman, N. (1987). Aristotle on friendship and the shared life. *Philosophy and Phenomenological Research, 47*, 589–613.

Sherman, N. (1997). *Making a virtue out of necessity: Aristotle and Kant on virtue*. CUP.

Shields, C. (2015). The science of soul in Aristotle's *ethics*. In D. Henry & K. M. Nielsen (Eds.), *Bridging the gap between Aristotle's science and ethics* (pp. 232–253). CUP.

Shields, C. (2009). The priority of soul in Aristotle's *De anima*: Mistaking categories? In D. Frede & B. Reis (Eds.), *Body and soul in ancient philosophy*. de Gruyter.

Shields, C. (2007). The peculiar motion of Aristotelian souls. *Proceedings of the Aristotelian Society, 81*, 139–161.

Shields, C. (1994). Mind and motion in Aristotle. In M. L. Gill & J. G. Lennox (Eds.), *Self-motion: From Aristotle to Newton* (pp. 117–134). Princeton.

Shields, C. (1990). The generation of form in Aristotle. *History of Philosophy Quarterly, 7*, 367–390.

Simplicitus. (1992). *Commentary on Aristotle's physics*. Duckworth. [*orig. c.* 500 AD].

Sisko, J. E. (1997). Space, time and *phantasms* in Aristotle, *De Memoria* 2, 452B7-25. *The Classical Quarterly, 47*, 167–175.

Slomkowski, P. (1997). *Aristotle's topics*. Leiden.

Smith Pangle, L. (2003). *Aristotle and the philosophy of friendship*. CUP.

Smith, A. D. (1996). Character and intellect in Aristotle's ethics. *Phronesis, 41*, 56–74.

Solmsen, F. (1957). The vital heat, the inborn Pneuma, and the Aether. *Journal of Hellenic Studies, 77*, 119–123.

Sorabji, R. (1983). *Animal minds and human morals: The origins of the Western debate*. Duckworth.

Sorabji, R. (1974). Body and soul in Aristotle. *Philosophy, 49*, 63–89.

Sorabji, R. (1972). Aristotle, mathematics, and colour. *The Classical Quarterly, 22*, 293–308.

Spiegel, N. (1975). The significance of "mimesis" in the light of Aristotle's doctrine of the four ontological Causes. *Revue Belge de philologie et d'histoire, 53*, 5–23. (*).

Sprague, R. K. (1991). Plants as Aristotelian substances. *Illinois Classical Studies, 16*, 221–229. (*).

Sprague, R. K. (1977). Aristotle and the metaphysics of sleep. *The Review of Metaphysics, 31*(2), 230–241.

von Staden, H. (2007). Physis and Technē in Greek medicine. In B. Bensaude-Vincent & W. R. Newman (Eds.), *The artificial and the natural: An evolving polarity* (pp. 21–50). MIT.

Stauffer, D. J. (2008). Aristotle's account of the subjection of women. *The Journal of Politics, 70*, 929–941.

Stohr, K. (2006). Manners, morals, and practical wisdom. In T. Chappell (Ed.), *Values and virtues: Aristotelianism in contemporary ethics* (pp. 189–211). OUP.

Stone, M. A. (1985). Aristotle's distinction between motion and activity. *History of Philosophy Quarterly, 2*, 11–20.

Strauss, D. F. M. (2013). Two remarkable contributions of Aristotle to the intellectual legacy of the west. *Phronimon, 14*, 61–78. (*).

Sweeney, L. (1960). L'infini quantitatif chez Aristote. *Revue Philosophique de Louvain, 58*, 505–528. (*).

Telfer, E. (1989–1990). The unity of the moral virtues in Aristotle's "Nicomachean Ethics". *Proceedings of the Aristotelian Society, 90*, 35–48.

Timmermann, J. (1995). Why we cannot want our friends to be gods. Some notes on "NE" 1159a5-12. *Phronesis, 40*, 209–215.

Torres, J. (2021). Philia: the biological foundations of Aristotle's ethics. *History and Philosophy of the Life Sciences, 43*. (*).

Tracy, T. J. (1969). *Physiological theory and the doctrine of the mean in Plato and Aristotle*. Mouton.

Tracy, T. J. (1986). Two views of soul: Aristotle and Descartes. *Illinois Classical Studies, 11*, 247–264. (*).

Tracy, T. J. (1979). Perfect friendship in Aristotle's Nicomachean ethics. *Illinois Classical Studies, 4*, 65–75. (*).

Tress, D. M. G. (1992). The metaphysical science of Aristotle's "generation of animals" and its feminist critics. *The Review of Metaphysics, 46*.

Tress, D. M. G. (1999). Aristotle against the Hippocratics on sexual generation; a reply to Coles. *Phronesis, 44*, 228–241.

Tymieniecka, A.-T. (2011). Inspirations of Heraclitus from Ephesus fulfilled in our new enlightenment. *Analecta Husserliana, 110*, 3–12.

Waerdt, V., & Paul, A. (1991). The plan and intention of Aristotle's ethical and political writings. *Illinois Classical Studies, 16*, 231–253. (*).

Vilkko, R., & Hintikka, J. (2006). Existence and predication from Aristotle to Frege. *Philosophy and Phenomenological Research, 73*, 359–377.

Vinci, T., & Scott Robert, J. (2005a). Aristotle and modern genetics. *Journal of the History of Ideas, 66*, 201–221.

Vlastos, G. (1945). Ethics and physics in Democritus. *Philosophical Review, 54*(6), 578–592.

Vogt, S. (n.d.). Semiotics of human body and character: Aristotle's logical foundation of physiognomics. *Paideia*. (Online) (*).

Waldron, J. (1995). The wisdom of the multitude: Some reflections on book 3, chapter 11 of Aristotle's politics. *Political Theory, 23*, 563–584.

Wardy, R. (1993). Aristotelian rainfall or the Lore of averages. *Phronesis, XXXVIII/1*, 18–30.

Wedin, M. V. (1994). Aristotle on the mind's self-motion. In M. L. Gill & J. G. Lennox (Eds.), *Self-motion: From Aristotle to Newton* (pp. 81–116). Princeton.

Wedin, M. V. (1989). Aristotle on the mechanics of thought. *Ancient Philosophy, 9*, 67–86.

Weinman, M. (2013). Education: The Ethico-political Energeia. In C. Baracchi (Ed.), *The Bloomsbury companion to Aristotle* (pp. 263–276). Bloomsbury.

Weiss, H. (1948). Aristotle's teleology and Uexküll's theory of living nature. *The Classical Quarterly, 42*, 44–58.

White, M. J. (1992). *The continuous and the discrete: Ancient physical theories from a contemporary perspective*. OUP.

White, M. J. (1993a). Aristotle on the non-supervenience of local motion. *Philosophy and Phenomenological Research, 53*, 143–155.

White, M. J. (1993b). The metaphysical location of Aristotle's Μαθηματικα. *Phronesis, XXXVIII*, 166–182.

White, M. J. (2004). The problem of Aristotle's "nous Poiētikos". *The Review of Metaphysics, 57*, 725–739.
Whiting, J. (1992). Living bodies. In M. C. Nussbaum & A. O. Rorty (Eds.), *Essays on Aristotle's De Anima* (pp. 75–92). OUP.
Whiting, J. (1988). Aristotle's function argument: A defense. *Ancient Philosophy, 8*, 33–47.
Wilson, M. (2013). *Aristotle's meteorologica: A more disorderly nature*. CUP.
Winslow, R. (2013). Aristotelian definition: On the discovery of *Archai*. In C. Baracchi (Ed.), *The Bloomsbury companion to Aristotle* (pp. 41–56). Bloomsbury. (*).
Winslow, R. (2009). On the life of thinking in Aristotle's *De Anima*. *Epoché, 13*, 299–316. (*).
Winslow, R. (2007). *Aristotle and rational discovery*. Continuum.

Select Bibliography of Modern Resources

Acton, H. B. (1952–3). Tradition and some other forms of order. *Proceedings of the Aristotelian Society, LIII*, 1–28.
op Akkerhuis, G. J. (2008). Towards a hierarchical definition of life, the organism, and death. In C. Vidal (Ed.), *The evolution and development of the universe* (pp. 169–192). ENS. (*).
Alexander, H. B. (1912). The conception of "soul". *Journal of Philosophy, Psychology, and Scientific Methods, 9*, 421–430. (*).
Ambrose, A. (1966). Unkowables and logical atomism. In *Essays in Analysis*. Allen and Unwin.
Anderson, P. W. (2011). On the "unreasonable efficacy of mathematics"—A proposition by Wigner. In *More and different: Notes from a thoughtful curmudgeon* (pp. 201–203). World Scientific.
Anderson, P. W. (1972). More is different: Broken symmetry and the nature of the hierarchical structure of science. *Science, 177*, 393–396. (*).
de Anna, G. (2015). Hylomorphism and substantial gradualism. *Revista Portuguesa de Filosofia, 71*, 855–872.
Anzaldo, A. A. (2009). Darwin's two hundred years: Is it not time for a change? *Ludus Vitalis, XVII*, 87–99. (*).
Atmanspacher, H. (2020). The Pauli–Jung conjecture and its relatives: A formally augmented outline. *Open Philosophy, 3*, 527–549. (*).
Atmanspacher, H. (2012). Dual-aspect monism à la Pauli and Jung. *Journal of Consciousness Studies, 19*, 96–120. (*).
Atmanspacher, H., & Primas, H. (2006). Pauli's ideas on mind and matter in the context of contemporary science. *Journal of Consciousness Studies, 13*, 5–50. (*).
Atmanspacher, H., & Primas, H. (1996). The hidden side of Wolfgang Pauli: An eminent physicist's extraordinary encounter with depth psychology. *Journal of Consciousness Studies, 3*, 112–126.
Babcock, G., & McShea, D. W. (2021). An externalist teleology. *Synthese, 199*, 8755–8780. (*).
Bacon, F. (1620). *Novum organum*.
Bacon, F. (1605). *On the advancement of learning*.
Badiou, A. (2013). *Philosophy and the event*. Polity.
Badiou, A. (2011). *Wittgenstein's antiphilosophy*. Verso.
Ball, P. (2011). A metaphor too far. *Nature*. (*).
Barbieri, M. (2013). The paradigms of biology. *Biosemiotics, 6*, 33–59.
Barbieri, M. (1985). *The semantic theory of evolution*. Harwood.
Barbour, J. (2003). The deep and suggestive principles of Leibnizian philosophy. *The Harvard Review of Philosophy, XI*, 45–58.
Barham, J. (1999). On the objectivity of the Scala Naturae. *Evolution and Cognition, 5*(1), 2–10.
Barham, J. (2007). The reality of purpose and the reform of naturalism. *Philosophia Naturalis, 44*, 31–52. (*).

Barnes, G. P. (2003). The paradoxes of Hylomorphism. *The Review of Metaphysics, 53*, 501–523.
Barrow, J. D., & Tipler, F. J. (1986). *The anthropic cosmological principle*. OUP.
Bauchau, V. (2006). Emergence and reductionism: From the game of life to the science of life. *Synthese, 331*, 29–40. (*).
Baudrillard, J. (2010). *The agony of power*. Semiotext(e).
Baudrillard, J. (1993). *The transparency of evil*. Verso.
Baudrillard, J. (1988). *Selected writings*. Polity.
Baudrillard, J. (2008). *Fatal strategies*. Semiotext(e). [*orig.* 1983].
Bechtel, W. (2011). Mechanism and biological explanation. *Philosophy of Science, 78*, 533–557.
Bechtel, W., & Richardson, R. C. (1998). Vitalism. In *Routledge encyclopaedia of philosophy*. Routledge. (*).
Benoist, J.-M. (1978). *The structural revolution*. Wiedenfeld and Nicholson. [*orig.* 1975].
Berdyaev, N. (1957). *The beginning and the end*. New York, Harper. [*orig.* 1941].
Berdyaev, N. (1943). *Slavery and freedom*. Geoffrey Bles. [*orig.* 1939].
Berdyaev, N. (1937). *The destiny of man*. Geoffrey Bles.
Berdyaev, N. (1936). *The meaning of history*. Everyman.
von Bertalanffy, L. (2014). *Problems of life: An evaluation of modern biological thought*. Martino. [*orig.* 1952].
von Bertalanffy, L. (1981). *A systems view of man*. Westview.
Bohm, D. (1990). A new theory of the relationship of mind and matter. *Philosophical Psychology, 3*, 271–286. (*).
Bohm, D. (1980). *Wholeness and the implicate order*. Routledge.
Bohm, D. (1979). [Interview with] David Bohm. In P. Buckley & D. Peat (Eds.), *A question of physics: Conversations in physics and biology* (pp. 124–150). Routledge.
Bohm, D. (1974). On the subjectivity and objectivity of knowledge. In J. Lewis (Ed.), *Beyond chance and necessity: A critical inquiry into professor Jacques Monod's chance and necessity* (pp. 121–110). Garnstone.
Bohr, N. (1934). *Atomic theory and the description of nature*. CUP.
Bohr, N. (1998a). Physical science and the study of religions. In J. Faye & H. J. Folse (Eds.), *The philosophical writings of Niels Bohr* (Vol. IV, pp. 155–160). Ox Bow. [*orig.* 1953].
Bohr, N. (1998b). Chemistry and the quantum theory of atomic constitution. In J. Faye & H. J. Folse (Eds.), *The philosophical writings of Niels Bohr* (Vol. IV, pp. 29–61). Ox Bow. [*orig.* 1932].
Bohr, N. (1998c). Address delivered at the second international Germanist conference. In J. Faye & H. J. Folse (Eds.), *The philosophical writings of Niels Bohr* (Vol. IV, pp. 189–191). Ox Bow. [*orig.* 1960].
Bohr, N. (1998d). Causality and complementarity. In J. Faye & H. J. Folse (Eds.), *The philosophical writings of Niels Bohr* (Vol. IV, pp. 83–91). Ox Bow. [*orig.* 1937].
Bohr, N. (1963). The connection between the sciences. In *Essays 1958–1962 on atomic physics and human knowledge* (Vol. 17, pp. 17–22). Wiley.
Bohr, N. (1933). Light and life. *Nature, 131*, 421–423. (*).
Bosanquet, B. (1914–15). Science and philosophy. *Proceedings of the Aristotelian Society, XV*, 1–21.
Bouquiaux, L. (2006). On a "mathematical neo-Aristotelianism" in Leibniz. *Synthese, 331*, 163–169.
Boutot, A. (1993b). *L'invention des formes: Chaos—Catastrophes—Fractales—Structures dissipatives—Attracteurs étranges*. Odile Jacob.
Bowler, M. (2008). *Heidegger and Aristotle: Philosophy as praxis*. Continuum.
Broackes, J. (2006). Substance. *The Proceedings of the Aristotelian Society, 106*, 131–166.
Brogan, C. (2022). Engineers uncover secret 'thinking' behind dandelions' seed dispersal. *Imperial*. (*).
Buchanan, B. (2008). *Onto-ethologies: The animal environment of Uexküll, Heidegger, Merleau-Ponty, and Deleuze*. SUNY.
Calhoun, C. (1995). Standing for something. *The Journal of Philosophy, 92*(5), 235–260.

Camus, A. (1953). *The rebel*. Penguin. [*orig*. 1951].

Canguilhem, G. (2008). *Knowledge of life*. Fordham Univ. [*orig*. 1952].

Carnap, R. (1937). *Logical syntax of language*. Kegan Paul.

Carnap, R. (1947). *Meaning and necessity: A study in semantics and modal logic*. Univ. Chic.

Cartwright, N. (1997). Why physics? In R. Penrose (Ed.), *The large, the small and the human mind* (pp. 161–168). CUP.

de Carvalho, O. (2011). *Mr Rorty and his fellow animals*. voegelinview.com. (Online) (*).

Cassirer, E. (1970). *An essay on man*. Bantam. [*orig*. 1944].

Cassirer, E. (2012). Form and technology. In A. S. Hoel & I. Folkvold (Eds.), *Ernst Cassirer on form and technology: Contemporary readings* (pp. 15–53). Palgrave. [*orig*. 1930].

Chaput, E. (2021). Hegel and Bichat on symmetry and life: Between physiology, aesthetics, and freedom. *Cosmos and History, 17*(3), 196–223.

Chomsky, N. (2011). *Human nature: Justice versus power: The Chomsky-Foucault debate*. Souvenir. [*orig*. 1971] (*).

Chomsky, N. (1972). *Language and mind*. Open Library.

Chomsky, N. (1959). A review of B.F. Skinner's *verbal behaviour*. *Language, 35*, 28–58. (*).

Chomsky, N. (1970). Recent contributions to the theory of innate ideas. In L. Hudson (Ed.), *The ecology of human intelligence* (pp. 79–89). Penguin. [*orig*. 1967].

Churchland, P. (1984). *Matter and consciousness*. MIT.

Clarke, A., & Rothery, P. (2017). Scaling of body temperature in mammals and birds. *Functional Ecology, 22*(1), 58–67. (*).

Comte, A. (1975). *Cours de philosophie positive*. Hermann. [*orig*. 1842].

Coppock, P. (n.d.). Pragmaticism and semiotics. *The Whitehead Encyclopedia*. (Online) (*).

Crane, T., & Mellor, D. H. (1990). There is no question of physicalism. *Mind, 99*, 185–206.

Davidson, D. (1982). Rational animals. *Dialectica, 36*, 317–327.

Davies, P. (2006). *The goldilocks enigma: Why is the universe just right for life?* Allen Lane.

Dawkins, R. (1973). *The selfish gene*. OUP.

Deacon, T. (2021). How molecules became signs. *Biosemiotics, 14*, 537–559.

Debord, G. (1957). Report on the construction of situations and on the international situationist tendency's conditions of organisation and action. *The Anarchist Library*. (Online) (*).

Deikman, A. J. (1996). I=Awareness. *Journal of Consciousness Studies, 3*, 350–356. (*).

Delbrück, M. (1970). A physicist's renewed look at biology—Twenty years later. *Science, 168*, 1312–1315.

Dennett, D. (2013). Kinds of things—Towards a bestiary of the manifest image. In D. Ross, J. Ladyman, & H. Kincaid (Eds.), *Scientific metaphysics* (pp. 96–107). OUP. (*).

Dennett, D. (2006). Higher-order truths about chmess. *Topoi, 25*, 39–41. (*).

Derrida, J. (1978). Force and signification. In *Writing and difference*. Univ. Chic.. (*).

Des Chene, D. (2001). *Spirits and clocks: Machine & organism in Descartes*. Cornell UP.

Descartes, R. (1985–1991). *The philosophical writings of Descartes* (Vol. I, II, and III). CUP. (abbreviated as "CSM").

Devitt, M. (2010). Species have (partly) intrinsic essences. *Philosophy of Science, 77*, 648–661.

Devitt, M. (2008). Resurrecting biological essentialism. *Philosophy of Science, 75*, 344–382.

Dewey, J. (1916). The pragmatism of peirce. *The Journal of Philosophy, Psychology and Scientific Methods, 13*, 709–715. (*).

Dobzhansky, T. (1974a). Two contrasting world views? In J. Lewis (Ed.), *Beyond chance and necessity: A critical inquiry into professor Jacques Monod's chance and necessity* (pp. 131–141). Garnstone.

Dobzhansky, T. (1974b). Chance and creativity in evolution. In F. J. Ayala & T. Dobzhansky (Eds.), *Studies in the philosophy of biology* (pp. 307–338). Macmillan.

Dolsenhe, O. (2022). The revival of structuralism: The periodic table of mental science. *Journal of Scientific Exploration, 36*, 311–319.

Driesch, Hans, Mind and Body: A criticism of psychophysical parallelism (, 1927).

Driesch, H. (1929). *The science and philosophy of the organism: The Gifford lectures delivered before the University of Aberdeen in the year 1907–1908*. Adam & Charles Black. [*orig.* 1908].
Duhem, P. (1969). *To save the phenomena: An essay on the idea of physical theory from Plato to Galileo*. Univ. Chic.. [*orig.* 1909].
Dummett, M. (1978). *Truth and other enigmas*. Harvard.
Dupré, J. (2017). Book review: D.M. Walsh: Organisms, agency and evolution. *British Journal for the Philosophy of Science*.
Ellis, B. (2002). *The philosophy of nature: A guide to the new essentialism*. Acumen.
Ellul, J. (1964). *The technological society*. Vintage.
Elsasser, W. M. (1998). *Reflections on a theory of organisms: Holism in biology*. John Hopkins. [*orig.* 1987].
Elsasser, W. M. (1981). A form of logic suited for biology. In R. Rosen (Ed.), *Progress in theoretical biology* (Vol. 6, pp. 23–62). Academic Press. (also online on www.vordenker.de) (*)
Elsasser, W. M. (1966). *Atom and organism: A new approach to theoretical biology*. PUP.
Esparza Urzúa, G. A. (2024). Body, illness and symbol in Ernst Cassirer. *Filozofia, 79*(4), 365–379. (*).
Favareau, D., et al. (Eds.). (2012). *A more developed sign: Interpreting the work of Jesper Hoffmeyer*. Univ. Tartu.
Favareau, D., et al. (2010). *Essential readings in biosemiotics: Anthology and commentary*. Springer.
Favareau, D., et al. (2015a). Symbols are grounded not in things, but in Scaffolded relations and their semiotic constraints (or how the referential generality of symbol scaffolding grows minds). *Biosemiotics, 8*, 235–255.
Favareau, D., et al. (2015b). Why this now? The conceptual and historical rationale behind the development of biosemiotics. *Green Letters: Studies in Ecocriticism, 19*, 227–242.
Feiten, T. E., Holland, K., & Chemero, A. (2020). Worlds apart? Reassessing von Uexküll's Umwelt in embodied cognition with Canguilhem, Merleau-Ponty, and Deleuze. *Journal of French and Francophone Philosophy, XXVIII*, 1–26. (*).
Feyerabend, P. (2011). *The tyranny of science*. Polity. [*orig.* 1992].
Feyerabend, P. (1999a). *Conquest of abundance: A tale of abstraction versus the richness of being*. Univ. Chic.
Feyerabend, P. (1981a). *Problems of empiricism: Philosophical papers* (Vol. 2). CUP.
Feyerabend, P. (1981b). Realism and instrumentalism: Comments on the logic of factual support. In *Realism, rationalism & scientific method: Philosophical papers* (Vol. 1, pp. 176–202). CUP.
Feyerabend, P. (1999b). *For and against method*. Univ. Chic.
Feyerabend, P. (1983). Some observations on Aristotle's theory of mathematics and of the continuum. *Midwest Studies in Philosophy, VIII*, 67–88.
Fodor, J., & Piattelli-Palmarini, M. (2011). *What Darwin got wrong*. Profile.
Fodor, J. (2000). *The mind doesn't work that way: The scope and limits of computational psychology*. MIT.
Fodor, J. (2004). Water's water everywhere. *London Review of Books, 26*, 17–19.
Fodor, J. (1974). Special sciences (or: The disunity of science as a working hypothesis). *Synthese, 28*, 97–115. (*).
Foucault, M. (2011a). *The courage of truth; The government of self and others II*. Palgrave. [*orig.* 1983–4].
Foucault, M. (2010). *The government of self and others*. Palgrave. [*orig.* 1982–3].
Foucault, M. (2005). *The hermeneutics of the subject*. Picador. [*orig.* 1981–2].
Foucault, M. (2008). *The birth of biopolitics*. Picador. [*orig.* 1978–9].
Foucault, M. (2001). *Fearless speech*. Semiotext(e). [*orig.* 1983].
Foucault, M. (1988). *Technologies of the self* (Amherst). [*orig.* 1982].
Foucault, M. (2016). *About the beginnings of the hermeneutics of the self: Lectures at Dartmouth College, 1980*. Univ. Chic.

Foucault, M. (2011b). *Human nature: Justice versus power: The Chomsky-Foucault debate.* Souvenir. [*orig.* 1971] (*).

Fox-Keller, E. (2010). It is possible to reduce biological explanations to explanations in chemistry and/or physics. In F. J. Ayala & R. Arp (Eds.), *Contemporary debates in the philosophy of biology* (pp. 19–31). Blackwell.

van Fraassen, B. C. (1999). The manifest image and the scientific image. In D. Aerts (Ed.), *Einstein meets Magritte: The White book—An interdisciplinary reflection* (pp. 29–52). Kluwer.

van Fraassen, B. C. (1994). The world of empiricism. In J. Hilgevoord (Ed.), *Physics and our view of the world* (pp. 114–134). CUP.

van Fraassen, B. C. (1976). To save the phenomena. *The Journal of Philosophy, 73*, 623–632.

Frankl, V. E. (1969). Reductionism and nihilism. In A. Koestler & J. R. Smythies (Eds.), *Beyond reductionism: New perspectives in the life sciences* (pp. 396–416). Hutchinson.

Gare, A. (2021a) (*)). Code biology and the problem of emergence. *Biosystems, 208*, 104487.

Gare, A. (2021b). Against posthumanism: Posthumanism as the world vision of house-slaves. *Borderless Philosophy, 4*, 1–56. (*).

Gare, A. (2013). Overcoming the Newtonian paradigm: The unfinished project of theoretical biology from a Schellingian perspective. *Progress in Biophysics and Molecular Biology, 113*, 5–24. (*).

Gare, A. (2019). Biosemiosis and causation: Defending biosemiotics through Rosen's theoretical biology. *Cosmos and History, 15*. (*).

Gare, A. (2008). Approaches to the question "what is life?": Reconciling theoretical biology with philosophical biology. *Cosmos and History, 4*, 53–77. (*).

Gare, A. (2000). Systems theory and complexity. *Democracy and Nature, 6*, 327–339. (*).

Gaukroger, S. (2016). Kant and the nature of matter: Mechanics, chemistry, and the life sciences. *Studies in History and Philosophy of Science Part A, 58*, 108–114.

Gilbert, S. F. (2003). The reactive genome. In G. B. Muller & S. A. Newman (Eds.), *Origination of organismal form; beyond the gene in developmental and evolutionary biology* (pp. 87–101). MIT.

Gnoli, C., & Poli, R. (2004). Levels of reality and levels of representation. *Knowledge Organization, 3*, 151–160. (*).

Gödel, K. (1992). *On formally undecidable propositions of principia Mathematica and related systems*. Dover Publications. [*orig.* 1930] (*).

Goodwin, B., & Solé, R. (2000). *Signs of life: How complexity pervades biology*. Perseus.

Gottfried, P. (2024). Showdown at Davos. *The European Conservative*. (*).

Greco, M. (2005). The vitality of vitalism. *Theory Culture & Society, 22*, 15–27. (*).

Grosholz, E. R. (1991). *Cartesian method and the problem of reduction*. OUP.

Hacker, P. (2006). Passing by the naturalistic turn: On quine's cul-de-sac. *Philosophy, 81*, 221–243. (*).

Hacker, P. (2002). Strawson's concept of a person. *Proceedings of the Aristotelian Society, 102*, 21–40. (*).

Hacker, P. (1979). Substance: The constitution of reality. *Midwest Studies in Philosophy, 4*, 239–262.

Haldane, J. B. S. (1926). On being the right size. *Harper's Magazine*. (*).

Haldane, J. S. (1931). *The philosophical basis of biology*. H & S.

Harries, K. (2007). The descent of the *logos*: Limits of transcendental reflection. In S. Crowell & J. Malpas (Eds.), *Transcendental Heidegger* (pp. 74–92). SUP.

Hassing, R. F. (2013). Laws versus teleology. In B. Kaldis (Ed.), *Encylopaedia of philosophy and the social sciences*. SAGE. (*).

Hassing, R. F. (2012). History of physics and the thought of Jacob Klein. *The New Yearbook for Phenomenology and Phenomenological Philosophy, XI*, 214–248. (*).

Hassing, R. F. (2004). The meaning and force of liberalism. In E. Banús & A. Llano (Eds.), *Present and future of liberalism*. Eunsa. (*).

Hassing, R. F. (1992). Animals versus the Laws of inertia. *Review of Metaphysics, 46*, 29–61. (*).

Hegel, G. W. F. (1975). *Hegel's logic*. OUP. [*orig.* 1830].

Hegel, G. W. F. (2007a). *Lectures on the proofs of the existence of god*. OUP. [*orig*. 1829].
Hegel, G. W. F. (2007b). *Lectures on the philosophy of religion (selected extract) in Lectures on the proofs of the existence of god*. OUP. [*orig*. 1829].
Hegel, G. W. F. (1871). The philosophy of Aristotle. *The Journal of Speculative Philosophy, 5*, 61–78. [*orig*. 1825–6].
Hegel, G. W. F. (1974). *Lectures on the history of philosophy* (Vol. II). OUP. [*orig*. 1840].
Heidegger, M. (1995a). *Aristotle's metaphysics Θ 1–3: On the essence and actuality of force*. IUP. [*orig*. 1931].
Heidegger, M. (1998a). On the essence and concept of Φυσις in Aristotle's *physics* B, I. In *Pathmarks* (pp. 183–230). CUP. [*orig*. 1939].
Heidegger, M. (1992a). *Plato's sophist*. IUP. [*orig*. 1924–5].
Heidegger, M. (1995b). *The fundamental concepts of metaphysics*. IUP. [*orig*. 1929–30].
Heidegger, M. (1984). *Early Greek thinking: The dawn of western philosophy*. Harpers and Row. [*orig*. 1950/7].
Heidegger, M. (1992b). *Parmenides*. IUP. [*orig*. 1942–3].
Heidegger, M. (2009). *Basic concepts of Aristotelian philosophy*. IUP. [*orig*. 1924].
Heidegger, M. (2008). *Basic concepts of ancient philosophy*. IUP. [*orig*. 1926].
Heidegger, M. (1992c). *The metaphysical foundations of logic*. IUP. [*orig*. 1928].
Heidegger, M. (1996). *The principle of reason*. IUP. [*orig*. 1957].
Heidegger, M. (1967). *What is a thing?* Regnery. [*orig*. 1935–6].
Heidegger, M. (1971a). *Poetry, language, thought*. Harper.
Heidegger, M. (1972). *On time and being*. Harper. [*orig*. 1962].
Heidegger, M. (1962). *Being and Time*. Blackwell. [*orig*. 1926].
Heidegger, M. (2013). Science and reflection. In *The question concerning technology and other essays*. Harper. [*orig*. 1954].
Heidegger, M. (1971b). The thing. In *Poetry, language, thought*. Harper. [*orig*. 1950].
Heidegger, M. (2002a). The origin of the work of art. In *Off the beaten track* (pp. 1–56). CUP. [*orig*. 1935–6] (On the origin of the work of art. In *Pathmarks*. Cambridge: CUP [*orig*. 1935–6] (2002)).
Heidegger, M. (2002b). The age of the world picture. In *Off the beaten track* (pp. 57–85). CUP. [*orig*. 1938].
Heidegger, M. (1973a). Metaphysics as history of being. In *The end of philosophy* (pp. 1–54). Harper. [*orig*. 1961].
Heidegger, M. (1973b). Sketches for a history of being as metaphysics. In *The end of philosophy* (pp. 55–74). Harper. [*orig*. 1961].
Heidegger, M. (1973c). Overcoming metaphysics. In *The end of philosophy* (pp. 84–110). Harper. [*orig*. 1954].
Heidegger, M. (1971c). Building, dwelling, thinking. In *Poetry, language, thought*. Harper. [*orig*. 1935–6].
Heidegger, M. (1985). *History of the concept of time*. Indiana. [*orig*. 1925].
Heidegger, M. (1998b). Comments on Karl Jasper's psychology of worldviews. In *Pathmarks*. CUP. [*orig*. 1919/21].
Heisenberg, W. (1958a). *Physics and philosophy: The revolution in modern science*. Unwin.
Heisenberg, W. (1961). Planck's discovery and the philosophical problem of atomic physics. In *On modern physics*. Clarkson.
Heisenberg, W. (1958b). The representation of nature in contemporary physics. *Daedelus, 87*, 95–108.
Heisenberg, W. (1979a). [Interview with] Werner Heisenberg. In P. Buckley & D. Peat (Eds.), *A question of physics: Conversations in physics and biology* (pp. 3–16). Routledge.
Heisenberg, W. (2020). *Le Manuscrit de 1942*. Éditions Allia. [*orig*. 1942].
Heisenberg, W. (1979b). The teachings of Goethe and Newton on colour in the light of modern physics. In *Philosophical problems of quantum physics* (pp. 60–76). Ox Bow. [*orig*. 1941].

Heisenberg, W. (1979c). On the history of the physical interpretation of nature. In *Philosophical problems of quantum physics* (pp. 27–40). Ox Bow. [*orig.* 1932].

Ho, M.-W. (1993). *The rainbow and the worm*. World Scientific.

Ho, M.-W. (2013). Review: Circular thermodynamics of organisms and sustainable systems. *Systems, 1*(3), 30–49. https://doi.org/10.3390/systems1030030. (*).

Ho, M.-W. (2009a). Quantum jazz: Liquid crystalline water music of the organism. *Subtle Energies & Energy Medicine, 20*, 37–52. (*).

Ho, M.-W. (2009b). How development directs evolution. In: *Evolution and the future conference paper*. (*).

Ho, M.-W. (2009c). Darwin's pangenesis, the hidden history of genetics & the dangers of GMO. *Isis* (*).

Ho, M.-W. (2004). *In search of the sublime*. Ratical.org. (*).

Ho, M.-W. (1998). Organism and psyche in a participatory universe. In D. Loye (Ed.), *The evolutionary outsider: The impact of the human agent on evolution, essays in honour of Ervin Laszlo* (pp. 49–65). Praeger. (*).

Ho, M.-W. (1997). Quantum coherence and conscious experience. *Kybernetes, 26*, 265–276. (*).

Ho, M.-W. (1996). The biology of free will. *Journal of Consciousness Studies, 3*, 231–244. (*).

Ho, M.-W. (1995). Bioenergetics and the coherence of organisms. *Neuronetwork WORLD, 5*, 733–750. (*).

Ho, M.-W. (1991). The role of action in evolution: Evolution by process and the ecological approach to perception. *Cultural Dynamics, 4*, 336–354.

Ho, M.-W. (1987). Evolution by process, not by consequence: Implications of the new molecular genetics on development and evolution. *International Journal of Comparative Psychology, 1*(1), 3–26. (*).

Ho, M.-W. (1994). What is (Schrödinger's) negentropy? *Modern Trends in BioThermoKinetics, 3*, 50–61. (*).

Ho, M.-W., & Popp, F.-A. (1993). Biological organisation, coherence, and light emission from living organisms. In W. D. Stein & F. J. Varela (Eds.), *Thinking about biology: An invitation to current theoretical biology* (pp. 183–214). CRC.

Ho, M.-W., & Saunders, P. T. (1980). Adaptation and natural selection: Mechanism and teleology. In S. Rose (Ed.), *Towards a liberatory biology: The dialectics of biology group* (pp. 85–102). Allison & Busby.

Ho, M.-W. (1979). Beyond neo-Darwinism—An epigenetic approach to evolution. *Journal of Theoretical Biology, 78*, 573–591. (*).

Ho, M.-W. (1976). On the increase in complexity in evolution. *Journal of Theoretical Biology, 63*, 375–384.

Ho, M.-W., & Ulanowicz, R. (2005). Sustainable systems as organisms? *Biosystems, 82*, 39–51. (*).

Hoffmeyer, J. (2008). *Biosemiotics: An examination into the signs of life and the life of signs*. SUP.

Hoffmeyer, J. (2011). Biology is immature biosemiotics. In C. Emmeche & K. Kull (Eds.), *Towards a semiotic biology: Life is the action of signs* (pp. 43–66). Imperial.

Hoffmeyer, J. (2010). God and the world of signs: Semiotics and the emergence of life. *Zygon, 45*(2), 367–390. (*).

Hoffmeyer, J. (1998). Surfaces inside surfaces. On the origin of agency and life. *Cybernetics and Human Knowing, 5*, 33–42. (*).

Hoffmeyer, J. (2002). The central dogma: A joke that became real. *Semiotica, 138*, 1–13. (*).

Hooper, S. E. (1948). Whitehead's philosophy: The world as *process*. *Philosophy, XXIII*, 140–160.

Hume, D. (1951). *Enquiries concerning the human understanding and concerning the principles of morals*. OUP. [*orig. c.* 1757].

Hume, D. (2000). *A treatise of human nature: Being an attempt to introduce the experimental method of reasoning into moral subjects*. OUP. [*orig.* 1740].

Hurlblut, W. B. (2006). Framing the future: Embryonic stem cells, ethics and the emerging era of developmental biology. *Pediatric Research, 59*, 4R–12R. (*).

Husserl, E. (1969). Universal teleology. *Telos, 4*, 176.

Husserl, E. (1970a). The origin of geometry. In *The crisis of European sciences and transcendental phenomenology: An introduction to phenomenological philosophy* (pp. 353–378). Northwestern Univ.. [*orig.* 1936].

Husserl, E. (1970b). Philosophy and the crisis of European humanity. In *The crisis of European sciences and transcendental phenomenology: An introduction to phenomenological philosophy* (pp. 269–299). Northwestern Univ.. [*orig.* 1936].

Husserl, E. (2013a). *Ideas II*. Isha. [*orig.* 1928].

Husserl, E. (2001). *Logical investigations* (Vol. 1). Routledge. [*orig.* 1900–1].

Husserl, E. (1999). *Cartesian meditations: An introduction to phenomenology*. Springer. [*orig.* 1933].

Husserl, E. (2013b). Letter to Albrecht, August 2 1917. In *Ideas II*. Isha. [*orig.* 1928].

Huxley, A. (1949). Letter to George Orwell "1984 v. Brave New World. *Letters of Note*. (Online) (*).

Huxley, A. (1980a). Art. In *The human situation* (pp. 180–194). Triad. [*orig.* 1959].

Huxley, A. (1980b). The ego. In *The human situation* (pp. 137–150). Triad. [*orig.* 1959].

Huxley, A. (1980c). Language. In *The human situation* (pp. 166–179). Triad. [*orig.* 1959].

Huxley, A. (1980d). The individual life of man. In *The human situation* (pp. 110–123). Triad. [*orig.* 1959].

Huxley, A. (1971a). Beliefs. In *Collected essays* (pp. 361–377). Harper. [*orig.* 1937].

Huxley, A. (1971b). Words and behaviour. In *Collected essays* (pp. 245–255). Harper. [*orig.* 1937].

Huxley, A. (1971c). The oddest science. In *Collected essays* (pp. 319–326). Harper. [*orig.* 1937].

Huxley, A. (1971d). Knowledge and understanding. In *Collected essays* (pp. 377–399). Harper. [*orig.* 1956].

Huxley, A. (1927a). Varieties of intelligence. In *Proper studies* (pp. 31–88). Chatto & Windus.

Huxley, A. (1927b). Education. In *Proper studies* (pp. 89–138). Chatto & Windus.

Huxley, A. (1927c). The substitutes of religion. In *Proper studies* (pp. 171–191). Chatto & Windus.

Il'in, I. (2010). *The philosophy of Hegel as a doctrine of the concreteness of god and humanity* (Vol. I & 2). Northwestern. [*orig.* 1918].

Ireland, T. (2022). Beauty: Synthesis of intellect and senses commentary on the biosemiotic fundamentals of aesthetics. *Biosemiotics, 15*, 47–55. (*).

James, W. (1913). *The varieties of religious experience*. Routledge. [*orig.* 1902].

James, W. (1975). *Pragmatism: A new name for some old ways of thinking*. HUP. [*orig.* 1907].

James, W. (1904). Does "consciousness" exist? *The Journal of Philosophy, 1*, 477–491. (*).

James, W. (1950). *The principles of psychology*. Dover. [*orig.* 1890].

Jaspers, K. (1958). *Myth and Christianity: An inquiry of the possibility of religion without myth*. Noonday Press.

Jaspers, K. (1959). *Truth and symbol*. Vision. [*orig.* 1947].

Jay, M. (1993). *Downcast eyes: The denigration of vision in twentieth-century French thought*. University of California Press.

Jolley, N. (1990). *The light of the soul: Theories of ideas in Leibniz, Malebranche, and Descartes*. OUP.

Jung, C. G. (2002). *The earth has a soul: C.G. Jung on nature, technology & modern life*. North Atlantic.

Jung, C. G. (1959a). The ego. In *Aion: Researches into the phenomenology of the self* (pp. 3–7). Routledge.

Jung, C. G. (1959b). The shadow. In *Aion: Researches into the phenomenology of the self* (pp. 8–10). Routledge.

Jung, C. G. (1933a). A psychological theory of types. In *Modern man in search of a soul* (pp. 85–108). Routledge.

Jung, C. G. (1933b). The stages of life. In *Modern man in search of a soul* (pp. 109–131). Routledge.

Jung, C. G. (1933c). Archaic man. In *Modern man in search of a soul* (pp. 143–174). Routledge.

Jung, C. G. (1933d). The spiritual problem of modern man. In *Modern man in search of a soul* (pp. 226–254). Routledge.

Jung, C. G. (1933e). The poet. In *Modern man in search of a soul* (pp. 192–199). Routledge.

Jung, C. G. (1960a). On psychic energy. In *The structure and dynamics of the psyche* (pp. 3–66). Routledge. [*orig.* 1928].

Jung, C. G. (1960b). A review of the complex theory. In *The structure and dynamics of the psyche* (pp. 92–104). Routledge. [*orig.* 1934].

Jung, C. G. (1960c). The structure and dynamics of the psyche. In *The structure and dynamics of the psyche* (pp. 139–158). Routledge. [*orig.* 1927].

Jung, C. G. (1960d). On the nature of the psyche. In *The structure and dynamics of the psyche* (pp. 159–234). Routledge. [*orig.* 1947].

Jung, C. G. (1960e). The soul and death. In *The structure and dynamics of the psyche* (pp. 404–415). Routledge. [*orig.* 1934].

Kant, I. (1987). *Critique of judgment*. Hackett. [*orig.* 1790].

Kant, I. (2002). *The critique of practical reason*. Hackett. [*orig.* 1788].

Kant, I. (1934). *The critique of pure reason*. Everyman. [*orig.* 1787].

Kant, I. (1949). *Fundamental principles of the metaphysic of morals*. Liberal Arts. [*orig.* 1785].

Kant, I. (1998). *Groundwork of the metaphysic of morals*. CUP. [*orig.* 1785].

Kapp, E. (2018). *Elements of a philosophy of technology: On the evolutionary history of culture*. Minnesota Press. [*orig.* 1877] (*).

Karaca, Ç. (2021b). Life's organization between matter and form: Neo-Aristotelian approaches and biosemiotics. *History and Philosophy of the Life Sciences, 43*, 55.

Kass, L. R. (1994). *The hungry soul: Eating and the perfecting of our nature*. Free Press.

Kauffman, S., & Gare, A. (2015). Beyond Descartes and Newton: Recovering life and humanity. *Progress in Biophysics and Molecular Biology, 119*(3), 219–244. (*).

Kenaan, H. (2002). Language, philosophy and the risk of failure: Rereading the debate between Searle and Derrida. *Continental Philosophy Review, 35*, 117–133. (*).

Kierkegaard, S. (1944). *Either/or*. Princeton. [*orig.* 1843].

Kierkegaard, S. (1936). *Philosophical fragments*. Princeton. [*orig.* 1844].

Kisner, W. (2008). The category of life, mechanistic reduction, and the uniqueness of biology. *Cosmos and History, 4*, 113–153. (*).

Kivi, R. (2019). How do elephants behave? *Sciencing*. (*).

Koestler, A., & Smythies, J. R. (Eds.). (1969). *Beyond reductionism: New perspectives in the life sciences*. Hutchinson.

Krois, J. M. (2004). More than a linguistic turn in philosophy: the semiotic programs of Peirce and Cassirer. *Sats—Nordic Journal of Philosophy, 5*, 14–33.

Kull, K. (2022). The biosemiotic fundamentals of aesthetics: Beauty is the perfect semiotic fitting. *Biosemiotics, 15*, 1–22. (*).

Kull, K. (2023). Interview with Gerd B. Müller on theoretical biology. *Biosemiotics, 16*, 381–394. (*).

Kull, K. (2020). Codes: Necessary, but not sufficient for meaning-making. *Constructivist Foundations*. (*).

Kull, K. (2015a). Semiosis stems from logical incompatibility in organic nature: Why biophysics does not see meaning, while biosemiotics does. *Progress in Biophysics and Molecular Biology, 119*, 616–621. (*).

Kull, K. (2010). Theoretical biology on its way to biosemiotics. In D. Favareau (Ed.), *Essential readings in biosemiotics: Anthology and commentary* (pp. 417–444). Springer.

Kull, K. (2015b). Evolution, choice, and scaffolding: Semiosis is changing its own building. *Biosemiotics, 8*, 223–234. (*).

Kull, K. (2003). Ladder, tree, web: The ages of biological understanding. *Sign Systems Studies, 31*(2), 589–603.

Kull, K. (2001). Jakob von Uexkül: An introduction. *Semiotica, 134*, 1–59. (*).

Kull, K. (2007). Biosemiotics and biophysics—The fundamental approaches to the study of life. In M. Barbieri (Ed.), *Introduction to biosemiotics: The new biological synthesis* (pp. 167–178). Springer. (*).

Kull, K. (1998). On semiosis, umwelt, and semiosphere. *Semiotica, 120*, 299–310. (*).

Kvasz, L. (2014). Newton as a Cartesian. In *Scientific cosmopolitanism and local cultures: Religions, ideologies, societies—Proceedings Athens 1–3 November 2012* (pp. 203–207). IHR. (*).

Kvasz, L. (2013). Heidegger's interpretation of mathematical science in the light of Husserl's concept of Mathematization in the *Krisis*. *Philosophia Naturalis, 50*, 337–363. (*).

Kvasz, L. (2012). Galileo, Descartes, and Newton—Founders of the language of physics. *Acta Physica Slovaca, 62*, 519–561. (*).

Kvasz, L. (2004). The invisible link between mathematics and theology. *Perspectives on Science and Christian Faith, 56*, 111–116. (*).

Kvasz, L. (2003). The mathematisation of nature and Cartesian physics. *Philosophia Naturalis, 40*, 157–182. (*).

Lakatos, I. (1970). Falsification and the methodology of scientific research programmes. In I. Lakatos & A. Musgrave (Eds.), *Criticism and the growth of knowledge* (pp. 91–196). CUP. [*orig*. 1965].

Lakatos, I. (1978a). Newton's effect on scientific standards. In *The methodology of scientific research programmes: Philosophical papers* (Vol. 1, pp. 193–222). CUP.

Lakatos, I. (1978b). The method of analysis-synthesis. In *Mathematics, science and epistemology: Philosophical papers* (Vol. 2, pp. 70–106). CUP.

Lakatos, I. (1962). Infinite regress and foundations of mathematics. *Proceedings of the Aristotelian Society, XXXVI*, 155–184.

Lanz, H. (1918). Infinity as method. *The Monist, 28*, 46–67.

Laughlin, R. B. (2005). *A different universe: Reinventing physics from the bottom down*. Basic Books.

Leggett, A. J. (1987). Physics on a human scale. In *The problems of physics* (pp. 111–143). OUP.

Leibniz, G. (1967). *The Leibniz-Clarke correspondence*. Man. UP.

Leibniz, G. Untitled essay. In C. I. Gerhardt (Ed.), *Die philosophischen Schriften* (Vol. IV, pp. 1875–1890). Wiedmann.

Leibniz, G. (1971). The twenty-four statements. In M. Heidegger (Ed.), *The end of philosophy* (pp. 49–54). Harper. [*orig. c.* 1690].

Leibniz, G. (1973a). New system, and explanation of the new system. In *Philosophical writings*. J.M. Dent. [*orig*. 1695–6].

Leibniz, G. (1989a). Preface to the new essays (1703–05). In *Philosophical essays*. Hackett.

Leibniz, G. (1989b). Against barbaric physics. In *Philosophical essays*. Hackett. [*orig*. 1686].

Leibniz, G. (1896). Appendix. In *New essays concerning human understanding*. Macmillan. [*orig*. 1702].

Leibniz, G. (1697). *The ultimate origin of things*. Inters.org. (*).

Leibniz, G. (1973b). Correspondence with Arnauld. In *Philosophical writings*. J.M. Dent. [*orig*. 1686].

Leibniz, G. (1965a). Critical remarks concerning the general part of Descartes' principles. In *Monadology and other philosophical essays*. Bobbs-Merrill. [*orig*. 1692].

Leibniz, G. (1989c). Letter to Johann Bernoulli 18 November 1698. In *Philosophical essays*. Hackett.

Leibniz, G. (1989d). Letter to Arnauld April 30, 1687. In *Philosophical essays*. Hackett.

Leibniz, G. (1989e). Letter to Molanus (?), On god and the soul, c. 1679. In *Philosophical essays*. Hackett.

Leibniz, G. (1989f). Letter to Arnauld Nov/Dec 1686. In *Philosophical essays*. Hackett.

Leibniz, G. (1989g). Letter to de Volder 20 June 1703. In *Philosophical essays*. Hackett.

Leibniz, G. (1989h). Letter to de Volder, 1704 or 1705. In *Philosophical essays*. Hackett.

Leibniz, G. (1989i). Letter to des bosses 29 May 1716. In *Philosophical essays*. Hackett.

Leibniz, G. (1955). Letter to Malebranche [?], June 1679. In A. Robinet (Ed.), *Malebranche and Leibniz*. Vrin.

Leibniz, G. (1973c). Metaphysical consequences of the principle of reason. In *Philosophical writings*. J.M. Dent. [*orig*. 1712].

Leibniz, G. (1973d). Monadology. In *Philosophical writings*. J.M. Dent. [*orig.* 1714].

Leibniz, G. (1973e). New essays on the human understanding. In *Philosophical writings*. J.M. Dent. [*orig. c.* 1704].

Leibniz, G. (1989k). Notes for Leibniz to des bosses; 5 February 1712. In *Philosophical essays*. Hackett.

Leibniz, G. (1989l). On freedom. In *Philosophical essays*. Hackett. [*orig.* 1689?].

Leibniz, G. (1965b). On the improvement of metaphysics. In *Monadology and other philosophical essays*. Bobbs-Merrill. [*orig.* 1694].

Leibniz, G. (1989m). Principles of nature and grace based on reason. In *Philosophical essays*. Hackett. [*orig.* 1714].

Leibniz, G. (1989n). Remarks on Arnauld's letter about my proposition that the individual notion of each person includes once and for all everything that will ever happen to him. In *Philosophical essays*. Hackett. [*orig.* 1686].

Leibniz, G. (1965c). What is nature? Reflections on the force inherent in created things and on their actions. In *Monadology and other philosophical essays*. Bobbs-Merrill. [*orig.* 1698].

Leibniz, G. (1989o). Discourse on metaphysics. In *Philosophical essays*. Hackett. [*orig.* 1686].

Lennon, T. M. (1993). *The battle of the gods and giants: The legacies of Descartes and Gassendi, 1655–1715*. PUP.

Lewis, C. S. (1943). *The abolition of man*. OUP.

Lewis, J. (Ed.). (1974). *Beyond chance and necessity* (pp. 27–50). Garnstone.

Lewis, T. (2013). Getting a head: How worms regenerate lost tissue. *Live Science*. (Online) (*).

Lewontin, R. (1991). *Biology as ideology: The doctrine of DNA*. Harper.

Lobkowicz, N. (1967). *Theory and practice: History of a concept from Aristotle to Marx*. Univ. Notre Dame.

Lockwood, M. (1989). *Mind, brain & the quantum: The compound "I"*. Blackwell.

Lohmar, D. (2019). The phenomenological method of eidetic intuition and its clarification as eidetic variation. In J. J. Drummond & O. Höffe (Eds.), *Husserl: German perspectives* (pp. 110–140). Fordham.

Lohmar, D. (2012). Ego and arch-ego in Husserlian phenomenology. In R. Breeur & U. Melle (Eds.), *Life subjectivity & art: Essays in honor of Rudolf Bernet* (pp. 277–302). Springer.

Lorenz, K. (1977). *Behind the mirror: A search for a natural history of human knowledge*. Methuen.

Lorenz, K. (1951). The role of gestalt perception in animal and human behaviour. In L. L. Whyte (Ed.), *Aspects of form* (pp. 157–178). Lund.

Lotman, Y. (2010). Excerpts from *universe of the mind: A semiotic theory of culture* (1990). In D. Favareau (Ed.), *Essential readings in biosemiotics: Anthology and commentary* (pp. 197–214). Springer.

Lotman, Y. (2013). On the dynamics of culture. *Sign Systems Studies, 41*, 355–370. [*orig.* 1992] (*).

Luther, M. (1520a). *An open letter to the Christian nobility of the German nation concerning the reform of the Christian estate*. (*).

Luther, M. (1520b). *Treatise on Christian liberty*. (*).

Lyotard, J.-F. (1991). *Phenomenology*. SUNY. [*orig.* 1954].

Macdonald, G. (1992). Reduction and evolutionary biology. In D. Charles & K. Lennon (Eds.), *Reduction, explanation, and realism*. OUP.

Manent, P. (1995). *An intellectual history of liberalism*. PUP. [*orig.* 1987].

Manent, P. (2013). Birth of the nation. *City Journal*.

Mayol, J. (2013). Why do elephants eat soil? *Metode*. (*).

Mayr, E. (1988). *Towards a new philosophy of biology*. Harvard.

Mayr, E. (2001). Ernst Mayr: What evolution is. *Edge*. (Online).

McCarthy, J. C. (1992). Parts, wholes, and the forms of life: Husserl and the new biology. In L. Hardy & L. Embree (Eds.), *Phenomenology of natural science* (pp. 135–156). Springer. (*).

Mercer, C. (2002). The Aristotelianism at the core of Leibniz's philosophy. In C. H. Leijenhorst, J. M. M. H. Thijssen, & C. H. Lüthy (Eds.), *The dynamics of Aristotelian natural philosophy from antiquity to the seventeenth century*. Brill. (*).

Merleau-Ponty, M. (1968). *The visible and the invisible*. Northwestern University Press.

Merleau-Ponty, M. (1962). *Phenomenology of perception*. Routledge. [*orig.* 1945].

Merleau-Ponty, M. (1967). *The structure of behaviour*. Beacon. [*orig.* 1942].

Merleau-Ponty, M. (2003). *Nature: Course notes from the Collège de France*. Northwestern University Press. [*orig. c.* 1956].

Merleau-Ponty, M. (1969a). The primacy of perception and its philosophical consequences. In *The essential writings of Merleau-Ponty* (pp. 47–63). Harcourt. [*orig.* 1964].

Merleau-Ponty, M. (1969b). The relations of the soul and the body and the problem of perceptual consciousness. In *The essential writings of Merleau-Ponty* (pp. 138–181). Harcourt. [*orig.* 1963].

Messerly, J. G. (1996). *Piaget's concept of evolution: Beyond Darwin and Lamarck*. Rowan & Littlefield.

Midgely, M. (2010). *The solitary self: Darwin and the selfish gene*. Acumen.

Midgely, M. (1985). *Evolution as a religion: Strange hopes and stranger fears*. Methuen.

Mohanty, J. N. (2009). *Lectures on consciousness and interpretation*. OUP.

Monod, J. (1971). *Chance and necessity*. Collins.

Monod, J. (1974). On chance and necessity. In F. J. Ayala & T. Dobzhansky (Eds.), *Studies in the philosophy of biology* (pp. 357–375). Macmillan.

Müller, G. B., & Newman, S. A. (Eds.). (2003). *Origination of organismal form: Beyond the gene in developmental and evolutionary biology*. MIT.

Murdoch, I. (1970a). On "god" and "good". In *The sovereignty of good* (pp. 46–76). Routledge.

Murdoch, I. (1970b). The idea of perfection. In *The sovereignty of good* (pp. 1–45). Routledge.

Murdoch, I. (1970c). The sovereignty of good over other concepts. In *The sovereignty of good* (pp. 77–104). Routledge.

Murdoch, I. (1962). Metaphysics and ethics. In D. F. Pears (Ed.), *The nature of metaphysics* (pp. 99–123). Macmillan.

Nagel, T. (2012). *Mind and cosmos: Why the materialist neo-Darwinian conception of nature is almost certainly false*. OUP.

Nagel, T. (1998). Reductionism and antireductionism. In G. Bock & J. Goode (Eds.), *The limits of reductionism in biology*. Wiley. (*).

Nagel, T. (1974). What is it like to be a bat? *The Philosophical Review, 83*, 435–450. (*).

Neander, K. (1991). The teleological notion of function. *Australasian Journal of Philosophy, 69*(4), 454–468. (*).

von Neumann, J. (1969). *Theory of self-reproducing automata*. Univ. Illinois. (*).

Niemann, H.-J. (2014). *Karl popper and the two new secrets of life*. Mohr Siebeck.

Noble, D. (2006). *The music of life: Biology beyond genes*. OUP.

Noble, D. (2017). *Dance to the tune of life: Biological relativity*. CUP.

O'Callaghan, C. (2009). Sounds and events. In M. Nudds & C. O'Callaghan (Eds.), *Sounds and perceptions: New philosophical essays*. OUP.

Oliver, A., & Smiley, T. (2013). *Plural logic*. OUP.

Özel, J., & Psarros, N. (2021). Conatus, the will to persist: An introduction. *Conatus, 6*, 9. (*).

Padial, J. (2015). *Umwelt* and *Innenwelt* in Hegel's philosophy of subjective spirit. *Fenomenologia, 13*, 123–132.

Pasnau, R. (2007a). Democritus and secondary qualities. *Archiv für Geschichte der Philosophie, 89*, 99–121. (*).

Pasnau, R. (2014). Epistemology idealised. *Mind, 123*. (*).

Pasnau, R. (2012). Mind and hylomorphism. In J. Marenbon (Ed.), *The Oxford handbook of medieval philosophy* (pp. 486–504). OUP. (*).

Pasnau, R. (2011). On existing all at once. In C. Tapp (Ed.), *God, eternity, and time*. Ashgate. (*).

Pasnau, R. (2007b). Mind and extension (Descartes, Hobbes, more). In H. Lagerlund (Ed.), *Forming the mind: Essays on the internal senses and the mind/Body problem from Avicenna to the medical enlightenment* (pp. 283–310). Springer. (*).

Pasnau, R. (2004). Form, substance, and mechanism. *The Philosophical Review, 113*, 31–88. (*).

Patočka, J. (1998). *Body, community, language, world*. Open Court. [*orig*. 1968–9].

Patočka, J. (2019). What is phenomenology? In J. J. Drummond & O. Höffe (Eds.), *Husserl: German perspectives* (pp. 84–109). Fordham.

Pattee, H. (1969). How does a molecule become a message? *Developmental Biology, 3*, 1–16. (*).

Pauly, D. (2000a). Goethean science: A phenomenological study of plant metamorphosis. *Biodynamics*, 21–25. (*).

Pauly, D. (2000b). Goethean science: A phenomenological study of plant metamorphosis—Part 2. *Biodynamics*, 19–30. (*).

Pauly, D. (2000c). Goethean science: A phenomenological study of plant metamorphosis—Part 3. *Biodynamics*, 24–26. (*).

Peirce, C. S. (2010). The logic of signs. In D. Favareau (Ed.), *Essential readings in biosemiotics: Anthology and commentary* (pp. 115–148). Springer.

Peirce, C. S. (1997). *The essential peirce [EP]*. IUP.

Peirce, C. S. *Collected papers [CP]* (pp. 1931–1958). HUP. (*).

Peirce, C. S. (1905). What pragmatism is. *The Monist, 15*, 161–181. (*).

Peirce, C. S. (1903). Lowell lectures. *Lecture, III*(MS 460) (*).

Perlman, M. (2010). Traits have evolved to function the way they do because of a past advantage. In F. J. Ayala & R. Arp (Eds.), *Contemporary debates in the philosophy of biology* (pp. 53–71). Blackwell.

Perner, J. (1991). *Understanding of the representational mind*. MIT.

Petruzzello, M. (n.d.). Why do leaves fall in autumn? *Britannica*. (Online).

Piaget, J., & Inhelder, B. (1969). The gaps in empiricism. In A. Koestler & J. R. Smythies (Eds.), *Beyond reductionism: New perspectives in the life sciences* (pp. 118–160). Hutchinson.

Pickard Winsor, M. (2006). Linnaeus' biology was not essentialist. *Annals of the Missouri Botanical Garden, 93*, 2–7.

Pieper, J. (1965). *Leisure the basis of culture*. Pantheon. [*orig*. 1947].

Popper, K. (2014). A new interpretation of Darwinism: The first Medawar lecture. In H.-J. Niemann (Ed.), *Karl Popper and the two new secrets of life* (pp. 115–128). Mohr Siebeck. [*orig*. 1986].

Popper, K. (1982). The place of mind in nature. In R. Q. Elvee (Ed.), *Mind in nature: Nobel conference XVII* (pp. 31–59). Harper & Row.

Popper, K. (1979a). Two faces of common sense: An argument for commonsense realism and against the commonsense theory of knowledge. In *Objective knowledge: An evolutionary approach* (pp. 32–105). OUP. [*orig*. 1970].

Popper, K. (1979b). On the theory of the objective mind. In *Objective knowledge: An evolutionary approach* (pp. 153–190). OUP. [*orig*. 1968].

Popper, K. (1979c). Of clouds and clocks. In *Objective knowledge: An evolutionary approach* (pp. 206–255). OUP. [*orig*. 1965].

Popper, K. (1979d). Evolution and the tree of knowledge. In *Objective knowledge: An evolutionary approach* (pp. 256–284). OUP. [*orig*. 1961].

Popper, K. (1974). Scientific reduction and the essential incompleteness of all science. In F. J. Ayala & T. Dobzhansky (Eds.), *Studies in the philosophy of biology* (pp. 259–284). Macmillan.

Popper, K. (1979e). A realist view of logic, physics, and history. In *Objective knowledge: An evolutionary approach* (pp. 285–318). OUP. [*orig*. 1966].

Popper, K. (1994a). Knowledge: Objective and subjective. In *Knowledge and the mind-body problem: In defence of interaction* (pp. 1–23). Routledge. [*orig*. 1969].

Popper, K. (1994b). The autonomy of world 3. In *Knowledge and the mind-body problem: In defence of interaction* (pp. 24–46). Routledge. [*orig*. 1969].

Popper, K. (1994c). World 3 and emergent evolution. In *Knowledge and the mind-body problem: In defence of interaction* (pp. 47–78). Routledge. [*orig*. 1969].

Popper, K. (1994d). Description, argument, and imagination. In *Knowledge and the mind-body problem: In defence of interaction* (pp. 79–104). Routledge. [*orig*. 1969].

Popper, K. (1994e). Interaction and consciousness. In *Knowledge and the mind-body problem: In defence of interaction* (pp. 105–128). Routledge. [*orig*. 1969].

Popper, K. (1994f). The self, rationality, and freedom. In *Knowledge and the mind-body problem: In defence of interaction* (pp. 129–142). Routledge. [*orig.* 1969].
Popper, K. (1978). Natural selection and the emergence of mind. *Dialectica, 32*, 339–355. (*).
Popper, K. (1945). *The open society and its enemies*. Routledge.
Putnam, H. (2004). *Ethics without ontology*. Harvard.
Quine, W. V. O. (1969). Natural kinds. In *Ontological relativity and other essays*. CUP.
Quine, W. V. O. (1960). *Word and object*. MIT.
Quine, W. V. O. (1953a). On what there is. In *From a logical point of view* (pp. 1–19). Harvard.
Quine, W. V. O. (1953b). Two dogmas of empiricism. In *From a logical point of view* (pp. 20–46). Harvard.
Quine, W. V. O. (1957–1958). Speaking of objects. *Proceedings and Addresses of the American Philosophical Association, 31*, 5–22. (*).
Rafelski, J., & Müller, B. (1985). *The structured vacuum: Thinking about nothing*. Deutsch. (*).
Rieppel, O. (2010). New essentialism in biology. *Philosophy of Science, 77*, 662–673.
Rieppel, O. (1990). Structuralism, functionalism, and the four Aristotelian causes. *Journal of the History of Biology, 23*.
Ritterbush, P. C. (1968). *The art of organic forms*. Smithsonian.
Robert, J. S. (2004). *Embryology, epigenesis, and evolution: Taking development seriously*. CUP.
Rocknak, S. (2015). Locke, Hume, and the treacherous logos of atomism: The eclipse of democratic values in the early modern period. *Notre Dame Philosophical Reviews*. (Online) (*).
Rorty, R. (1979). *Philosophy and the mirror of nature*. Princeton.
Rorty, R. (1982). Mind as ineffable. In R. Q. Elvee (Ed.), *Mind in nature: Nobel conference XVII* (pp. 60–95). Harper & Row.
Rosen, R. (1991). *Life itself: A comprehensive inquiry into the nature, origin, and fabrication of life*. Columbia.
Rosen, R. (2000). *Essays on life itself*. Columbia.
Rosen, S. (2002). Austin and ordinary language. In *The elusiveness of the ordinary: Studies in the possibility of philosophy*. Yale.
Royce, J., & Kernan, F. (1916). Charles Sanders Peirce. *The Journal of Philosophy, Psychology and Scientific Methods, 13*(26) (*).
Ruse, M. (1985). *Sociobiology: Sense or nonsense?* (2nd ed.). Reidel.
Russell, B. (1953). *The impact of science on society*. Simon & Schuster.
Russell, B. (1926). *Our knowledge of the external world*. Allen & Unwin. [*orig.* 1914].
Russell, B. (1959). Logical atomism. In A. J. Ayer (Ed.), *Logical positivism*. Free Press. [*orig.* 1924].
Russell, B. (1945). *A history of western philosophy*. Simon & Schuster.
Russell, B. (1929). On the notion of cause, with application to the free-will problem. In *Our knowledge of the external world* (pp. 247–256). Norton. (*).
Russell, B. (1963). Logic as the essence of philosophy. In R. Jager (Ed.), *Essays in logic from Aristotle to Russell*. Eaglewood Cliffs. [*orig.* 1914].
Russell, B. (1943). *An outline of intellectual rubbish: a hilarious catalogue of organized and individual stupidity*. Haldeman-Julius. (*).
Russell, B. (1956). On denoting. In *Logic and knowledge*. Allen & Unwin. [*orig.* 1905].
Russell, B. (1937). *The principles of mathematics*. CUP. [*orig.* 1903].
Saunders, P. T. (2003a). The organism as a dynamical system. In W. Stein (Ed.), *Thinking about biology* (pp. 41–63). CRC.
Saunders, P. T. (2003b). Bricks without straw: Darwinism in the social sciences. *Theoria, 18*(3), 259–272. (*).
Saunders, P. T. (1993). Alan turing and biology. *IEEE Annals of the History of Computing, 15*(3), 33–36. (*).
Schrödinger, E. (1958). *Mind and matter*. CUP.
Schrödinger, E. (1951). *Science and humanism: Physics in our time*. CUP.
Schrödinger, E. (1967). *What is life? The physical aspect of the living cell*. CUP. [*orig.* 1944].

Schrödinger, E. (1980). The present situation in quantum mechanics: A translation of Schrödinger's "Cat Paradox" paper (trans. John D. Trimmer). *Proceedings of the American Philosophical Society, 124*, 323–338. [*orig.* 1935].

Schuhmann, K. (2019). Intentionality and the intentional object in the early Husserl. In J. J. Drummond & O. Höffe (Eds.), *Husserl: German perspectives* (pp. 141–162). Fordham.

Schumacher, E. F. (1973). *Small is beautiful: A study of economics as if people mattered*. Blond & Briggs.

Scruton, R. (2012). *The face of god: The Gifford lectures 2010*. Continuum.

Scruton, R. (2007, June 12). Richard Rorty's legacy. *Open Democracy*. (Online) (*).

Sebeok, T. A. (2001). Biosemiotics: Its roots, proliferation, and prospects. *Semiotica, 134*(1–4), 61–78.

Sehon, S. (2005). *Teleological realism: Mind, agency, and explanation*. MIT.

Sellars, W. (1963). Philosophy and the scientific image of man. In *Empiricism and the philosophy of mind* (pp. 1–40). Routledge.

Senatore, M. Teleotheology: Derrida and the Aristotelian foundations of structuralism. *Philosophy Today: An International Journal of Contemporary Philosophy, 63*(1), 175–194. (*).

Shields, J. (2019). Why some animals eat rocks to aid in digestion. *How Stuff Works*. (*).

Shingleton, A. W. (2010). Allometry: The study of biological scaling. *Nature Education Knowledge, 3*(10), 2.

Shouse, B. (2002). How Geckos Stick on der Waals. *Science*. (*).

Skinner, B. F. (1971). *Beyond freedom and dignity: A stunning new plan to alter human behaviour*. Vintage.

Skinner, B. F. (1957a). *Verbal behaviour*. Appleton.

Skinner, B. F. (1957b). The validity of the author's *Verbal Behaviour*. In *Verbal behaviour* (pp. 453–460). Appleton.

Skinner, B. F. (1957c). The verbal community. In *Verbal behaviour* (pp. 461–470). Appleton.

Skinner, B. F. (1988). Philosophy of behaviourism: An informal talk about behaviour and its determinants. *Youtube*. (Online) (*).

Slatman, J. (2005). The sense of life: Husserl and Merleau-Ponty on touching and being touched. In *Life and individuation* (pp. 305–324). Mimesis. (*).

Smiley, T. (1982–3). The schematic fallacy. *Proceedings of the Aristotelian Society, LXXXIII*, 1–18.

Sober, E. (1980). Evolution, population thinking, and essentialism. *Philosophy of Science, 47*(3), 350–383.

Sorel, G. (1969). *The illusions of progress*. UC. [*orig.* 1910].

Soontiëns, F. J. K. (1991). Evolution: Teleology or chance? *Zeitschrift für Allgemeine Wissenschaftstheorie, 22*, 133–141.

Stengers, I. (2012). Reclaiming animism. *e-flux*. (*).

Straus, E. W. (1967). An existential approach to time. *Annals New York Academy of Sciences, 138*(2), 759–766. (*).

Straus, E. W. (1952). The upright posture. *The Psychiatric Quarterly, 26*, 529–561.

Strauss, D. F. M. (2006). *Infinity and wholeness: The mutual distinctiveness of multiplicity and wholeness*. (Online) (*).

Strauss, D. F. M. (2003). The "Copernican turn" of biology in the 20th century. *Acta Academica, 35*(3), 49–76. (*).

Strauss, D. F. M. (2007). Did Darwin initially develop a theory of evolution in the biological sense of the word? *South African Journal of Philosophy, 26*(2), 190–203. (*).

Sturrock, J. (2003). *Structuralism* (2nd ed.). Blackwell. [*orig.* 1986].

Sullivan, L. (2023). Seed dispersal: Deciding when to move. *eLife*. (*).

Suppes, P. (2014). [Commentary on] Dagfinn Føllesdal. In C. E. Crangle et al. (Eds.), *Foundations and methods from mathematics to neuroscience* (pp. 265–267). CSLI.

Suppes, P. (2011). Why the effectiveness of mathematics in the natural sciences is not surprising. *Interdisciplinary Science Reviews, 36*, 244–254. (*).

Taguchi, S. (2023). Ur-Ich in Husserl. In N. de Warren & T. Toadvine (Eds.), *Encyclopedia of phenomenology*. Springer.

Taylor, C. (1971). Interpretation and the sciences of man. *The Review of Metaphysics, 25*(1), 3–51. (*).

Thom, R. (2016). *To predict is not to explain: Conversations on mathematics, science, catastrophe theory, semiophysics, and natural philosophy*. Thombooks. [*orig.* 1991].

Thom, R. (1991a). *Prédire n'est pas expliquer*. Eschel.

Thom, R. (1990a). *Semiophysics: A sketch: aristotelian physics and catastrophe theory*. Addison-Wesley.

Thom, R. (1975). *Structural stability and morphogenesis: An outline of a general theory of models*. W.A. Benjamin. [*orig.* 1972].

Thom, R. (1999). Aristote topologue. *Revue de synthèse, 4*, 39–47. (*).

Thom, R. (1991b). Matière, forme et catastrophes. In M. A. Sinaceur (Ed.), *Penser avec Aristote* (pp. 367–398). Toulouse Erès.

Thom, R. (1991c). Aristote et l'avènement de la science moderne: la rupture galiléenne. In M. A. Sinaceur (Ed.), *Penser avec Aristote* (pp. 489–494). Toulouse Erès.

Thom, R. (2010). From animal to man: Thought and language. In D. Favareau (Ed.), *Essential readings in biosemiotics: Anthology and commentary* (pp. 337–376). Springer.

Thom, R. (1989). Causality and finality in theoretical biology: A possible picture. In J. Casti & A. Karlqvist (Eds.), *Newton to Aristotle: Toward a theory of models for living systems (Mathematical modeling)* (pp. 39–45). Birkhäuser.

Thom, R. (1988a). Les intuitions topologiques primordiales de l'aristotélisme. *Revue Thomiste, 88*, 393–409.

Thom, R. (1988b). Structure et fonction en biologie aristotélicienne. In *Biologie théorique*. Solignac.

Thom, R. (1990b). Homoéomères et Anhoméomères en Théorie Biologique d'Aristote à Aujourd'hui. In D. Devereux & P. Pellegrin (Eds.), *Biologie, Logique et Métaphysique chez Aristote* (pp. 491–511). CNRS.

Thom, R. (1986). Organs and tools: A common theory of morphogenesis. In J. L. Casti & A. Karlqvist (Eds.), *Complexity, language, and life: Mathematical approaches* (pp. 197–230). Springer.

Thom, R. (1984a). Techniques, sciences et technologies: une classification catastrophiste. *Séminaire de Philosophie et Mathématiques*. (*).

Thom, R. (1984b). *Stabilité structurelle et morphogénèse*. InterÉditions.

Thom, R. (1980a). *Modèles mathématiques de la morphogénèse 2e éd.* Bourgois.

Thom, R. (1980b). L'explication des formes spatiales: réductionnisme ou platonisme. In *La morphogénèse de la biologie aux mathématiques: Actes de trois colloques organisés par l'Ecole Pratique des Hautes Études* (pp. 71–79). Maloine.

Thom, R. (1977). Structural stability, catastrophe theory, and applied mathematics: The John von Neumann lecture, 1976. *SIAM Review, 19*, 189–201. (*).

Thompson, D.'. A. W. (1917). *On growth and form*. CUP.

Thompson, D.'. A. W. (1911). Magnalia Naturae: Or, the greater problems of biology. *Science, 34*, 417–428.

Toulmin, S. (1957). Contemporary scientific mythology. In A. MacIntyre & R. Gregor Smith (Eds.), *Metaphysical beliefs*. SCM.

Turing, A. M. (1952). The chemical basis of morphogenesis. *Philosophical Transactions of the Royal Society of London, 237*, 37–72.

von Uexküll, J. (1992). A stroll through the worlds of animals and men: A picture book of invisible worlds. *Semiotica, 89–4*, 319–391. (*).

von Uexküll, J. (1982). The theory of meaning. *Semiotica, 42–1*, 25–82. [*orig.* 1940].

von Uexküll, J. (1926). *Theoretical biology*. Kegan Paul. [*orig.* 1920].

VandeWall, H. (2007). Why water is not H_2O, and other critiques of essentialist ontology from the philosophy of chemistry. *Philosophy of Science, 74*(5) (*).

Vinci, T., & Scott Robert, J. (2005b). Aristotle and modern genetics. *Journal of the History of Ideas, 66*, 201–221.

Vivien, C. J., et al. (2016). Evolution, comparative biology and ontogeny of vertebrate heart regeneration. *npj Regenerative Medicine*. (*).

Voegelin, E. (2004). *Science, politics & gnosticism: Two essays by Eric Voegelin*. ISI. [*orig.* 1968].

Voegelin, E. (1952). *The new science of politics*. UC.

Vörös, S. (2021). Minding the body: From corporeal mind to minded corporeality. *Phainomena, 30*, 5–34.

Walsh, D. M. (2018). Objectcy and agency: Towards a methological vitalism. In D. Nicholson & J. Dupré (Eds.), *Everything flows: Towards a processual philosophy of biology* (pp. 167–185). OUP. (*).

Walsh, D. M. (2010). Not a sure thing: Fitness, probability, and causation. *Philosophy of Science, 77*, 147–171.

Walsh, D. M. (2006). Evolutionary essentialism. *British Society for the Philosophy of Science, 57*, 425–448.

von Weizsäcker, C. F. (1988). *The ambivalence of progress*. Paragon. [*orig.* 1977].

Whitehead, A. N. (1933). *Adventures of ideas*. Penguin.

Whitehead, A. N. (1926). *Religion in the making*. CUP.

Whitehead, A. N. (1929a). *The function of reason*. PUP.

Whitehead, A. N. (1938a). Importance. In *Modes of thought* (pp. 1–19). Free Press.

Whitehead, A. N. (1938b). Expression. In *Modes of thought* (pp. 20–41). Free Press.

Whitehead, A. N. (1938c). Understanding. In *Modes of thought* (pp. 42–64). Free Press.

Whitehead, A. N. (1938d). Perspective. In *Modes of thought* (pp. 65–85). Free Press.

Whitehead, A. N. (1938e). Forms of process. In *Modes of thought* (pp. 86–104). Free Press.

Whitehead, A. N. (1938f). Civilised universe. In *Modes of thought* (pp. 105–126). Free Press.

Whitehead, A. N. (1938g). Nature lifeless. In *Modes of thought* (pp. 127–147). Free Press.

Whitehead, A. N. (1938h). Nature alive. In *Modes of thought* (pp. 148–170). Free Press.

Whitehead, A. N. (1938i). The aim of philosophy. In *Modes of thought* (pp. 171–174). Free Press.

Whitehead, A. N. (1916–17). The organisation of thought. *Proceedings of the Aristotelian Society, XVII*, 58–76.

Whitehead, A. N. (1967). *Science and the modern world*. Simon & Schuster. [*orig.* 1925].

Whitehead, A. N. (1929b). *Process and reality: An essay in cosmology*. Macmillan.

Whyte, L. L. (Ed.). (1951). *Aspects of form*. Lund.

Wigner, E. (1967). The unreasonable effectiveness of mathematics in the natural sciences. In *Symmetries and reflections: Scientific essays of Eugene P. Wigner* (pp. 222–237). IUP. [*orig.* 1959].

Williams, R. (2019). Single-celled organism appears to make decisions. *The Scientist*. (*).

Wilson, C., & Schildknecht, C. (1988). The cogito meant "no more philosophy": Valéry's Descartes. *History of European Ideas, 9*, 47–62.

Winslow, R. (2014). Biological meaning. *Epoché, 19*, 65–85. (*).

Wittgenstein, L. (1922). *Tractatus logico-philosophicus*. Kegan Paul.

Woese, C. (2004). A new biology for a new century. *Microbiology and Molecular Biology Reviews, 68*, 173–186. (*).

Wolfram, S. (2017). Are all fish the same shape if you stretch them? The victorian tale of on growth and form. *Wired*. (Online) (*).

Woodfield, A. (1976). *Teleology*. CUP.

von Wright, G. H. (1971). *Explanation and understanding*. Cornell.

von Wright, G. H. (1997). Progress: Fact or fiction. In A. Burgen et al. (Eds.), *The idea of progress* (pp. 1–18). de Gruyter.

Zeeman, E. C. (1976). Catastrophe theory. *Scientific American, 234*, 65–83. (*).

Index

P. N. Jackson, *Aristotle on Meaning in the Living World*, Biosemiotics 30, https://doi.org/10.1007/978-3-032-00602-8

GPSR Compliance

The European Union's (EU) General Product Safety Regulation (GPSR) is a set of rules that requires consumer products to be safe and our obligations to ensure this.

If you have any concerns about our products, you can contact us on ProductSafety@springernature.com

In case Publisher is established outside the EU, the EU authorized representative is:

Springer Nature Customer Service Center GmbH
Europaplatz 3
69115 Heidelberg, Germany

Batch number: 09187159

Printed by Printforce, the Netherlands